65

T0180592

Information Systems Methodology
Proceedings. Venice 1978

Edited by G. Bracchi and P. C. Lockemann

Springer-Verlag
Berlin Heidelberg New York

Lecture Notes in Computer Science

Lecture Notes in Computer Science

Edited by G. Goos and J. Hartmanis

65

Information Systems Methodology

Proceedings, 2nd Conference of the
European Cooperation in Informatics,
Venice, October 10–12, 1978

Edited by G. Bracchi and P. C. Lockemann

Springer-Verlag
Berlin Heidelberg New York 1978

Editors
Giampio Bracchi
Instituto di Elettrotecnica
ed Elettronica
Politecnico di Milano
Piazza L. da Vinci 32
I-20133 Milano

Peter Christian Lockemann
Fakultät für Informatik
Universität Karlsruhe
Zirkel 2
D-7500 Karlsruhe 1

Library of Congress Cataloging in Publication Data

European Cooperation in Informatics (Organization).
   Information systems methodology.

   (Lecture notes in computer science ; 65)
   Bibliography: p.
   Includes index.
   1. Management information systems--Congresses.
I. Bracchi, Giampio, 1944-   II. Lockemann, P. C.
III. Title. IV. Series.
T58.6.E9 1978   658.4'03   78-12358

AMS Subject Classifications (1970): 00A10, 68-02, 68A05, 68A10,
68A50, 68A55, 90-02, 90B99, 93A05
CR Subject Classifications (1978): 2.10, 2.11, 2.2, 3.50, 4.33, 4.34, 4.35,
4.6, 6.0, 8.1

ISBN 3-540-08934-9 Springer-Verlag Berlin Heidelberg New York
ISBN 0-387-08934-9 Springer-Verlag New York Heidelberg Berlin

Printing and binding: Beltz Offsetdruck, Hemsbach/Bergstr.
2145/3140-543210

EDITORS' PREFACE

In the last decade a new and flourishing activity has developed within
organizations: the design and operation of computer-assisted information
systems.
The nature of information systems tasks and the large numbers of individuals involved in them create challenging problems for computer specialists, administrators and management personnel.

As in all evolving disciplines, information system analysts and designers debate the question of their profession as an art or a science.
We strongly believe that the area of information systems has passed beyond
the stage of an art: from modest beginnings as an empirical art in the
Sixties, the area has developed into a broadly based, highly interdisciplinary science and technology that draws on the resources of many
diverse fields, ranging from informatics through engineering, economics
and the behavioral sciences. The development of larger and larger information systems has stimulated new research activities, it has forced
the application areas to a more precise analysis of their own needs and
institutions, and their social and economic impact has led to the beginnings of legislative control.

The ultimate goal of all these efforts is to provide a set of generally
accepted, widely applicable methods and techniques to specify the information needs, predict the effects of computer-based information systems, design them, analyze their operational effectiveness and evaluate
them within the context of an organization. A wealth of methods and
tools has already been developed or has been adapted from other areas
in informatics and from other fields. However, results about information system methodologies are presently widely scattered in various
journals and conference proceedings. This fragmentation produces difficulties in communication among interested persons and in integration
of interdisciplinary experiences.

The second conference of the European Cooperation in Informatics on
'Information Systems Methodology' that was held in Venice, October 10-
12, 1978, has brought together for the first time a wide range of information system experts, from theoreticians through system analysts
and designers to users, from the academic world through manufacturers
to industry and government.

This book contains the papers selected for the conference, covering
subjects such as information system planning, analysis of user needs,
specification tools, data modelling, software systems development, implementation and simulation techniques, parallel processes, man-machine
interface, operation and evaluation of information systems, relationships among information systems, information technology, organizations
and the society.

Elsewhere the reader will find the names of the Program Committee members and the Organizing Committee members, who fulfilled their role
with admirable dedication. In selecting the papers to be presented at
the conference, the program committee had the help of many wellknown
specialists and we wish to thank all of them for their contributions.
Considerable support and encouragement were provided by the ECI board.

We also want to thank the sponsors, supporters and cooperating societies who made this conference possible:

> Association Française pour la Cybernétique,
>   Economique et Technique (AFCET)
>
> Associazione Italiana per il Calcolo Automatico (AICA)
>
> British Computer Society (BCS)
>
> Gesellschaft für Informatik (GI)
>
> Nachrichtentechnische Gesellschaft im VDE (NTG)
>
> Nederlands Genootschap voor Informatica (NGI)
>
> Association for Computing Machinery (ACM)
>
> Association of European Operational Research Societies (EURO)
>
> European Economic Community
>
> European Research Office
>
> IBM Scientific Centers of Italy
>
> Assicurazioni Generali

We hope that this book will be useful to the information systems
community at large.

        Giampio Bracchi   Peter Lockemann

October 1978

MEMBERS OF THE ECI

Association Française pour la Cybernétique,
Economique et Technique (AFCET)

Associazione Italiana per il Calcolo Automatico (AICA)

British Computer Society (BCS)

Gesellschaft für Informatik (GI)

Nachrichtentechnische Gesellschaft im VDE (NTG)

Nederlands Genootschap voor Informatica (NGI)

CONFERENCE ORGANIZATION

Conference Chairman

Giorgio Sacerdoti,
Olivetti, Ivrea

Program Committee

Peter Lockemann, chairman
Universität Karlsruhe

Giampio Bracchi, vice-chairman
Politecnico di Milano

Rudolf Bayer
Techn. Universität München

Victor Chaptal de Chanteloup
Université de Paris VIII

Claude Girault
Université de Paris VI

A. Th. Handstede
Philips, Eindhoven

R.J. Lunbeck
Eindhoven University of Technology

Sidney Michaelson
University of Edinburgh

R.D. Parslow
Brunel University, Uxbridge

Claudio Rossetti
Olivetti, Ivrea

Organizing Committee

Luciano Lippi, chairman
IBM Italia, Venezia

Guido Abiuso
Assicurazioni Generali, Venezia

Antonio Lepschy
Università di Padova

Mariagiovanna Sami
Politecnico di Milano

Carlo Tedeschini-Lalli
Italsiel, Roma

Acknowledgement

Every paper submitted was carefully reviewed by three referees.
Besides the reviewing done by the members of the Program Committee,
invaluable help was provided by the referees listed below. Their
assistance is gratefully acknowledged.

J. Arsac

M. Adiba

M. Atkinson

W. Augsburger

P. Azema

C. Ballée

F.L. Bauer

R. Beaufils

A. de Beer

H. Biller

N. Bjørn-Andersen

G. Blain

P. Bosch

A. Bosman

W. Brauer

S. Braun

K. Brunnstein

J.A. Bubenko

C. Daquin

G. Dathe

G. Degli Antoni

C. Delobel

M. Diaz

H.-D. Ehrich

A. Endres

C. Floyd

M. Galinier

F. Gebhardt

M. Gedin

H. Genrich

G. Giorgi

G. Goos

B. Grassmug

J. Griese

R. Gunzenhäuser

C. Hackl

T. Härder

Y. Hebrail

L. Heinrich

P. Heyderhoff

E. Holler

P. Hoschka

H. Hummel

M. Italiani

U. Kastens

P.J.H. King

J.W. Klimbie

H. Klimesch

P. Köppe

E. Konrad

K.-D. Krägeloh

C. Landrault

P.E. Lauer

K. Lautenbach

J.L. Le Moigne

M. Lundeberg

F. Maero

A. de Maio

F. Marra

A. Mastellari

H.C. Mayr

K. Mehlhorn

P. Mertens

P.L. Modotti

CONTENTS

Information System Planning and Analysis

Information Systems and Organization

ANALYSIS OF USER NEEDS

B. Langefors
Administrative Information Processing,
University of Stockholm
S-10691 Stockholm,Sweden

1. The Problem

The design of information systems, IS, presents challening problems
to the computer specialists or IS designers. They are also aware of
this and are eager to meet this challenge. Consequently computer
people are anxious to have the clients or users specify their requirements for the system as soon as possible. They do not expect the
clients to present their needs in very precise terms and, hence, they
try to interpret the requirements into "computer language". In doing
so the computer people formulate their conceptions of the requirements according to their own ideas of the solutions.

Unfortunately, the computer people are not aware of the fact that to
specify the requirements for the IS is at least equally difficult as
to design the IS. Indeed computer system specialists tend to ignore
the fact that the IS is just a subsystem of a larger system which it
is to serve. This ignorance is serious. To develop the specifications
for the IS means to design the main structure of the System of which
IS is but a component. And the System is not a data processing system,
it is a living organization containing people and different subsystems only few of which are data systems. It follows that the System
design task is not a task of computer specialists.

It is quite some time since the recognition of the analysis of user
needs - and of the System needs - as an explicit project activity
was beginning to attract the attention of some researchers. One and
a half decade ago I wrote [Langefors 1963-1]:

> "...great risks for a system to grow up, that will process large
> masses of data which are not used for decisions - and yet ignore
> important data - ...

> What is sorely needed in this area is a systematic ... technique
> for establishing the real needs for information within an organi-
> zation. Thus it has to define the information needed, its volume,

the time intervals at which it is required, and that at which it
is available, the data from which it can be produced, and the process - or alternative processes - needed for its production and the
form for presentation of the results ...

An analysis of an information system must therefore, in the first
place, be hardware independent. It has to provide an abstract definition of the system itself."

Likewise, a decade ago, Ackoff pointed out the need for more careful
analysis of the user needs and for systematic approaches [Ackoff 1967].
Ackoff also pointed out the problem of collecting and processing masses
of data which are not used:

"It seems to me they suffer more from an
overabundance of irrelevant information."

Also Ackoff emphasized the fact that crucial questions may be asked
by people who know the organization but do not know much about computers:

"The recommendation was that the system be redesigned as
quickly as possible...

The questions asked of the system had been obvious and simple
ones. Managers should have been able to ask them but - and this
is the point - they felt themselves incompetent to do so. They
would not have allowed a handoperated system to get so far out
of their control."

Main Problems

Some Basic Problems

. Who are the users?
. Who does the analysis?
. Why are systems not made use of?
. Which system, what problem, opportunity, or need?
. How combine overview, comprehension, user meaningfulness
  with concreteness and precision in details, in design
  documentation?

User Needs

. Needs for better organization, management, job design
. Organizational change, learning & development
. Needs for information, information service, computeraided data handling
. Social and personal needs

Problems with the information

. Relevance of information, for whom?
. Who are able to share the same data?
. How make sure the data provide the right information?
. Information for Everybody - what design implications?
. Information administration and data administration?

Who are the users?

Often when we talk about the users we have in mind all those who are affected by the system in some way, without having a designer's interests in the system. This is, for instance, quite reasonable in cases where the main question is about dominance of the experts or when one is concerned with other user/designer controversies. But the people affected are of many distinct kinds and are affected in distinct ways, often in ways such that the term "user" appears quite inadequate. Therefore, to study "user needs" or effects of systems upon people it is important to identify either the distinct kinds of affected people or the distinct ways people are affected. The latter way has several advantages and can be pursued by listing distinct roles vis-à-vis the information systems. Then any person or work group can be analyzed into what roles he or it plays and to what extent.

Some roles are

. System sponsors - people who initiate and finance systems projects, because they estimate this is in the interest of the organization, regardless of whether or not they will be serviced directly by the system. Corporate management and the concerned line manager are in this role

but any employee may be too, depending on
the power structure at work.

. Information consumers - people who are aided in their work, deci-
                          sion-making or operating activities, by
                          being serviced with information.

. Information suppliers - those who have to provide information to
                          the system.

. System operators -   those who have to manipulate the equipment,
                          for instance interact with the systems on
                          terminals.

. Other employed people - who are affected by the functioning of the
                          system, for example workers whose work is
                          scheduled by the system.

. Other affected people - who are not employees of the organization
                          owning the system, for instance citizens
                          vis-à-vis a computerized taxation system.

It is important to recognize that one person may simultaneously have
several of these roles. Thus a manager may operate a terminal to
feed in data for a decision model and receive results at the terminal.
He is then an information consumer and supplier and is also, part time,
a system operator. He may also be one of the system sponsors. It is also
important that the needs of these distinct roles are distinct.

Normally an IS is developed for the purpose of supporting information
consumers. People in the other roles are, typically, affected by the
system not because they wanted to use the system - being no information consumers - but because it is required as part of their job in
order that the system will work. It is natural, though unfortunate,
that the needs of theese people have attracted less interest. It is
unfortunate because if the system was not designed with the purpose
of serving them in their job, their jobs may become most affected.
However, more recently the needs of these people have begun to be
taken seriously into consideration.

Because many people are affected in the distinct roles they have, the effects on all the non-information consumer roles are complex and not yet well understood. It is not only how the system affects these people but also how they will experience these effects in the long run that must be clarified. Nevertheless these problems do not seem to be essentially distinct from other job design problems and it seems likely that there will, fairly soon, be a reasonable knowledge among system architects - and among the people themselves - how to handle these user needs.

The needs of the information consumers, though they have been investigated more early, appear to be the most problematic ones and they are likely to challenge us as researchers still for a long time. Because they are also the central problems of any IS design, they will be given the main focus in this presentation. Also, remember that I am talking of roles, not of persons, and any user, or affected person, may perhaps improve his situation through becoming also an information consumer. If so he will also profit from any advances we make in determining the needs of information consumers.

Who does the analysis?

Traditionally, computer programmers have expected the users to specify their needs. When this did not work system analysts entered the scene. In this way we had two kinds of specialists: the computer specialists - programmers and computer system analysts on the one hand and management scientists on the other hand. Often the same person tried to be all these three experts. The specified needs became more a reflection of what these experts believed to know about users' needs, perhaps after quick interviewing, than they reflected the real needs. Somewhat later, around 1965, it became common to have user representatives as members of the project teams. They were supposed to know the user needs. It turned out that the "system analysts", though often recruited from the user lines and trained by the computer manufacturer, behaved more as a kind of computer system experts than as real analysts of user needs. Their documentation was not understood by the users. As a consequence, the user did not know what the system would be doing and be requiring until the system was implemented.

Nowadays the insight is spreading that to determine IS user needs -
at least those of the information consumers - requires a search-learn-
ing process through which the users develop an understanding of what
their real needs are and what new opportunities have become available
to them. As a consequence, the users emerge as the key persons in the
task of analysing user needs. Information analysts appear to have an
important part in aiding the users in learning and analysing, in doing the documentation in a way understandable to the users, while
efficient as a design input for the data system design stages to follow, and in doing feasibility estimates.

It should be obvious that the new way of doing the analysis of IS
users needs, that we have found necessary, raises questions of new
methods. It also becomes clear that the analysts should have a social
science and infology grounding more than one of a computer science or
data processing background.

Why are the systems not made use of?

A frustating experience, to managers and system designers alike has
been that the systems are not used or only used in their most trivial
aspects. Expected gains have not been attained. Cost have greatly exceeded
estimates, various difficulties have been encountered. Why do people
not use the systems? Are they not enough motivated? Is this because
they did not participate realistically in the design? Are the wrong
goals defined for the systems? Are the data - and the system documentation - not accessible and understandable? Is the system architecture, the software, the hardware, unsuitable? Which system, what
problem, opportunity or need? Before system design is taken on, one
must decide which system to address. How should the system's border-
line be drawn. How integrated with the organization? What problems
should be attacked? Perhaps no problem but a new opportunity? What
are the important needs?

How combine overview and comprehension with concreteness and precision in details? The term system is associated with things that have
many, interdependent components. Accordingly, overview over the whole
is difficult and is one of the most typical problems of system design.
It is also a foremost difficulty with analysing user needs. How do

they mutually interact? By what methods can one reach overview and combine this with systematic detailed analysis?

User Needs

The service from an IS may be needed for several reasons. The above list of user roles gives an indication in this respect. Better information makes possible better decisions. Better information service allows new organization structures, more effectively coordinated/directed local autonomy and may make for more convenient office work - though sometimes the opposite occurs.

With the present, rapid changes in economic and social environment organizations need to change, learn, and develop. Information service might be designed to aid in those aspects.

Social and personal needs are also, potentially, served by an appropriately designed IS.

Problems With Information

Relevance of Information. This is a problem not merely of what is relevant from an obejctive or "scientific" point-of-view, though this is often the only kind of relevance mentioned. The infological (or conceptual) view on IS has taught us that information must be "subjectively relevant" in order to become used. For instance, user's cognitive styles need be reflected. How may we go about to satisfy the two relevance requirements simultaneously?
Who are able to share the same data? IS (and data bases) presuppose that distinct people may use the same data but how may data be subjectively relevant to distinct people? What determines "shareability" of data?
How make sure the data correctly represent the intended information, to the intended users? This is the central infological question.
Information for Everybody. Not only democratic demands call for making the information available to everybody, also organizational efficiency and creativity makes this desirable. What can be done in this direction?

If the information is made available to everybody then it becomes
natural to consider "everybody's needs" in designing the IS. How
might this be approached? What additional information content will
become relevant?

Information administration. To determine shareable information and
shareable data and to care for system integrity and security there
must be a central IS administration. To handle this by a special management may hurt the subjective relevance. How can both needs be com-
promised?

It seems likely that local, decentralized systems will be common.
They will contain local data that do not need to be standardized or
centrally available. But some data are likely to have to satisfy corporate requirements and must be controlled by a central body. How
may one delimit such requirements to such data only, for which they
are justified?

Problems with User Involvement

User involvement is important and needs to be improved in all applied technology projects but in the connection with IS design this
is especially significant. Thus to define information needs one must
be able to specify the information to have. This specification must
be stringent in order to provide a basis for data design and computer programming. The stringency easily leads to unintelligibility for
the various users. This is one factor that hampers user involvement.

The information needs and the interrelations between them makes for
an extremely complex information structure or abstract information
system involving many elements and linkages. In order not to get lost
one must introduce abstract, overall information system models so,
in addition to stringency, the users also will have to cope with abstraction if they shall have a real influence upon the IS design.

Another problem is that the term 'user' refers to many distinct roles.
In other words many distinct groups of people need to be involved
and as they are affected in distinct ways this causes distinct problems all of which affects the analysis work and methodology.

2. Available Results

2.1 Project Stages and Problem Areas of IS-design

Projects have since long been structured into stages and phases such
as

>Feasibility study
>Main Study
>System Design
>Implementation
>Operations, Maintenance, and Assessment

but this does not give much guidance as to what-to-do-when, in IS
design. What problems are brought up in the Feasibility Study? What
structure and content has Main Study?
In a systematic study of the problem categories involved in IS design and their interrelations, it was found that some distinct problem areas (or method areas or topic areas) are always involved and
they present distinct kinds of problems and call for distinct skills.
This result offered improved insight into how to organize and staff
IS projects and what to teach and do research on [Langefors 1963-1,
1969]. It provided knowledge of various classes or problems involved
in every IS project and established a framework for developing of
methods and doing research on the distinct problem areas. Groups of
tasks which require similar background can be assigned to common project activities and to designers/analysts with the appropriate skill
and experience.

It was found that because an IS has the purpose of serving as a component in a larger system, its Object System, OS, the specification
of the IS cannot be produced directly but must result from an earlier
stage in which the main (or top-level) "subsystems structure" of the
OS is designed. Importantly the main OS design stage is not a data
system design and, hence, not properly delegated to the data system
department (which should therefore not be referred to as the "Systems
Departments"). In this way a specific problem area or topic area Object Systems Analysis & Design (OSA) was identified. The reader may
alternatively interpret OSA as "organization systems analysis" if
he so prefers.

Only after some object system analysis and design activities have
identified some object system functions can the information needed
of these functions be determined. This can be done in computer-in-
dependent terms which has significant consequences for methods development and user participation. Thus emerged the area <u>Information
Analysis & Design (IAD)</u>.

The areas OSA and IAD constitute the <u>infological</u> realm of IS design
problems. Following them, logically, are the areas within the <u>data-
logical realm</u>. Below the problem areas are presented in a tableau.
The datalogical realm, which is less interesting in this context is
quite condensed in the tableau, Fig. 2.1.1.

One important characteristic of the framework of problem areas is
that after each completed activity in one area more knowledge has
been obtained which is relevant for the cost and feasibility estimates in previous areas. This should be exploited for re-assessment
of previously set goals. In the tableau feed-back arrows are shown,
to point out this.

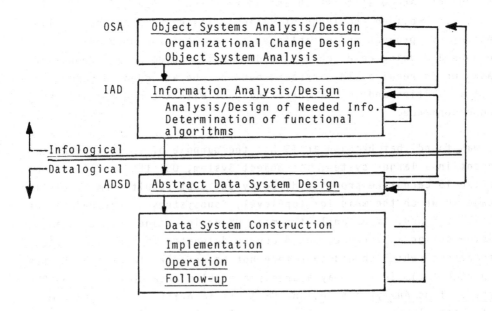

Fig 2.1-1

Problem Areas & Project Stages

Advantages_of_these_concepts

. Distinct classes of project tasks are recognized and problem
  areas are identified.

. More systematic research and methods development can be done

. Better knowledge of the skills required in distinct project stages
  and project activities was gained - henceforth neglected problems
  and methods needs got detected

. Important insights into how IS projects might be more efficiently
  structured and managed were obtained.

Especially

The areas Object System Analysis/Design OSA and Information Analysis/
Design (IAD) became recognized as full-fledged work areas with their
own needs for methods and skills, greatly distinct from computer
science and data processing. Research in OSA and IAD got started and
practical experience (for one decade, now) was acquired. These are
the infological areas.

Thus the introduction of problem areas started the work on Analysis
of User Needs.

The informatological (nowadays: infological) approach to IS design
began to be developed, emphasizing the distinction between infological problems (OSA,IAD) concerned with people and their work and information use and the datalogical problems concerned with formal representation and computers.

Conclusions

Some object system design, change, and analysis should be done before information analysis, data design and programming (organization,
people, jobs).

Change process, social interaction, learning, and identification of
human values are crucial to the system design, as much as is technology.

The necessarily unstructured change processes must result in stringent design specifications that lead to control of the technical design - a clash between two "scientific cultures or paradigms". A
challenge to human communication. Formal methods can only do part
of the task.

As distinct social groups, with distinct values, will prefer distinct changes, the project problem arises as to whether there can
be one single change project involving all groups or there should
be distinct change projects which are then merged. Conflicting designs would then be developed and a negotiation process would follow.

Before it is determined what the real problem is, the boundaries
for the actual system are undefined, hence the set of users is undefined.

Some problems:

- How to balance distinct user needs against each other in specifying design goals?
- How overview or comprehend the distinct needs and the effects of
  design decisions?
- Abstractions are necessary for overview, concrete details are
  needed for construction.

System design, being a social change process, may change power and
will always be sensitive to conflict situations.

Problem Areas are not Identical with Project Stages

There is a strong, logical precedence order between the problem areas.
Functions in the OS must be identified before their needs for information can be determined and this must be done before the data and programs can be designed that may satisfy these needs. But this does not
mean that all of OSA must be done before any part of IAD can get
started and so forth. For example OSA followed by IAD can be done on
a crude level and then be succeeded by an OSA activity followed by
a new IAD activity on the next finer level and so forth. Also, after

13

concrete, detailed data system construction - for specific hardware -
has been done, one may need to do some OSA work to modify the job design for some system operators. This will be done in order to adapt
the job procedures to the operating characteristics of the computer
system.

Other Work on IS Design Framework

A Framework For Information Systems Development which has many similarities to the above framework of problem areas was postulated in
[McFarland,Nolan,Norton,1973]. It, likewise, differs from the traditional project frameworks by talking about the content of the distinct areas. The similarity with our framework is illuminated by the
following quotation as well as the display of the framework, presented below.

  "With a subject as broad and encompassing as information systems,
  expediency requires that the whole be subdivided into manageable
  parts. If this subdivision is performed properly it becomes possible to address each part somewhat independently, for the dual
  purpose of understanding and generalizing. The overall framework
  is then used to integrate the parts into a meaningful whole."

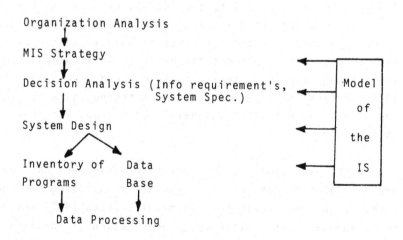

Fig. 2.1-2 The Framework of McFarland,Nolan, and Norton,1973

The similarity speaks for itself. I only want to point out that "System Design", as used here, had better be named "Data System Design".

It is of interest to notice that recent data base work also has begun to recognize distinct levels of the data base schema. Although this has a different background it is related to our problem area structuring and the framework by [Senko,1976] shows interesting similarity to ours. Senko talks of System Name-Based and User Name-Based levels and the Access Path Level, the Encoding Level and the Physical Device Level.

2.2 Infological and Socio-Technical Approaches to User Needs

The infological view is that because data are handled for the sole purpose of utilizing information in an organization and because many distinct data formats may be used to represent the same information it is necessary to describe information in a way which is accessible to the users and is independent of computer technology. The information specification states what is required or desired from the data processing. It ignores how the data system works, internally.

The infological view led to the introduction of specific, early analysis stages where the goals/desires of users are studied and identified, separated from the intricate technical construction work that will subsequently implement the desires. As a consequence, the infological approach [Langefors 1963,1966, Lundeberg 1968,1970, Nissen 1972, Sundgren 1973] subordinated the technological construction to social and human factors and developed methods and tools for making this possible.

The recent development of the database management has begun to address infological aspects for instance in defining information in terms of abstract relations [Codd 1970, Senko 1976] and developing query languages that allow the specification of the information to be retrieved in infological, datastructure-independent terms. Often, user needs may be defined in terms of such query statements.

The Socio-Technical Approach

The infological approach is, clearly, a sociotechnical approach. It
was developed from a computer science and systems engineering background. Else sociotechnical approaches have been advocated and developed by distinct social scientists, for distinct kinds of systems.
Mumford and Ward [Mumford and Ward,1968] presented a sociotechnical
approach to data systems design. They propose that two distinct lists
of design alternatives be worked out, one list of alternatives developed from the point-of-view of the users (the "system operators" mainly, in our terminology). The other list should be developed from a
computer efficiency point-of-view. Then both lists are compared and
a design alternative is selected which is as good as possible from
both points-of-view. This scheme, like other sociotechnical proposals by social scientists, does not employ the strict subdivision into stages. This is natural because the social scientists were not
aware of this kind of development in the IS engineering area. The
sociotechnical approach, by putting the social and the technological
drafting at equal levels implies a compromising between the human
and the computer aspects. Contrary to this, the infological approach
gives a priority to the human aspects, requiring from the datalogical designers to satisfy (realizable) infological specifications (or
to feed back realizability problem information). It should be noted
that the sociotechnical model described here was published a decade
ago and that development is going on in the area. The Socio-Technical approach was also discussed recently in [Boström and Heinen 1977].
See also my discussion letter to MIS Quarterly regarding that article
[Langefors 1978].

Object Systems Analysis and Social Change Design

The infological approach started by taking information definition as
a basis for data and program construction. As a consequence the un-
avoidable, subjective human conditions for interpreting data and utilizing information came out as a fundamental analysis/design basis.
In this way the problem area of Object Systems Analysis/Design emerged.
The people, the stations, the functions in the OS (the organization)
to be serviced, were analyzed as to what was in existence and what
was desired for the future [Langefors 1963-1]. It was proposed by

Nissen [Nissen 1972] that the techniques for precedence and component
analysis developed in the information analysis area ought to be applied also in OSA, after suitable amendments offered by Nissen. This
turned out to be a fruitful approach and has come to be a standard
part of infological systemeering (IS analysis/design). Its advantages
are illuminated in [Lundeberg 1976].

An important development in OSA has been the study of IS development
as a social change process by the ISAC group in Stockholm [Lundeberg
1977] and by Høyer in Oslo [Høyer 1975]. This work has made it increasingly clear that the OSA - and, as a consequence the overall
System Design task - must be arranged so as to begin with a social
change and learning process involving the future users as the main
actors. In view of these findings (which are now accumulating) "analysis of user needs" is, fundamentally, a social learning and creating
process and it is my impression that this should be integrated as a
central area preceding all systemeering (or IS design) projects. With
this insight it is no longer reasonable to go on as if one would be
believing that the users know their needs or - still worse - as if
computer experts or management science experts would know. The practical experience associated with this insight is still meagre, but
what has been attained is quite convincing.

It is important to point out that the insights from our work on the
early stage of systemeering as a social change process clarify that
both the unstructered social processes and the well-structured integration and documentation of their results are necessary and worthy
of being developed further. Thus the bridging of the two emerges as
the central infological systemeering issue.

2.3 Structured Information Analysis

The infological perspective made it clear that data and algorithms
are to be determined based on the needs for information - knowledge -
provided to people through data. At the basis is therefore not scientific or logical laws but intuitive, non-logical user view of their
work and its needs. No unique solutions can be expected and purely
formal methods are inapplicable. But unaided intuition gets lost in

complex systems problems. The best that can be hoped for is a structured approach which identifies semiseparate subdesign tasks and
achieves an integration between them. Crude pictures alone permit
total views but decisions regarding the whole thus comprehended must
be formulated as a precise framework which strictly controls subsequent, detail design activities. Thus the latter essentially fills
in detail content into the meshes of the framework. A <u>structured
design</u> approach is thus obtained. A structured infological systemeering approach has been developed based on a few principles

1   The <u>information product</u> users want from the system determines
     the design task (together with goal information users want to
     prescribe).

2   Information <u>precedence analysis</u>, step by step, starting at user
     determined information products determines where to collect the
     information or from which other information to derive it and
     identifies the derivation processes.

3   The <u>internal design</u> of the information derivation processes can
     be done <u>separately</u> from the information precedence analysis.

4   To gain overview, crude pro-concepts of information may be iden-
     tified first and then <u>component analysis</u> can be performed to
     identify finer pro-concepts, components, until <u>e-concept</u> (ele-
     mentary information kinds) and e-messages (elementary messages)
     are identified. Precedence analysis can be performed on any pro-
     concept level, as deemed suitable, as well as on the elementary
     level.

5.   Through the successive refinement of the IS structure which is
     obtained during the structured design, successively more is spe-
     cified about the information processes and about the data but at
     the elementary level, process design (algorithm design) is com-
     pleted.

The steps 1,2, and 3 were developed in [Langefors 1963-1 and 1966].
Steps 4 and 5 were formulated in [Langefors 1969] and further developed in e.g. [Lundeberg 1969] and [Langefors 1973].

The structured information analysis (and object systems analysis)
based on the precedence and component analysis allows a precise
guidance and a precise documentation to support any intuitive,
social learning analysis/design process. The precision is obtained
without using any computer language which would have hampered the
understanding by various users. It is stringent enough to permit
performance calculations on design alternatives, employ computer aid
to documentation and data and program design and form a systematic
basis for subsequent design tasks. The structure obtained can
directly be represented through system matrices, decision tables
and list structures but graphical representation is usually preferred by the users.

Structured IS analysis/design "automatically" produces crudelevel
structured programming specifications so that a direct continuation
to program design is available. The same is true for data base
design.

A very brief and simplified illustration of a structured information
analysis work is given in the "information graphs" Fig 2.3-1 and
2.3-2. It is assumed that the relevant users have decided to look
into a prospect of an IS for goal steering of the sales activities.
It is emphasized by the method that the organization should be careful to select "all" relevant people to participate in the study.
Information analysts aid them in getting into a creative, social
interaction and in doing the documentation. First one decides, we
assume for the illustration, that "Operating Sale's Goals (g,t)"
are to be proposed by the IS. This is represented by a rhomboid
(representing an information entity), which is placed below (for
output) the large box, "O", symbolizing the IS to be explored. The
indication (g,t) is used to suggest that the goals will be set for
a series of product groups "g" and a time period "t". Operating
Sale's Goal (g,t) is a pro-concept, that is, an information kind
which is undefined, so far, except for its name and the meaning it
suggests. It is still concrete enough to allow the analysis/design
team to do a first step of precedence analysis. This they do by
raising, and debating, the question of what one would need to know
in order to determine realistic and efficient Sale's goals, that is,
what information precedents are relevant. The question of how to

derive the goals from its precedents is postponed. The team starts
by listing a large number of candidate precedents and after lenghty
discussions, analysis and estimating they decide to use only two
precedents, the (pro-concepts) Sales Forecast (g,t) and Other
Corporate Goals. The latter is assumed to be provided to the IS
from outside which is indicated by placing it above the system box
"0" (zero). "0" is used to tag this box in order to indicate that
it represents the whole of the IS being analyzed. "Sales Forecast
(g,t)" is placed inside the box to symbolize that it will be defined
within this system. The line "3" is part of the bundle of lines
symbolizing the precedence relations. It will later be associated
with a subsystem "3" intended for producing Operating Sales' Goals
(g,t) from its precedents. The design of "3" is postponed.

The rest of the graph illustrates the remaining precedence analysis
for the IS (System "0"). The graph represents the whole IS as is
directly seen by the team from the fact that all information to be
produced by the IS has been provided with all its specified
precedents and the initial precedents (above the box) are all
assumed to be available from outside. The team now makes a careful
"workability" check on the system as specified, before going on to
a more detailed analysis.

Note that at the present, crude (pro-concept) level it would not be
possible to define the algorithms involved so this analysis/design
was made possible by separating and postponing the internal design
of the process from the IS structure design.

A more detailed analysis/design can be built systematically on the
top graph "0" by addressing, one by one, the subsystems associated
with the line "1", "2", and "3". This, again, requires the team and
(possibly other teams) to engage in creative work but one is guided
by having the precise tasks of starting with component analysis of
the information product of one subsystem at a time and then repeat
the precedence analysis on all the components. This work is
illustrated in Fig. 2.3-2, for the subsystem 3. It can be seen
that the team decided to define three components for the sale's
goal. They are shown as subboxes within Operating Sale's Goal. They
happen to be e-concepts, elementary information each occurrence of

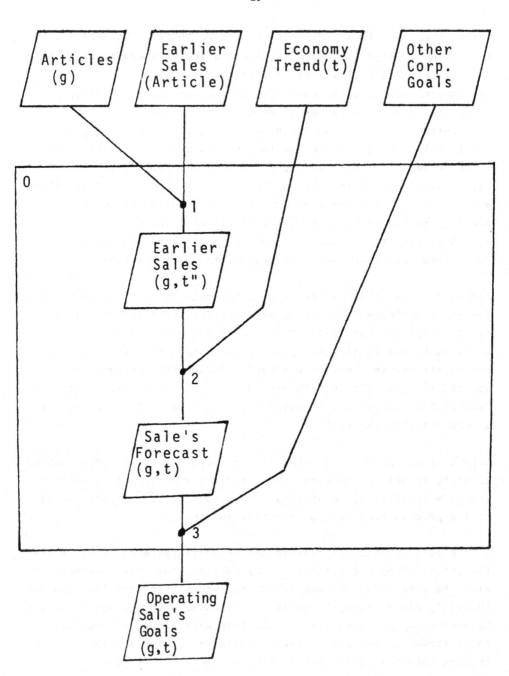

Fig. 2.3-1

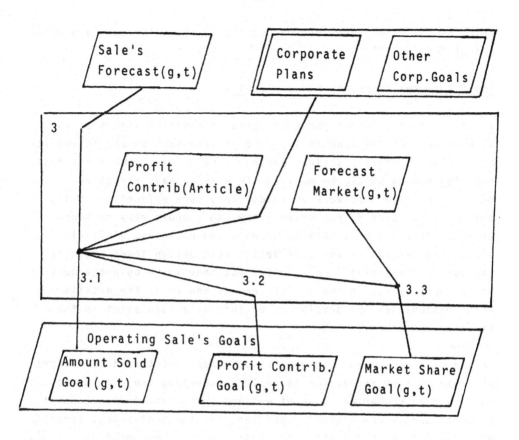

Fig. 2.3-2

which is one elementary message possessing one value for each value
set of "g" and "t".

The Structured Method is Always Partial

It should be clear that the structured information analysis, as
illustrated, can be done in any kind of organization and leaves it
open to be separately decided whether more or less democratic design
decision-making should be applied. This is to say, structured
methods are like a network of roads. They make effective driving
possible. But they do not drive. They can support many distinct
kinds of driving, many driving methods and policies. Similarly,
structured methods of analysis design help design teams organize
the work and integrate the results. But they leave open the question
of who participate in the design team or how to do the details of
work. Methods need be developed for this as a supplement to the
structuring methods.

In the case of information graphs, or other structure descriptions
the details left aside, for separate documenting are e.g. quantitive
data such as number and size of message instances of an e-concept,
time, frequency and control conditions for the processing, semantic
material for the e-concepts and algorithmic or mathematical details.

The fact that details may be represented separately, and piece by
piece, implies that the way they are represented and documented is
not very critical - as long as there are clear references to points
in the structure description.

The structure documentation must be abstract in order to provide
overview and comprehension. It may therefore be desirable to add
concrete material such as concrete layout of reports or terminal
screen pictures to show how specified information (messages) may be
communicated between people and system.

The structured documentation has also the advantage of lending itself to computerized documentation and computer aids to analysis
and design.

2.4 Other Structured Analysis Approaches

Some of the published work on the structured systemeering approach
described in Section 2.3 in addition to the reference given above
are [Lundberg 1976,1977], [Nissen 1972,1976].

The structuring in the approach discussed above was characterized
by stringent linking of unstructured chunks of work and by providing
systematic linkage both horizontally (precedence analysis) on the
same level of detail and vertically (component analysis) from cruder
to finer levels. Postponing of local design tasks was also a key
feature. Other approaches which attempt at systematic work provide,
of course, some structuring but not any integration of all the
mentioned characteristics. Also most structured approach proposals
are emphasizing the program design efficiency whereas the above
work is giving at least the same emphasis on identifying user needs.

One structured approach which has some similarities to the above,
though not taking up all the mentioned aspects is HIPO (Hierarchy
plus Input-Processing-Output) [Stay 1976]. It is more program-
focussed.

One can see from the name already that HIPO also defines vertical
and horizontal relationship ("hierarchy", Input-Output). One can
also see that it does not postpone process definition and that it
has not used the abstracting from "process-input-output" to information precedence relations. On these grounds it appears less
developed conceptually. Also HIPO does not employ the strict
distinction between data and information that has been found so
important for the obtaining of a clear-cut structuring of the
design task.

Some quotes from Stay's article will have to suffice here, to give
an idea of what HIPO is or attempts.

> "The HIPO design process is an iterative top-down activity in
> which it is essential that the hierarchy chart and the input-
> process-output chart be developed concurrently, so as to create
> a functional break-down."

". User understanding and agreement on functional content are
    made easier.

 . Missing or inconsistent information is identified early.

 . Functions are discrete and are therefore more easily
    documented and, if necessary, modified."

The quotes illustrates the (partial) similarities in approach and
itentions. It appears that development along the lines described
earlier is beginning in more and more places.

2.5 Other Information Analysis Approaches

The infological, structured approach described earlier did not
assume that users - or specialists - do know their needs. The
approach was directed to guide in defining and documenting known
needs but also - most importantly - to perform creative work in
finding out unknown needs. Most other approaches presume that users
do know the information (or the data) that they need.

Some of the more developed other analysis approaches are

 .   Data Analysis, Looking at existing data flows
 .   Decision Analysis, Looking at decision processes that need
     to be done
 .   Systematics, as developed by Grindley

An important, different approach is the experimentation with quickly
implemented prototype systems.

Data Analysis and Decision Analysis appear as special and limited
cases of information precedence analysis while ignoring the
vertical (component analysis) aspects entirely.

Data Analysis [Munro & Davis 1977] ascertains information requirements by examining all reports, files, and other information
sources. Managers are asked to state any additional information
requirements. Data for which there is no perceived need are
eliminated. As compared with structured, infological analysis this

appears as a very limited special case. The organizational learning aspects, the vertical structuring and the formalized documentation aspects seem to be missing.

Decision Analysis [Ackoff 1967] [Mc Farlan,Nolan,Norton 1973], similarly to the infological precedence analysis, takes the needs for information and decisions as the starting points for the analysis. Contrary to precedence analysis it studies the decision process in its detail steps in order to identify the information required. Likewise contrary to the infological approach it restricts itself to the needs of managers. Because in the infological work it was detected early [Langefors 1963-1] that the information needs are not determined by procedure steps but by dependencies in reality, the Decision Analysis model reminds on the very earliest infological models (before 1963). Again, no formally structured tools appear to be developed in the Decision Analysis.

Systematics, as developed in [Grindley 1966] has several similarities with the infological analysis/design approach. Thus to obtain a precise definition of the information the concept of "element" is introduced, corresponding to e-concepts and e-processes in the infological approach. But the methodological separation between the information and the processes is not employed in Systematics. There also is no counterpart to the pro-concepts and the component analysis. Systematics, like the infological approach, develops semi-formal description techniques. Decision tables are used as a central description tool.

One of Grindleys starting points was an observation from a Diebold study, around 1964, in which it was found that programmers spent less than half of their time writing and testing programs. The rest was spent on various System Queries such as

1. Changes made by the analysts to original specifications
2. Misunderstandings between analysts and programmers
3. Errors in specifications
4. Omissions in specifications

Many of the propositions in Systematics are in complete agreement with fundamental infology. For instance the proposition that two essentially distinct tasks go into any IS design work:

I   Consideration of user needs
II Consideration of computer needs.

Experimentation with Prototype Systems

A different way to determine user needs is to directly implement
extremely simplified prototype systems. Through ignoring performance
problems and employing extremely simple design one can quickly
implement application systems, using miniature data banks. One can
thus allow the users to experiment with the system and then modify
and extend the system successively, according to user demands.
Afterwards, when the users feel satisfied one can redesign to
achieve acceptable performance. In some application performance is not important so the prototype can be used operationally.
This way of developing IS is sometimes advocated as an alternative
to analysis. It seems to me that it is more appropriate to regard
this approach as another way of doing the analysis. It is essentially
similar to the infological approach in some fundamental aspects,
such as involving users actively and realistically in the specification work.

The idea to use simplified fast implementation for "incremental
system implementation" was proposed by Bubenko and Källhammar in
the CADIS project at the Royal Institute of Technology, Stockholm,
as a part of a development of Computer-Aided Design of Information
System [Bubenko,Langefors,Sölvberg,1971, article by Bubenko and
Källhammar]. This kind of ideas have appeared in different places,
for instance [Bally,Britten, and Wagner,1977]. One interesting,
recent development of this kind has been done by Staffan Persson,
Stockholm School of Economics. He is using a small minicomputer
which he takes along when visiting companies in Scandinavia, by
car or by plane. He offers to implement a prototype of any system
in three days, sometimes. He programs on-line while querying the
users. These can then immediately experiment with the system and
find out desirable modifications. Typically, one finds that
questions arise early which may need several weeks for decision-
making. It is important to identify such problems before large-scale

implementation has been finished. This happens, typically, also in infological analysis work where then no programming at all has had to be done.

It is my estimate that the structured infological analysis is superior in some aspects, where abstraction and comprehension is at issue while the prototype experiment is superior in making the operative situation of the user concrete and live. It seems that the two approaches supplement each other, which ought to be done research on.

2.6 Documentation and Analysis/Design Aids

All systematic or structured analysis/design methods need be supplemented with some documentation/description methods. In the IS analysis/design area some one-dimensional languages have been proposed. Examples are the system algebra language (supplemented with Algol or COBOL statements) [Langefors 1966] and PROBLEM STATEMENT LANGUAGE (PSL) [Teichroew 1972]. PSL is, in the ISDOS project, supplemented with PSA, Problem Statement Analyser which checks the statements for consistency and some kinds of completeness. The 1-dimensional languages are poor system-languages in the sense of not aiding overview very well. Their advantage seems to be expected to be that they can be interpreted and analyzed by computers. But then one should not ignore the fact that 2-dimensional descriptions are also quite easy to input to a computer.

Two-dimensional languages are useful for systems descriptions because they employ both dimensions of the document paper to better aid overview, which is one of the central problems with complex systems. System matrices, precedence and incidence matrices were introduced for IS descriptions both for needs analysis and design computations, based on system algebra, in [Langefors 1963-1, and 1966]. These matrices describe the structure of the system whereas decision tables were available for system logic descriptions [Langefors 1970, 1970 version of 1966], [Grindley 1966].

As was observed by Jäderlund in 1974, decision tables may be treated
as special parts of the IS incidence matrix so that a combined
matrix, the Process Control Matrix, [Jäderlund 1976] was obtained
which describes, concurrently, the IS structure and the process
control logic, see also [Langefors and Sundgren 1974].

As we have already seen, formalized graph presentations have been
developed and used extensively as documentation tools for user needs
analysis, both information analysis and object system analysis
(organizational analysis). This kind of presentation seems to be
preferable to users though a systematic comparison with matrices
has not been undertaken so far. It is often held against the use of
graphs that they cannot be input to computers. Such statements,
however, are built upon ignorance. It is rather easy to input formal
graph information to computers as may be concluded from the facts
that such graphs are isomorphic to incidence matrices. A little more
difficult is to produce graphs by computers but reasonable solutions
do exist.

2.7 Computer Aids to Needs Analysis

It was pointed out early [Langefors 1963-1,1966] that formalizing IS
analysis/design concepts to improve understanding would also have
the advantage of allowing the use of computers as an aid to IS
design but also to the analysis of user needs. The possibility of
computer aid to IS design has aroused a great deal of interest in
many places. It is a question, though, whether we have not underestimated the other kinds of knowledge that we need to have before
it will be worthwhile to use computer aided needs analysis. As
always, it is easy to jump to formalizing and automating too much
too early.

3. Some Practical Experiences

3.1 Early Experience

It has long been known that data processing systems often were
failures and even those seen as reasonably successful were
questionable in many respects. Costs and time for implementation
were typically twice the estimated figures, or more, and gains were
half. Users failed to use the systems, maintenance was extensive and
endless. This sort of experience was published widely in the 1960's.
As late as in the early 1970's one could still see reports of this
kind by (still) surprised and disappointed authors. One such work
[Dew & Gee 1972] studied the use made of budget information. They
report for instance

> "in a study of four manufacturing enterprises with "well
> established budget control systems" 60 middle managers were
> queried. It was found that 32 ignored completely their budget
> information and 12 made only limited use of it."

3.2 Experiences with Infological Analysis Applications

Already in the early application work (mid 1960's) some of the
intended advantages were clearly realized:

- . users did find the documentation possible to understand
  and appreciated the absence of computer language, for
  instance, users have by themselves corrected mistakes in
  system graphs
- . thanks to the precision of description users detected
  mistakes and asked for changes - typically 10 to 20
  changes were done on the earliest graphs - and, thus,
  faulty implementation was avoided
- . users learned about their own work through participating
  in the analysis - they stated they had reached an overview
  of their own work and its place in the organization, never
  possible before.

. it was easy to replace analysts, thanks to the structured
documentation. This is important in all large projects.

Later on (in the 1970's) some "finer points", not visible to the
researchers earlier, have come out:

. formal training should be provided to the users before
project starting

. users should be encouraged to take control in the analysis
work and do some of the documentation, themselves

. users, from many organizations, have become rather
enthusiastic about the infological analysis approach,
for instance, several of them have emphasized the importance that universities continue research and teaching in
the area

. all kinds of users, not only managers, have taken part,
for instance, in some cases the factory workers were able
to specify requirements for economic information that would
illustrate the "value added" of their work in the company
profit

. in some cases it has been possible to mail the documentation
to the computer programming team who then mailed the
programs to user representatives who took care of the
implementation.

As already mentioned, experience lead to expanding on the "social"
part of the information analysis and object system analysis/design
and organizational change:

. several projects falter early because of social conflicts
between distinct user groups, a conclusion has been that a
social feasibility study ought to be done before other
feasibility studies. No project should be started unless
the social environment is cooperative

. in research projects where the users took over most of the
analysis work positive results have been most certain

- in projects where teams of users for doing the social change
  design have been established, operative valuable organizational innovations have been created and implemented in the
  system

- it has become clear that "information analysis" has to be a
  creative, learning process, not just a mapping of known
  needs

- users tend to gain power through working with the structured
  analysis; this may generate conflicts with other users who
  are brought into the project later on - hence it seems
  important to bring in all people from the beginning
  [Lundeberg 1977].

3.3 Some Observations from Empirical Research

Problem with User Representatives. Efforts at really involving users
in IS design projects has been taken since long. It used to be found
that this didn't work. We know now that it couldn't with traditional,
technology dominated analysis methods. More recently user representatives have been more involved, in some cases. This seems to be
more the case with on-line systems, especially with local, minicomputer projects. A disturbing finding, reported for instance by
John Kjaer and by Peter Neergaard from distinct projects at the
Copenhagen School of Economics, has been that if user representatives do get to function in design teams then they quickly get
regarded as technological experts and, thus cease to be really
representative. I have, myself, seen this happen, also, at different
places.

One-Line Users - System Operators

Many of the systems studied in recent research have been on-line
systems. There is a clear shift in findings in that whereas batch
systems used to be problematic from both user aspects and business
aspects, studies of on-line systems have regularly shown results
where positive user reactions dominate the overall estimates. On
the other hand it is also regularly found that people who work for

long stretches of time have physical problems of various kinds
caused by unsuitable design of terminals, desks and chairs. Job
rotation, daily, has been found important for these reasons.

IS Influence upon Social Communication. The design of modern IS
tends to replace man-to-man communication by man-to-computer
terminal communication. This may reduce social contacts in a
serious way which has been pointed to by some researchers. However,
recent studies have also shown that IS may have the opposite effect
in encouraging distinct system users to take up contacts with each
other that they did not have before [John Kjaer,1977,Copenhagen
School of Economics, personal communication].

4.   Ongoing Work

The study of user needs, both as a separate research field and as
a specific problem area in IS projects is quite recent, as we have
seen, and not too much is going on or known to be. Indeed it is
hoped the present conference will contribute to improving the
situation in this respect. Only a few indications can be presented
here.

Communication Aspects in IS design

The structured infological approach and its emphasis on user understanding, hierarchic descriptions and graphical aids is clearly
oriented towards communication but that term is not emphasized in
published work. Communication has many associations with IS design
but three may deserve special attention: (1): the design documents
as vehicles for communication between users, analysts, programmers,
operators and maintenance staff, (2): the system design process as
a communication process and (3): the use of data processing systems
as communication systems.

Empirical research on description techniques from an interspecialist
communication perspective was reported in [Willoughby and Arnold].
I hope work of this kind is going on and I think more of it would
be desirable. Work on methods concerned with human communication to
be included in IS design work is presently being pursued at the
Royal Institute of Technology, Stockholm [Lundeberg 1977].

The hypothesis that it is useful for the study of existing IS to
regard them as instruments for human communication was forwarded
and pursued by Nissen [Nissen 1976]. It seems promising also from
the needs analysis point-of-view. Nissen points out that the use of
a formalized language restricts the subject that can be communicated.
Computerization restricts further. It is important, when formalized
channels are introduced, so as to replace informal communication,
to determine the communication that becomes ruled out by the change
and to introduce appropriate means of compensation.

Human Information Processing Aspects

The infological approach emphasizes that data must be designed with their use in human information processing in mind. This is the essence of distinguishing between data and information. This area is, of course, closely related to that of the human communications approach. Recently, much work has been started in this area. Ongoing work is indicated by some recent publications. Barkin, at the University of Utah, and Dickson at the University of Minnesota [Barkin and Dickson 1977] report on empirical work, on hypothesis testing, regarding information system utilization as dependent on the cognitive style of the decision-makers. Similar studies are reported in [Lucas 1975].

Dialectical Needs Analysis

We pointed out that structured methods partition the total analysis and design task into subtasks. The boundary conditions for subtasks and the linkage between them is handled precisely by the structured method while the content of each subtask is left entirely open by the structured method. Various methods and various decision power systems can use the same structuring methodology but lead to choice of distinct subtask methods. The use of creative learning processes within subtasks was mentioned. An alternative method to be used for subtask activities is "the Dialectical Method". This is thus not an alternative to structured methods but, rather, one possible supplementary method to be considered for some subtask activities.

Dialectical approaches to systems design have been advocated and worked on by Churchman and followers, such as Ian Mitroff [Mitroff, 1972] and, apparently, is still going on. They point out that information and epistemology are closely related, a position which is very similar to that of infology. Philosophers have discussed various systems of epistemology or inquiring systems. Different kinds of information are associated with each of them, Mitroff asserts. Churchman and Mitroff present a list of distinct systems of inquiry

Leibnitzian:		purely formal, deductive
Lockean	:	experiential, inductive, consensual, empirical
Kantian	:	both formal and empirical, reconciles Leibnitzian and Lockean
Hegelian	:	conflict, synthetic design, the strongest possible conflict on any issue
Singerian	:	synthetic, interdisciplinary, attempt to integrate scientific, ethical, and aestetic modes of thought.

Compared with the infological framework the Leibnitzian corresponds to the datalogical, Lockean corresponds to the empirical, data collection aspect and the Kantian, Hegelian and Singerian correspond to various aspects of the infological concept of "interpreting structure" or frame of reference. Thus the Churchman and Mitroff list does not take us outside infology but, rather, provides additional content inside it. An example is the dialectical, Hegelian approach as suggested by Mitroff. On every important issue one should find two experts with opposing views (corresponding to possessing distinct data interpreting structures in the infological sense). They should be "equally creditable". These experts should perform a strongest possible debate on the issue and be watched by the decision-maker (the users we would say) who should create his own synthesis. Clearly this is one proposal for trying to handle the infological problem, mentioned earlier, of how to reconcile subjectively relevant and "objectively" relevant information specifications. Let me end this account of Mitroff's work with a quote from him [Mitroff 1972].

"... the search for a common, consensual position may be ill-fated, if not outright dangerous. It may be more important to have a methodology that allows one to make sense out of direct confrontation than to eliminate confrontation by either suppressing it or whishing it away."

Empirical Research to Evaluate Needs Analysis Methods

When alternative analysis methods become available it becomes
interesting to evaluate their respective advantages and disadvantages. This turns out to be an extremely complicated task.
It appears to me that we do not yet know what criteria might be
relevant for the comparison. Nevertheless we ought to get started.
Hence the work taken up in this direction in [Munro and Davis 1977]
is welcome beginning. Munro and Davis try to compare two variants
of information needs analysis, the "data analysis" and the "decision
analysis". They formulate a number of hypotheses and design an
experiment to collect data for testing them.

REFERENCES

Ackoff, R.L., 1967, Management Misinformation Systems, Management Science, Vol. ü4, No 4, December 1967.

Bally, Britten, and Wagner, 1977, The Prototype Approach to Information Systems Design and Development, Information and Management, Vol.1, No 1, pp 21-26.

Barkin, S.R., and Dickson, G.W. 1977, An Investigation of Information System Utilization. Information and Management 1(1977) North Holland Publ.

Boström, R.P. and Heinen, J.S., MIS Problems and Failures: A Socio-Technical Approach, Part I MIS Quarterly, September 1977, Part II MIS Quarterly, December 1977.

Bubenko, J., Langefors, B., Sölvberg, A. (eds), 1971, Computer-Aided Analysis and Design, Studentlitteratur, Lund, Sweden.

Codd, E.F., 1970, A relational Model of Data for Large Shared Data Banks, CACM 1970.

Dew & Gee, 1973, Management Control and Information , MacMillan,London.

Grindley, K. 1966, Systematics - a Non-Programming Language for designing and Specifying Commercial Systems to Computers. The Computer Journal 9(2)Aug. 1966.

Grindley, K. 1975, Systematics, A new Approach to Systems Analysis, Mc Graw-Hill 1975.

Høyer, R., 1975, Systemeering and Social Reality: Formal and Informal Aspects of Administrative Control, In Lundeberg, M., Bubenko,J.(Eds), Systemeering 75, Studentlitteratur, Lund, Sweden.

Jäderlund, Chr.,1976, "Systematik" in Data/4-1976, pp 35-42.

Langefors, B., 1963-1, Some Approaches to the Theory of Information Systems BIT 3 (1963) 229-254, Copenhagen, Denmark.

Langefors, B., 1963-2, Toward Integration of Engineering Data Processing and Automatization of Design, In Vistas in Information Handling The Augmentation of Man's Intellect by Machine (P.Howerton,ed) Spartan Books.

Langefors, B., 1966, Theoretical Analysis of Information Systems (THAIS) Studentlitteratur, Lund, Sweden.

Langefors, B., 1969, Management Information System Design, IAG Quarterly 2 (No 4).

Langefors, B., 1970, Integrated Control by Information System, Effectiveness and Corporate Goals, in Management Informations Systeme, Grochla and Szyperski (eds) Betriebswirtschaftlicher Verlag, Dr Th. Gabler GmbH, Wiesbaden 1971.

Langefors, B., 1973, Control Structure and Formalized Information Analysis in an Organization in Grochla & Szyperski (eds) Information Systems and Organizational Structure, Walter de Gruyter, Berlin, New York 1975.

Langefors, B., 1978, Discussion of [Munro and David 1977], in MIS Quarterly, September 1977.

Langefors, B. and Sundgren, B. 1974, Information Systems Architecture, Mason/Charter, New York

Lucas, H.C., 1975, Why Information Systems Fail Columbia University Press, New York.

Lundeberg, M., 1976, Some Propositions Concerning Analysis and Design of Information Systems, TRITA-IBADB-4080, Royal Institute of Technology, Stockholm. Doctorate thesis.

Lundeberg, M., 1977, Utilization of New Information Systems Development Methods in Practice - Perspectives and Prospects, Information Processing 77, B.Gilchrist, Editor, IFIP, North-Holland Publ.

Mitroff, I.I., 1972, Dialectial Inquiring Systems: A New Methodology for Information Sciences, Journ. of the Am. Soc. for Information Science - Nov.-Dec. 1972.

Mumford, E., and Ward, T.B., 1968, Computers: Planning for People, London: B.T. Batsford

Munro, M.C. & Davis, G.B., 1977, Determining Management Information Needs ... MIS Quarterly/June 1977.

Nissen, H.-E., 1972, A Method for the Description of Object Systems in Information Systems Work, TRITA-IBADB-4403, Royal Institute of Technology, Stockholm.

Nissen, H.-E., 1976, On Interpreting Services Rendered by Specific Computer Applications, Liber/Allmänna Förlaget, Stockholm (Doctorate thesis) Royal Institute of Technology, Stockholm.

Senko, M.E., 1976, Specification of Stored Data Structures and Desired Output Results in DIAM II..., Lockeman and Neuhold (eds), Conf. on Systems for Large Data Bases, Brussels 1976.

Stay, J.F., 1976, HIPO an Integrated Program Design, IBM System Journal No 2 1976.

Teichroew, D., 1972, A Survey of Languages for Stating Requirements for Computer Based Information Systems, Proc. FjCC, p. 1203-1224.

Willoughby, T.C., and Arnold, D.C., Communication with Decision Tables, Flowcharts and Prose, (The copy I have does not have ref. to the publication. I believe it is the ACM publication Data Base. The authors were with the department of Accounting and Mgmt Inf.systems College of Bus.Adm., The Pennsylvania State University, University Park, Pennsylvania) 1971?

A PARTICIPATIVE APPROACH TO FORWARD PLANNING AND SYSTEM CHANGE

John Hawgood
Frank Land
Enid Mumford

Automation Benefit Appraisal Consultants Ltd.
Durham DH1 4DY, England

(Most of this paper is based on an article commissioned from us by the UNESCO journal
"Impact of science on society" for its issue (Vol.28, No.3, July-September 1978),
which is devoted to the theme "Computers and social options".   Permission of UNESCO
to use this material here is gratefully acknowledged.)

ABSTRACT

Forward planning and system design can be improved through the participation of the
people who will eventually use the system, if they can be provided with the necessary
skills.   The authors have developed and tested a framework for participative forward
planning and aystem design which includes a set of analytic and design procedures:

1   Variance analysis, for identifying operational deficiencies

2   Job satisfaction analysis, for measuring the lack of fit between
    employees' actual and preferred work situations

3   Future analysis, to identify significant opportunities and
    development goals

4   The BASYC approach to multi-attribute utility analysis, to
    compare the desirability of alternative courses of action
    (possibly using "fuzzy logic")

5   Socio-technical systems design, to bring human and technical
    factors to bear simultaneously on system improvements.

* * * *

John Hawgood, University of Durham, Durham DH1 3LE, England
Frank Land, London School of Economics, London WC2A 2AE, England
Enid Mumford, Manchester Business School, Manchester M15 6PB, England

INTRODUCTION

It is gradually being recognised that systems design, like management in general, is primarily about people. Many past failures of computer-based information systems can be directly attributed to two complementary causes:

1 A lack of knowledge of human needs and motivation on the part
   of technically oriented systems analysts and designers

2 A lack of technical confidence on the part of general and
   departmental managers which makes them reluctant to intervene
   in design decisions

The long term answer must lie in the education of specialists and users, but two present day trends provide a means for improving the situation. These are, first, the movement towards employee participation in forward planning and work design and, second, the trend away from conventional money based cost-benefit analysis and towards multi-attribute utility analysis which takes account of all the different types of advantages and disadvantages associated with any proposed change. The authors have developed a method which combines these two elements in a formal yet flexible participative framework in such a way that a number of related techniques can be applied as and when needed. Our approach is based on four important value judgements:

1 That the financial, human and technical factors in system
   design can and should be treated compatibly

2 That everyone affected by a system change can and should be
   considered in planning it

3 That employees at all levels can and should design their own
   systems

4 That the overall approach to systems design and development
   should be based on the principle of reducing uncertainty

In our view the process of forward planning, including systems design, implementation and evaluation, should be carried out by two types of team. The first team is a steering group which sets the basic organisational objectives and constraints under which the new system is to be developed. It will generally include the managers of affected departments, official representatives of trade unions and other major interests affected by the new system. The second team is responsible for the detailed systems design and consists mainly of representatives of the department where the new computer system is to be introduced. This team will define the scope of the local problem, scan the environment for new opportunities or changing constraints, analyse deficiencies in the current system, define development goals in consultation with other groups likely to be affected and provide decision makers with assessments of the likely impacts of alternative strategies on these goals. Finally it will design and test the selected system and the work organisation and task structures associated with it.

Such a team may include members from the traditional EDP department, but their function will be to transfer the required skills to other members rather than to carry out the design themselves.

BACKGROUND

Economic analysis of information systems

The introduction of computers into an organisation's information system generally requires substantial resources.   In order to justify the use of such resources, the organisation has to set-off the development costs against the net benefits the changes bring to the organisation.   Benefits may accrue to the organisation directly through a reduction in costs or an increase in production, or indirectly in that the changes in the information system enable the users of the system to perform their functions more effectively.   Further, some of the benefits of the changed system have an effect on the organisation through the changed behaviour of the customers or the improved image of the organisation in the community.   Thus the concept of a net benefit to the organisation as a measurement of the benefits accruing to the organisation over the life-time of the project, less the cost of operating the changed system over that time, contains a number of practical and theoretical difficulties (1).

In practice there have been a number of different approaches to solving the problem of evaluating the worthwhileness of systems changes.   At one extreme (2), organisations will only accept systems change projects if it is possible to demonstrate that the change will result in directly measurable savings.

At the other extreme (3,4,5), cost benefit analysis techniques are suggested which attempt to provide an economic value measured in money terms for all costs and benefits, whether these be measurable through the normal accounting system or not.

A new approach has recently begun to take favour.   This recognises that an organisation has many goals and that any goal may be regarded as having a different value by different groups within the organisation.   Some goals may appear to be conflicting; for example, it may be desirable to have both an efficient system (operating at minimum costs) and a flexible system.   It is difficult for a system to be both optimally efficient and very flexible.

Multi-objective, multi-criteria decision methods are being developed all over the world (6,7,8) and have found a number of applications in the evaluation of computer-based information systems (9,10,11).

The approach to evaluation chosen here is based on this approach, which has been shown (12,13) to overcome some of the problems associated with other techniques of analysing the expected value to the organisation of making a change in the information system.

The planning and management of change

This paper has two objectives related to the management of change. First, it seeks
to legitimate a value position in which the future users of computer systems at all
organisational levels play a major part in their design. The argument here is that
people have a moral right to influence the organisation of their own work situations
and that if this right is conceded then there are likely to be both job satisfaction
and efficiency gains: job satisfaction gains because the group are better able to
diagnose their own job satisfaction needs than any outside group of specialists; efficiency gains because the people who are in the change situation are likely to have an
excellent knowledge of day-to-day work problems and can make useful contributions to
the solution of these. Also, it is hypothesised that they will be committed to operating efficiently a work system which they have designed (14,15).

The second objective of the paper is to persuade groups concerned with the design of
computer systems to set specific job satisfaction objectives in addition to the usual
technical and operational objectives. Here we argue that unless job satisfaction
objectives are made explicit, and the computer system and associated organisation of
work designed to achieve these, then the human consequences of a new computer system
will be unpredictable because they have not been consciously planned for. The result
can be that the new system will have undesirable human consequences such as a routini-
sation or deskilling of work, or other features that are not welcomed by the user.
Employees in the user department may then respond in a negative way, refusing to operate the system or ensuring that it runs at low efficiency, and, in addition, absenteeism
and labour turnover may increase.

Our belief is that greater user involvement together with clear job satisfaction objectives will assist the successful planning, design and implementation of computer
systems. Ozbekhan, discussing theories of planning, suggests that the raison d'etre
of any planning is to change the environment in a manner that is smooth, timely and
orderly so that a dynamic evolution that is in line with our ideas of organisational
progress can be achieved (16). He also points out that because the objective of
planning is change it is likely to be a threat to people, unless the different groups
associated with the change believe that they are participants in the change process.
We would support this view.

Figure 1 sets out the authors' perceptions of the risks associated with the traditional
approach to the design of computer systems in which planning and design responsibility
rests with a group of EDP specialists.

If a group believes that it is threatened by another group it is likely to draw together
to show a collective identity and to introduce group norms directed at emphasising
group unity and solidarity. This response enables it to reduce uncertainty by introducing behaviour patterns which are seen as helpful in protecting the interests of the

43

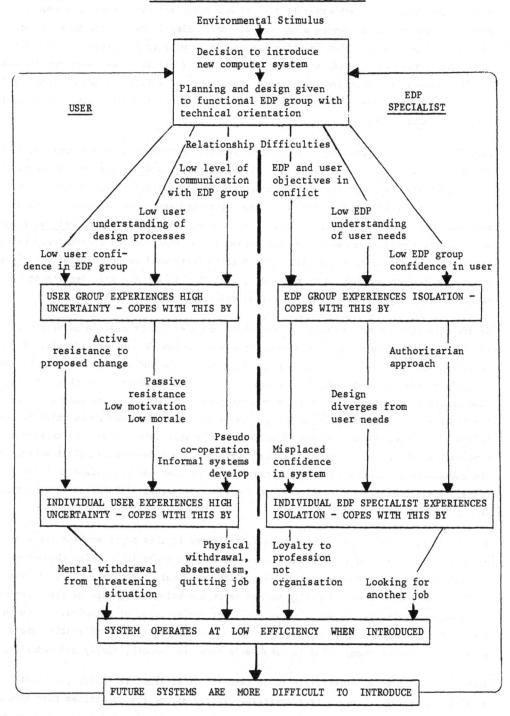

Figure 1   SYSTEMS PLANNING AND DESIGN
The risks of the traditional approach

group. If the response of the threatening group is one of increased pressure on the threatened group to conform, conflict will increase. It has been suggested that negative actions feed upon each other (17). If one group acts with hostility towards another, then this will provoke a counter action. Also if two groups have an interdependent relationship - that is, they are unable to work in isolation and depend upon each other's services for the successful completion of a task - then hostility is likely to be increased, for they are forced into a constant and irritating association. All of these statements seem applicable to the traditional relationship between EDP specialists and user groups.

Planning and design approaches that tend to generate rather than reduce conflict place the organisation in a risk situation. It may incur _financial_ risks through introducing expensive computer systems which operate at low efficiency, or reduce job satisfaction and increase labour turnover. Many firms underestimate the cost of labour turnover, which can be very high (18). The firm may also incur _organisational_ risks, for a poorly functioning department may spread dissatisfaction and inefficiency through departments which interact with it. In addition there will be human relations risks. A new computer system that is introduced against the wishes of a user department has the potential to produce serious industrial relations problems (19).

We are therefore suggesting that the technical and user groups associated with the introduction of a new computer system are unlikely to have a complete identity of interest, and may have major conflicts of interest. The EDP specialists will be keen to optimise the use of a technology which they know and understand and this can lead them to design systems which have a high technical competence but are poor at catering for human needs, such as a desire for job satisfaction. The user group which has no active role in the design process is unlikely to be able to challenge the technical knowledge of the specialists and this can force it into a dependency relationship, and the subsequent acceptance of a computer system which does not adequately meet its needs. This produces low commitment to the system together with increased resistance to any future change.

A participative design approach such as we recommend in this paper enables the user group to identify its own needs and interests, to turn these into design objectives and to see that these receive equal weight with technical objectives. This is the basis of the socio-technical design method described below, which has as its objective the design of work systems so as to secure joint optimisation of technical and human needs (20). Figure 2 sets out our view of the advantages of a participative approach as a facilitator of change that in Ozbekhan's terms is 'smooth, timely and orderly'.

Such an approach enables conflicts to be resolved or at least recognised, so that evasive action can be taken, or a consensus arrived at, on the objectives that should be attained through the new computer system. Because planning and design responsi-

Figure 2   SYSTEMS PLANNING AND DESIGN

The advantages of the approach described
in this paper

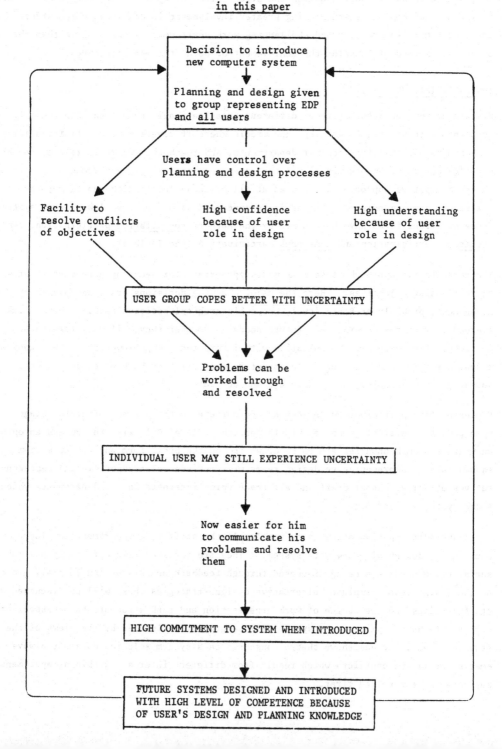

bilities are shared between technical experts and users, competences are shared and
each group can learn from the other. Perhaps most important of all is the fact that
user values on participation and on the organisation of work can be catered for. Today
many groups of employees are seeking greater involvement in decisions which affect
them and a more interesting and challenging work environment. We believe that the
use of our method will assist the achievement of both of these objectives.

THE PARTICIPATIVE PROCESS

Systems design can occur at three different organisational levels: the top where it is
concerned with strategic planning, the middle where it covers system definition for a
number of divisions, functions or departments, and the bottom where it relates to the
detailed design of an organisational subsystem serving a single department or function.
The participative approach can be used at all of these levels although it may take a
different form depending on whether it is concerned with higher or lower level systems.
These different forms have been named by the authors consultative participation, representative participation and consensus participation (see Table 1).

The consultative approach is seen as most appropriate for securing agreement on strategic planning objectives. Here the major planning decisions are taken by senior
management, probably at Board level, whose hierarchical position enables them to take
a broad view of the enterprise's future needs. However they will only take these
decisions after extensive consultation with interested groups lower down the organisational hierarchy and a consultative structure must exist or be created so that this
sounding out of opinion can be thorough and accurate.

Representative participation is seen as appropriate at the system definition stage
when powerful interest groups at middle management level will wish to express an opinion
on where system boundaries are to be drawn and on the broad form any future system
should take. Representative design teams will include selected or elected representatives of all grades of staff and all trade union interests in the departments which
a new system will affect.

Consensus participation attempts to enable all the staff in a department to play a
part in the design of a new work system. They are involved when efficiency and job
satisfaction needs are being diagnosed through feedback and discussion in small groups.
As the design team formulates alternative design strategies these will be discussed at
staff meetings and the choice of work organisation and task structure to be associated
with the technical part of the system will be greatly influenced by the views of the
staff. Experience has shown that a consensus on a system solution does not always
emerge easily and conflicts which result from different interests within a department
may have to be resolved first.

PROCEDURES TO ASSIST PARTICIPATION

Variance analysis

An important prerequisite to designing an improved system of work is a detailed knowledge of those weak parts of the existing system which produce operational problems.
The method used to obtain this is known as variance analysis and was developed by
Professor Louis Davis of the University of California, Los Angeles (21).   A design
group which uses this approach will examine in detail all the different operations
which a department or section undertakes and note areas where variances are prone to
occur.   By variance is meant a tendency for the work system to deviate from a desired
standard or specification.   This tendency arises as a result of some problem associated with the work process itself in its normal operation.   Variance analysis is not
concerned with temporary problems such as machine breakdown or with human errors which
are a result of inadequate training.   It concentrates on system weaknesses associated
with the organisation of work operations.   An important objective of the method is
to identify clearly those key variances that significantly affect the ability of a
work system to pursue its major objectives.   These variances are often found at the
boundaries of a system, for example, where the work of one department interacts with
that of another and there are problems of co-ordination.

Once variances have been identified they must be examined in detail in order to determine the following:

    1 where the variance originates
    2 where it is observed
    3 where it is controlled (corrected)
    4 who has responsibility for this control
    5 what he does to control it
    6 what information he requires to restore control and where he gets it from
    7 possible alternative control mechanisms (of which a computer may be one)

Variances are frequently controlled not where they originate, but later by supervision.
A good design principle is to ensure that corrections are made as close to the source
of the variance as possible.

Many variances interact with others, thus causing a set of problems that affect the
efficiency of the work process from the input to the output stage.   A variance matrix
is a useful method for showing this interlinking.   An example of a simple variance
matrix for a department dealing with customer orders is set out as Figure 3.   It shows
how variance 1, orders which incorrectly identify goods required, and variance 3, too
many orders for staff to handle, have the greatest impact on the total work system.
Particular care must therefore be taken when redesigning the technical and human parts
of the work system to get these two variances under better control.   The computer
can be of assistance here.

Figure 3   VARIANCE MATRIX

Receipt of orders	1 Orders which incorrectly identify goods required				
		2 Orders without proper customer identification			
Transmission of orders to warehouse	1	2	3 Too many orders for staff to handle		
	1		3	4 Errors in goods requested	
Goods sent from warehouse to customer	1		3	4	5 Wrong goods sent to customer
		2	3		6 Goods sent to wrong customer
UNIT OPERATIONS	Receipt of orders	Transmission of orders to warehouse	Goods sent to customer		

This example shows some of the variances associated with a line department with one functional activity. The same approach can be used at different levels in the organisation, for example, identifying the variances of a research and development department or a planning activity.

Alternatively, a single manager can be used as the unit of analysis and his personal variances identified. This last approach would prove useful when designing information systems.

A major part of the socio-technical design task is to eliminate system variances or enable these to be more effectively controlled. If a computer system is being introduced then its ability to assist the elimination or control of variances is one measure of its efficiency.

Job satisfaction analysis

Job satisfaction is defined by Mumford as the FIT between what an individual or group is seeking from the work situation and what they are receiving from it, in other words the FIT between job needs and positive expectations and the requirements of the job (22). Job satisfaction is seen as being achieved when three kinds of needs are met in the work situation. These are personality needs, competence and efficiency needs, and needs associated with personal values. It can be argued that an improvement in job satisfaction should always be made a design objective and a design group concerned with job satisfaction should be able to answer the following questions:

Needs associated with personality

1 Knowledge needs: To what extent does the existing organisation of work meet
   the needs of the group of employees which the design group represents for work
   that fully uses their knowledge and skills, and to what extent does it provide them with the opportunity to develop their knowledge and skills further?

2 Psychological needs: To what extent does the existing organisation of work
   meet the needs of employees for recognition, responsibility, status, advancement, esteem and security?   Does it also give them a sense of achievement?(23)

Needs associated with competence and efficiency in the work role and the successful performance of work activities

3 Support/control needs: To what extent does the work situation provide employees with the kind of support services which enable them to carry out their job
   efficiently?   These support services include the information and materials
   necessary to work at a high level of competence, supervisory help, and encouragement and good working conditions.   We are here making an assumption
   that an efficient and supportive work environment increases job satisfaction.

   To what extent also does the way work is controlled through checks and audits
   fit with employee ideas and wishes on how their work should be controlled?
   The level and structure of wages and salaries will be an important part of
   the control system.

4 Task needs: To what extent does the way in which work is organised and jobs
   designed meet employee needs for the following?

   a) the opportunity to use a variety of different skills and different
       levels of skill.

   b) the opportunity to achieve targets, particularly quality targets,
       and to obtain feedback on how well these targets have been achieved.

   c) Autonomy.   The opportunity to take decisions, exercise choice and
       exert a degree of control over what is done and how it is done.

   d) Task identity.   The opportunity to undertake work which is viewed
       as important, which is organised in such a way that the work of
       one group is clearly separated from the work of other groups, and
       which has a reasonably long task cycle so that an employee can look
       back with pride on the way in which he has solved a particular work
       problem or carried out a challenging set of tasks (24).

Needs associated with employee values

5 Ethical needs: To what extent does departmental management, and senior management also, treat employees in the way they think they should be treated?   This

applies particularly to issues such as communication, consultation and opportunities for participation in decisions which affect employee interests.

This job satisfaction information can usefully be collected by questionnaire provided
that three important criteria are met:

(a) The information is collected in such a way that employees are
convinced of its confidentiality.

(b) Aggregate data derived from analysis of the questionnaires is
given to everyone who completed a questionnaire.

(c) Questionnaire data is discussed with all employees who complete
a questionnaire in small groups. This will check its accuracy,
provide an understanding of the reasons for high and low satisfaction and get employees involved in improving the design of
their own work organisation.

Variance analysis and job satisfaction analysis will provide essential diagnostic data
for gaining an understanding of existing efficiency and job satisfaction problems.
In addition there is a need to identify future needs, so a 'future analysis' is also
required.

Future analysis

The time taken to design, construct and implement a computer-based information system
will depend on the scale of the planned change and the resources available to make
the change. Even quite small systems changes take substantial time and resources,
whilst the implementation of major systems may require a period of several years.
Time cycles of three to five years are not uncommon for the period between the start
of an information systems project and its implementation.

To recover the cost of development, to cover the operating costs and to provide an
adequate return on the capital employed, the system might be expected to be operated
for a number of years before it is replaced by a new system.

Suppose it takes three years from project start to project implementation and a further
five years of systems operation before the advent of a new technology makes it advantageous to design a new system, which itself takes two years to design and implement.
Then the system which is being planned at A in Figure 4 will have to meet the requirements of the organisation not only three years later when it is implemented, but also
through its lifetime of seven years. In other words, those who design the system
must be capable of foreseeing the needs of the organisation ten years later.

Figure 4   SYSTEM LIFE CYCLE

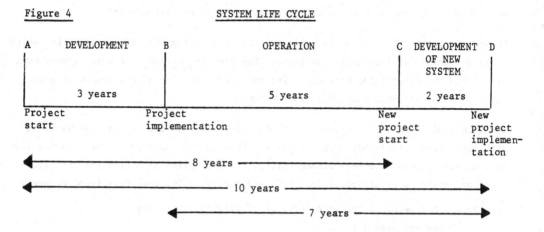

Many types of changes may have an impact on the information system; for example, the
development of new technologies, revisions in the organisation's structure, alterations
in the organisation's management style, changes in the attitude of employees to work,
new laws, changes in the economic climate, or alteration in the organisation's scope
or function.

In order to design a system which meets the future as well as the present requirements
of an organisation, the systems designers have to:

1 predict the kind of changes which could occur over the expected life
   time of the system

2 predict the extent to which the kinds of changes outlined above will
   have an impact on the jobs the systems have been designed to carry
   out.   In other words, how sensitive is the system to changes?

Because it is inherently impossible to build a completely flexible, completely portable
system capable of coping with any change, we have to define in some way the extent of
change it is possible to accommodate.

The further the designers look into the future the more they are faced with uncertainty
regarding the changes that may occur and the potential impact on the system they are
designing.   At some point in the future the uncertainty is so great that the system
designers cannot conceive of any design which can cope with a possible range of circumstances.   That point of time is called the forecasting horizon.

In practice the forecasting horizon will vary from organisation to organisation and
will vary in different epochs.   At times of high technological change, such as now,
with the coming availability of new micro-computer technology, the forecasting horizon
is closer to the present than at any other time.   A similar effect is noted at times
of economic instability.   Some organisations have a very stable environment and hence
can forecast with reasonable certainty over quite long periods, others live in a much

more dynamic or uncertain world and forecast for a short period ahead.

No systems designers can plan to build systems on an assumption of a system life which could go beyond the forecasting horizon. The planning horizon for a new system must lie within the forecasting horizon. The expected life time of the system then is related to the planning horizon of the system.

The traditional method of designing computer-based systems has not in general dealt with the future uncertainty adequately, and the actual as opposed to the expected life time of the systems have been disappointing. The problems stem from the division of function between user and specialist EDP departments. The user department manager:

a) is not aware of the inherent lack of flexibility of the computer-based system,

b) does not realise the sensitivity of the system to different kinds of changes,

c) regards many aspects of future policy as being outside the scope of discussion and, in many cases, regards questions of future policy as confidential.

Hence, he makes no special effort to predict the kind of changes which might occur.

A specialist designer, in concentrating his efforts on finding a technical answer to meet the immediate problem, is unaware of the dynamic nature of his environment and the extent of uncertainty about the future. The level of communication between user and specialist regarding the future tends to be low and unstructured.

The solution lies with a structured approach to future analysis. The two groups involved in the design process - the steering group and the design team - join together in the first instance to attempt to define the forecasting and planning horizon for the system. At this stage the scope of the new system has already been defined and the groups have some rough idea of the expected life of the system. Changes can affect the system by a change in the system's logic, or by a change in activity levels.

The structured process involves, first, drawing up a list of factors relevant to the system which are subject to change, and second, assessing the likelihood of the factors changing significantly over the expected life time of the system.

This completes the diagnostic stage of our approach. The next step is to set objectives and evaluate alternative strategies for meeting the needs identified in the variance, job satisfaction and future analyses.

Benefit assessment

Our approach to multi-attribute utility analysis, which we call BASYC (Benefit Assessment for System Change), expresses the benefits of alternative policies to different groups of people in terms of relative progress towards the goals important to them. The BASYC approach is intertwined with the other procedures described in this paper, but its key concepts can be displayed simply:

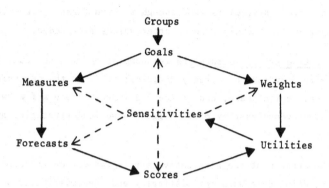

The sequence is shown without an exit to emphasize that a decision "emerges" after a number of cycles of discussion held by members of the design team with their consti- tuents, with other interest groups and with their steering committee. The discussions are guided and disciplined by the formal structure of utility calculation, but as the diagram shows, the <u>sensitivities</u> are central to the pattern.

The word "sensitivity" is used in two senses here; both in its mathematical sense - in which it means the extent to which a value would change if one or more of the input variables changed - and in its conversational sense - drawing attention to the effect of proposed policies on the feelings of the people involved.

The detailed steps the design team will carry out in the benefit assessment sequence are as follows, when using <u>numerical</u> measures at the starred points:

<u>Identify interest groups</u> which are not already represented in the team (customers, owners, managers, taxpayers, ...), paying particular attention to any subgroups that might suffer from proposed changes.

<u>Shortlist the measurable goals</u>, relevant to the situation being studied, which are most important to all the interest groups (including here both groups which have direct representation on the design team and those which have indirect repre- sentation). The techniques of variance analysis, job satisfaction analysis and future analysis will provide a great deal of the information required to do this.

<u>Assign a weight*</u> to each goal for each group to represent its importance to that group in relation to the other goals; this should be done in consultation with members of the group. Job satisfaction must always be included as a goal for internal groups. (For calculation of utilities these weights are converted to

percentages so that the total weight of all goals for any one goal is 100.)

<u>Estimate current measures</u>* (often several are required for each goal) and set
target measures for the planning horizon date; again all three of the analytical
techniques will provide information for these measures. Guesswork (by people
who know the current system) is good enough for the first cycle - later, sensitivity analysis will determine where better values are needed.

<u>Define alternative systems strategies</u> for comparison with the current system (as
it will be after any changes already decided); this is the beginning of the
socio-technical design phase, and at the preliminary stage a few "wild" ideas
are quite advantageous as they often suggest practical strategies not generated
otherwise.

<u>Forecast</u>* the effects of the alternative strategies on the measures at the plan-
ning horizon, doing this both optimistically and pessimistically as outlined in
the future analysis section above. Again the sensitivity of recommendations to
assumptions will determine how much effort to put into refining forecasts.

<u>Score</u>* each strategy for its success in meeting each goal as seen by each group
using a scale running from -10 (change very much worse than existing system)
through 0 (same as existing) to +10 (change very much better than existing system). Note that the comparison is between the changed system at the planning
horizon and the existing system <u>also at the planning horizon</u>. Different optimistic and pessimistic scores corresponding to different trends are often needed.
Often the scores will be the same for all or most of the groups, but this needs
checking as groups' ideas on what constitutes "success" may differ. This scoring
operation needs human judgement; it is not mechanical.

<u>Calculate utilities</u>* by multiplying each percentage weight (for each goal for
each group) by the corresponding score (for each goal for each group for each
strategy) and adding over all shortlisted goals. This is a purely mechanical
operation, conveniently done by computer. The result will be (optimistic and
pessimistic) utilities for each strategy as seen by each group. The numbers
obtained are not important in themselves - what we are seeking is the ranking
order of the alternative systems as seen by each group and, more important, the
sensitivity of the ranking order to changes in goals, weights, measures, scores
or assumptions.

<u>Consider the sensitivities</u> carefully to decide what further investigations are
required; often it can be seen that some strategies can be discarded, and occasionally all agree that the right decision has emerged - but more often a new or
compromise strategy will require examination in a further cycle before an acceptable solution is found. The actual decision when "enough is enough" must always

remain with management - or with the steering committee if this power has been delegated to it.

It must be emphasized that a number of the stages of benefit assessment are likely to be negotiating processes as the representatives of different interests on the design team press for priority to be given to those goals and strategies most likely to further the needs and expectations of their constituents. One advantage of the method is that these issues are brought out into the open.

When a decision to proceed does emerge, the team will usually go on to the full socio-technical design process, described below. This may itself be iterative - with initial cycles corresponding to design of prototype systems or part-systems, followed by BASYC cycles to check on likely benefits as compared to those estimated in the assessment stage.

The BASYC approach has been applied in two major projects: comparison of alternative computerisation schemes in savings banks (25), and benefit assessment in public libraries (26), the latter being an application not involving computer systems design at all.   Two short-cut methods were developed for use in the library project: MINIBASYC for initial training and SEMIBASYC for simple real problems (27).

Fuzzy approach to benefit assessment

One of us (JH) is now engaged, with Janet Efstathiou and Vladislav Rajkovic, in applying the concepts of "linguistic variables" and "fuzzy algorithms" (28) at the starred points in the BASYC approach.   The use of numerical weights, measures, scores and utilities is abandoned - partly because of the spurious precision which numbers convey to the statistically unsophisticated participant, but more because some important theoretical obstacles concerning orthogonality of goals and aggregation of utilities are thereby surmounted - or, rather, the obstacles simply disappear.

In this new variation, weights may be expressed as "very important", "rather important", "not important", ..., measures in terms appropriate to individual goals (e.g. "comfortable", "prompt", "cheap", "accurate"), and comparative benefits as "much more acceptable", ..., "much less acceptable".

The final comparative rankings will be derived direct from the linguistic measures without use of "partial utilities", by fuzzy algorithms chosen by the design team (within the rules of fuzzy logic) to represent the decision processes of the different interest groups, real members of these interest groups being consulted about this.

Socio-technical systems design

A participative approach to work design means that the employees of a department or their representatives construct a new form of work organisation which is based on a diagnosis by them of their own needs. There are a number of philosophical approaches to work design which such design groups may want to consider. The two most frequently used are job enrichment and the socio-technical approach. Job enrichment focusses on the job of the individual worker and tries to build up this job in such a way that it increases in interest, responsibility and challenge. The job may be extended by adding to it preliminary activities such as setting it up and acquiring the necessary materials, or completion activities such as final quality inspection and the rectification of errors, tasks which previously have been carried out by other individuals. The aim of job enrichment is to improve the relationship between the individual and his work.

The socio-technical method was originally developed by the Tavistock Institute in Great Britain and this takes a very different approach. The concept of a socio-technical system is derived from the premise that any production system requires both a technology, a process of transforming raw materials into output, and a social structure linking the human operators both with the technology and with each other. A socio-technical system is any unit in the organisation composed of a technological and a social sub-system having a common task or goal to accomplish (29). If we are concerned with clerical systems based on the use of a computer, the technical system will consist of the tools, techniques and procedures used for processing the raw material of information. The social structure is the network of roles, relationships and tasks which interact with the technical system. The purpose of the socio-technical systems approach is to produce technical and social structures which have a high capacity to achieve technical and social goals, and which reinforce each other in the achievement of these goals.

Socio-technical analysis incorporates a logical analysis of the technical components of the work system (machines, procedures, information) and the grouping of these into 'unit operations' (21). Unit operations are logically integrated sets of tasks, one set being separated from the next by a change of state in the input or product. For example, in a Purchase Department the tasks of preparing accounts data for a computer, putting the data into the computer and correcting errors is a logically different set of activities from matching accounts with goods received notes and investigating discrepancies. Work design which uses a socio-technical approach identifies unit operations and allocates one or more unit operations to each work group. The work group then has the responsibility for allocating tasks amongst its members and for training its members so that eventually each individual is competent to carry out all tasks.

The analysis of the social part of the work system consists of analysing the role relationships within the system, in other words who needs to work with whom for the system to function efficiently. In addition an analysis is made of the job satisfaction needs of individuals in the department, using the theoretical framework described earlier.

An important aspect of the socio-technical approach is the notion of 'control'. A further analysis of the technical part of the system is carried out using the variance analysis technique already described. Operating efficiency is then improved by giving each work group the responsibility for eliminating and correcting variances which occur within the set of unit operations for which it is responsible. Job satisfaction is also improved through handing over this control function to the group. The group requires a set of problem solving skills to enable them to handle variances successfully and the acquisition of these involves an enhancement of the knowledge of individual group members. It is believed that responsibility for, and ability to solve problems increases the interest of work and provides a sense of achievement.

Final steps

Once our new system has been designed it then has to be implemented and its success evaluated. Both of these stages are complex and demanding, and there is no space to discuss them in this paper. We would however suggest that the creation of a small prototype system is a wise safeguard as it enables the selected system design to be tried out and modified before it has wide scale implementation (30). When the time for evaluation of the new system arrives we have found that an excellent check of its technical efficiency is the extent to which it has eliminated or gained better control over those variances identified at the diagnostic stage without introducing new ones. A second measure is the extent to which job satisfaction has improved through the creation of a better fit between employee job needs and expectations and the requirements of their jobs on our five satisfaction variables.

CONCLUSION

The approach set out in this paper stems from the authors' belief that 'participation' should be associated with systems planning and design. This belief is based first on the value position, that people affected by new technical systems should have a right to influence the design of these systems and, second, on the practical proposition that the future users of systems will possess a detailed knowledge of both the efficiency requirements of the situations in which they work and their own job satisfaction needs. They will, at the same time, have a knowledge gap in that they are unlikely to have much experience of the analytical and synthesising skills that are required in systems design. The authors have attempted to fill this gap by developing a set of simple, structured procedures that will take a newly formed design team from an initial diagnosis which covers efficiency, job satisfaction and future contingency needs, through the setting and weighting of objectives and the development of alternative strategies, to a choice of system and its socio-technical design. Table 1 shows these different procedural steps used at different organisational levels and with different forms of participation.

Table 1

THE PARTICIPATIVE APPROACH AT DIFFERENT ORGANISATIONAL LEVELS

	STRATEGIC PLANNING	SYSTEM DEFINITION	SUBSYSTEM AND WORK DESIGN
SCOPE	Whole organisation	Division or function	Department or subfunction
TIME SCALE	5 years or more	1 to 5 years	2 years or less
FORM OF PARTICIPATION	Consultative*	Representative*	Consensus*
TASK	To identify systems likely to be needed under different circumstances, and make outline contingency plans for system development.	To specify overall system architecture, interfaces, hardware and software compatibility requirements, etc., within agreed strategy.	To design system modules and work procedures within overall concept defined at higher levels
DESIGN TEAM	3 or 4 senior managers or Board members plus 2 or 3 internal or external consultants.	5 or 6 operational staff selected by senior manager responsible for division or function, plus technical experts as advisers.	5 or 6 elected or selected departmental staff, plus department manager if he/she wishes, technical expert and trainer/consultant.
DIAGNOSIS	Check performance of current system against original design objectives and forecasts. Identify relevant economic, social and technological trends.	Analyse operational and job satisfaction deficiencies of current system (variance analysis, job satisfaction analysis).	Identify and measure local variances between system performance and agreed social and technical targets.

59

GOAL SETTING	Identify new opportunities and adjust priorities of different elements in broad aims of organisation.	Define system development objectives relevant to different interest groups within broad priorities laid down at strategic level.	Interpret system goals as detailed local social and operational targets, in consultation with steering group.
ASSESSING ALTERNATIVES	Define main alternative strategies and forecast their impact on broad aims under different conditions.	List all feasible system options and forecast their effects on system goals important to interest groups.	Consider possibilities for local variations within overall system parameters and compare likely measures.
CONSULTATION	Discuss goals and options with representatives of major interest groups and with other study teams; present to Board for decision.	Discuss goals and options with "constituents" and with other study teams; present to decision maker for choice of system.	Discuss goals and options with all departmental staff and with other study teams; obtain staff agreement to prototype trials.
PLANNING AND DESIGN	Work out details of chosen strategy or contingency plans and check forecasts under different conditions.	Outline recommended system and work out details of modular structure and interfaces; recheck forecasts.	Design and implement prototype modules and/or work procedures, observe effects and redesign accordingly.
EVALUATION	Present to Board and, if accepted, to Unions, other study teams and relevant external agencies; report reactions to Board.	Collate results of local trials, check their compatibility with overall design and important goals and present to decision maker.	Check redesigned procedures and subsystems against the full list of operational and social targets and present to staff for agreement.

* In each case the form of participation in the adjoining column may also be appropriate. For example, a consensus approach might not be feasible in a department with more than fifty staff. A representative approach might be preferable. Similarly, if for any reason a representative design group cannot be created, a consultative approach can still be used.

REFERENCES

1 LAND, F.F. Criteria for the evaluation and design of effective systems, Economics of Informatics, A.B. Frielink ed., North Holland, 1975

2 ZIMA, K.P. Project-Wirtschaftlichkeitsrechnungen bei der ESSO AG, Proceedings of the Informationsforum "Die Wirtschaftlichkeit von Informations- und Kommunikationssystemen", BIFOA, 1977

3 MARSCHAK, J. Economics of information systems, Journal of the American Statistical Association, March 1971

4 LEVEN, P.H. On decision and decision-making, Systems and Management Annual, 1974, R.L. Ackoff ed., Petrocelli Books, 1974

5 FLOWERDEW, A.D.J. & WHITEHEAD, C.M.E. Problems of measuring the benefits of scientific and technical information, Economics of Informatics, A.B. Frielink ed., North Holland, 1975

6 ZALANY, M. Introductory notes on conflict dissolution, Proceedings of ORSA/TIMS Joint National Meeting, Philadelphia, 1976

7 DUZMOVIC, J.J. Criteria aggregation technique for evaluation, optimisation and selection of computer systems, University of Belgrade, March 1976

8 ZANGEMEISTER, C. 'Nutzwertanalyse', in The Economics of Informatics, A.B. Frielink, ed., North Holland, 1975

9 GENKINGER, P.F. Implementing 'Nutzwertanalyse', in Economics of Informatics, A.B. Frielink, ed., North Holland, 1975

10 SIBLEY, E.H. Evaluation and selection of data base systems, Database Systems Infotech State of the Art Report, 1975

11 BUSCH, U. & KROPP, I. Kombination von Monetaren Wirtschaft-Lichkeitscrechnungen und Nutzwertanalysen, Proceedings of the Informationsforum "Die Wirtschaftlichkeit von Informations- und Kommunikationssystemen", BIFOA, 1977

12 LAND, F.F. Evaluation of systems goals in determining a decision strategy for a computer based information system, The Computer Journal, Vol.19, No.4, November 1976

13 HAWGOOD, J. Quinquevalent quantification of computer benefits, Economics of Informatics, A.B. Frielink ed., North Holland 1975

14 MUMFORD, E. Towards the democratic design of work systems, Personnel Management, Vol.8, No.9, September 1971

15 SUSMAN, G. A socio-technical analysis of participative management, Praeger, 1976

16 OZBEKHAN, H. Towards a general theory of planning, in E. Jantsch (ed.), Perspectives of Planning, OECD, 1969

17 BLAKE, R., SHEPARD, H. & MOUTIN, J. Managing intergroup conflict in industry, Houston, Gulf Publishing, 1964

18 GUSTAFSON, H.W. 'Force-loss lost analysis', paper presented to the AWV-Fachseminar: Das Human Kapital der Unternehem, Bonn, September 1974. Quoted in A. Hopwood 'Towards assessing the economic costs of new forms of work organisation', Accounting, Organisations and Society, 1977

19 MUMFORD, E. & PETTIGREW, A. Implementing strategic decisions, Longmans, 1975

20 DAVIS, L. Job satisfaction research: the post industrial view, Industrial Relations, Vol.10, 1971

21 TAYLOR, J.C. The human side of work: the socio-technical approach to work design, Personnel Review, Vol.4, No.3, Summer 1975

22 MUMFORD, E. Job satisfaction: a method of analysis, Personnel Review, Vol.1, No.3, Summer 1972

23 HERZBERG, F. Work and the nature of man, London, Staples Press, 1966

24 COOPER, R. How jobs motivate, Personnel Review, Vol.2, Spring 1973

25 LAND, F.F. Evaluation of systems goals in determining a design strategy for a computer based information system, Computer Journal, November 1976

26 HAWGOOD, J. Participative assessment of library benefits, Drexel Library Quarterly, July 1977

27 HAWGOOD, J. How to apply participative planning to your library policy change, Automation Benefit Appraisal Consultants, Durham, 1977

28 ZADEH, L.A. Outline of a new approach to the analysis of complex systems and decision processes, IEEE Trans. on Systems, Man and Cybernetics, Vol.SMC-3, No.1, January 1973

29 ROUSSEAU, D. Technological differences in job characteristics, employee satisfaction, and motivation: a synthesis of job design research and socio-technical systems theory, Organisational Behaviour and Human Performance, 1977, Vol.19, pp.18-42

30 EARL, M. Prototype systems for management information and control (to be published shortly)

A PARTICIPATIVE APPROACH TO SYSTEMS ANALYSIS : AN ACTION RESEARCH

IN THE LOCAL GOVERNMENT

C.Ciborra *   G.Gasbarri **   P.C.Maggiolini ***

ABSTRACT

In order to present a participation-oriented systems analysis method,
the rationale of users' participation is discussed considering the recent trends in systems analysis techniques from an organizational
point of view and putting forward a typology of organization and information systems based on the "integration" criterion.

The local government appears to be an organization which controls a
"hypointegrated system" composed of differently goal-directed socio-economic units operating on a portion of land. For such a system the participation issue in systems analysis gains a crucial importance.

The description of an action research in the local government shows a
possible approach to the analysis and design of Information Systems
and Organizational Systems in a hypointegrated context.

*   Istituto di Elettrotecnica ed Elettronica del Politecnico di
        Milano, P.zza Leonardo da Vinci, 32 - 20133 Milano, Italia.

**   Independent Consultant and Lecturer, C.so Plebisciti, 15 -
        Milano

***   Centro Studi Ingegneria dei Sistemi di Elaborazione delle In-
        formazioni, CNR - Milano
        Dipartimento di Sistemi dell'Università della Calabria -
        Cosenza.

1. PRELIMINARY REMARKS

An information system (IS) may be defined as the set of methods, procedures, formal and informal techniques and technologies for collection, storage, processing and distribution of information.

The IS is part of the organizational systems. Both entities form the real field on which system analysts act and to which the computer-based IS is applied.

Systems analysis and design are activities directed to analyze/descri be the present IS and OS, and build a new IS (and, in a subordinate and often implicit way, a different OS). Both activities may be labeled as "model building" processes: a first model is the outcome of the analysis phase, a means for the diagnosis and evaluation of the real IS; the second model (the newly designed IS) is implemented through hardware and software media, new procedures, new rules etc., and becomes a part of the real organization.

Schematically, the analyst's work, which covers the life cycle of the IS (shown in fig.1), is a dialectical process of abstraction, model building and action.

System analysis techniques are formalized "packages", which support the analyst's job: they usually contain a check-list of the activities to be performed in the different stages of system development, some documentation techniques and, more recently, computer-aided tools.

2. RECENT TRENDS IN SYSTEM ANALYSIS TECHNIQUES: PROBLEMS AND EXIGEN-CIES FROM AN ORGANIZATIONAL POINT OF VIEW

In what follows only the analysis stage of system development is discussed. The analysis phase may be thought of as a process of perception and selection of the real organizational and "informational" processes, followed by the description in terms of a formal model and, through abstraction and classification activities, by the building of an "infological" model, which will be the input for the design and implementation of the "datalogical" model.

As far as the analysis is concerned, the system analysis techniques suggest, more or less extensively, what is worth to be examined in the OS and IS, how to analyze it, how the relationship between IS and OS should be looked at, what is the interaction between the analyst and the rest of the organization (top management, users, workers, etc.).

Many authors (for example Couger (1973), Golvers (1976)) have pin-
pointed the crucial role of the analysis stage, both from the functional and economic point of view. The functional importance is due to
the following factors:

- mistakes in the requirements definition are often detected only very late in the development of the system: this fact may influence
  negatively the effectiveness of the system;

- the analysis produces a "reference model" for the evaluation of the
  efficiency/effectiveness of the new IS;

- during the analysis, forms of interaction are institutionalized
  between the specialists and the users,which have an influence on
  the whole design/implementation and usage of the new system.

The economic relevance derives from these facts:

- the time spent in the analysis may reach one third of the total
  system effort. Couger (1973) writes that, passing from the first to
  the third computer generation, the analysis phase (study of the existing system and evaluation of requirements) has increased its importance (expressed in percentage of total life-cycle time) from 5%
  to more than 20%;

- in order to eliminate mistakes (made during the analysis) from the
  new system,costs may be one hundred times higher than those derived
  from an effective analysis.

Comparing the increasing importance of the analysis stage on one hand
and the evolution of the system analysis techniques (from ADS/3/to
BISAD/4/,SOP/5/,BSP and ISDOS/6/) on the other, one may still notice
gap between the two and the urgent need for further research in this
area.

In fact, while the present trend is characterized by an effort of au-
tomatizing the description/documentation and design of formal models
(info-and-data-logical;see as an example the ISDOS project), relatively little effort is put on the specification of the   crucial
areas/problems in the IS and OS, in order to build a model which is
not going to be useless   although extremely formalized and coherent.

Golvers /2/ describes the present unsatisfyng situation in this way:

- there is no effective methodology for the mapping of the real system
  into a descriptive model, functionally oriented to the implementation
  of the new IS;

- there is a confusing coexistence of extremely formalized descriptions

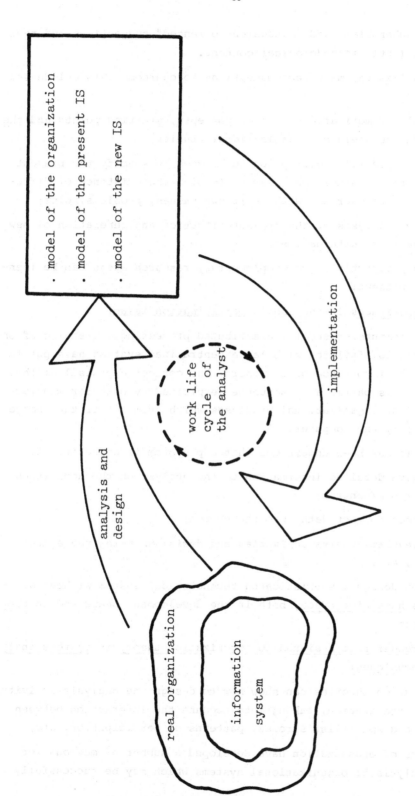

Fig. 1 — Work life of the system analyst

- model of the organization
- model of the present IS
- model of the new IS

work life
cycle of
the analyst

analysis and
design

implementation

real organization

information
system

of local subproblems and "management oriented" suggestions and reci-
pes, with little methodological content.

Sol summarizes the main reccomandations to overcome this cul-de-sac
/7/:

- rational and motivated choice of concepts, starting points and the
  ories in the field of organizational studies;

- choice of possible description techniques to specify the present
  and the target situation, bearing in mind the existence of different users'categories, with their own values, problems,roles;

- effective analysis of the present situation and generation of new
  organizational alternatives.

In such a perspective, two complementary research areas can be identified at present:

a) the experiences in the user's participation issue.

New small computer-based IS, distributed processing, spreading of on-
line terminals, failures with large centralized systems have put in
evidence the importance of the users. The correct appraisal of the
user's role is partly a consequence (exploited by computer manufac-
tures,too) of organizational processes,which stem out of the change
introduced by the computer.

Briefly, it has been understood that Edp system is effective if:

- users give detailed information to the analyst on the activities
  they perform/control;

- users feed correct data into the system;

- users regulate theirs activities and decisions using the system
  outputs, etc.

One cannot design a sophisticated technological system without designing its human interface, both in the development phase and in the
utilization.

b) the broader role assigned to organization theory in systems analysis techniques:

- organization theories can show how to design the analysis activity
  within organization, taking into account the interaction between
  users' and specialist's roles, patterns of participation, etc;

- students of organization have developed a number of methods for
  the analysis of organizational systems which may be successfully

included in or adapted to Edp analysis tecniques;

- various organizational models have been proposed and implemented,
which may be a reference for the introduction of Edp technology.
Up to now, i.e. in the current techniques, organizational models
are implicity assumed as costants in the change process.

Works and experiences that are worth to be mentioned as examples of
such new trends are those of E. Mumford /8/, who has applied a participative work design to the Edp environment, the suggestion of I.
Mitroff /9/ and B. Hedberg /10/ about "dialectic information systems",
the new approach of the Swedish School represented by the "user-oriented system design" /11/ and, finally, the workers'participation implemented by the Scandinavian trade Unions (see K.Nygaard) /12/.

In such melting pot of proposals and experiences, a rigourously defined methodology has not yet appeared: just now the participation
issue and the ideas derived from organization theories are being explored and tested. In this turbulent search process naive attempts
of manipulation, social control, misuses of organizational concepts
have not been avoided.

The normative aspects of participation are often emphasized, without
providing the means for effective (real) participation, for example
in the definition of information requirements bound to real exigen-
cies.

3. ANALYSIS TECHNIQUES FOR HYPOINTEGRATED SYSTEMS

Often, the user-oriented approach is just a means to make the user responsible for the new system which has been designed according to criteria stated by someone else (top management and/or specialists). This
formally participative approach may be effective when the users are
few, when the system to be automatized is limited and cleary goal-directed.

Our impression is that such approaches, even in their newer look, are
condemned to failure, if they are applied to "imperfectly organized
systems" (1), systems where various users' categories are present. This
seems to be the case of urban, regional socio-economic information systems.

Such an impression stems from the assessment of Edp applications in the
public government (local and central): there are plenty of sectorial
computer-based IS, with no co-ordination between them and a surprisin-

gly low automation degree.

Most probably, there is a link between the problem of imperfectly
organized systems and the participation issue. Such a link may be
identified considering the structural motives or factors that support
the user's participation in specific organizational structures during the system analysis and design process.

The automation of activities and procedures, or the collection and
storage of data for management decisions (data base) require a process of formalization of the laws that govern decision processes and activities within the organization (processes that take place outside
the organization are somehow internalized and can be treated as the
formers). There are two possibilities: formalization exists already,
and the problem is to transfer it into an Edp-oriented form; or,
formalization can only be reached through a bargaining process with
those who own the process local memory (i.e. the users) (2). In the
latter hypothesis, the analysis, design and implementation of an effe
ctive IS require that the users agree to placing the memory of the
processes they control at the analyst's disposal. When the new system
runs, the users must also participate and cooperate in order to fill
in those "gaps" or "fissures" between the implemented Edp model and
the real OS.

User-oriented systems analysis techniques try to take into account
(sometimes uncounsciously) the power bargaining game (memory transfer)which
takes place during the implementation of a new model within the organization.

A typology of information and decision systems - Having shown the stru
ctural bases of participation, it is necessary to turn to the different types of organizational systems, in order to find out in what
situation a fully participative approach is absolutely essential to
create an effective IS.

J.L. Le Moigne /13/ has suggested the following typology of systems,
based on the integration degree of their parts :

- hyperintegrated systems (e.g., the family)

- mesointegrated systems (e.g., the entreprise, the purposive administration)

- hypointegrated systems (e.g., local communities, a region, a country,
  etc.).

The mesointegrated systems are purposive organizations, such as an industrial entreprise: each of its departments is a purposive unit, whose objective is to execute a part of the purchase-production-sale cycle; the whole entreprise has, as main goàl, profit, (in the short and/or long term) etc. In such systems the organizational structure that controls and regulates the subsytems (departments, functions, individuals) is formally defined, and so is the autonomy of each subsystem. The system and subsystems dynamics is observable, at least to a certain degree; the "memory" of each department is formally located within the organization and it is composed of: procedures, methods, tasks, quantitative and qualitative information (on production or administrative processes) etc. The organizational texture and the set of its objectives and subobjectives(to a certain degree) reflects selectively the socio-economic processes of the environment and internal needs of coordination.

A regional or local community are hypointegrated for the following reasons: each socio-economic unit (agricultural,industrial, service or commercial units) produces goods or services, which may be not integrated with the objectives of the other units operating on the same territory, although adequate to the survival of the specific socioeconomic unit. Thus, the texture of the whole region appears to be fragmented from the point of view of the coordination of the goals/ objectives of the various socio-economic units.

Furthermore, in such systems there are observability problems: their "memory" is fragmented,departimentalized, "distributed"; there are many decision points, with a high degree of autonomy and a low or non existent degree of coordination.

The"organizational structure" of hypointegrated systems does exist but is imperfect;for example functional and economic bindings between agriculture, manufacturing industry, local government, services and so on may be identified but such links, which form the "texture" of the systems, are often difficult to observe and control. In such a context an effective IS is difficult to implement using traditional techniques.

One should consider that systems analysis has been developed taking as reference model mesointegrated systems (see for example the Hearle and Mason's approach to the analysis and design of the IS in a local government /14/). Up to now, little effort has been put in the analysis and design of IS in hypointegrated systems, where the"organizatio nal structure" is a problem, i.e. something to be reached and not something to be simply described in terms of functional analysis. The action research reported below does not provide a fully developed systems analysis methodology, but it shows an interesting experiment of participation-oriented analysis of the IS and OS in the Public Administration environment, taking into account some of the concepts discussed above.

4. MOTIVES OF INTEREST OF THE CASE OF LOCAL GOVERNMENT INFORMATION SYSTEMS

The Public Administration and, more specifically, the local Government environment, are very interesting from the point of view of system analysis techniques and Edp, at least in the Italian context. In fact,

- the public administration is in the middle of a deep crisis: it is unable to change its structure, its way of operating and its relationships with the outer society, in a turbulent socio-economic environment. Its "texture" is more and more not coherent with the changing texture of society. Edp technology has been proposed and considered as a means to rationalize the old procedures, to redesign the organizational structure in order to build up regulative capabilities on economic and social processes;

- the diffusion of Edp in the Public Administration has meant the automation of routine procedures (not always in the interests of the, citizen-user) and the reinforcement of the pre-existing bureaucratic structure; often it has also meant a considerable economic waste or under-used facilities;

- the philosophy which has guided the introduction/application of the computer in larger and smaller authorities has been that of high centralization with the exclusion of users and citizens;

- a public agency (e.g. a local government) must intervene in processes which are extremely loose, fuzzy, difficult to reach, control and regulate: this fact has not had any consequence on the development of appropriate system analysis and design strategies;

- according to our hypotheses, in such hypointegrated system the users (citizens, economic operators, public managers, etc.) should play an active role as sources/receptors of information and as "organization-builders".

5. SYSTEMS ANALYSIS IN THE LOCAL GOVERNMENT

An IS, dedicated to the regulation of a socio-economic system characterized by its extension on a territory, should be able to identify
the interdependences between subsystems, and the relationships between them and the regulating administrative superstructure.

Different plans of analysis may be identified: firstly there is a
physical or geographic plan which indicates the location of socio-
economic activities on the territory. Secondly, there is a "logical
plan" which indicates the aggregations of interrelated activities.
This plan may represent the location of the "memory" of socio-economic processes. Then, there is the "bureacratic plan", which gives
the formal scheme of the administrative structure of the local government, subdivided into goal-seeking departments, agencies, etc.

Finally, there is the political system (local parliament or council)
which should govern the whole administrative structure and the socio-
economic system underneath.

To analyse and design such a multilevel structure, trying to preserve a certain degree of coherence between the "texture" of territory
and the regulation superstructure, it is necessary to:

- identify the problems that are tackled or avoided by the present
  administrative structure (comparing the extant functions of the
  administration with the socio-economic processes in the environment);

- check the regulation and control variables at disposal, i.e. the
  means to intervene on the territory (laws, commercial and house-
  building plans, urban plans, etc.);

- identify the legal constraints to the intervention of the local
  government, and the economic and human resources at disposal;

- define the means that can be used to study the territory system:
  surveys, other forms of data-generator units, such as general meetings of users-citizens who operate on the territory; specific departments within the local government, etc.

- istitutionalize forms and mechanisms of participation for the users,
  as feeders and actors of the new system: such mechanisms should aim
  to collect and unify the "distributed" memory of each operator on
  the territory. Only the part of memory which may be of collective
  interest will be shared.

These steps are preliminary conditions to start a process of"building the organization" of the hypointegrated system. The participation issue appears to be not only a democratic ideal but, more rudely, a constraint for the effectiveness of the new IS and OS.

The model-building process (core of the analysis phase) starts from the local perception of socio-economic phenomena carried out by the local user (citizen) and the mutual share of the local knowledge acquired, together with an activity of stimulus carried out by the local government itself.

This down-top approach, in which the definition of users, socio-econo mic units, operators etc. is a problem not a datum, is radically dif ferent from the current approach of the socio-economic data base for the local government.

Implicitly assuming that the territory is a hypointegrated system, the Edp specialists think that a general-purpose data base will be the right solution for the information needs of a local government: all the data at disposal are put in the data base; every user can pick out the relevant information for him, etc.

But data-base approach is an apparent solution. In the data base only general statistics are usually included (of little meaning for the local user), because it is difficult for the user to pre-define what information is relevant to his decisions; the hypointegrated system being loosely connected and/or scarcely purposive,each operator is unable to define its information needs a priori.

The effort, on the contrary, should consist in starting a process of organization and purpose definition in the hypointegrated system; only later it will be useful to think about possible data-base functions and structure.

5.1. The steps of the analysis

The analysis of the hypointegrated systems is an iterative and participative process, that can be split into the following phases:

1. Identification of the users and their main goals

2. Identification of the economically important subsystems (the socio-economic units)

3. Identification of the interdependences between socio-economic units

(drafting of the "logical map" of the territory)

4. Identification of information sources (statistics, opinion leaders, social groups) and design of the procedures to collect missing data and information about the socio-economic units

5. identification of the main mutual influences between units, trying to quantify the interdependences

6. Research on the administrative structure and definition of organizational alternatives coherent with the "texture" of the territory identified by the logical plan.

A detailed description of each step applied in local government of Otranto (South Italy) is contained in /5/. Further remarks are made here to pin-point some aspects in relation to the hypointegrated texture of the system, and to the more general problem of systems analysis:

i) the definition of the users and their objectives is of major importance as a first step of the analysis. In fact, users have an active information role in structuring the organization of the hypointegrated system. A possible list of users is the following:
   - the elected political men (the city major, the city council, etc.)
   - members, chief Executives, officers belonging to the various public agencies and to the administrative structure
   - the socio-economic operators, some of whom may be opinion leaders
   - the citizenship: users of public services, organized users (trade unions), workers,etc.

   The user's participation is,at the same time,an instance of democracy and a choice of method in analysing and designing the new IS. More specifically, the analysis should become a collective learning process, concerning the socio-economic reality of the geographic area in which users live and work.

ii) The analysis of the territory system takes place by decomposing it into subsystems according to criteria of economic and functional importance. Another criterion is given by the relative responsibility of the social actors who are going to intervene on a specific subsystem.

iii) The logical plan, that is the output of the preceding phases, is related to a possible logical structure of the database: but this does not mean that its function is Edp oriented! On the contrary,

the logical plan is a proposal in order to start a process of
information collection and of socio-economic unit description.
Its degree of definition increases as the collective consciousness
of the people about socio-economic phenomena increases.

Each socio-economic unit is represented as a "box" in the logic
plan. Usually such "boxes" represent important economic activities
(agriculture, families, labour force, industry, trade, etc.). Each
"box" in connected with the others according to the perceptions
and knowledge of the users. Each activity is then composed of
"segments" : for example, the "box" "Agriculture" can be subdivided into the following segments: hunting and fishing, zoothechny
growings, etc.; the "box" "Tourism" into the segments: international holiday resorts, hotels, private rooms, etc.

For each "segment" the following aspects should be identified and
quantitatively analyzed:

- economic flows (revenues, turnover, etc.)

- functional relationships (with other segments)

- structures (geographic location, number of employees, type of
  products, etc.)

The graph of the "logical map" serves as a schema of reference ,
i.e. as a grid in order to aid the collection of data from different sources. Different methods of collection are employed.

The starting point obviously, should be a consultation of data
from ISTAT (National Institute of Statistics).

Information about socio-economic units is more easily found from the
principal operators in each sub-system (economic operators, opinion
leaders, etc.) who can furnish personal opinions which are, however,
based on specific experiences.

Such opinions can be considered hypotheses to be verified by ad hoc
inquiry on the field.

When estimates (even qualitative ones) are obtained from various
opinion leaders, it is possible to compare them in order to identify both the trend of particular local socio-economic phenomena and
the consequent problems involved.

The opinions collected and the results of partial inquiries are
then presented for discussion in popular assemblies on specific

themes corresponding to the more important socio-economic units.

The process of information enrichment brought by popular participa
tion makes it possible to transform the logical map into an oriented flow graph (were the arcs which join the various blocks are
in some way quantified).

On the graph it is possible to identify priviledged itineraries
or "circuits" according to their socio-economic importance.

At first the nature of the relationships is of a qualitative kind;
in order to organise an intervention it is essential that quantitative structure be found, even, if necessary, by carring out ad
hoc inquiries.

5.2. The control superstructure

The local authority is, as any bureaucracy, a meso-integrated organizational system.

Each office (service) is relatively identifiable on the basis of its
objectives, functions, degree of autonomy, "memory". But the loose
structure of the phenomena which take place outside the authority
is often not coherent with the purposive structure of the local
government.

And it may happen that the public bureacratic structure becomes some
sort of "opaque filter" between the socio-economic processes on the
land and the elected representatives (i.e., the political decision-
makers in the council) (see Fig. 2).

Such structure becomes an opaque filter between territory and the
political council (Fig.2). On one hand land use planning requires
integrated services, mainly related to the important "chains" identified, on the other the council needs an integrated view of relevant
phenomena in order to implement effective decisions. The identification of the chains can be a first means to pin-point the activities
which need financial support, further investigation and information,
evaluation of consequences etc.

In the local government considered, the first outcome of the action
research has been the redesign of the administrative structure according to the oriented network resulted from the analysis, taking
into account the juridical constraints.

The redesign is a first step towards the creation of the new IS.

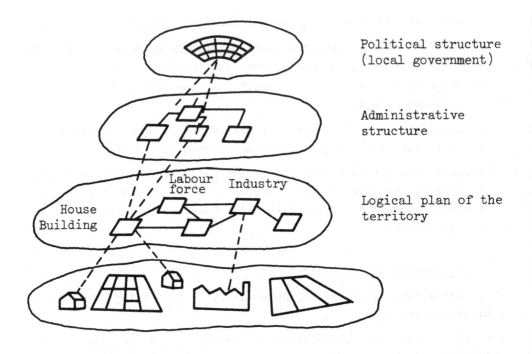

Political structure
(local government)

Administrative
structure

Logical plan of the
territory

Geographic structure of the territory

Fig. 2 - Model of land unit.

6. CONCLUSIONS

Systems analysis techniques need to include participation strategies, in order to implement effective computer-based information systems.

Besides ideological or marketing-oriented justifications, there seem to be structural reasons for user's participation: during the analysis and design phases the "memory" of the members of the organization has to be made explicit, formalized and partially transferred into the computer programs; in the usage phase users must co-operate to "feed" the systems and to fill in the gaps which inevitably occur after the introduction of the new system.

Moreover, participation appears to be crucial for the case of imperfectly organized (hypointegrated) social systems. Creating a new information system in a hypointegrated social system is a fairly different task from introducing a computer in a firm, for in the former structural relationships among system parts are poorly defined.

The local government and the socio-economic processes which take place on its land form a hypointegrated system.

Many purposeful subsystems (firms,families,schools,hospitals,transports,etc.) operate on the territory of the local government, the definition of the relationships among the subsystems and between these and the administrative structure is an open problem.

In such a context advanced information systems are usually conceived according to a general purpose data base approach.

For example, a socio-economic information system for the local authority is designed as a data base, which owing to its general purposes, should satisfy the needs of different users. In this way the variety and the fuzzy texture of the hypointegrated system and of its objectives set is taken into account.
A different (down-top) approach is being tested with an action-research experiment carried out in the local authority of Otranto.

In order to define a new IS, at first the imperfect organization of the hypointegrated system is improved through a learning process of knowledge acquisition, memory exchange, and participation among the socio-economic actors who live and work on the land of the local government.

"Organization of a system is an activity that can be carried out only by purposeful entities" (Ackoff /22/), and in this case such entities are officers, associations, employers, trade unions, individuals, citizens of the local community. The act of organizating is a process of participation. The analysis and design of both the organizational and information system are an outcome of the participation process.

Information system analysis in a hypointegrated context appears to be first of all a social system design activity.

NOTES

(1) According to R.L.Ackoff,in an imperfectly organized system,"even if every part performs as well as possible relative to its own objectives, the total system will often not perform as well as possible relative to its objectives" /22/.

(2) The concept of memory has been used by various authors in the context of organizational and IS studies. J.G. March and H.A. Simon /16/ refer the importance of the organizational memory to the innovation process within an organization: each time an innovation is introduced, the most important source of organizational proposals is composed of the solutions and procedures "stored" in the memory of organization members.

According to J.L.Le Moigne /17/ the memory contains models and programs related with the transformation system of the organization. E.S. Dünn /18/ writes that the memory of an IS, applied to whatever social process, is composed of the individual and organizational "history", decision rules, programs and models, the goals and values of individuals and groups. Finally, the notion of "track" by S.Beer /19/ should be recalled and the examples of organizational memory proposed by M. Landry /20/. The concept of memory applied to the analysis of the consequences of Edp technology or decision-making has been presented in /21/; in such a context organizational memory seems to be a component of the "real power" within organization.

BIBLIOGRAPHY

/1/ J.D. Couger, Evolution of Business System Analysis Techniques, Computing Surveys (ACM), 5, Sept. 1973, pp.167-168.

/2/ L. Golvers, Structured Analysis and Design of Business Information Systems, Int.rep. University of Bruxelles, May 1977.

/3/ NCR, A study guide for Accurately Defined Systems, Dayton, Ohio, 1968.

/4/ Honeywell, BISAD-1 and -2, 1973.

/5/ IBM, Study Organization Plan, doc.n C 20-8075, White plains, New York, 1961.

/6/ D. Teichroew, E.W. Hersey, Computer-aided Structured Documentation and Analysis of Information Processing Requirements, Int. doc. University of Michigan, Aug. 1976.

/7/ H.G. Sol, SIMULA(tion) in the Development of Information Systems, Lect.not., CREST-Course on MIS ,Stafford, July 1977.

/8/ E.Mumford, Procedures for the Institution of Change in Work Organization, IILS Symposium ,Moscow, Feb. 1977.

/9/ I.I. Mitroff, Dialectis in System Design, Lect.not., CREST-Course, Stafford, 1977.

/10/ B.Hedberg, Information Systems for Alternative Organizations, lect.not., CREST-Course, Stafford, 1977.

/11/ B. Sundgren, User-Controlled Design of Information Systems, paper, CREST-Course, Stafford, 1977.

/12/ K. Nygaard, O.T. Bergo, The Trade Unions-New Users of Research, Personnel Review, vol. 4, n. 2, Spring 1975.

/13/ J.L. Le Moigne, L'émergence d' une nouvelle discipline: les sistémes d' information, proc. AICA Congress, Milan, Oct. 1976.

/14/ E.R.F. Hearle, P.J. Mason, A Data Processing System for State and Local Government, Prentice-Hall, Englewood Cliffs, 1963.

/15/ C.Ciborra, G.Gasbarri, P.C. Maggiolini, Systèms d' Information et Systèmes d' Organisation Hypointegrés: une Méthodologie de Modélisation et de Maitrise de Systèmes Hypointegrés fondée sur la Participation, AFCET Congress, Nov. 1977, Versailles.

/16/ J.G. March, H.A. Simon, Organisations, Wiley, New York, 1958.

/17/ J.L. Le Moigne, Les systèmes de décision dans les organisations, PUF,Paris, 1974.

/18/ E.S. Dünn, Social Information Processing and Statistical Systems: Change-Reform, Wiley, 1974.

/19/ S.Beer, The Brain of the Firm, Penguin Press, London 1972.

/20/ M. Landry, L' Information et ses problèmes de gestion dans l' organisation: une perspective globale, Cost and Management, Jul. Aug. 1976 ,pp. 29 - 39.

/21/ E. Bartezzaghi, C. Ciborra, A.De Maio, P. Maggiolini,
P. Romano, Information Systems and Organizational Empirical
Findings regarding the introduction of Computers in Manu-
facturing Firms, IFIP Congress-77, Toronto, Aug. 1977.

/22/ R.L. Ackoff, Towards a system of systems concepts, in Management Science, vol. 17 n. 11, Jul. 1971, pp. 661-671.

/23/ IBM, Business Systems Planning, New York, 1975.

TOOLS FOR HANDLING HUMAN AND ORGANIZATIONAL PROBLEMS OF COMPUTER-BASED INFORMATION SYSTEMS

FRANK KOLF
HANS-JÜRGEN OPPELLAND
DIETRICH SEIBT
NORBERT SZYPERSKI

Betriebswirtschaftliches Institut für
Organisation und Automation
Universität zu Köln

D-5000 Köln-Lindenthal

A B S T R A C T

The last decade has confronted system designers with the problem that
they have to consider the requirements and restrictions of human and
organizational context in system development processes. There are lots
of tools for supporting the technical design activities but until now
project management is lacking effective and efficient tools for handling
the human and organizational problems in system design.
The paper describes some tools which were developed in the research
project PORGI (Planning model for the Organizational Implementation)
and are currently tested in CBIS-projects in two German companies.

1. Organizational implementation of computer-based information
   systems (= CBIS)

Literature on the development of CBIS as well as activities of
system analysts and system designers have been and still are dominated by many attempts to solve technological questions (concerning computers, programs, data bases, methods, models etc.). There
exists a continuously growing number of tools to support these
activities (e.g. techniques for program design, interactive programming support systems, documentations methods and systems etc.).

Roughly since 1970 some large organizations (companies etc.) try
to achieve computer-based management information systems especially to support management planning and control processes. These
systems not seldom perform multiple functions and/or serve users
in multiple areas of the organization. They provide users with
access to data and analytical capabilities which previously were
not available to them. Trying to develop such systems a special
kind of "implementation" problems originated, which system designers used to attribute to the fact, that "users" have only insufficient technological knowledge and therefore are not able to make
optimal use of CBIS.
More and more it became clear that introduction and usage of CBIS
in existing organizational environments was confronted with large
difficulties comparable to those which originate with implantations
of artificial organs into the human body. In many cases potential
user departments put the new systems under a boycott. As a consequence diligent system analysts invented an additional phase in
the development process, the core of which was "user-training".
In parallel they were active to find out necessary modifications
of user behavior and of the organizational structures and tried
to convince management to perform those modifications. But then
new difficulties arouse in form of feed-back-effects to the technological solutions designed in earlier phases of the development
process. In many cases "trained users" were no longer satisfied
with the system, they themselves had defined before as they had
experienced learning processes. They changed their demands and
needs. New incompatibilities between the future system and its
environment arouse.

The date of system-delivery to the user had to be postponed, planned budgets could not be held, this being only few out of a large number of frustrations which hit users and system analysts participating in these situations.

Some consequences hat to be drawn from these experiences. The traditional understanding of "organizational implementation" and of its position in the process of system development had to be modified:

- objects of system development are not only the technological components, but simultaneously and equally the organizational and personal components of an information system (as a socio-technical system)
- the core of system development is to produce a fit between technological, organizational and personal system components. System development can only be successful, if activities to produce this fit, to integrate system components, start with the beginning of the design phases. Organizational implementation is understood here as this central process of integrating, mutually adjusting system components.

From another point of view: system development is a special form of "planned change or organizational innovation", a term which is influenced by social science thinking.

This approach is not very new. Many authors have already mentioned this point. There is a lot of publications especially on empirical studies dealing with personal and organizational aspects of system development processes:

a) Several papers stress the enormous importance of factors like
  - user participation[1],
  - top management support[2],
  - overcoming user resistance to change[3].

Authors of these papers always have only relatively small samples at hand on which they base their statements. Most of their conse-

---

1) e.g. Little = Models =
   Carter/Gibson/Rademacher = Critical Factors =
2) e.g. Radnor/Rubenstein/Bean = Integration =; Radnor/Rubenstein/Tansik = Implementation =
3) Huysmans = Joint Consideration =

quences is not discussed.

b) Some studies concentrate on "impacts": impacts of CBIS on individuals or small groups[1], impacts of certain technological components or special forms of CBIS on decision behavior[2], or impacts of CBIS on organization structure[3]. Most of these investigations represent mono-causal explanations, i.e. impacts studied are attributed solely or mainly to the introduction of CBIS or traced back exclusively to changes of technological system-components.

c) In a third kind of studies the relations between certain factors or groups of factors and success of CBIS-implementation are analysed[4]. The problem here is the variety of factors investigated. Studies are concentrating on different groups of factors, on different types of information systems and different organizational and personal settings of CBIS[5]. In many cases results of these studies are contradictory. You find a large number of single statements without any help how to coordinate them.

d) Many studies have a retrospective orientation. Relations between factors are analysed ex-post. Empirical data are mostly collected after system development has been finished. There are only few statements on the process of system development. This leads to the fact, that system people don't find much information in these studies which could help them directly with their daily work.

We think that implementation research in the field of CBIS as a special form of technological research should produce tools to improve the effectiveness of system development activities. Besides that it should generate empirically founded statements how to use these tools in an individual context. It should stress important attributes of organizational implementation in different situations.

---

1) Dyckman = Additudinal Study =; Shaw = Group Dynamics =;
   Ladd = Group's Reaction =; Schultz/Slevin = Organizational
   Validity =; Sorensen/Zand = Improving =; Vertinsky/Barth/
   Mitchell = Social Change =
2) Hedberg = Man-Computer =; Chervany/Dickson = Information Over
   Load =
3) Whisler = Impact of Computer =; Lucas = Performance =; Bean/Neal/
   Radnor/Tansik = Correlates =
4) Ginzberg = Implementation = gives a very comprehensive survey
   on such factor-studies;
5) Harvey = Factors =; Powers/Dickson = MIS =; Lucas = Systems =;
   Bean/Neal/Radnor/Tansik = Correlates =

The management of system development is context-dependent. General rules and general statements must be substituted by specialized combinations of methods and techniques which take account of special conditions of implementation cases.

These and other objectives are pursued in research project PORGI, which is sponsored by the German Federal Ministry for Research and Technology. Within this project we try to generate instruments for CBIS development

- which help to measure the fit (or mis-fit) respectively the degree of integration between technological, organizational and personal system components (= specialized methods for analysis/diagnosis)
- which help to produce a better fit between system components by concentrating on these fit problems.

A multi-stage research and design process is provided for this project:

Stage 1
Design of a first global version of a descriptional framework, which contains all important factors by which O.I.-situations may be characterized.

Stage 2
Empirical research: interviews with 50 experts in the field of information systems (22 cases). One purpose of these interviews was to examine the descriptional framework and to detail it in such a way that it is able to map all factors which are named by the experts as being important factors in their individual O.I.-cases. A second purpose was to generate context-dependent hypotheses about how implementation processes have been performed in different organizational settings, which tools have been or not been used to produce fit between system components.

Stage 3
Classification of typical implementation situations, problem and process profiles of these situations based on the expert interviews.

Stage 4
Development of techniques and methods to measure implementation situations, to diagnose problems and to derive specialized recommendations to overcome these problems.

<u>Stage 4 a (actual state of project PORGI)</u>

Application of these techniques and methods in a certain number of cases in order to examine and continously improve them in the practice of system development. At the moment the research team is participating in CBIS-projects in two German companies. The same purpose of examination and methodological improvement is pursued in a working group consisting of several German CBIS-experts belonging to different institutions and the members of the PORGI-team.

<u>Stage 5</u>

Collection and comprehension of all tools in a "handbook for organizational implementation of CBIS", which will support practitioners in system development and will be the basis for training courses performed to disseminate this kind of know-how.

The next chapter gives an overview about all tools prepared in project PORGI. These tools are then portrayed in some details during chapters 3 - 7. Chapter 8 shows the procedural interactions between these tools when used in system development. Chapter 9 points to some important implications and limitations of the approach. The appendix presents some examples of PORGI-tools.

2.   Synopsis of the PORGI-Tools

The concept of the PORGI-Tools for the support of handling the
human and organizational problems in the development of CBIS consists of five elements :   (See figure 1)

- Descriptional Framework (DF) for implementational situations
- Procedural Scheme (PROC) as an example for the sequence of
  design activities which has to be adapted to the design phases
  in a specific context. The design activities are associated
  with relevant organizational implementation (OI) activities and
  tools for handling these activities.
- A Pool of Methods (METH) for the support of OI-activities, especially those instruments, tolls and methods for analyzing and
  diagnosing implementational situations which are developed by
  the PORGI-team.
- A Pool of typical implementational problems (PROB), which have
  occured in typical situations.
- A Pool of Design Concepts (CON) consisting of procedure-oriented
  and system-oriented design-recommendations.

The following pages will give a closer look at some of these five
components.

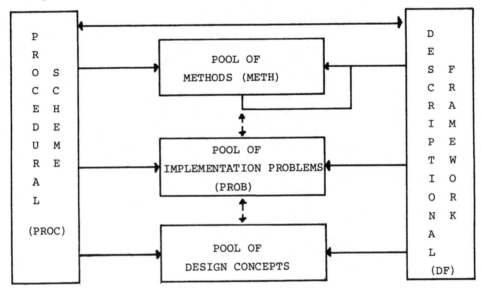

Fig. 1 :   Synopsis of the PORGI-Tools

3. The Descriptional Framework (DF)

We have already mentioned that all components have to be oriented
at the categories of a common descriptional framework.
That framework serves as a common terminological basis and not as
an explanatory model. The DF contains all those principal elements
which constitute specific implementational situations. That does
not mean that all elements have to occur in a specific situation.
According to the underlying implementation philosophy we have to
differentiate two areas of analysis :

- analysis of the fit between the system components "man, organizational structure, information technology and the task" to
  be supported;[1]   (System Fit)
- analysis of the design process (the project organization, project
  management etc.)   (Process Fit)

The principal categories of the DF are grouped into four modules[2]

- Personal System
- Task
- Information Technology
- Organization Structure

These principal categories are abstracts from concrete situations.
In a specific context, design situation etc. these categories have
to be specified in terms of that individual situation. For instance we don't speak in global terms of "the user" but we differentiate functional criteria which enable us within a specific
situation to describe those people which have to be considered as
"users", e.g. those who have to deliver inputs to the system, and
will receive ouputs from the system, their educational background,
their job record, needs and so on.
The following pages give an idea of that principle by showing a
part of the module "Organization Structure".

---

1) These components are oriented at Leavitt's categories Man,
   Structure, Technology and Task. See Leavitt = Change =
2) For the structure of the DF see Szyperski/Kolf/Claus/Oppelland/
   = Implementation = and especially Kolf/Claus/Oppelland
   = Beschreibungsmodell =

5   Organization Structure
...
54   Project Organization

541   Coordination Instance(s)   e.g. project commission
                                                    (incl. number&belonging department)
5411   Combination/Formation
54111   hierarchical   . top management
                                                    . middle management
                                                    .. offical expert, referee etc.
                                                    .. planning/operating department (line/staff)
                                                    .. organizational   "
                                                    .. data processing   "
54112   functional   .. other

5412   Recruitment   . time (available)
54121   recruitment criteria   .. qualification (capability, knowledge)
                                                    .. interest
                                                    .. delegation   . by position
54122   recruitment procedure   .. personal preference . election
                                                    . part time
5413   Occupation of (working) time   . full time
        (with coordination task)   . initiating
5414   Competence   . (global) project control
                                                    .. balancing interests
                                                    .. resource allocation planning
5415   Internal Structure   . functional differentiation
                                                    . hierarchical   "
                                                    . temporal   "

5416   Procedural Standards   . events   . point of time
54161   cause of meeting   . time statements
54162   frequency/duration   . existent
54163   standards of decision preparation   . majority rule (incl. power of veto etc.)
54164   standards of decision making   . authoritarian decision
...   . obligatory concensus
                                                    . no standard

Part of the DF-MODULE "Organization Structure"

4. The Procedural Scheme (PROC)

The Procedural Scheme consists of a sequence of well known design activities. Associated with the different design activities are implementational aspects and specific tools to support the analysis of these aspects. Like the other PORGI-tools the procedural scheme also has to be adapted to a specific situation, e.g. the sequence of the design activities may be different in different situations, or the number of separate activity blocks may vary depending on the project's complexity, budget, involved manpower etc. The conception of that procedural scheme has been guided by the following principles :

- Based on our definition of the organizational implementation as a process of securing the well-tuned design of the system components Man, Task, Structure and Information Technology there have to be planned specific activities which enable us to verify that tuning from the start-up-phase of the project.
- The design activities have to be supplemented by specific analytical steps. By performing repeated status analyses of all the system components we get information about the design status of the different components and FIT of these components. The evaluation of actual and planned status indicates specific (potential) problem areas early in the design process which have to be counteracted.
- Those who are concerned with system design and for system usage have to be involved in the degree of their individual concern. Therefore we have to plan activities which ensure that the organization of the design process besomes a design variable too.

The following abstract from the PORGI-Procedural-Scheme shows the realization of these principles. The nature and function of the analytical tools mentioned in the last column will be explained in chapter 5.

Design Activities	OI-Aspects	Analytical Tools
(1) Problem Definition	Identification of the "interested" people	
- Identification and Description of the problem	- Who has the problem?	PORGI-Checklist PROB
- Information about the problem	- Who articulates the problem?	PORGI-Descriptional Framework
- Evaluation of the problem	- Who participates in the evaluation (priorities etc.) of the problem?	PORGI-Checklist BTR-1.1 Group-evaluation methods (e.g. BASYC)
- Decision about the realization of Problem Analysis	- Who decides upon the realization of the Problem Analysis?	PORGI-Questionnaire BEW
(2) Problem Analysis	- Analysis of problematic areas in/between the system components · Man · Task · Structure · Information Technology	PORGI-Questionnaire 1-69 PORGI-Checklist ZIE PORGI-Checklist INN
- Analysis of the different elements of the problem and its context	- Identification of the interests of the people concerned	PORGI-Criteria-List EIG-1; AUS-1
- Analysis of possible causes	- Group consensus about the required characteristics of the different system components	Group-evaluation-methods
- Specification of the requirements for the system components		PORGI-Checklist PROB

Design Activities	OI-Aspects	Analytical Tools
(3) Discussion of possible problem solutions		
– Generation of alternative solutions for the system components	– Identification of those who are concerned and those who participate	Group-oriented design-methods in PORGI-METH
– Generation of alternative solutions for the integration of the system components	– Analysis of the existing FIT	PORGI-Checklist ZIE
	– Securing the FIT regarding new problem solutions	PORGI-Questionnaire INN
		PORGI-Criteria-List EIG-1; AUS-1
(4) Feasibility Study		
– Analysis of the need for change in the existing status of the system components	– Analysis of the required innovatorial step of the different system components regarding · degree (complexity, size, etc.) · past innovatorial experience · evaluation	PORGI-Checklist INN
– Check of possible restrictions (What is changeable, what is not?)		PORGI-Questionnaire BET
– Evaluation of feasibility	– Securing the FIT between system components in the problem solution	PORGI-Questionnaire FIT
– Formulation of the project application	– Planning of project organization and phases	PORGI-Evaluation-Scheme
		PORGI-Pool of potential implementation problems (PROB)
		Group-Evaluation-Methods

94

Design Activities	OI-Aspects	Analytical Tools
(5) Organizing the Project		

– Institutionalization of the
  · Control level
  · Project management level
  · Task Forces
– Planning of
  · needed resources
  · design activities
  · milestones
	– Identification/Determination of people to be involved
	– Determination of coordination and decision rules
	– Determination of principles for the documentation of the system and the project
	– Task assignment according to
	  · qualification
	  · motivation
	  · availability
		PORGI-Questionnaire 70-110
		PORGI-Analytical-Scheme
		PORGI-Questionnaire BTR
		PORGI-Checklist BET
		PORGI-Pool of potential implementation problems (PROB)
		Cooperation models from PORGI-CON

(6) Design Concept – Coarse Version		

– Analysis of relations between the components of the existing system
– Generation of alternative solutions for system components
– Evaluation of alternatives and selection of a specific alternative
– Design of concepts for the system components
	– Identification/Determination of people to be involved
	– Planning of FIT between
	  · Man
	  · Task
	  · Structure
	  · Information Technology
	– Analysis and evaluation of FIT
		PORGI-Checklist BTR
		PORGI-Checklist/Questionnaire BET
		PORGI-Checklist/Questionnaire FIT

Design Activities	OI-Aspects	Analytical Tools
(7) Design Concept – Coarse Version - Detailed design of chosen alternative - Specifications for the realization of system components - Planning of realization, e.g. planning of programming, planning of personnel development (training, promotion, etc.)	- Identification/Determination of people to be involved - Analysis/Planning of FIT between the system components	PORGI-Planning-Checklists PORGI-Questionnaire FIT PORGI-Evaluation-Scheme PORGI-Checklist ZIE PORGI-Checklist INN PORGI-Criteria-List EIG-1; AUS-1

5. Pool of Methods for System Development

5.1 Classification of methods

Most development methods have in <u>common</u> that they apply systematic
procedures for solving problems of (information) system development
but <u>differ</u> in the degree of specialization, applicability and design efforts.
They are methods for

- . collection,
- . description,
- . analysis and diagnosis,
- . forecasting,
- . design and model,
- . planning and control,

the characteristics, structure, state, behavior, performance etc.
of models, systems, elements and design-processes in conception
and reality.

<u>Specialized methods</u> (or special purpose methods) are those, which
are mainly applicable in specific areas and to selected objects.
We think it is helpful to differentiate two groups of specialized
methods :

(1) <u>Specialized methods for technical system development</u> :

- . assume well defined problems; intend to solve the problem
  efficiently from an engineering point of view,
- . do not explicitly consider the social and organizational
  environment.

(2) <u>Specialized methods for socio-technical system development</u> :

- . concentrate on "wicked" problems of system development and
  consider social and organizational environment,
- . focus on integrative aspects of system development,
- . conceive design and implementation as a necessarily social
  interaction process,
- . assume definition of system objectives as depending on
  social learning processes and therefore changing over time.

In addition we can find different types of methods which focus on
different developmental aspects, i.e. regarding the group of

(1) specialized methods for <u>technical</u> system development, especially for the
   (a) <u>design of system components</u>
   (b) <u>integration of system components</u>
   (c) <u>management of the system development process</u>
   and regarding the group of
(2) specialized methods for <u>socio-technical</u> system development those for
   (a) <u>system design and implementation</u>
   (b) <u>system development process support</u>.

In order to improve a better mutual understanding here we add a summary of our method classification with some <u>examples</u> for their explanation.

A. General (basic) methods

   1) <u>description methods</u>
      - graphical description (network-structures, structograms, flow charts, etc.)
      - matrix presentation
      - profile presentation
      - verbal presentation

   2) <u>collection methods</u>
      - observation
      - interview
      - questionnaire
      - document analysis

   3) <u>analysis/diagnosis methods</u>
      - regression analysis
      - correlation analysis
      - ABC-analysis
      - utility analysis (Nutzwertanalyse)

   4) <u>modelling/design methods</u>
      - simulation
      - optimization
      - stepwise refinement
      - morphological methods

B. Specialized methods for system development

    1)   methods for technical system development

        a) methods for design of system components
            - structured programming
            - Jackson diagram
            etc.

        b) methods for integration of system components
            - Kölner Integrationsmodell (KIM)
            - AEG-Telefunken-Modell
            etc.

        c) methods for the support of administrative aspects
            of system development processes
            - PAC II
            - IFA-PASS
            - ORGWARE

    2)   methods for socio-technical system development

        a) methods for system design and implementation
            - ETHIC (Mumford)
            - BASYC (Hawgood, Land, Mumford)
            - MIRA/SPIRA (Taggart, Tharp)
            - PORGI-System FIT

        b) methods for the support of socio-technical aspects
            of development processes
            - PORGI-process FIT

Regarding to the PORGI approach there is a special need for process-oriented methods, which enable all those who are presumed users of the system in managing the social interactions in developmental processes.

As a first step PORGI intends to attain the development of appropriate instruments to achieve this socio-technical design support. The following chapter (5.2) presents some of these instruments.

5.2 Instruments, Tools and Methods for Analyzing and Diagnosing Implementational Situations

The above mentioned Procedural Scheme (PROC) for all design activities points to such aspects, which have to be considered in the

view of what we call <u>Organizational Implementation</u> (see chapter 4,
the middle column). With these different aspects are associated
instruments, tools and methods for their analysis and diagnosis,
which will be combined to the planned Implementation Handbook (IHB).
Problem-oriented questionnaires, checklists, lists of criteria etc.
support the diagnosis and analysis of constellations of variables,
which are relevant to organizational implementation activities in
this specific situation.

As an example for problem-oriented questionnaires we offer those
named "<u>SITUATIONAL DIAGNOSIS</u>" (BET, BTR) and "<u>METH</u>" (ORG-INT), wich
are suitable for the organizational implementation (O.I.) activities "<u>Planning of process-FIT and system-FIT</u>" according to the
system development step <u>(6) Design-Concept-Coarse-Version</u> (see the
special marks in the above presented Procedural Scheme and the explanation in chapter 5 and the appendix).

In the appendix you will find parts of these PORGI-questionnaires,
which intend to enable those, who are concerned with the system
development, to analyze the fit of the outlined system design concept and the perceived design and implementation process by themselves. These questionnaires are developed under certain assumptions
of cause/effect-relations, which are based on empirical data of
the PORGI-expert-interviews and design experiences of the PORGI-
team-members.

As our empirical experiences indicate a special effect of questionnaire application can be seen in a "participation-by-answering-
effect", which influences the development process and therefore its
success positively. Beyond that the questionnaire offer the chance
to include users' demands, evaluation and attitudes adequately in
system design.[1]

As an example of a PORGI-<u>checklist</u> we present the checklist PROB
for the system development step <u>(2) Problem analysis</u>. This checklist "<u>PROBLEM DEFINITION (PROB)</u>" intends to ensure the consideration of all relevant situative aspects for the definition of

---

1) See e.g. Mumford = Job Satisfaction =; and = Procedures =;
   Hawgood = Evaluation =; Hawgood/Land/Mumford = Comparison =

system and system development objectives and problems.

The following list gives additional examples for instruments, which are developed in PORGI and which are now in an on-going process of empirical test and refinement in practical development projects :

. Checklist for the identification of those members of the organization who are concerned with the development process and/or
  will be system users;
. Checklist and questionnaire for the definition of project and
  system objectives;
. Checklist and questionnaire for the identification of those who
  are concerned with changes through system development and the
  expected concequences of these changes for their responsibility,
  autonomy of decision-making, control and other working conditions;
. Questionnaire for identification of actual and required participation in system development activities;
. Questionnaire and checklist for identification and analysis of
  the required/planned innovatiorial step and its importance regarding the system components :
  Task, Personal System, Information Technology and Organization
  Structure, e.g. task content, procedures, communication;
. Checklist and questionnaire for the evaluation of the FIT between the system components by those who are concerned.

Checklists and questionnaires are instruments, which can be handled by the project management or special implementors according to the different steps of system development. The questionnaires are ad-dressed to those members of the organization who are concerned with system design, implementation and use.

6. Pool of Implementation problems (PROB) for the contextual
   analysis and diagnosis

This pool consists of specific implementation problems which can
be identified by a situational profile. These profiles describe
those empirical situations in which that specific implementation
problem has occured repeatedly, e.g. results or consequences of
specific actions in the design process which deteriorate the intended implementation success. These problems lead to economically
and/or socially insuffient results of the system design process,
process of system usage. Descriptions of these implementation problems use the categories of the descriptional framework and are
based on those problems which have been identified in our expert
interviews to be typical for specific situations.

Additionally it contains theoretically or empirically based implementation problems found in literature, case studies and own design
experiences of the members of the PORGI-team. Objective of the diagnosis is the comparsion of the characteristics of the individual
situation and the situative profile of the implementation problems
of PROB in order to identify possible problem areas in the actual
situation. Follow-up activities focus on the analysis of possible
causes for that problem and possible solutions (see the CON-pool
in the next chapter).

7. Pool of Design Concepts (CON)

The result of the contextual analysis and diagnosis is the knowledge of specific implementational problems (which probably deteriorate the intended implementation success) and possible causes for
these problems. Based on this result the PORGI-Design-Concepts
support problem solution by helping to select appropriated implementational actions - appropriate for avoiding expected problems
or reducing their possible consequences. But the design concepts
should also be helpful to handle unexpected problems or reduce
their consequences.

The choice of appropriated implementational actions is supported
by offering
- general design principles and
- situative design proposals,
which can be seen as sufficiently tested and confirmed by the system development praxis. The description of these design concepts
is also oriented at the categories of the descriptional framework,
thus their situative applicability could be better evaluated in
actual situations by the project manager, implementor or other
people responsible for the system development.

The design concepts contain as possible implementational actions
proposals to correct (or confirm) social and organizational behavior and its rules regarding the different aspects of
- organization and procedures of design and
  implementation processes, e.g.
  .. project planning, project organization
  .. participation of those who are concerned
  .. consideration of changes in the definition
     of project and/or system objectives
  .. improvement of implementational activities
     initiated by differences between actual
     and planned situations
- design of system components under consideration
  of their fit to accepted
  .. objective requirements
  .. subjective demands and needs of those
     who are concerned with the system
     development and use.

8. Procedural Interactions of the PORGI-Instruments

The interaction of the proceeding chapters has been to give an
imagination of that kind of tools and instruments which have been
developed by the PORGI-team and are currently under test in practical DP-Projects in industrial firms.
The following comments will show up briefly the procedural interactions of these instruments in the course of a design project.
(See fig. 2)

The procedural scheme (PROC) works as a directory which tells us
what next steps should be undertaken and what kind of implementational aspects should be considered. The details of such an
OI-aspect (e.g. (5.1.1) Participation) are compiled in the Dis-
criptional Framework (DF), in that example the elements of the
group 1113 Individual evaluation of participation.
The Pool of Methods (METH) offers for the analysis of these OI-
aspects specific analytical tools, in the example the PORGI-Ques-
tionnaire BET (see the abstract in the appendix).
After having gathered facts about the specific situation by use
of the PORGI-tools the problem-oriented analysis of that situation
is being performed by searching the pool of implementation problems
(PROB). Objective of that search is to find out whether the situative profile just gathered is similar to one of those situative
profiles in the problem pool in which specific implementation problems have repeatedly occured.
Then the pool of Design concepts (CON) is to be searched whether
it contains a possible solution with a similar situative profile,
that means a solution which has proven to be useful in similar
situations.

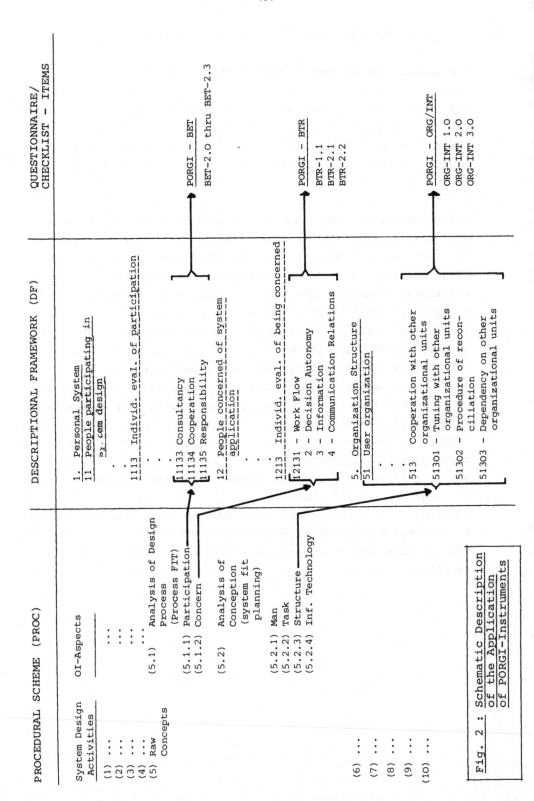

Fig. 2 : Schematic Description of the Application of PORGI-Instruments

9. Implications and limitations of the PORGI-approach

. The approach is focused on the integration of technological,
organizational and human system components as one of the main
tasks of CBIS-development. The authors think that the performance
of this task is crucial to system success. Knowledge about how
to achieve this integration up to now is of a fragementary nature. PORGI tries to coordinate and organize the existing know-
how, to enrich it with additional tools and bridge the existing
gap between research and system development, which has hindered
the perception of valuable empirical findings by practitioners.

The main advantage of the required O.I.-activities will be that
the risk of unseccessful systems and unsuccessful system development decreases. On the other side it must be clear that O.I.-ac-
tivities cause additional expenses. It starts with the training
of highly qualified system people, which are able and willing
to take the role of "implementors", people, which will become
responsible for the implementation activities shown in chapter
4. This may require additional manpower compared to the traditional procedures of system development.
In some cases there will be a personal union of the project
management function and the O.I.-function, in other cases (e.g.
large projects) there might be a division of labour between project managers and specialized implementators. In some cases
(e.g. smaller projects) the O.I.-function might be performed by
external consultants.

Some technologically oriented system designers apply a special
conception or strategy, which might be called "release of bombs"-
strategy. It can be characterized as follows: In their view personal and social behavior of future users and system specialists
is so complex and multidimensional that it cannot be measured
and controlled. They feel that the utility of extending the area
of measureq and controlled variables is lower than the cost of
research necessary to achieve such an increase of know-how plus
the cost of know-how-introduction and -application.
Therefore they concentrate on the efficient production of technological system components, leave organizational implementation
to the future user and hope that the user will succed in understanding and adapting the new system to his individual situation.
As the user sometimes is not prepared to face and to solve such

problems the delivery of new systems might have the same effects
as the "release of bombs". Nevertheless the adherents of this
strategy don't see efficient ways to avoid these effects.
- We think that we cannot restrain anybody of applying such a
strategy, but maybe using O.I.-methods like those developed in
project PORGI could lower the risk of such releases.

. Another objection against this approach could concentrate on
the fact, that transparency of system development processes is
rigorously increased. It might be that this is not in accordance
with the management style of some organizations. If management
style cannot be changed to accept more transparency we would not
recommend to make a practice on this approach. On the other side:
Most system development processes are already understood as
"bargaining processes". Applying tools like those proposed here
will lead to rationalization of these processes.

. The PORGI-approach is focused on the development of single
"local" (sub-)systems. Long-range and global system planning
problems (like definition of priorities of competing projects)
are not discussed here. This does not mean, that such a participative approach would not be applicable to long-term global
systems planning. It only means that you need specialized tools
to support such planning processes.

. The traditional "classical" activities of system design concentrate on the question: which technological alternative is the
best (most efficient) one to solve a given, predefined task?
Concentration of PORGI-tools on planning and control of the fit
between technological and organizational/human system components
supposes that this question has been or will be answered during
system development. The PORGI-approach does not support answers
to this difficult and in many cases very unclear question (for
which a large amount of research always will be needed). But
this limitation has been chosen deliberately. The PORGI-approach
does not intend to substitute existing system design approaches,
which concentrate on the questions of optimizing technological
solutions for predefined tasks, but complement them in respect
to certain factors extremely relevant to system success.

. Chapter 1 of this paper has related O.I.-problems to computer-
based management information systems. Today it has become clear,
that M.I.S. were among the first examples, where such problems

were encountered, but management orientation is neither the only nor the main reason for producing O.I.-problems within system development. We find such problems within those classes of system development, where we try to support relatively weekly structured but complex information processing tasks with CBIS and where the users of CBIS are not able (for different reasons) to create these CBIS themselves. In many organizations relatively simple and highly structured tasks have already reached a high degree of computer support. Therefore we feel that systems with O.I.-problems will dominate the future of system designers and implementors.

Organizational Implementation of CBIS has many different - last not least cognitive and political - dimensions. The tools proposed here do not cover all these dimensions and the problems stemming from them. The PORGI-approach directs attention to some of these problems and tries to find methods better adjusted to the complexity of reality. Continous improvement of these methods during a longer period (much longer than the duration of project PORGI) is necessary. Applications of the tools within organizations should be structured as systematic learning processes.

108

10. References

Bean, A.S.; Neal, R.D.; Radnor, M.; Tansik, D.A. = Correlates =
    Structural and Behavioral Correlates of Implementation in
    U.S. Business Organizations. In: Schultz/Slevin = Implemen-
    tierung =, S. 77-132
Carter, Deane, M.; Gibson, Harry L.; Rademacher, Robert A.
    = Factors = Zum Situationsansatz in der Organisations-
    forschung. In: ZfO, 46. Jg. (1977), Nr. 2, S. 67-74
Chervany, N.L.; Dickson, G.W. = Informations Overload =
    An Experimental Evaluation of Information Overload in a
    Production Environment. In: Management Science, Vol. 20,
    1974, S. 1335-1344
Dyckmann, T.T. = Attitudinal Study =
    Management Implementation of Scientific Research: An Atti-
    tudinal Study. In: Management Science, Vol. 13, 1967,
    No. 10, B612-B620
Ginzberg, M.J. = Implementation =
    A Detailed Look at Implementation Research. Report CISR-4,
    Sloan School of Management, MIT, Working Paper 753-4,
    November 1974
Harvey, Allan = Factors =
    Factors making for Implementation Success and Failure.
    In: Management Science, Vol. 16, 1970, No. 6, S. B312-B321
Hawgood, J. = Evaluation =
    Evaluation and management of computer-based systems: an
    inter-disciplinary approach. Information Processing 71.
    Amsterdam 1972
Hawgood, J.; Land, F.F.; Mumford, E. = Comparison =
    Comparison of alternative strategies and systems.
    In: Datenverarbeitung im Europäischen Raum. Wien 1972,
    S. 283-287
Hedberg, B. = Man-Computer =
    On Man-Computer Interaction in Organizational Decision-
    Making. A Behavioral Approach. Gothenburg 1970
Huysmann, J.H.B.M. = Joint Consideration =
    The Implementation of Operations Research: An Approach to
    the Joint Consideration of Social and Technological Aspects.
    New York, 1970
Ladd, D.E. = Group's Reaction =
    Report on a Group's Reaction to the "Researcher and the
    Manager": A Dialectic of Implementation. In: Management
    Science, Vol. 12, 1965, No. 2, B24-B25
Leavitt, H.H. = Change =
    Applied Organizational Change in Industry. Structural,
    Technological, and Humanstic Approaches. In: March, J.G.
    (Ed.): Handbook of Organizations. Skokil, Ill., 1965,
    S. 1144-1170
Little, J.D.C. = Models =
    Models and Managers: The Concept of a Decision Calculus.
    In: Management Science, Vol. 16 (1970), No. 8, S. B466-B483
Lucas, H.C. = Performance =
    Performance and the Use of an Information System.
    In: Management Science, Vol. 21, No. 8, 1975, S. B908-B912
Lucas, Henry C. = Systems =
    Why Information Systems Fail. New York, London 1975
Mumford, Enid = Procedures =
    Procedures for the Institution of Change in Work Organization.
    Participative work design: A Contribution to democracy in the
    office and on the shop floor. Unpublished Paper. Manchester
    Business School, 1976

Mumford, Enid = Job Satisfaction =
    Job Satisfaction. A Study of Computer specialists. London 1972
Powers, Richard F.; Dickson, Gary W. = MIS =
    MIS Project Management: Myths, Opinions and Reality. In:
    California Management Review, Vol. 15, 1973, No. 3, S. 147-156
Radnor, M.; Rubenstein, A.H.; Bean, A.S. = Integration =
    Integration and Ultilization of Management Science Activities
    in Organizations. In: Operations Research Quarterly, Vol. 19,
    1968, No. 2, S. 117-141
Radnor, M.; Rubenstein, A.H.; Tansik, D.A. = Implementation =
    Implementation in Operations Research and R & D in Government
    and Business Organizations. In: Operations Research, Vol. 18,
    1970, S. 967-991
Schultz, R.L.; Slevin, D.P. = Organizational Validity =
    Implementation and Organizational Validity: An Empirical
    Investigation. In: Schultz/Slevin = Implementing =,
    S. 153-182
Shaw, Marvin E. = Group Dynamics =
    Group Dynamics: The Psychology of Small Group Behavior.
    New York 1971
Sorensen, R.E.; Zand, D.E. = Improving =
    Improving the Implementation of OR/MS Models by Applying
    the Lewin-Schein Theory of Change. In: Schultz/Slevin
    = Implementing =, S. 217-236
Whisler, T.L. = Impact of Computers =
    The Impact of Computers on Management. 1970

11. <u>Appendix :</u>   <u>Parts of PORGI-Tools</u>

A. Situational Diagnosis : BET-2.0 thru BET-2.2

B. Situational Diagnosis : BTR-1.1

C. METH : ORG-INT-1.0 thru 3.0

D. Checklist Problem Definition
    And Analysis : PROB-1.0 and 3.0

2.0 <u>Conceptual Design of the System</u>

R P C I N *

2.1 What kind of participation have
you experienced ?

☐ ☐ ☐ ☐ ☐

2.2 What kind of participation would
have been appropriate ?

☐ ☐ ☐ ☐ ☐

2.3 If you would like to participate
<u>more intensive</u>, what reasons do
you see ?

   (1) The consideration of my
       know-how/knowledge would
       be helpful   ☐

   (2) The consideration of my
       requirements is justified
       and necessary   ☐

   (3) To avoid disadvantages for
       my own position my parti-
       cipation is necessary   ☐

   (4) Other :

*
R = Ratification
P = Responsible performing
C = Cooperation (by giving suggestions,
                consultancy etc.)
I = Receiving information about
N = Not involved

| PORGI | SITUATIONAL DIAGNOSIS | BET - 2.0 |

2.4 If you would like to participate <u>less intensive</u>, what reasons do you see ?

    (1)   The consideration of my know-how/ knowledge is not necessary ☐

    (2)   I don't have requirements to that specific system to be considered ☐

    (3)   I don't have the time available ☐

    (4)   Other :

2.5 If you would like to have participated in a different way (more or less intensive), have you tried to accomplish it ? ☐ Yes ☐ No

2.6 If you would like to have participated in a different way but <u>not tried</u> to accomplish it, what reasons were responsible ?

    (1)   I didn't get any information about that project ☐

    (2)   I don't have available

        a. the required time ☐
        b. the required know-how ☐

    (3)   I don't have good experiences by active participating or starting initiatives for it ☐

    (4)   Would look like criticism for my superiors ☐

    (5)   Wish to participate <u>less intensive</u> would look like an admission of incompetence or lack of interest ☐

| PORGI | SITUATIONAL DIAGNOSIS | BET - 2.1 |

(6) Wish to participate more intensive
would look like "lust for power or
promotion" ☐

(7) Other :

2.7 If you have tried to accomplish a different kind
of participation but you didn't succeed, what
were the reasons ?

(1) I don't know the reasons ☐

(2) The consideration of my know-how/
experiences ...

a. is not necessary ☐
b. would be helpful but not
realizable with available
time ☐

114

1.1 For which of the below mentioned task-oriented
    aspects do you expect or know changes induced by
    the system design and what importance do they
    have for you ?

Changes are of
none
or   Don't
minor   major   know the
         importance   importance

a. task assignment   ☐   ☐   ☐
b. work flow   ☐   ☐   ☐
c. decision autonomy   ☐   ☐   ☐
d. responsibility   ☐   ☐   ☐
e. information   ☐   ☐   ☐
f. quality of own work ☐   ☐   ☐
g. Other :

1.2 For which of the below mentioned person-oriented
    aspects do you expect or know changes induced by
    the system design and what importance do they
    have for you ?

Changes are of
none
or   Don't
minor   major   know the
         importance   importance

a. personal image   ☐   ☐   ☐
b. influence   ☐   ☐   ☐
c. chances for
   promotion   ☐   ☐   ☐
d. financial
   implications   ☐   ☐   ☐
e. personal well-being ☐   ☐   ☐
f. work load/stress   ☐   ☐   ☐
g. Other :

| PORGI | SITUATIONAL DIAGNOSIS | BTR - 1.1 |

1. What is the official hierarchial structure of that
   department/division etc. ?
   (if possible take plan of organization)

2. Which meetings, conferences etc. exist in your
   department ?

   2.1 Name :   (1) _____

                 (2) _____

                 (3) _____

   2.2 How often do they take place ?

No.	frequency
(1)	
(2)	
(3)	

   2.3 Who takes part ?

No.	Name	Department

   2.4 What is the subject of these meetings ?

No.	Subject area

PORGI	M E T H	ORG-INT 1.0

2.5 Of which character are these meetings ?

    - Exchange of information .... No.
    - Discussion of proposals .... No.
    - Elaborating of problem solutions .... No.
    - Common decision-making .... No.

2.6 Are you content with the realization of
these meetings ?

   ☐ Yes

    No, because

    ☐ too seldom
    ☐ too often

    ☐ wrong persons
    ☐ irrelevant/peripher problems
    ☐ without obligation

    Other : _____

3. Regarding the planning task to be supported by the
new system who in your department

3.1 is intensively occupied with operative
activities ?

3.2 represents proposals against other
departments ?

PORGI	M E T H	ORG-INT 2.0

4. Is the <u>preparation</u> of a plan proposal being performed

    ☐ <u>solely</u> in <u>your</u> department

    ☐ <u>mainly</u> in <u>your</u> department and other departments

        ☐ delivers specified inputs

        ☐ will be informed about the proposal

    ☐ <u>mainly</u> in <u>other</u> departments and your department

        ☐ delivers specified inputs

        ☐ will be informed about the proposal

5. Are <u>you</u> involved in the

    5.1 preparation of plan proposals ?

        ☐ Yes   ☐ No

    5.2 discussion of plan proposals ?

        ☐ Yes   ☐ No

    5.3 ratification of plan proposals ?

        ☐ Yes   ☐ No

| PORGI | M E T H | ORG-INT 3.0 |

The following checkpoints have to be considered in
problem definition and analysis :

1. Characteristics of the problem situation

    1.1 Characteristics of task performance
        (See PROB-3.0, 3.1)

    1.2 Characteristics of the technological support
        of task performance
        (See PROB-5.0)

    1.3 Characteristics of the personal system
        (See PROB-4.0)

    1.4 Characteristics of the organizational
        structure
        (See PROB-6.0, 6.1)

2. Members or organizational units,

    2.1 who/which are concerned by that problem

        -   task assignment
        -   work flow
        -   decision autonomy
        -   responsibility
        -   information

        (See Checklist BTR and analytical scheme)

    2.2 who/which have articulated that problem

    2.3 who/which initiated the study
        (See Role structure in PROB-6.0)

PORGI	CHECKLIST PROBLEM DEFINITION AND ANALYSIS	PROB - 1.0 CONTINUED

Characteristics of the problem situation	possible causes
System component : planning TASK	
(1) Time structure of task	
• Planning horizon	too short; lack of long-range perspectives
• Length of planning process	too long; insufficient possibility to answer changes in market conditions
• Initiative for planning process	unsystematic ad-hoc planning
(2) Level of task	
• Problem-level	only operative planning, no integration with strategic planning
• Considered units	isolated unit plan, integration with other units is insufficient handled
(3) Complexity of task	
• Number of different activities	too high; not to handle with available methods
• Intellectual requirements	too low; permanent coordination problems with other units performing similar activities
• Degree of detail	
(4) Work load	too big; permanent performing in over-time
	too low; people feel no "challenge"

PORGI	CHECKLIST PROBLEM DEFINITION AND ANALYSIS	PROB - 3.0 CONTINUED

ORGANISING FOR COMMON SYSTEMS

by

N.L.   Richards

Systems Development Manager,

Gold and Uranium Division,

Anglo American Corporation of South Africa Limited

SYNOPSIS

It is argued that to design large, common information systems and
implement them in several locations needs a special organisation.
Most sites where such systems are to be installed have different
organisation structures which makes the development process difficult.   Internally, most Information Systems Departments do not
have the control procedures in place which will give adequate fore-
warning that projects of this type are not producing an acceptable
systems solution.   This paper describes the way that the Gold
Division of Anglo American have tackled these and associated
problems.

1. INTRODUCTION

In this paper I will describe the methodology that we use to
develop large Information Systems, together with some problems
that have been experienced by ourselves and some other major
South African Companies. The companies who have been cited,
albeit anonymously, are, likewise large, capital intensive
firms which are geographically distributed into semi -
autonomous sites.

Large being a relative description of size, requires me to
give a few statistics about The Anglo American Corporation of
South Africa.

Anglo American Corporation is the head of an international
group of mining, industrial and investment companies which
it administers in South Africa and elsewhere. We have close
associations with De Beers Consolidated Mines Limited,
Charter Consolidated Limited (in London), Engelhard Minerals
and Chemicals Corporation (in USA) and Hudson Bay Mining and
Smelting Company (Canada).

The Group's main interests are, however, in Southern Africa
where income is derived from gold (44%), diamonds (17%), copper
(9%), industrial companies (16%), coal (3%), other mining
activities (2%), finance (6%) and platinum (3%).

Anglo's Gold and Uranium Division is responsible for the
management of our ten major gold mining companies operating in
South Africa. Production from these mines amounts to about
40% of South Africa gold production and 30% of world production.

This is an impressive operation. The gold mines which are
spread over a wide geographical area employ one hundred and
fifty thousand people, have a capital expenditure each year of
$200 million and working costs of $630 million.

Yet the evolution of the Division's use of computers has
been very slow.

Table 1 shows, however, that since 1976 a build-up is certainly now under way:

Year	Hardware Installed	Operating System	Principal systems in Development	Principal systems in Operation
1970	1 x IBM 360/40	DOS	-	˙Stores ˙Payroll
1972	2 x IBM 360/40	DOS	-	˙Stores ˙Payroll
1974	2 x IBM 370/135	DOS	-	˙Stores ˙Payroll
1976	1 x IBM 370/158 1 x IBM 370/135	OS/VS	˙Personnel Management	˙Stores ˙Payroll
1977	IBM 370/158MP	OS/MVS	˙Personnel Management ˙Mine Planning ˙Geological Information ˙Materials Control	˙Stores ˙Payroll

TABLE 1 :   EVOLUTION OF DATA PROCESSING IN GOLD DIVISION

The main reason for this acceleration is that in 1975/76 the Division became dissatisfied with the way the computer was being used as a Management Information tool. It therefore undertook a BSP study (Business Systems Planning) whose purpose was to:

- create the basic systems theory that would be used in the Division

- review the Management's requirements for Information

- define the Division's business processes

- develop a strategy for providing required information

-       create an organisation responsible for carrying it
        through

-       define the system which had the highest development
        priority

The study took more than seven man years to complete. It involved
a large number of interviews with people from all levels of
management who were asked a series of searching questions about
their Information needs and the benefits that would accrue if they
got them. I do not intend to dwell on the BSP methodology. Your
friendly IBM salesman would be only too willing to describe it. I
would say, though, that as a method for developing a long term
Information Systems plan, it is very useful. It certainly gave us
a highly committed Top Management and a wide visibility on the
mines but, for reasons that are internal to an Information Systems
Department, I believe BSP should only be used by a company who is
able to ensure a tight quality control over the manner in which the
study is carried out. A company which has effective project control
standards in operation would be ideal. Unfortunately, we did not
have this absolutely essential discipline, and consequently, as I
shall describe, we created many problems for ourselves.

2.   SYSTEMS THEORY

I will describe the basic systems theory that we have adopted
because I believe, and my company believes, that without a similar
credo, an information systems man will never implement a successful
system. Even though we systems people are trying to be a practical
profession, we must nevertheless always remember that a systems
theory is necessary. "The theory inspires and informs the practice,
and the practice confirms and modifies the theory." [1]
The theory that we are using evolved from reading many authorities.
One book in particular which I found very instructive, especially
in the way that it concerned effective relationships with Users,
was by Oliver Wight.[2] I have borrowed themes from this book quite
often in this paper for which I acknowledge my debt.

Our theory can be reduced to 8 basic rules, which are shown in
Table 2 :

Rule 1:	The User is responsible for realising the business improvements that Information Systems can produce.
Rule 2:	The User owns the system.
Rule 3:	Information Systems must only be considered as a means to an end - they are not an end in themselves
Rule 4:	The User must be heavily involved in Information Systems development.
Rule 5:	The User must be continually reviewing the development of the system to ensure that he can accept responsibility for systems success.
Rule 6:	The User must understand the system and come to depend on it.
Rule 7:	Unnecessary embellishments are not allowed.
Rule 8:	The only way to determine the success of a system is to show that the operation improved after using it. Therefore evaluation criteria are essential.

TABLE 2:   THE 8 RULES TO ENSURE SYSTEMS SUCCESS.

It can be seen that we intend to keep our Users in the vanguard of the systems development effort and the systems people where they should be - as catalysts and technical supporters. We believe that, unlike process control systems which are able to regulate such processes as oil refining and electricity distribution, the systems that Gold Division produces, Information Systems, will never control our business. We believe that only people will do that. Information Systems are purely a "tool of the trade" of a manager. Therefore it is logical to presume that only the User will ever get the improvements from an Information System. He should, as part of his management responsibility, be always designing improvements to the way he runs his function. The computer can often help him to attain these improvements by giving him better information faster, but the fact remains that the responsibility for determining what he wants, for developing these improvements, and for obtaining the desired results belongs to the Users - not the systems people. We systems

people, for our part, must never see ourselves as the specifiers of
the Users systems requirements. It is quite possible that, when
developing a system, we will initiate an idea and try to have it
included in the design, but, if the User manager, who really has
the responsibility, does not consider it acceptable whether or not
this is due to his lack of perception or our lack of business appreciation, we must accept his decision. Similarly, we must never
add that "bell" or "whistle" which, though not essential, would
complete the symmetry of our design. This is certain to put the
whole system at risk by placing it on to the next level of complexity.
Our approach has, on the face of it, little to attract a systems man.
Gone is the Sir Galahad image he has enjoyed up till now. Indeed,
when a system is seen to be a success the User is quite likely to
think that it was all his own work. Therefore, if this approach to
Systems Development is to be used, a large responsibility is placed
on Information Systems management to recognise the good systems man.
This recognition is something that must be actively sought out,
because the more successful the systems man is, the more the success
is apparently due to somebody else.

3.   INFORMATION SYSTEMS ORGANISATION

The next logical step in describing our methodology is to consider
four things:

-    The type of system that we develop

-    The way the systems inter-relate

-    The benefits of designing for business processes

-    The I.S. organisation structure that we chose to
     produce and run our systems.

3.1 SYSTEM TYPES

We categorise our Information Systems into four categories (as
shown in Table 3).

-    Central Systems - developed by a single project team for
     a User who is centrally based (e.g. at Head Office), who
     is solely responsible for the co-ordination of the multiple
     users and for submitting the input, and who therefore acts
     as a central authority. The system is identical at
     all sites. In our case, our Stores and Payroll systems
     are run on this basis.

- Common Systems - developed by a single project team for Users based on every mine, where there is no central submission of input but where a Head Office manager has functional control over the system. The system is identical for all mines.

- Shared Systems - developed by a single project team for a User based at one mine. When the system is proved to be a success at this mine, it is offered to the others on a "no guarantee and no maintenance" basis. There is no central submission of input, nor has Head Office functional control. A mine which accepts such a system is responsible for its own maintenance and therefore, gradually, the system becomes unique to a particular mine.

- Local Systems - developed by a single project team, for a User based at one mine. The system is of a local nature which no other mine wishes to use. Shared systems start out like this.

Type of System	No. of Development Sites	No. of Installation Sites	No. of Maintenance Sites	End User
1. Central	Single	Single	Single	Multiple
2. Common	Single	Multiple	Single	Multiple
3. Shared	Single	Single (leading to Multiple)	Single (leading to Multiple)	Single (leading to Multiple)
4. Local	Single	Single	Single	Single

TABLE 3: TYPES OF SYSTEM DEVELOPMENT IN MULTI-LOCATION COMPANIES

The systems which we have identified as being part of our long term strategy belong, to the Central Common and Shared categories and are always developed using a standardised approach. Our choice of preferring the systems which allowed a standardised approach was deliberate, notwithstanding the disadvantages of this method. We felt that the advantages (see Table 4) outweighed the disadvantages (see Table 5) and that, by organisational means, we could overcome the major drawbacks.

1.	Easier to develop - less time to design - more use of commercial packages.
2.	Easier to train - system can be seen in operation at other sites, training programme becomes increasingly refined.
3.	Better communication between sites - all on same system.
4.	Is the accepted standard tool - no more unique approaches
5.	No more arguments about which way is best.
6.	Maximises use of scarce development resources (especially D.P. Personnel)

TABLE 4:   BENEFITS OF STANDARD APPROACH TO SYSTEMS DEVELOPMENT

1.	Computer gets blamed for the standards that are needed anyway
2.	Systems have a Head-Office-imposed image.
3.	Individual needs are sacrificed.
4.	Systems tend to be compromises.
5.	Designers have no intimate familiarity with local problems.
6.	Slow response to Users need for operational system.
7.	No identification with the objectives of local management - divisional need catered for only.
8.	Difficult to use because of an in-built rigidity.
9.	Common systems are often impractical in local circumstances
10.	Operating level not in sympathy, so difficult to implement.

TABLE 5:   DISADVANTAGES OF STANDARD APPROACH TO SYSTEMS DEVELOPMENT.

128

3.2 HOW OUR SYSTEMS INTER-RELATE

As part of the BSP study, we produced:
- a list of required systems in priority order and
- a conceptual network diagram of how they inter-connected.

Figure 1 is a sample extract from the total I.S. Network showing the systems with which we have initially been concerned.
As can be seen, we believe that systems only interface by transferring data. Therefore, if one controls the data, one controls the interface.

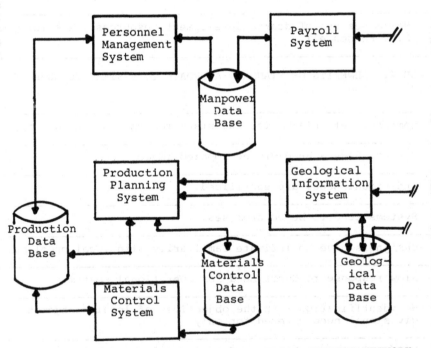

FIGURE 1:   EXTRACT FROM INFORMATION SYSTEMS NETWORK

3.3 DESIGNING FOR BUSINESS PROCESSES

A few companies are fortunate in that their remote locations have similar organisation structures. This gives standard job titles and standard job descriptions. It does not occur very often though.
In most companies, as is the case with our own mines, the locations are semi-autonomous. They each have a different history, make

different products, or have different management styles, and,
consequently, have evolved different organisation structures. Even
where job titles might be similar, the job content is almost certain
to be different. This, on the face of it, makes the design of
common systems impossible - there being no common denominator.
However, there is a technique which copes with this situation. We
discovered it in the BSP study. In fact, to my mind, it was the
best technique to come out of BSP. It was to design systems around
business processes.

A company has many business processes. In Gold Mining, we have
the Exploration Process, where geologists search for prospective
sites for boreholes. When a mine is to be established there is the
New Investment process which arranges shaft sinking, etc. After
the mine is in operation, there are a series of processes such as
Exploitation, Recovery, Administration, Materials, etc. Each process
can be broken down to sub-processes. Materials sub-divides into
Inventory Management, Maintenance, Requisition, Issue, etc. When
we get to the lower levels, it matters little what the organisation
structure of the various sites at which the process is performed
looks like, or what the job title is of the person who performs each
of the various activities. The process is, or could be made to be,
the same. This then is the basis on which we design our systems.
Of course it means that implementation is different for each mine,
but as I shall describe, we feel that this is another example where
we can solve the problem by organisational means.

3.4 INFORMATION SYSTEMS ORGANISATION

The objective of our Information Systems Department is to produce
and run information systems. As already stated, we have identified
and given a priority to all currently required systems. These have
been documented in the form of a conceptual network, an extract of
which was shown in Figure 1. In order to carry out our objectives,
we have a functional organisation as shown in Figure 2.

This shows that the line functions produce and run the systems and
the two staff functions exercise control over the line. This type
of organisation is standard and well known except, possibly, for
Box 2 - Systems Integration - which I will describe in more detail.

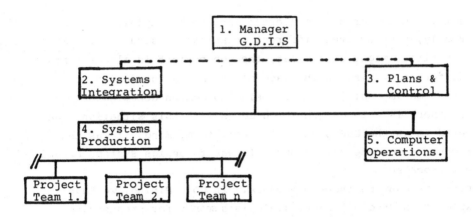

FIGURE 2: GOLD DIVISION INFORMATION SYSTEMS FUNCTIONAL
ORGANISATION.

The purpose of Systems Integration is to provide staff control by
safeguarding the integrity of the:

- SYSTEMS NETWORK
    - by keeping the I.S. Network up-to-date
    - by initiating studies in areas where the Network
      could be expanded.
    - by ensuring that present and future systems
      interface with each other.
- DATA
    - by reviewing data collection methods, validation,
      security, classification and storage (this to
      include the design of logical data bases).
    - by resolving any conflicts in the areas of data
      ownership and data usage.
    - by being the administrator of the Data Dictionary.
- COMMON (DIVISION-WIDE) CODES.

We consider this to be a function of major importance and yet there
are not many appointments to this type of position. Many companies
to whom we spoke in Europe and America expressed the wish that they
could have established a Systems Integration function when they were
embarking on a network of systems as we are currently doing.

4.   PROJECT MANAGEMENT

Consider a few statistics:

- The divison employs 130,000 black mine workers whose
  average tour of duty lasts eight months, giving an

annual turnover of over 200,000 people.

- There are over 1,000 white supervisors responsible for developing their sets of production plans each month. A set consists of one production plan together with six supporting resource requirement plans.

- There are over 7,000 survey measurements which have to be taken monthly and then plotted on maps and translated into production and other statistical figures.

- There have been 2,500 surface boreholes and over 50,000 underground boreholes drilled in the Witwatersrand Basin, where our mines are situated. None of this data is currently available in computer files.

This serves to show that the systems we are undertaking are major systems which will take several years to develop and implement. We therefore have a series of directives which the manager of a project must follow.

4.1 DIRECTIVES TO PROJECT MANAGERS

- All projects must conform with the standard phases shown in Figure 3.

- All projects which will take more than one year to develop must be broken down into "deliverables" - firstly in order to produce a faster return on the investment in development and secondly to keep the project in the User's eye.

- A deliverable - which is the General Design, the production and the testing of a subset of the complete project - must take no longer than six months to complete.

- Formal documentation must be produced before a phase can be considered complete.

- Review approvals must be formally obtained before the project can proceed to the next phase.

This being a practical paper, I must admit to a problem with the six month yardstick for the "deliverables". This has been difficult to achieve in practice. Maybe this is because we are relatively in-experienced in breaking down a project into "deliverables" or

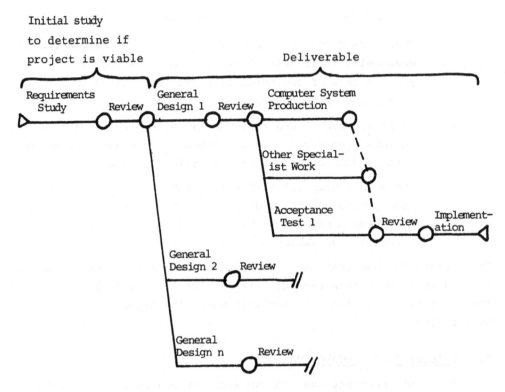

FIGURE 3: STANDARD PHASES USED IN INFORMATION SYSTEM PROJECTS

because we have encountered true exceptions where deliverables cannot
be used. In our Personnel Management System, our first deliverable
included not only the collection of basic data about each of our
mining workers but also included the creation of the infrastructure
on which the system would operate. This included the acquisition
and installation of a IBM 370/158 MP, the upgrading of the operating
system to MVS, the installation of a pilot terminal network on one
of our mines, the creation of several new organisations (e.g. Network
Control, Data Base support, On Line systems support), and the formation of various User review bodies. In our Materials Control
System, our first deliverable was to rewrite our existing stores
system. This was found to be completely undocumented - save for
program listings - and had been designed so that we could not
isolate a meaningful portion of the system to rewrite. It was all
or nothing. Except for these two experiences, however, it seems
that the other projects, and even later deliverables in these
projects will be able to use the deliverables concept to good effect.

4.2 PROJECT MANAGER

Before an Information System can be produced and implemented, two
important questions must be answered:

- What does the Project need to do?

- How are we goinç to do it?

The people able to answer these questions have completely different
backgrounds. From our System theory, we hold the User responsible
for the specification of requirements. Therefore in the Gold Division,
the Requirements Study is manaçed by a senior User Manager (see
Figure 4), who will have been selected because of the esteem in
which he is held by the Divisional Executive and by his peer group
on the mines. This is a commitment which the Division willingly
gives, and it has already appointed this calibre of person to the
four requirements studies that we have so far performed.

- For Materials Control - the General Manager of the Central
  Stores - this took three months.

- For Geology - the Chief Exploration Geologist - this
  took four months

- For Production Planning - a Section Manager from one of
  the mines - this took ten months.

- For Payroll - a company secretary from one of the mines -
  this took three months.

Such a person would be too valuable to lead the project for its
duration so we exchange project managers after the Requirements Study
- when the first question would have been answered - and appoint an
experienced systems project manager for the remainder (see Figure 5).
This project manager defines how the project will satisfy the given
requirements, and, providing approval is obtained, proceeds to
produce the system.

4.3 PROJECT CONTROL PROCEDURES

The project control procedures that we use are not rigid. Guidelines
are given for what is the minimum that should appear in the documentation at the end of each phase, but we consider the Project Managers
to be highly responsible people who should be allowed to use their
initiative and not be tied by meaningless rules. In the Requirements
Study, the User Project Manager must, amongst other things, set out

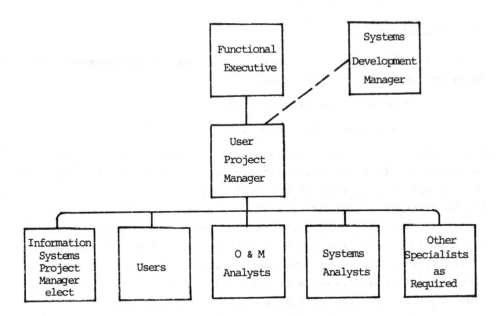

FIGURE 4:   ORGANISATION STRUCTURE OF REQUIREMENTS STUDY PHASE

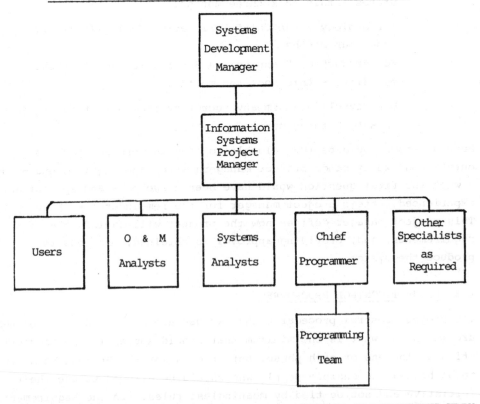

FIGURE 5:   ORGANISATION STRUCTURE OF PROJECT AFTER APPROVAL OF REQUIREMENTS STUDY.

the sequence of deliverables, giving a budget and a delivery date
for the first one. This commitment has to be supported by a detailed
plan and it is against this plan that the Systems Project Manager is
controlled. There are no required monthly progress reports as this
type of communication is given at informal progress sessions with the
Systems Development Manager. The Development Manager does produce
a consolidated report on all projects for Divisional Management, but
this is really a public relations vehicle.

4.4 PROJECT TEAM MEMBERSHIP

In designing the membership of the project teams who will design and
implement our systems, we took account of the recent experiences of
three project teams employed by a South African Utility. Each team
used a different approach. They were:

- Fully respresentative project team
- Head Office Staff project team
- Head Office Staff development team and a fully represent-
  ative test team

The first approach was used by the developers of a Stock Recording
and Costing System. This was a Common system. They assembled a
project team under the leadership of the Head Office Inventory
Controller. The team consisted of the Inventory Controllers of each
of the seven Divisions and an experienced Systems Analyst who knew
Inventory Control theory, but who had no practical experience of
storekeeping. This team of nine people realised that they were too
many to do detailed design work, so they formed a working group com-
prising of the Project Leader and the Systems Analyst. The other
team members received working papers containing the latest proposals
in time for them to be read and checked against local circumstances
before the regular monthly formal team meeting where they would be
discussed and, if agreed, incorporated. This Utility covered the
whole of South Africa and it was decided that the monthly meeting
would be held on a rotating basis in each of the major centres. This
was more expensive than holding it in Johannesburg because, travel
being more difficult, it necessitated the expense of all members save
the host staying as a group overnight at a hotel. However the
camaraderie built up during the course of this project became almost
tangible. The system took four months to develop and, because it had
to be implemented in eighty stores, it took six months to implement.

Each divisional representative became the implementation co-ordinator for his division; the Systems Analyst supplied his team of four programmers to visit each store to explain the input forms and the reports, and he himself acted as central control; and the system went in with no major problems on time and on cost. There were many minor problems but, because of the team spirit, these were accepted and solved in a positive fashion.

The second approach was used by the Payroll project. This was also a Common system, and like the Stores, had to be implemented throughout the Republic. This team, which was set up at the same time as the first and consequently could not draw on its experience, included only Head Office members. The team, when it wanted something accepted by the Divisions, had the Personnel Manager issue a Head Office order to to amend the appropriate procedure manual. The Project Leader, a senior member of the Personnel department, accompanied by his exper-ienced Systems Analyst, would also occasionally tour the Divisions to inform the paymasters of what was being developed for them. During these tours, because they could not carry all the details of the design with them, certain local queries were never satisfactorily answered. The first design was completed on time in four months but, during implementation at the project's pilot site, so many local anomalies were found that the team had to abandon its implementation and return to the design stage. This had side effects because the changes necessary were added on to the system like poor quality main-tenance, rather then designed into it. Documentation suffered, and this aggravated the situation still further. In the end, the development phase took three years and the system was never really satisfactory to anybody.

The third approach was used by the Accounting project. This was a Central System, controlled by a senior accountant at Head Office. When operational, it was proposed that the Divisions would submit their accounting transactions via their local computer (this Utility had a mini computer in every main centre of the Republic) for processing on the central hardware. There was to be a relatively complex month end closing procedure and, because the teleprocessing software was fairly primitive, any error found in an accounting batch caused the whole batch to be rejected. The project development team consisted solely of Head Office personnel. In this case it was consciously decided that the people in the Divisions were not really of a calibre

to be able to do development work. Progress reports on what the project team were considering were given at the quarterly Accountants
meeting, but suggestions for improvement were not sought. Development took fifteen months and finished on 12 December. Implementation
date was to be 1 January. December in South Africa is the start of
the summer holiday season. At the November Accountants meeting, the
Divisions were invited to nominate a four-man acceptance test team,
just to ensure that the system would work. In December, four Divisional accountants sat down to test the system. With the aid of the
Accounting project's Systems Analyst, they learned how to do an
acceptance test and then they set about putting a series of possible
transactions through the system. The shortcomings of the design
started to become apparent following the probing test data submitted
by the test team. They dredged up from the depths of their experience
all the tricky conditions that had ever happened to them, and in
practically every case they scored a hit. All leave was cancelled
and the development team and the test team sat down to rectify
matters. Fortunately they had until the end of January to get things
operational and, with much overtime and much praying, the deadline
was met and the system implemented without any of the problems being
visible to the company at large.

The membership of our project teams is a compromise between the first
and the third approach just described. We have Users from the mines
on the project teams and have other independent User involvement to
ensure the project is as widely acceptable as possible - as will be
described in the next section.

4.5 USER INVOLVEMENT

The last aspect of our organisation that I shall describe is User
involvement. I have already mentioned that Users are heavily involved
as members of the project team throughout the life of the project.
In addition there are three other major ways where we have User
involvement.

The first of these is the Executive Committee. This is a committee
representing senior Divisonal management. It is chaired by the
Managing Director of the Orance Free State mines. Its membership
includes director level personnel, and it has the responsibility,
amongst other things, for initiating common (Divisional) projects
and for approving the movement of each project from phase to phase.
The project managers, when approval is required, give a short

presentation on the project to date to the members. Really though, the Executive Committee procedures are only an exercise in communication because, being senior executives, it is impossible to expect the members to review in any detail the documentation produced. This is why another responsibility that the Committee has is to appoint personnel in the Division who will do the detailed review of each project's work. This leads us to the second type of involvement - Divisional reviewing.

In practice, as many different people as possible are appointed to review different projects. This has the effect of increasing the visibility of our projects and directing the attention of various specialists to the task of reviewing. We have found this aspect of our methodology a very great success. Project managers really have to prove the worth of their systems to get approval to go to the next phase. This has the effect of motivating the systems people to produce a high quality product whose logic is faultless and which overwhelmingly proves to be a sound business decision. In our recent experiences, the various review bodies have:

- suspended the Payroll project as not having a sound enough business case to continue.

- refused to allow the Personnel Management System past its general design phase until a better system of security backup was devised.

- increased the scope of the Materials Control project to include an additional class of material.

- decreased the benefits claimed for one of the systems.

The third type of involvement is implementation. As stated previously our systems in Gold Division are designed for business processes, and not for organisations. As the organisation on each mine could be different, a group of people are appointed with the task of implementing the system on their mine. This group is given training in the tasks expected of them and are constantly guided by the project team. They have to allocate the responsibilities for the system functions, build the facilities required, authorise the ordering of the equipment, carry out the training, do a test run of the system etc. These groups have been established on all our mines and the job is being tackled with great responsibility.

5. PROBLEMS

Naturally, we have had problems in putting a few of these ideas into
practice. Most of these occurred on the Personnel Management system.
The reason why this system was particulary badly favoured was because
it was the system identified by the BSP study as the first system to
be developed. The methodology of BSP provided for the Requirements
Study of the first system to be done in parallel with the rest of
the study, which in effect meant before the project control procedures
had been put into place. This particular Requirements Study was
done to nowhere near an adequate level of detail. The scope and
objectives of each of the deliverables were not defined and the
time table was far too optimistic. The I.S. department had not yet
been organised in a way to give adequate supervision to a project
manager who, it soon became apparent, was not sufficiently experienced to cope with this very ambitious project. The project, to
be fair, was enough to extend even the most experienced project manager.
Everybody was having to learn new techniques - the South African Post
Office was extended in implementing the data communcations network,
data communications was new to Gold Division, new software and hardware was being installed and data bases were to be used extensively
for the first time.
The project began to slip and the slide into chaos was getting faster
because the Divisional reviewers would not allow the project past its
General Design checkpoint. Information Systems management was able
only with difficulty to check the deterioration sufficiently to take
corrective action. This needed great strength of character, because
at no time did we pretend to our management that everything was going
well. They were informed of the truth and because of that we
continued to have their support. We were able to appoint an outside
consultant with proven experience in similar projects to lead the
project. The previous project manager now reported to the new man.
New functional specifications of the first deliverable were documented
from scratch, and the Requirements Study phase was re-opened. The
project is now heading in the right direction once again. It has
slipped by a year but much of that was due to the optimistic plan
that was originally set. On the other projects, we have had far
fewer problems. This I believe is because they were only initiated
after we had organised ourselves. One experience in Project Management that would be of interest is the following story as described
to me by the I.S. Development Manager of a large South African

company with several subsidiaries. This company was to implement a
Management Accounting System. The accounting transactions were to
come from existing resources control systems such as stores and payroll etc. Because the problem was so large, a project manager was
appointed for the central accounting system and a different project
manager was appointed for each of the interface systems which were to
produce the transactions. Implementation date was to be 1st July
1975 and the time allowed for development was six months which was
more than adequate. Everybody made it - except the project manager
of the Materials Control interface. Seeing that he was allowed six
months, and having the interest of his User at heart (the Chief
Buyer, unfortunately, not the Chief Accountant) he added a few "bells"
and "whistles" that helped the Buying Department. He installed a
part of a Purchasing package, but because this used CFMS files, he
also had to introduce the IMS Bridge. Somehow or other this was
not stopped by I.S. management. By the time April came, the project
was in its 90% state and continued that way through May and June.
I.S. management was in a dilemma. It could not abandon the whole
exercise because the company was going to switch to its new accounting system on the 1st July. Ultimately, the Materials Control
interface was delivered five months late. This meant that no
materials costs could be reflected in the books of account during
this time. Even by the end of the first year of running the new
system, the initial five months backlog had only been reduced to
three months. The direct effect of this misguidedness was that it
took the company more than a year to be able to manage its affairs
again, and it was only because it had a virtual monopooly of the
market that it avoided having to cease trading. The project
manager "left" and yet I feel that he was not the main person in
error. An Information Systems organisation must ensure that this
type of occurrence is never able to take place.

6. CONCLUSION

In this paper I have attempted to deliver a practitioner's view of
the organisation that must be in place before large Information
Systems projects can be developed and implemented successfully. I
have described the organisation that we have developed in the Gold
Division of Anglo American and also a few situations experienced by
ourselves and by other developers of large systems in South Africa.
I have covered many points, and I will repeat the two main ones.

Firstly systems are means by which a User manager organises his
department in order to achieve his objectives. Making amendments
to these systems is his prime responsibility as a manager. Therefore, even if he chooses a computer solution it is still his
responsibility to design it, to test that it is correct and to make
it work.

Secondly, the most important word in project development is
control - especially quality control. This control should not be
exercised in an oppressive fashion as would be the case if a
supervisor were to be continually peeping over the shoulder of a
Project Manager. Nevertheless the control, which we believe should
be in the form of reviews, should be entrenched in the organisation
so that, unless it satisfies the reviewers, a project should not
be allowed to continue.

ACKNOWLEDGEMENTS

I would like to record my appreciation to the Management of the
Gold Division of the Anglo American Corporation of South Africa
for permission to publish and present this paper.

BIBLIOGRAPHY

1.   Corporate Planning: A Practical Guide
      John Argenti - George Allen and Unwin Ltd, London.
2.   The Executives New Computer
      Oliver Wight - Reston Publishing Co. Inc. Reston, Virginia.

VALUES, TECHNOLOGY AND WORK

by

Enid Mumford

Reader in Organisational Sociology
Manchester Business School, England.

Introduction

This paper describes some of the results of research carried out by the author
into the design of a number of large scale computer systems affecting white collar
employees.   The objective of the research was to establish the extent to which the
values of computer technologists, top management and user management and clerks
influenced the design objectives that were set for these systems and their human
consequences.

Six systems were studied: three of these were large government applications,
two were in industry and one in a large international bank.

Framework of analysis

Any useful and meaningful interpretation of research data requires a framework
for analysis which will assist the description and understanding of the research
situations.   The analytical framework for this paper has been derived from the
work of Talcott Parsons[1].   Parsons provides a framework for classifying action,
or goal directed behaviour, in terms of the meaning of that action for the people
concerned and its consequences for a particular system.   The four components of
this model are 'goal attainment', 'adaptation', 'integration' and 'pattern maintenance'.   Actions directed at the attainment of goals will directly reflect the
values of the social system in which the action takes place.   Adaptive behaviour
facilitates the attainment of these goals, while integrative acts ensure that a
system subjected to change either maintains its equilibrium or reverts to a state
of equilibrium once the change is completed.   Pattern maintenance acts are those
that help sustain the new pattern of behaviour once it is adopted[2].

Goal attainment

Goal attainment covers the setting of goals and the actions that contribute directly to the realisation of these goals.   If goals have been defined in technical terms, then goal attainment will be the processes of technical systems design.   Goal attainment usually involves cooperation, conflict and the use of power.   It can also involve the manipulation of people's perception, so that the new is seen as preferable to the old.[3]

Adaptation

Adaptation is said to have taken place when a group or unit is able to interact successfully with its environment, in the sense that it gets from this environment the inputs which enable it to meet its needs, and is able to return to the environment those things of which it wants to dispose.   Adaptation is the process of moving from one state of integration or equilibrium to another and the means by which this process can take place smoothly and successfully.   Facilitators of adaptation will include values, attitudes, programmes and incentives that make change acceptable and understandable.   Adaptation in most organizations is a political and negotiating process in which the interests of the individual are likely to be subordinated to the interests of the group, and in which the interests of the less powerful groups may be subordinated to the interests of the more powerful.

Integration

Integration is the action taken, once goals have been attained, to restore a situation to a state of equilibrium.   This involves bringing the different components of tasks, technology, people and structure together into a viable and reasonably stable relationship.   Successful integration implies a positive relationship between the new tasks which are a product of the new technology and the needs of the employees who have to perform these tasks.

Pattern maintenance

Once a new system has been introduced, the essential function of pattern maintenance is to maintain its stability by making it acceptable to the values and norms of the groups associated with it.   This requires that users view the system positively and have confidence in its ability to provide the kind of environment they need to achieve job satisfaction.

Values and Change in Six Organizations

The six organizations in this research all had different approaches to the

144

introduction of computer systems. Three of them, one of the government departments, one industrial firm and the international bank set precise human goals related to job satisfaction at the start of the design process. Two of these, the
firm and the bank, used a participative approach in which clerks played a major
role in the development of a new form of work organization into which the computer
system would be embedded.

The diagram below shows the strength and nature of values in each organization.
It can be seen that those organizations which set job satisfaction goals have more
'Y' oriented managers and systems designers than those which set business or technical goals only. The values of managers and systems designers are also stronger
than those of the clerks.

<u>The strength and extent of values in each organization</u>

X indicates a preference for tightly structured
   and controlled forms of organization

Y indicates a preference for flexible, self
   managing forms of organization

C indicates a neutral attitude

Organizations with business
and technical goals only.

Organizations with job
satisfaction goals.

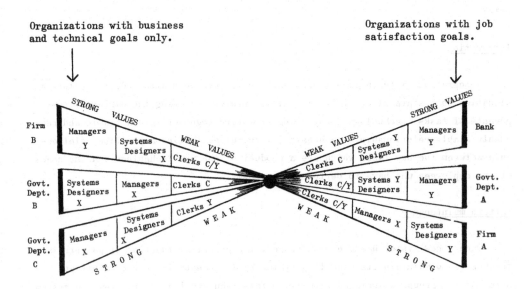

The best value fits are in

a) Government Dept. A
b) International Bank

The worst are in

a) Government Dept. C

It was hypothesised that a good fit between the values of all the groups associated with a major change would lead to change consequences which are approved of. The data supported this.

The Value Fit

Organizations which did
not set job satisfaction goals

Government Department C

Poor fit on values between the clerks and management/systems designers.

System consequences

Computer seen as routinizing work.

Government Department B

Poor fit on values between the clerks and management/systems designers.

Computer seen as disadvantageous in busy offices.

Firm B

Good fit on values between the clerks and management. Less good fit between these two groups and the systems designers.

Computer liked in one department, disliked in another.

Organizations which set job satisfaction goals

Government Department A

Good fit on values between all three groups.

Computer system approved of. System designed to improve job satisfaction.

Firm A

Good fit on values between clerks and systems designers. Less good fit between these two groups and management.

Computer system approved of. Initiative to associate new system with new form of work organization came from systems designers.

Bank

Good fit on values between clerks and managers and systems designers responsible for design of on-line system.

Computer system approved of. Initiative to associate new system with new form of work organizations came from management and systems designer.

Goal attainment

Change provides an opportunity for making choices and the kinds of choices that are made provide an insight into the values of the different groups involved in the change process. For example, values may lead to problems being defined in such a way that choice opportunities are limited, or conflicts of values and interests may give the choice process a dynamic which makes outcomes difficult to predict.

But values will not be the only influence on the kinds of goals that are set, these will also be constrained by the nature of the decision making environment. This will include the number and diversity of the groups involved, the efficiency of information systems in providing the decision group with relevant, broad based, and up to date information, and the extent to which internal and external organizational environments are subject to change.[4]

All the six organizations in this research claimed to have set very precise goals at the start of their projects. The majority of these goals were directed at the attainment of economic, rational or technical objectives. The intellectual model used to attain goals was close to that recommended by Steiner.

> Planning is a process that begins with objectives; defines
> strategies, policies and detailed plans to achieve them;
> which establishes an organization to implement decisions
> and includes a review of performance and feedback to intro-
> duce a new planning cycle. [5]

This was particularly true of the government departments whose extremely large systems involved the setting up of a network of groups to take and implement decisions. The formulation of clear objectives appeared to give a sense of order, clarity and purpose to the design process although it may have had the disadvantage of inhibiting the exploration of alternative futures once the movement towards goals had begun.[6]

Listed on the next page are the principal goals set by the six organizations, the environmental influence that stimulated the goal and an assessment of whether the goal was successfully achieved. All of these goals, with the exception of the economic goal that the system must pay for itself, were a response to some perceived discrepancy between the existing situation and the situation as an influential group in the organization would like it to be.[7] This discrepancy acted as a stimulus for change. An interesting feature of the goal setting process was that technical goals did not dominate. Sackman has described very critically how the programmers and systems analysts of the fifties and sixties were technically blinkered. He says,

> Most programmers tended to view accommodation
> for users as costly and disconcerting in the
> more urgent business of building programs and
> making them work with limited resources. [8]

In contrast to this technical perspective, the strongest influence in the six

Goal	Environmental influence	Attainment of goal
Rational/Economic/Technical goals	**External**	
Reduce staff numbers.	Tight labour market meant staff difficult to recruit and costly.	Yes. Considerable staff savings achieved in government departments. Some savings elsewhere.
	Internal	
System must pay for itself.	Treasury and Board of Director requirement.	Yes, in government departments. Probably yes elsewhere.
	External	
Increase efficiency.	Customer requirement for faster more accurate service.	Yes, orders, claims, etc. processed more quickly.
	Internal	
Increase control.	Organizational requirement for better control.	Yes, through better information.
Acquire technical knowledge.	New technical equipment on market.	Yes, through designing on-line systems.
Socio-technical goals	**Internal**	
Increase job satisfaction (three organizations).	Management/system designer values. High labour turnover.	Yes, in organizations which set job satisfaction goals.
Remove man-machine barrier (one organization.)	User dislike of previous batch system + systems designer values.	Yes, in Firm A.
Moral goals		
Provide equitable service to public.	Organizational values + public pressure.	Yes, in government departments.

organizations was the wish to achieve what were seen as important business requirements; the reduction of costs and the improvement of efficiency. In those organizations which set social or socio-technical goals, the needs of user management and clerks were given high priority, although not at the expense of efficiency. This business and social orientation was a consequence of a number of factors.

First, computer systems are easier to design today and therefore technical constraints are fewer.   Second, the power situation has altered as a result of a
wider spread of knowledge through increased use of computers.   Effectively computer specialists have lost their monopoly of EDP information.   In the government
departments the systems designers were recruited from staff who had previously
worked as administrators in the areas which were computerised, and therefore had
considerable knowledge of and sympathy for the needs of the user, at least at
managerial level.   The two firms and the bank were introducing their second stage
of computerisation and the systems designers were all conscious that their design
responsibility covered the setting and meeting of business as well as technical
goals.   They therefore worked in close association with user management throughout
the design process.

Another feature of the goal attainment process was that the goals were all
attainable, and all attained, suggesting perhaps that innovators who are going to
be later judged on the success of their actions do not make explicit objectives
which are difficult, uncertain, and may not be achieved.   The principal features
of the goal attainment process can be summarized as follows.

1. Goals did reflect organizational values and were associated with the attainment of business and human rather
   than technical objectives.

2. A number of important goals represented the values of
   organizational sub-groups.   For example, the socio-
   technical goals related to increasing job satisfaction
   originated with management in government department
   B and in the bank, and with the systems designers in
   Firm A.

3. Those who participated directly in the early stages of the
   systems design process, for example, user management, were
   able to influence the nature of the goals set.

4. The goals that were formally set and made explicit were
   those that could be attained without difficulty. This enabled the systems to be seen as a 'success' when they were
   evaluated after implementation.

The choices associated with the design of the social part of the systems are
of particular interest.   In the three organizations which did not set any positive
human goals for their own staff, government departments B and C and Firm B, the
reasons given for not doing so were as follows.

1) Job satisfaction was not seen as the responsibility of the
   systems design team.   It was perceived as the responsibility
   of the user manager or the local office.

2) In the government departments human goals were related to the needs of the public, not the needs of staff.

3) The participation of user management in systems design led to priority being given to managerial goals such as efficiency and security. There was a belief that greater work flexibility could threaten the attainment of these.

4) Systems designers saw clerks as compliant and unlikely to protest.

5) Clerks were viewed primarily as resources facilitating the operation of an administrative system.

In each of the three organizations which set specific human goals great thought was given to different ways of organizing work and to the creation of interesting and challenging jobs. In government department A this was done entirely by the systems designers and management. In Firm A design alternatives were worked out by the systems designers assisted by the author and with considerable consultation with the clerks. The knowledge acquired by the clerks during this process led to them rejecting the job enrichment solution recommended by the systems designers and substituting their own, based on the creation of autonomous groups. In the bank, total responsibility for the design of the social system was given to a representative group of user clerks who also chose an autonomous group form of organization.

The fact that social as well as technical goals had been set had two consequences for the systems designers. First, it made the systems design process more complex, requiring a knowledge of social as well as technical systems design. Second, it introduced the constraint that technical design had to be very flexible so that the technical part of the system did not inhibit any desired social solution. Three conclusions can be drawn from an examination of the consideration given to human design alternatives. These are that, in the six organizations,

1) If a human goal such as an increase in job satisfaction was accepted and made explicit then considerable efforts would be made to attain it.

2) If no precise human goals were set then little attention would be paid to human needs in the systems design process.

3) If the user, particularly the user clerk, participated in or controlled the design of the social system then goals which were important to users, such as job satisfaction, were likely to be attained.

Adaptation

Any major change process is a test of an organization's philosophy. It will

show whether the philosophy has adequately prepared the organization for change, and
if it assists the successful management of change. The organization has to be
capable of introducing and legitimizing new objectives, procedures and attitudes
and of creating a structure and a set of levers to enable it to do this. It requires what Gross has called an 'institutionalized capacity to build other institutions' which in turn requires 'the development and maintenance of a network of
supporting groups'[9]. Edstrom has suggested that when large-scale computer
systems are being designed there will be a great deal of top management intervention
for this group will wish to influence the outcome of a change that is going to have
considerable organizational impact.[10]

The structures set up to assist adaptation in our six organizations are shown
below. It can be seen that Edstrom's hypothesis is supported, for the three government departments had senior management steering committees whose role was the
formulation of policy and the evaluation of proposals from the systems design
groups.

Adaptation Structures

Adaptation structure	Organizational level of membership	Adaptation function
Government departments		
Steering Committee.	Senior management.	Change policy.
Liaison Committee.	Senior systems designers, Users, Trade Union, Board.	Change policy.
User Committees.	Management.	Development of new procedures.
Design groups with user members.	EDP, Management.	System transform- ation.
Implementation or Conversion Committees or teams.	Management, Special Group.	System transform- ation.
Consultative (Whitley Committee).	Clerks.	Problem identifica- tion once system was implemented.
Industry and Bank		
Steering Committee.	Systems designers, User management, Personnel department, Trade Union.	Change policy and progress.

Adaptation structure	Organizational level of membership	Adaptation function
Industry and Bank		
Progress Committee.	EDP, Management	Decide priorities, monitor progress.
User Committee.	EDP, Management Clerks.	Development of new procedures.
Job satisfaction working parties.	Clerks.	Design of work organization.
Implementation working parties.	Clerks.	System transform- ation.

Each organization was careful to involve its user management in all aspects of the change process. In government department C, and to a lesser extent in government department A, management itself formed the design team and so adaptation came as a result of direct participation in system development. Government department B's system was designed by a centralized systems group and care had to be taken to ensure that the user's view was represented through the cooption of users onto this team. In the firms and the bank every effort was made to design the kind of system that management wanted and this policy stimulated managerial enthusiasm for the change. Managerial participation was formalized through the setting up of user committees with responsibility for developing new administrative procedures to fit the new level of technology. These efforts to involve management were in sharp contrast with the situation ten years ago when the author was studying the introduction of computer systems in other organizations. A common policy then was for the EDP group to design a system and 'sell' it to management. It was believed at that time that management's ignorance of computers meant that it had nothing to contribute to systems design. But, whereas participation at management level was considerable, this was not true at the level of the clerks, except in those organizations which set job satisfaction goals.

The creation of a support network of committees did appear to assist adaptation in the six organizations. These committees provided opportunities for influence, for negotiation, for learning, and for taking responsibility. They worked best when the members were not restricted to a passive role of question or comment but were given specific tasks so that they could actively contribute to policy or design. Each organization formed groups of this kind at management level. Two of the organizations which set job satisfaction goals also created representative groups of clerks.

There are many strategies which one group can use to ensure the compliance or cooperation of another.   There is compliance through deference, because a group knows its place; or through trust because there is confidence that the right thing will be done.   There is compliance through the understanding which comes from good communication.   There is compliance through negotiation and through shared control and participation in decisions.   Which of these strategies is used will be determined by power, by expediency and by values.   The chart below shows the strategies used in the six organizations.

Adaptation Strategies

	Communication with	Consultation with	Participation in decisions	Control by
Government Departments				
Dept. C	All groups.	User management.	User management.	User management.
Dept. B	All groups.	User management representatives.	User management representatives.	EDP group, user management representatives.
Dept. A	All groups.	User management, staff representatives.	User management, staff representatives.	User management.
Firms and Bank				
Firm B	All groups.	User management, section leaders.	User management.	EDP/User management.
Firm A	All groups.	All groups.	All groups.	EDP/User management.
Bank	All groups.	All groups.	All groups (clerks through representatives.)	All groups (of some aspect of design)

There are four arguments for involving lower level groups such as clerks in a more fundamental way than merely communicating to them what is going to happen. The first is a values argument which states that people have a moral right to control their own destinies and that this applies as much in the work situation as elsewhere.[11]   The second is an expediency argument and states that activities are ultimately controlled by those who perform them, and that people who do not have a say in decisions may decide to repeal the decisions of others as soon as

those others leave the scene.[12]   The third relates to the location of knowledge
and states that the experts on operational factors such as task design are the
people who do these jobs.[13]   The fourth argument is that involvement acts as a
motivator.[14]   Firm A and the bank were probably most influenced by the third
argument, with the first also carrying some weight; government department A by the
first and fourth arguments.

An important function of adaptation structures and strategies is to reduce
the ambiguity and equivocality in the change process and to give people a sense of
control, knowledge and purpose; an understanding of where they are going and what
they are trying to achieve.   The six organizations created a comprehensive network
of groups to assist adaptation, whose roles varied from the creation of policy to
responsibility for specific tasks.

Integration

Adaptation has been defined as the movement from one integrated set of relationships to another.   Integration is when a number of parts of a system are
mutually adapted to each other and contributing to the successful functioning of
the system as a whole.   These parts or units can be people or things or a combination of the two as in the concept of a socio-technical system.   Integration is
made difficult by the fact that organizations are complex entities that function
on several levels and contain groups whose different interests are often in conflict.[15]   Some writers have argued that integration was only possible when employees did not question the logic of the bureaucratic organizations in which they
worked whereas today, not only are organizations rational, rational habits of mind
have also infected the work force and they are now questioning what they previously
accepted.[16]   Others believe that integration has never been possible; that there
are no social systems that can be perfectly integrated with their cultural systems
and with the personality systems of their members, that the values institutionalized in a social system will always be somewhat inappropriate to some parts of
that social system.   It has also been argued that too tight an integration is not
even desirable as it removes the stimulus for change.[17]

When technical change has been introduced, successful integration requires
the bringing together of people, technology and structure into a viable and stable
relationship.[18]   This has to be done within the context of an organization
with values and goals as shown in the diagram on the next page.   The relationship
between these four variables needs to be stable but it should not be static.   Organizations should be able to respond to new pressures from their environments while
at the same time maintaining a state of equilibrium or being able to make adjust-

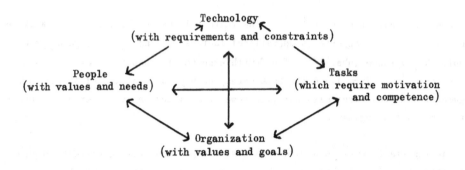

Technology
(with requirements and constraints)

People
(with values and needs)

Tasks
(which require motivation
and competence)

Organization
(with values and goals)

ments which restore equilibrium if internal relationships are badly disturbed.
The introduction of a new computer system is likely to affect each of the variables
in the diagram.   A new level of technology will bring with it a new man-machine
relationship incorporating both opportunities and constraints.   Because tasks are
altered by technology, the task structure of departments using the computer will
be affected.   New tasks mean that new demands will be made of people and this
will affect job satisfaction positively or negatively depending upon whether the
new situation meets their values of what is desirable work.   In turn, the technology, tasks and people variables will interact with an internal organizational environment which provides a structural context for the achievement of the organization's purpose, and this interaction may start the looping process again by making
new demands of technology.   A lack of recognition of the importance of integrating
these variables when introducing computer systems is seen as an important source
of their failure.

In this research the author concentrated on examining two aspects of integration.   The first was the extent to which technology and tasks were brought together into a set of interesting and stimulating activities which met the criteria
associated with good job design.   The second was how well this combination of
technology and tasks was integrated with the values and needs of employees and
produced feelings of job satisfaction.

Technology, values, and task structure

There is now considerable evidence that most employees prefer jobs that provide
an opportunity for the use of discretion, meaning by this using judgement, making
choices, taking decisions.   The exercise of discretion gives them a feeling of
personal identity and responsibility which contrasts with routine work which can
induce a sense of worthlessness or anonymity.   One way of providing discretion
is through organizing employees into multi-skilled, self managing groups with
responsibility for handling an important set of activities.   Such groups are often
called 'autonomous'.

155

The three organizations, which set job satisfaction goals all created autonomous group structure for their staff; and these ensured opportunities for the use
of considerable discretion, together with responsibility for meeting the needs of
a specific group of customers or clients. If our hypothesised relationship between values and action is correct then the values of the managers and systems
designers in these three organizations ought to support this kind of work structure
and we have evidence that this was the case.

The values of the managers and systems designers in the
organizations which set job satisfaction goals compared
with the values of those in the organizations which did
not set such goals.

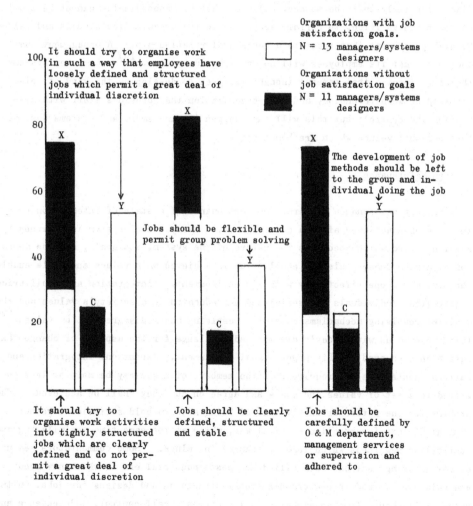

(All statistically significant at 5% level)

It can be seen that no managers and systems designers in organizations which did not set job satisfaction goals are on the Y part of the scale on the left and centre charts, and only 9% on the right hand chart.   In contrast a majority of managers and systems designers in the organizations which did set job satisfaction goals are on the Y side on two of the three questions.   In the organizations which did not make job satisfaction a system objective no effort was made to remove routine work through associating new work structures with the new computer system; and in one of them the computer led to a considerable increase in routine.

Fox has pointed out that what is being discussed here is the degree of confidence and trust between management and subordinates.   Routine jobs with low discretion imply that the manager believes that his subordinates cannot be trusted, of their own volition, to perform according to the organization's goals and values. He also points out that trust is a reciprocal relationship.   A management that does not trust its employees will not be trusted by them.[19]   Computer systems therefore require that careful thought is given to developing the kind of relationship between technology and work organization that provides staff with discretion and control; but this will only happen if management and systems designers have personal values which are humanistic.

Pattern Maintenance

Pattern maintenance requires the continuance of a state of integration once this has been achieved after a major change.   Every social system is governed by a value system which specifies the nature of the system, its goals, and the means for attaining these goals and equilibrium is achieved when values and goals enable the system to cope effectively with its environment.   Integration and equilibrium require that individuals are socialized and educated in a society's values and that conflict resolving mechanisms exist for resolving major disagreement on values.[20] This presents us with today's dilemma and challenge for the aspects of change that have been discussed in this paper - goal attainment, adaptation, integration and pattern maintenance all require that the members of a society be more or less committed to a set of values and goals and agree on how they shall be achieved.   The problem for the future is surely that of being able to hold societies together without violent insurrection or civil war, while at the same time assimilating and controlling new ideas which lead to changes in values.   Some commentators see us as now entering what has been called the post-industrial era in which new needs and values will lead to new organizational structures and designs for jobs. Cultural values will change from an emphasis on achievement, self-control, independence and endurance of distress to self actualization, self-expression, interdependence and a capacity for joy.[21]

Conclusions

The evidence from this study shows that in the six organizations that were investigated, values did influence the objectives, design and assimilation of computer systems. In three organizations the values of the groups responsible for systems design did not include the humanization of work. Because different values were given priority, the opportunity to associate a new technology with a new and more satisfying form of work organization was not taken advantage of. The unsatisfactory results of this policy showed up most clearly in government department C where greater routinization of work contributed to the low job satisfaction of intelligent young clerks.

In the organizations which set job satisfaction goals, in contrast, the introduction of a new level of technology was recognized as an opportunity to increase job satisfaction and this was done for humanistic as well as practical reasons. These organizations not only made job satisfaction a systems objective, they gave the clerks themselves a major role in the design of the new work organization which would be associated with the new computer system.

Schon believes that there are two powerful pressures in society today, these are the call to Return and the call to Revolt. The first tells us to return to the old ways and standards, to try and retrieve the vanished stability. The call to Revolt says that there is increasing alienation in society and that we must develop new values and objectives.[22] The practical problem for those who, like the author, believe that technology can be an instrument for human well being rather than a threat to this, is how to assist the development of these new values.

There are many suggestions on how a change of values is and can be made. Kuhn has put forward the scientific 'crisis' theory, which suggests that when existing paradigms cease to solve problems, they will be replaced with new ones.[23] The strategy here would be to draw attention to this kind of failure. Parsons believes that a system's inherent tendency to adjust when discrepancies occur between its needs and the external environment will keep it in a state of stability, providing there is value consensus on what these needs are.[24] This suggests that an emphasis should be placed on the development of shared values of a kind different from those of the past. Hedberg maintains that an increasing awareness of change consequences can be achieved through making values explicit.[25] Fox's view is that low trust relationships between management and employees should be replaced by high trust relationships, and that providing employees with opportunities for the use of discretion is a manifestation of this trust.[26]

158

All of these diagnoses and suggestions are insightful and helpful and will be increasingly debated in the future. In this paper the emphasis has been on the role of computers as a vehicle for change and the author's belief is that this technology can be used as an agent for 'technological bonding' so that it contributes to an enhancement of the quality of working life. Computers and automation can be an agent for good or bad, our values will determine what we regard as good or bad.

References

1. Parsons, T. and Shils, E. (1951) Towards a General Theory of Action; Harvard University Press.

2. Menzies, K. (1976) Talcott Parsons and the Social Image of Man, London: Routledge and Kegan Paul.

3. Neal, M.A. (1965) Values and Interests in Social Change. Englewood Cliffs, N.J.: Prentice Hall.

4. Friedmann, J. (1967) "A conceptual model for the analysis of planning behaviour', Administrative Science Quarterly. Vol.12 (2) 225-252.

5. Steiner, G. (1969) Top Management Planning, New York: Macmillan.

6. McCaskey, M.B. (1974) 'A contingency approach to planning: planning with goals and without goals'. Academy of Management Journal. Vol.17 (2) 281-289.

7. Vickers, G. (1973) Making Institutions Work, London: Associated Business Programmes.

8. Sackman, H. (1971) Mass Information Utilities and Social Excellence, New York: Aurbach.

9. Gross, L. (1967) Sociological Theory: Inquiries and Paradigms: New York: Harper and Row.

10. Edstrom, A. and Nanges, L. (1974) 'The implementation of computer based information systems under varying structural conditions'. Paper presented to Altorg Conference, Gothenburg.

11. Mumford, E. (1978) 'The design of work: new approaches and new needs'. Paper presented at IFAC Conference, Enschede, Holland, 1977.

12. Hedberg, B. Nystrom, P. and Starbuck, W. (1976) 'Camping on Seesaws: prescriptions for a self-designing organization', Administrative Science Quarterly. Vol.21 (1) 41-62.

13. Edstrom, A. (1974) Op. Cit.

14. White, J.K. and Ruh, R.A. (1973) 'The effects of personal values on the relationship between participation and job attitudes'. Administrative Science Quarterly. Vol.18 (4) 506-514.

15. Touraine, A.   (1974) <u>The Post-Industrial Society</u>, London: Wildwood House.

16. Fitzgerald, T.H.   (1971) 'Why motivation theory doesn't work'. <u>Harvard Business Review</u>. July-August.

17. Gowler, D. and   (1972) 'Occupational role development, part one'.
    Legge, K.   <u>Personnel Review</u>. Vol.1. (2) 12-27.

18. Leavitt, H.   (1958) <u>Managerial Psychology</u>, University of Chicago Press.

19. Fox, A.   (1974) <u>Beyond Contract:Work, Power and Trust Relations</u>. London: Faber.

20. Smelser, N.   (1959) <u>Social Change in the Industrial Revolution</u>. London: Routledge and Kegan Paul.

21. Davis, L.E.   (1971) 'Job satisfaction research: the post-industrial view', <u>Industrial Relations</u>, Vol.10. 176-193.

22. Schon, D.A.   (1967) <u>Technology and Change</u>. London: Pergamon Press.

23. Kuhn, T.S.   (1962) <u>The Structure of Scientific Revolutions,</u> The University of Chicago Press.

24. Parsons, T.   (1964) <u>The Social System</u>. New York: Free Press.

25. Hedberg, B.   (1975) 'Computer systems to support industrial democracy' in E. Mumford and H. Sackman (Eds.) <u>Human Choice and Computers.</u> Amsterdam: North Holland.

26. Fox, A.   (1974) Op.Cit.

A FRAMEWORK FOR THE ANALYSIS OF THE

RELATIONSHIP BETWEEN BUSINESS ORGANIZATION EVOLUTION

AND

BUSINESS INFORMATION SYSTEMS EVOLUTION

M. RICCIARDI

INFORMATION SYSTEMS

MARKETING DIRECTOR

IBM ITALY

SEGRATE - MILANO

ABSTRACT

This paper describes a matrix which put into evidence the tight "correlation" existing
between the EDP Development and the Organization Development of a Company.

The matrix is obtained linking together two models : the Greiner's representation
of the five phases of the Business Organization Development and the Nolan's hypothesis
on the stages of EDP growth.

The result is a method used as framework to evaluate the information systems status
and define its main development strategies to support the Business Organization
Evolution.

161

I N D E X

1.   BUSINESS ORGANIZATION DEVELOPMENT PHASES (GREINER)   (1)

2.   EDP DEVELOPMENT STAGES (NOLAN)   (2)

3.   SYNTHESIS BETWEEN MODEL 1 AND MODEL 2 TO ALLOW:

-   A DIAGNOSIS OF I.S. STATUS RELATED TO THE BUSINESS ORGANIZATION
    STATUS

-   A DEFINITION OF I.S. ARCHITECTURE AND MAIN DEVELOPMENT STRATEGY
    COHERENT WITH THE EVOLUTION OF THE BUSINESS ORGANIZATION.

REF. :

(1)   Larry E. GREINER
      Harvard Business Review, Vol. 50 July-August '72, p. 76, nr. 4

(2)   Richard NOLAN

      "Managing the Four Stages of EDP Growth"
      Harvard Business Review, Vol. 52 January-February '74, p.   , nr. 1

      "Thougts about the fifth stage" - Working Paper -
      Harvard Univ. - Oct. 1975

      "The Data Resource Hypothesis - Business Needs a New Bread of EDP Manager"
      Harvard Business Review, March-April 1976, Vol. 54, p. 125, nr. 2
      page 123.

1.   BUSINESS ORGANIZATION DEVELOPMENT PHASES

1.1   THE FORCES AND THE FACTORS DETERMINING THE DEVELOPMENT
       OF A BUSINESS

Larry Greiner, in his referenced paper, points out that the internal history of the
organization may influence the strategies, the feasibility of organizational changes
(new structures) and, as a consequence, the Business Development even more
than the external historical forces (like the market place).

He identifies the major factors determining the Business Development in:

1.   THE AGE   (of the Business)

2.   THE DIMENSIONS

3.   THE PERIODS OF DEVELOPMENT

4.   THE PERIODS OF REVOLUTION

5.   THE TREND OF DEVELOPMENT OF THE INDUSTRY SEGMENT.

1.2.   THE 5 PHASES OF THE DEVELOPMENT

Figure 1 A) reproduces the Greiner's Organization Development Model, which shows
the sequence of the five development and crisis phases, the main characteristic of
which are described in detail in Greiner's referenced paper and are summarized
in the first six lines of the matrix of fig. 2: Business Aspects versus Organization
Phases.
The main conclusion and recommendations arrived at by Greiner are the following:

- Understand in which stage of development or of crisis the company is

- Each phase leaves its mark of strenght & experience

- The tensions provide ideas & awareness for changes

- Recognize that the spectrum of solutions is limited

- Leave the present structure before the revolution becomes too much violent

- The evolution is not automatic: but it is struggle for survival

- Realize that each solution will generate new problems

- Analyze problems in a wide historical perspective

- The historical events determine in a sure way what will happen a long time after

- Analyzing better the history we can better put into perspective our future work

If we extend Greiner's analysis to the Information Systems of an organization we could add to the preceding some further considerations related to the following basic question: "which is the type of information system more appropriate for each organizational phase?"

In general we can consider obvious that the information systems support requested by the business is different in each phase and also that the EDP Management System and Style have to change as the Business Organization and therefore the EDP rôle change.

More specifically, adding a new line to the Business Aspects versus Organization Phases Matrix we can consider that:

- The Information Systems produced when the company organization is in phase 1 (creativity) are normally structured to support Operational/Accounting Aspects such as billing, payroll etc. .....

- In Phase 2 (authority) the demand is still for Information Systems of operational type to support a style of Management mainly concerned with control as opposed to standard of production/cost.

- The tendency to delegation which characterizes the phase 3 implies that the Business demand is toward Functional Information Systems able to support not only the operational activities but also the Management of the different functional and geographical units.

- The Phase 4 cannot practically be put in place whithout having implemented Information Systems integrated enough to be able to support the Company Planning and Measurement Processes which are essential for implementing the coordination of the Business Resources and Organization.

- The characteristics of the Phase 5 (Team work etc.) ask for a kind of Information System Cross Level and Cross Functions (based on Business Process) in a mature DB/DC environment.

2. EDP DEVELOPMENT STAGES
    (4 and 6 STAGES HYPOTHESIS)

In 1973 Professor Nolan discovered that the EDP Budgets for a number of companies,
when plotted over time from initial investment to mature operations, forms an
s-shaped curve.

The changes of this curve correspond to the main events (often crises) in the life
of the EDP Function that signal important shifts in the way the Information System
Resources are used and managed.

Nolan was able to single out four stages of the DP evolution, which he later expanded
into six stages, each of which is characterized by a particular aspect of the Information System growth process, and which he calls:

  I: Initiation
  II: Proliferation
  III: Control
  IV: Integration
  V: DB/DC Architecture
  VI: Maturity

Figure 1 B is a plot of the first five stages of Prof. Nolan's hypothesis v. EDP
Budget Expenditure.

Furthermore he recognized remarkable similarities between the problems which
arise and the Managment Tecniques applied to solve them at each stage, despite
variations among industries and companies, and different ways of usage of the
DP Equipment.

The main characteristics of each stage are summarized in the Matrix: I.S. Aspects
v. I.S. Development Stages of Figure 3.

3.   SYNTHESIS BETWEEN GREINER'S MODEL AND NOLAN'S MODEL

3.1.   EVOLUTION MATRIX

If we put together the two models as in Figure 1, we obtain a matrix which
represents all the possible combinations between organizational phases and
DP stages.
The logic of this matrix is to link together a model of development of the Leading
System (in this case the entire Company) and a model of develompment of the
subsystem under study (in this case the Information System) .
The matrix hase, horizontally, the scale of time of the DP Development Stages,
and vertically the Company Organization Phases.

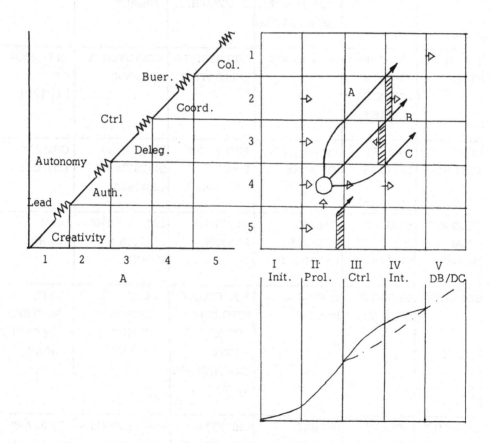

FIG. 1 - EVOLUTION MATRIX - COMPANY ORGANIZATION DEV. V. DP DEVELOPMENT

BUSINESS ASPECTS VERSUS ORGANIZATION PHASES

ORGANIZ. PHASES / BUSINESS ASPECTS	I CREATIVITY	II AUTHORITY	III DELEGATION	IV COORDINATION	V TEAM
MAIN GOAL OF THE MGT.	PRODUCE & SELL	EFFICIENCY OPERATIONS	ENLARGE THE MARKET PLACE	CONSOLIDATE THE ORG.	PROBLEMS SOLUTIONS & INNOV.
STRUCTURE	WITHOUT	CENTRALIZED BY FUNCT. (SPECIALIZED)	DECENTRALIZ. GEOGRAPHIC.	STAFFS & GROUPS	CREATION
TOP MANAG. STYLE	INDIVIDUA-LISTIC & BUSINESS ORIENTED	AUTHORITA-RISM	ORIENTED TO DELEGATION	COMMITED TO CONTROL	IN FAVOUR OF PARTICIP.
CONTROL SYSTEMS	MARKET RESULTS	STANDARDS OF WORK	REPORTS & PROFIT CENTRES	PLANNING & INVESTMENT CENTRES	COMMON OBJECTIVES
RECOGNIT. FROM THE MANAG.	SHARE IN THE PROPERTY	SALARY INCREASE PER MERIT	INDIVIDUAL AWARDS	PART OF THE PROFITS & STOCKS	TEAM AWARDS
COMMUNIC.	FREQUENT & INFORMAL	FORMAL & IMPERS.	NOT FREQ./ REPORTS/ LETTERS/ PHONE CALLS/SHORT VISITS	- PLAN REVIEWS - BY LINE - BY FUNCT.	- FREQ. MEETINGS - COMMON WORK
INFORMAT. SYSTEMS	OPERAT. (BILLINGS, PAYROLLS)	OPERAT. ORIENTED TO ACC. CONTR.	ORIENTED TO FUNCTIONAL UNITS MGT.	- DP CENTRAL - IS ORIENT. TO INTEGR.	DB/DC APPL. TO THE DAY BY DAY ACTIVITY

I.S. ASPECTS VERSUS I.S. DEVELOPMENT STAGES MATRIX

I.S. STAGES \ I.S. DEVEL. STAGES	I INITIALIZAT.	II PROLIFERATION	III CONTROL	IV INTEGRATION	V DB/DC ARCHIT.
SYSTEMS LEVEL	OPERAT.	OPERAT.	FUNCTIONAL	GENERAL	CRSS LEVEL & FUNCTIONS
APPLICATIONS	ACCOUNTING	OPERAT. CONTROL	FUNCTIONAL UNIT MANAG.NT	FUNCT. UNIT MGT/ BUSINESS MGT	PROCESS MGT.
DP PEOPLE SPECIALIZAT.	COMPUTING PROGRAMM.	VARIETY OF APPLICAT.	VALIDITY & CONTROL	DB/DC MGT SYSTEMS IS PLANNING	DATA RESOURCES MGT
ORGANIZATION	INSIDE THE FIRST USER DEPART.	ANALYSTS-PROGRAMM. ASSIGNED TO AREAS	DP DEPARTM. SEPARATED STEERING COMMITTEE	BECOMES SEPARATED FUNCTIONS	THE DIRECT DATA RES. FUNCTION IS CREATED
MANAGEMENT STYLE	PERMISSION	ORIENTED TO SELL	CONTROL-ORIENTED	SYSTEMS & RESOURCES PLANNING	DB/DC ARCH. COHERENT WITH CY STRATEGIES
PLANNING	EXPENSES BUDGET	EXPENSES BUDGET	COST CONTROL FOR HW & APPLICAT	3-4 YEARS PLAN FOR HW, SW, RESOURC. METHODS	DATA RES. FUNCTION PLANNING
CONTROL	VERY LOW	FLEXIBLE	TIED . PRIORITIES . BUDGETS . STANDARDS . PROJECTS	SOPHISTICATED EFFECTIVENESS VS. PLANS	MEASUR. OF DATA RES. FUNCTIONS VS. CY PLAN

3.2.   SOME CONSIDERATION ON THE EVOLUTION MATRIX AND EXAMPLES ON ITS USE

The Matrix is a scheme which:

- Represents an attempt to put in relation the Development stage of the Company macro organisation & the DP development stage identifying the point in the matrix which represents the combination of the 2 stages.

- Stresses the concept that the judgement upon the DP development stage has to be "relative" being dependant on the stage in which the Company is at that moment.

- Makes available a first diagnosis tool to evaluate if the stage and the development strategies (architecture) of the Business are coherent with the status and perspectives of the Company Organization.

- Allow to use a common and synthetic language.

The organizational phase in which the Company is when it starts the utilization of the DP Systems can influence its first reaction regarding DP. In general we can observe that:

- If the Company is in the Phase 1 (Creativity) the DP can be developed until Stage 2 (proliferation) but it is very difficult to reach the Stage 3 (Control) being the control completely out of the prevalent mentality of the people.

- The Phase 2 (Authority) can be a stable Phase & therefore the DP can be developed until the Stage 3 & more.

- The Phase 3 (Delegation - Decentralization) makes it difficult or impossible to realize the Stage 4 (Integration) because the company philosophy in this Phase is exactly the contrary.

- The Phase 4 (Coordination) requires a DP Development close to the Stage 4 (Integration) because the effective implementation of the organizational phase needs an architecture of the I.S. able to make available an effective Planning Measurement & Control System.

- The Phase 5 characterised by team work requires available DB/DC Systems supporting the teams with information complete from an horizontal & vertical point of view.

In reality the evolution of the DP stages is combined with the Company development phases.
The path in the Matrix becomes a line and each stage has problems depending also on the past history of the development.

In general, starting from a given square of the Matrix we can have 3 major types of development :

- DELAYED : DP is following passively the organizational evolution (path as A)

- SYNCHRONISED : DP development has well understood the main business tendencies and mantain available and is consistent enough with organizational development (path as B)

- ADVANCED : DP is adopting a development strategy consisting in anticipating I.S. availability to prevent (or reduce) next crisis (Path as C).

If, for example, the Company is moving from an authority phase (2) to a delegation Phase (3) and ask the DP to move from the DP Stage 2 (the Proliferation) to the stage 4 (Integration) (Path C), that will represent for the DP Management a kind of commitment very difficult to fulfill both for technical reasons (skip of Stage 3) and organizational reasons (delegation Phase is normally not the right condition for I.S. Integration).

This strategy can only be implemented if the company Top Management is well aware of the difficulty and the spectrum of problems that may be encountened, hires the right DP manager, anticipates the delegation phase carrying on investments, commitments and part of the actions normally typical of the following coordination phase (like the creation of Planning and Control Procedures, investment centres ....)

Another case which corresponds to many practical example is when a Company in Phase 2, having the DP in Stage 3 is evolving its organization toward delegation.

It may happen that the EDP function is mantained centralized so that all the decentralized functions will be served by the same EDP Organisation which is, in this case, in condition to progress as foreseen in the Nolan stage Model.

If, viceversa, even EDP function is impacted by the delegation process and single organizational units put in place their own EDP Organization, we may assist to the regression of these new EDP Environment to the Stage 1.

In this case it is easy to foresee a spectrum of problems going from discrepancies and difficult relations between the decentralized EDP Units and what remains of the central one, up to the loss of the EDP Development and Management Experiences already present in the Company.

Another typical example is when EDP persists in remaining in a proliferation (2) Stage characterized by operational Systems when the main problem of the Company is to solve the control crisis.

Often this situation happen when the EDP Management Style is too much oriented toward technical aspects of EDP and is unable to understand the major Business Requirements connected with the Macroorganization Evolution.

In this cases there is an evident lack of communications between the Company Top Management and the EDP Management, there are no appropriate I.S. planning methodology and no organizational structures and procedures able to create a common understanding.

Typically this situation is the true reason which, in many cases, explains the existance of strong tensions and disagreements between EDP and USERS.

The lesson coming from these examples is the necessity to first well analize and understand if EDP is really well supporting the development of the organization, and then define EDP Development Strategies coherent with the business objectives, before discussing any specific solution and technology.

INFORMATION SYSTEMS MANAGEMENT THROUGH OFFICE AUTOMATION :
AN ORGANIZATIONAL AND SOCIAL PERSPECTIVE

Jean-Paul De Blasis

Centre d'Enseignement Supérieur des Affaires

78350   Jouy-en-Josas, FRANCE

ABSTRACT

A number of concurrent technological and organizational developments
in the office automation area offer an excellent opportunity to greatly
improve information systems management. This paper will attempt to show
how information systems control could be taken over by managers through
office automation. The emphasis is first on showing how office automation can contribute in designing manageable and controllable information
systems from offices. Next, some organizational implications are discussed including possible structural changes and roles of the involved
actors. Finally, a research design approach is outlined. It provides
a practical framework to tackle implementation issues taking into account some critical constraints generated by office automation systems,
notably from an organizational and social perspective.

This work was partially supported by the Centre National de la Recherche
Scientifique (CNRS) and the Institut de Recherche d'Informatique et
d'Automatique (IRIA) under ATP contract.

INTRODUCTION

Attention has recently been focused on office automation systems with most experts agreeing that the "data" processing world and the "text" processing world will ultimately converge.[1,3,8,20]

Office automation systems resulting from this evolution have the potential to profoundly effect organizations in different domains at various levels.

Because of the technological and organizational implications that office automation systems have in such areas as communications, database access, distributed processing, etc., the design of information systems within many organizations is going to be greatly affected. Recent technological developments made with more emphasis on ultimate end users - non EDP specialists - indicate that corporate management and office support staff will be faced with revolutionary changes accompanying office automation. However, they are going to be able to better understand and relate to those systems than in the past with the traditionnal EDP systems.

For example, it has often been stressed that, in order to keep the control of their information systems, managers have to be at least as attentive as they are with any manual system they supervise. In fact, it is astounding to note the frequency with which managers become apparently disinterested in their system and abdicate control to staff specialists.[2,7,15] Current research efforts in the office automation area - notably Xerox[18], Citybank[23], Wharton[5,16,17], etc. - make an effort to put technological capabilities in each office in such a form that secretaries and managers can use them. This approach of preserving "the culture of people" over "the computer culture", will enable a greater involvement of managers in their information system operations and policies.

The intention of this paper is to show how information systems control could be put back in managers' hands through office automation. The emphasis is on organizational and social issues that have been brought about by the relatively new concept of office automation. These issues will undoubtedly need the attention of managerial skills.

OFFICE AUTOMATION AND INFORMATION SYSTEMS

To date, within organizations, data processing and text processing
- the most developed side of office automation so far - operate to a
great degree independently of one another. However, an office can be
viewed as the center of what goes on in an organization as far as
information generation and communication.[14,25] Usually a great deal of
the information eventually stored and processed by the corporate EDP
department originates in an office or an other. For this reason a
number of information specialists and executives in several
organizations are thinking about eventually closing this loop.
Basically three technological and organizational developments permit
to argue that office automation systems present a reasonably feasible
route to the implementation of more meaningful information systems.
These concern the areas of data and text capture, data base
integration, and communication networks.

Data/Text capture.

Information specialists have recommended for some time that in order
to improve accuracy, the data entry function must move back to the
source level i.e.the corporate office where most data and documents
are generated. Therefore it is an excellent source date capture point.
An office automation system is a feasible solution because when some
text is typed, it is also digitized for storage on a magnetic medium.
Such a system makes possible the indexing and referencing of material
prepared in the office at the time of the original text entry. This
tends to improve accuracy, timeliness, completeness, etc..., all
essential criteria for building up sound information systems.

Data base integration.

Another use of office automation systems is the indexing and
referencing of documents prepared externally which enter the office
files. These systems will also assist in the maintenance of local
files as well as in the eventual integration of some stored material
in centrally controlled data base.[11] There already exist many
commercially available text editing terminals linked to a shared
processor for economies of scale which for instance have access to
common data bases containing customers' names and addresses and
standard letter paragraphs. Another growing area of information

storage and retrieval is the use of microforms by various
organizations such as insurance companies, banks and utilities where
special filing applications may require it. Microfilm and microfiche
systems can easily become part of office automation by allowing for
the entry and query of file index information from a terminal. The
microform is then displayed on a reader that can eventually deliver a
hard copy.

Communication networks.

Recent technological breakthroughs in telecommunication such as the
programmable switchboard PABX (Private Automatic Branch Exchange)
allows the linking together of various administrative functions for
the automated office.[4,12] A PABX may also provide communications
facility to link one office to another, to the corporate EDP department
or to outside computer service vendors. The programmable switchboard
may also be used in a variety of applications integrating both voice
and data communications in having models shared by office terminals.
It enables incoming messages from the outside to be stored in one
office system or another for later delivery ; it allows office
dictation to be recorded on voice store-and-forward equipments for
later transcription by the text processing center ; it optimizes
telephone line usages within offices ; it provides toll accounting as
well as traffic analysis to allocate phone budgets, etc. A typical
"Office of the future" may use an office automation system where
document generation, text editing, fascimile transmission, database
system, microforms, photocomposition and programmable switchboard are
used together.

Some technological implications.

The introduction of office automation systems in organizations is most
likely to bring about changes not only in the traditional office
structure, but also in the structure of the organization as a whole.

When EDP was introduced in organizations, its impact was initially
most felt in accounting applications, and it did not have the broad
effect on people that office automation systems, which reach almost
everyone, are expected to have. Though, there is a large consensus
among experts who feel that the user's level of technological
knowledge with office automation systems will not reach that of

today's data processing because it will not be needed.[1,21,24] Most
manufacturers are right now working for developing an easier-to-use
"friendly" technology for such "naive" users as managers and office
personnel. However, given today's state of technology, it is very
difficult ot produce an office automation system that simply can be
installed in an organization and be directly used by office staff
other than in relatively simple ways.[22] For that, and for some years
to come, new roles will have to be defined associated with people who
are able to take bits and pieces of the office automation technology
to put together reasonably good ad hoc solutions to get things
working. Presently, it is difficult to hand that role to someone who
does not understand in principle the editing language or the way
information is stored in the machine for example.

The problem as pointed out by a professional, is that "only a
handful of computer specialists have the remotest idea of the
information storage and retrieval problems in typical offices. The
problem is that computer people know how to think numbers, but they
don't know how to think words".[20]

Another question concerns the suggestion made earlier to key
data/text once at the text processing station, then storing it for
later retrieval or transmitting it. Such a procedure implies the
support of a digital technology. One may argue that "offices do not
work digitally ; they work with analog images on paper".[20] In fact,
most of the work going on in the office automation field and
commercially available concentrates on letting people produce paper
faster and in higher volume. Thus it appears that it will be a
disastrous solution in the long term. Hopefully, this tendency will
be reversed in order to concentrate more on ways for minimizing and
in many instances eliminating paper as a communication medium. Some
research initiated on graphics, image digitizing and holographs as
opposed to character representation may provide some satisfactory
answers.[21]

ORGANIZATIONAL IMPLICATIONS

Depending on an organization's primary activity, the principal
incentive for implementing an office automation system may come from
different areas : text processing, telecommunications (electronic
mail, computerized conferencing, facsimile transmission, etc.), or
even from the printing/publishing technologies. Initially, the
responsibility for such services is likely to be given to an
administrative manager, not to the data processing manager who will
still be in charge of EDP and data communications. However, with the
increasing development of functions associated with office automation
systems, many believe that all these activities will be grouped under
the responsibility of one executive, perhaps a V.P. for information
and communication services.[8] Reporting to this executive - staffed
with a manager for information administration - might be three
different groups as shown in Figure 1. Each group is responsible for
the following area :

1. In staff, the information manager is the data administrator
   responsible for the content of data bases, microform supports,
   etc., coordinating groups within the department and interacting
   with users in the whole organization to meet their information
   needs.

2. In line, the three departments are in charge of : a) administrative
   services : general office services utilizing non-electronic
   support ; b) office automation : correspondence services (text
   processing, graphics, printing, publishing, duplicating
   technologies), communication services (verbal electronics, voice
   and dictation equipments, PABX, etc.) ; and finally c) data
   processing : all data electronics related services, staffed with a
   "Data Base Administrator" responsible for the data base technology.

Although this type of organizational structure is apparently not
going to be implemented within the near future, there is definitely
a trend among experts to see the MIS or EDP function going under an
executive for Administration, rather than the Controller.[6,26] At one
point, it may well also happen that a great majority of people
reporting to Administration will be DP professionals.

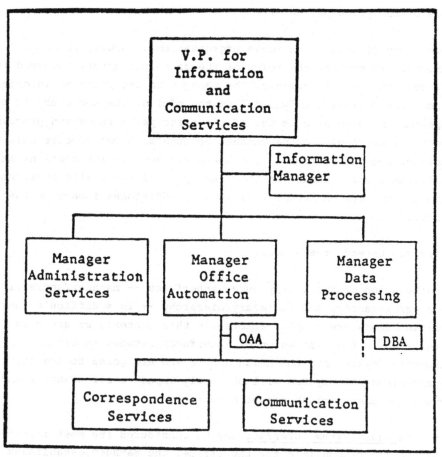

OAA : Office Automation Expediter

DBA : Data Base Administrator

Figure 1 - Organizational structure for Information
          and Communication Services.

Role of the principal actors.

Before any discussion on these organizational issues, it seems
important to outline the roles of the principal actors involved in
offices and their interfaces. The manager is the point of interest
since s/he defines the "values" in the office. The secretary is also
important because s/he is one of the principal information processing
resources available to the manager because s/he can process relatively
unstructured information.[13] The hierarchy and its interactions at
different levels is also an important part of the office environment.
Finally, the system itself will have an additional impact on the
office.[6]

Interface between the actors.

Consideration of the purely human interfaces -- manager/secretary,
manager/hierarchy, and secretary/hierarchy -- is a difficult task
because humans are complex actors. In this context, we are more
interested in the discussion of interfaces between an office
automation system and the human actors who are going to use it.
Nevertheless, the system will obviously impact their behavior and
some human observations can be made :

- the manager/system interface may be considered the most important.
  As the one responsible for the office, the manager establishes
  values and rules reflecting his/her style and personality. The
  system should in turn be able to reflect this style and personality
  if it is to be accepted. Personalization here is an important
  characteristic that must be part of the system, thus contributing to
  a more "human" computer system. In OASIS, an office automation
  system being developped at CESA[9], personalization is built in with
  such features as flexible report or schedule formats at different
  levels of details or times. A notion of "history of use" is being
  built in OASIS to remember the manager's useage familiarity with
  the system as part of the personalization effort. Therefore the
  system will "learn" and adapt to the user's degree of sophistication.
  So far, very few systems have been designed and actually run by
  managers themselves, with various degrees of success depending most
  of the time on the resistance to change and cooperation of managers
  involved in those experiments.[2,18,23]

- The secretary/system interface is very dependent on the scope of
responsibility given to the secretary. It is assumed that
secretaries will be primarily responsible for operating the system.
Consequently, the interface should be simple, requiring as little
abstraction and learning as possible. Here again, some office
automation systems - as the one under development at the Wharton
School (W.O.A.P.)[17] - provide some kind of results personalization.
By setting parameters, the W.O.A.P.* system is able to recognize who
it is "used by", and who it is "used for". But perhaps the key
problem here concerns who will get the job done. Most managers lack
typing skills and since we deal with an interactive system using
terminals with typewriter keyboards, there are physical and
psychological constraints for them. Of course, the primary data
input necessary to drive the system will center on the secretary.
It is conceivable that budget, schedule or planning supports
modules for example will typically be run by the manager
him/herself. This is important in maintaining the feeling that the
systems are built in such a manner that only a "clerical"
intermediary is required to use it.

- In the hierarchy/system interface, one of the most important
considerations is privacy of information, the protection of it
and the control of access to it. The answer is neither provided in
granting complete access nor in preventing it completely, but in
defining some level of access in between. Depending on the
hierarchy, some people may have access to detailed financial
information for example, while others may be restricted to some
summary data. In other words, an office automation system must be
able to provide some control mechanism allowing intermediate access
stages. Another advantage here is the communication facility that a
system may offer by cutting out several levels of hierarchy. For
instance, if a manager wants to call a meeting of his subordinates
with OASIS, he can review their schedules and set a date and time
without having to go through the whole hierarchy of managers and
their secretaries. A manager can also leave a message to a high-
ranking officer of an organization through electronic mail. Such
systems offer a definite advantage in situations where time is a
vital constraint.

---

* WOAP stands for Wharton Office Automation Project. The author who
participated in this project named it also OASYS in a previous paper[17]. It is not to be confounded with OASIS, another project described later which is conducted in CESA by B. Savonet and the author.

Impact of interactive computing systems on people.

To date office automation systems have had some definite impact on the
people who have used them. Following in an outline if the principal
areas of observation.

- the interactive nature of most office automation systems has an
  important impact on people's motivation. For example, to provide
  information through a system "asking questions" is a better accepted
  activity than filling out pre-printed forms. A system waiting for an
  answer appears to be more likely to get one. The "conviviality" of the
  system may be a good means of coping with the resistance to change
  by users whose jobs may be affected.

- an interactive terminal also provides a sort of "impersonality"
  which is likely to assist communications throughout the organization
  because of the non-censorship feeling it offers to users.

- finally, since it is possible for non-trained users to have a working
  system within an hour, they immediately feel rewarded for their invested time. Therefore the system is appealing to them and they are
  more willing to learn about it and continue to use it.

A RESEARCH DESIGN APPROACH

At CESA, we have been setting up a research design approach to organizational and social implications of office automation, and we are currently implementing it.

First established in early 1976 under the name OASIS[9] it was an initial research program aimed at developing operational tools to be used in an office environment. A number of office tasks were selected in cooperation with secretaries involved in this process and a number of software modules were implemented accordingly. So far, they have delt with :

1 - text processing (documents creation and preparation, personalized repetitive letters, automatic bibliography maintenance, etc..)

2 - scheduling of events (planning for certain dates and times, coordination of schedules, agenda maintenance, etc..)

3 - filing (preparation and maintenance of some office files such as resumes, publications, names and addresses files, document flows, etc..)

4 - communications (electronic mail and message system).

The OASIS system as shown in Figure 2 is in constant improvement. Modules are being modified, suppressed or added according to the needs and comments expressed by the various actors involved in its design and operations. For the most part, the system is being developed in the BASIC language and run on a PRIME computer from various terminals (visual displays and printers, all meeting office quality requirements).

182

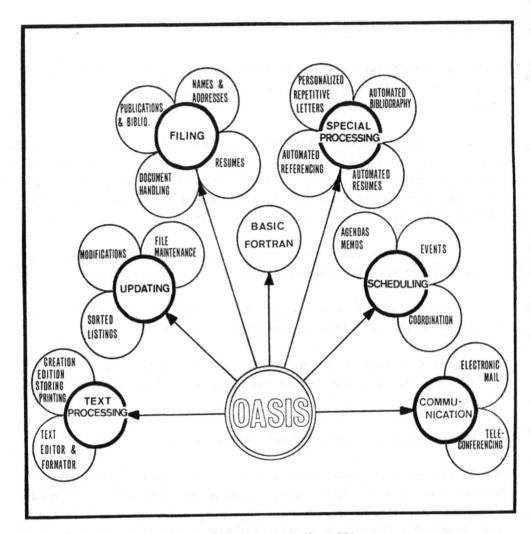

Figure 2 : Modules in CESA's Office
Automation System.

Objective of the research design.

The main objective of the research team working on office automation
at CESA is not software tool development although modules in OASIS
are constantly improved. Actually, the objective is to evaluate the
effect that our office automation system has in technical, human,
economical and organizational areas by means of a pluri-disciplinary
approach. This is what the MATHIAS[10] project is about. Some well-
identified office functions have been selected in actual offices
within CESA and other external organizations to insure that results
obtained in one environment are general enough to be transferable in
other environments.

Working hypotheses.

Our research team is currently investigating the following hypotheses :

1 - Office automation systems seem to offer a number of advantages
    over traditional manual office systems : what kind of experimental
    measures could we use to scientifically confirm those expected
    advantages ? For example, what criteria in terms of volumes
    processed, time spent, number of people, cost and economic
    consequences, etc. may impact the "productivity" related to the
    various categories of work processed in an office environment
    such as text processing, scheduling, communications, etc. ? Is it
    also possible to measure inconveniences in the same way ?

2 - Because of the automation of an increasingly large number of
    office tasks, functions of the social actors concerned, i.e.
    secretaries, managers, hierarchy, organization, are going to
    evolve toward a larger decentralization of responsibilities while
    the tendency is to increase the centralization of processing. Is
    it really so, and by what criteria may this be appreciated ?

3 - The eventual redefinition of some functions is going to affect
    the equilibrium of the social cell represented by an office in its
    organizational environment. By what evidence may one evaluate this
    lack of balance ? What consequences will this have from a human
    standpoint ? Is it possible to avoid them by using an appropriate
    implementation program ?

4 - Education needs of the office staff to recent computerized
technologies and to new working conditions and procedures is
going to be an important element affecting the success of the
office automation effort. Are there some satisfactory strategies
to tackle this education problem ? How to organize it ? What
about the other actors ?

5 - The technological changes introduced in an office environment
are not neutral, i.e. they are going to generate negative
reactions. In fact, the automation process is just starting in
some offices. So far, traditional tools in use in most offices
are writing tools (typewriters and copiers), simple and
monofunction that contrast with some integrated, multifunctions
of today's office automation system (writing - transmitting -
storing - accessing, etc.). This distinction seems important
because widely accepted individual technologies such as the
telephone or the writing tool must be differentiated from a
"system" tool. An office automation "system" is able to handle
several administrative functions and therefore is in a position
of modifying many working habits and many traditional
organizational frameworks. On what technical criteria should those
systems be built in order that they be least "disturbing" ?

6 - When dealing with the problem of justifying office automation
systems, most of the time one is faced with issues relative to
the improvement of the office work productivity. We all know
that the productivity concept carries a lot of emotional
reactions. Is it possible to conceive the organization of
administrative work with production as the only objective ? Are
we attempting to transpose in the office the spirit of scientific
organization which is reflected in most working conditions in
the industry ?

7 - A distinction is possible between offices according to major
tasks they perform :

a) the production offices which handle very structured tasks
such as heavy typing jobs in banks or insurance companies.

b) the support offices which handle unstructured and very
diversified duties - mostly administrative - such as scheduling,
telephone answering, filing, light-typing, etc.

The assistance brought by office automation systems to those two office environments may be very different. However, it is not radically different from one office to another or from an activity sector to another so long as those offices belong to the same category. How to experimentally demonstrate this assertion ?

RESEARCH METHODOLOGY

The problem with conducting this kind of empiric research is in making
sure that results are pertinent and transferable to other environments.
For this reason, it is necessary to take the prototype office
automation system away from the experimental setup and to transfer it
to real "operational" offices choosen for pilot studies. Besides
testing the transferability of results, this allows for an
appreciation of practical difficulties in installing the system. A
number of observations may be made with regards to the acceptance of
the system by the office staff, the degree of information/education
needed, the organizational constraints, etc. In particular, the pilot
study should try to determine quantitative measures of "productivity"
and to evaluate qualitative factors affecting the impact of this
technology on office environments under investigation.

A typical pilot study may be organized according to the four following
steps :

Step 1 : Systems analysis of the office cell and comparison with the
         traditional organization.

         - study of the office in terms of systems approach : its
           environment, structure, functions, actors involved,
           "culture", management style and personality, etc. The
           objective of this initial phase is an attempt to construct
           a system model of the office in order to better appreciate
           the various parameters that may affect its behavior -
           notably from a pluri-disciplinary approach.

         - this systems analysis phase should be conducted along with
           a detailed study of the various office functions in its
           traditional, non-automated functioning. During this study
           a number of measures should be made on the office tasks as
           far as volumes, delays, complexity, processing, frequencies,
           etc. according to the categories of office functions
           previously defined. This will allow for the establishment
           of a comparison basis with automated functions introduced
           in the next phases.

Step 2 : Introduction of the office automation technology.

This phase provides for the installation, test and
implementation in the office of a technology suited for the
automation of some selected office functions. It may be
initiated concurrently with the first step provided that
information/formation has been given to the office staff
prior to any installation so that "negative" reactions may be
avoided as much as possible.

Step 3 : Automation of some office functions and measures.

- during this phase some office functions and/or tasks are
  selected for automation according to results of the previous
  phases. Criteria for selection may include feasibility from
  a technical standpoint, opportunity from a processing
  standpoint, cost constraints, human impact, etc.

- from that juncture commences a period for quantitative and
  qualitative evaluations of the selected activities.

- then comparisons are made between measures taken during the
  traditional manual operations and those taken in the
  automated mode. Here the major difficulty is to evaluate
  precisely the qualitative aspects from one mode of
  operation to the other.

Step 4 : Validation of the results transferability into other
environments.

This phase is concerned with the validation of results
obtained and recommendations made as a result of the transfer
of the experimental conditions in another office environment.
An analysis of the same tasks performed in step 1 is
necessary to allow for comparison and validation of the
results. The same variables should be measured and then
compared. Significant differences and gaps should be
analyzed and conclusions made with regards to the working
hypothesis 7.

A final report should present results of the pilot study
with the following emphases : the implication of a
technological introduction on the evolution of office
functions and tasks, on human reactions due to new working
conditions, on psycholosociological constraints due to
changing working habits of office staff, and on economical
and organizational consequences. Particular attention should
be given to technical specifications that the office
automation tools should meet in this kind of office
environment.

CONCLUSION

A number of concurrent facts lead us to argue that an opportunity to greatly improve information systems management is offered by recent office automation developments.

Organizations have just discovered the enormity of office automation problems by looking at their increasing costs. The ballooning clerical costs of the office are being given more attention in the budgeting and planning cycles that every company undergoes on a formal or informal basis, thus generating an important market of potential customers for equipments and services.[22] It is believed that organizations are willing to make the necessary investments that can increase productivity and significantly reduce the time from idea to printed or recorded material ready for communication and distribution.

From another standpoint, organizations are increasingly recognizing that information may be compared with a resource that needs to be managed.[19] They also realize that the flow of corporate information is going to be considerably affected with the arrival of office automation technologies.[16]

Also, the declining hardware costs are going to assist in the wide dissemination and use of technological tools in offices, and "all kinds of applications which seem exotic today are going to become feasible and attractive".[12]

In the long run, office automation and data processing are going to move along together as their technologies gain in similarity, and it was argued that this may provide a basis for more meaningful information systems. However, any success in that area presupposes that managers of the traditional DP functions will become involved in office automation problems.

A first step in the design of an automated office is provided by OASIS and similar office automation systems under development. However, they usually are developed in somewhat idealized environments because users are relatively familiar with the technology involved, and costs are not yet well controlled. Therefore, it is not clear whether such systems working in some environments will work when transfered in other environments. Because of the tremendous potential for change brought by such office automation systems, the organizational implications cover a wide range of areas. The traditional office structures and the various actors involved are first to be affected, and this is a very sensitive issue in most organizations.

The MATHIAS project previously described provides a research methodology which constitutes a reasonable framework to tackle office automation implementation problems. The pilot study approach such as discussed may proved to be essential in taking in account the technological, human, social, organizational and economic constraints for a successful implementation and that in itself is really a challenge for managers.

REFERENCES

1. Burns, J. Christopher, "The Evolution of Office Information Systems", Datamation, April 1977, pp. 60-64.

2. Carlisle, James H. "Evaluating the Impact of Office Automation on Top Management Communication" Proceedings of the National Computer Conference vol. 45, AFIPS Press, June 1976, pp. 611-616.

3. Caswell, Stephen A., "Word Processing Meets DP", Computer Decisions (10:2), February 1977, pp. 52-56.

4. Cerf, Vinton G. and Alex Curran, "The Future of Computer Communications", Datamation, May 1977, pp. 105-114.

5. De Blasis, Jean-Paul, "OASYS-Problem Approach and Overview of the System" DSWP 75-10-05, The Wharton School, University of Pennsylvania, October 1975, 21 p.

6. De Blasis, Jean-Paul, "An Interactive System for Office Automation : Some Organizational Implications", IRIA Colloques, (IRIA 78150 Le Chesnay France), January 1976, pp. 389-398.

7. De Blasis, Jean-Paul, "Management Information Systems : A Current Appraisal", DSWP 76-06-02, The Wharton School, University of Pennsylvania, June 1976, 89 p.

8. De Blasis, Jean-Paul, "Office Automation Systems : Another Possible Route to M.I.S.", Proceedings of the International Symposium on Technology for Selective Dissemination of Information, IEEE Press, 76 CH1114-8c, September 1976, pp. 41-50.

9. De Blasis, Jean-Paul and Bernard Savonet, "Projet OASIS : Organisation et Assistance des Systèmes Informatisés de Secrétatiat", CESA Pub., 78350 Jouy-en-Josas, France, April 1977.

10. De Blasis, Jean-Paul, "Projet MATHIAS-Modifications et Aménagements des Tâches Humaines dus à l'Informatique dans l'Automatisation des Secrétariats", ATP-CNRS/IRIA, CESA Pub., 78350 Jouy-en-Josas, France, April 1977, 31 p.

11. De Blasis, Jean-Paul and Thomas H. Johnson, "Database Administration : Classical Pattern, Some Experiences and Trends", Proceedings of the National Computer Conference, vol. 46, AFIPS Press, June 1977, pp. 1-7.

12. Ferreira, Joseph and Jack M. Nilles, "Five-Year Planning for Data Communications", Datamation, October 1976, pp. 51-52.

13. Husbands, Bernard, "Centralized Secretarial Services", Journal of Systems Management, August 1977, pp. 23-27.

14. Mayfield, Henry L., "Improving Corporate Information Services in an Automated Word-Processing Network", Proceedings of the National Computer Conference, vol. 46, AFIPS Press, June 1977, pp. 443-448.

15. Mintzberg, William, "The Manager's Job : Folklore and Facts", Harvard Business Review (53:4), July-August 1975, pp. 49-61.

16. Morgan, Howard L., "Office Automation Project : A Research Perspective", Proceedings of the National Computer Conference, vol. 45, AFIPS Press, June 1976, pp. 605-610.

17. Ness, David N., "Office Automation Project : Overview", <u>DSWP 75-05-03,</u> The Wharton School, University of Pennsylvania, May 1975, 51 p.

18. Newman, William, "An Approach to Office Automation Systems Design at XEROX-PARC", Oral Presentation, IRIA Seminar, Le Chesnay, France, October 1977.

19. Nolan, Richard L. (Ed), <u>Managing the Data Resource Function</u>, West Pub. Co. 1974, 394 p.

20. "Office of the Future", <u>Business Week</u>, June 30, 1975.

21. Peacock, James, IDC Corp., "Information Processing and the Office of Tomorrow", <u>Fortune</u>, October 1977, pp. 41-109.

22. Strassman, Paul A., "Stages of Growth", <u>Datamation</u>, October 1976, pp. 46-50.

23. White, Robert B., "A Prototype for the Automated Office", <u>Datamation</u>, April 1977, pp. 83-90.

24. Wohl, Amy D., "What's Happening in Word Processing", <u>Datamation</u>, April 1977, pp. 65-74.

25. Wynn, Eleanor H., "Office Conversation as an Information Medium", XEROX-PARC, Office Research Group, Palo Alto, December 1976, 31 p.

26. Yasaki, Edward, "Toward the Automated Office", <u>Datamation</u>, February 1975.

METHODS AND TOOLS FOR INFORMATION SYSTEMS DESIGN

S. KRAKOWIAK

IMAG, Université de Grenoble

B.P. 53 38041 GRENOBLE-Cedex

ABSTRACT

The purpose of this survey is to present, in a comprehensive manner, some important
concepts that influenced the design of information systems in the last few years.
Emphasis is placed on recent progress in design methods, and on the development
of tools that may be used to apply these methods.

Some aspects of recent computer-implemented systems for assistance to requirements
analysis and system design are examined. The paper then reviews some advances in
the design of data and control structures. The impact of the abstract data type
concept and its use in system design is analyzed. Recent progress in the control
of parallel process cooperation is finally presented, with reference to distributed
systems.

1. - ARCHITECTURAL PRINCIPLES

A *system* may be defined, in general terms, as a set of interacting components. In
a man-made (as opposed to natural) system, these components are designed to operate
together towards some defined objective or purpose. A component of a system may be
an elementary object, or a system in itself (in which case it is called a subsystem).

Information processing is a global term for the set of operations (input, output,
transmission, storage, retrieval, transformation,..) that may be applied to data*.
The purpose of an *information system* is to provide a support for a variety of information processing tasks (technical, clerical or managerial) that are required
by an organization. Such a system is not closed, i.e. it interacts with an environment which is made up of physical objects and human users. This environment is
responsible for information exchange with the system, but also for various kinds
of unwanted interference.

---

* In the sense of the IFIP guide to concepts and terms in Data Processing
  (Gould 71) : "a representation of facts or ideas in a formalized manner capable
  of being communicated or manipulated by some process".

In the above general characterization of a system, they keywords are "interacting" and "purpose". "Interaction" means that a major part of the designer's activity must be concerned with the proper definition and management of the relations between the parts of a system. Decomposition, modularization and interface definition are one side of this activity; synchronization between parallel processes is another aspect.

"Purpose" means that a system, or a part of it, has a specific function that must be clearly defined and stated. Proper specification is a necessity if the design and development process is to be kept under control.

Computer programming is often referred to as an art rather than a scientific activity (Knuth 74a). Such reference applies more generally to information systems design. A number of factors account for this situation :

- User needs and requirements are ill-defined; and when defined, they are often mutually conflicting.

- Information systems have long lives and interact with a changing and complex environment; therefore, they are subject to constant modification.

- Large information systems are very complex creations, which cannot in general be completely mastered by a single person's mind.

Therefore, it is quite characteristic that information systems design is often compared to such activities as architecture or city planning, which are design activities with a long history. Alexander's book, "Notes on the synthesis of form" (Alexander 64) (which is mainly concerned with city-planning, although it defines a very general approach to the design of complex systems) is often quoted in relation to program and information systems design. Alexander analyses the transition from "unselfconscious" to "selfconscious" design. In the first attitude, design principles are unstated and transmitted by tradition; in the latter one, the design process relies on a wealth of explicit methods. Another fruitful source of inspiration is the methodological approach followed by Polyà in his book "How to solve it" (Polyà 71.) : e.g. the imbedding of a problem in a (well chosen) more general solvable problem, and the identification and reuse of already available results. As humorously pointed by Hamming in his 1968 Turing Lecture (Hamming 69), we are "standing on each other's feet" rather than on other people's shoulders.

In the rest of this survey, we shall try to give a review of some methods and tools that are currently being used to help the information system designer in his task.

The intellectual aids of the system designer are now well identified and we shall only recall them briefly :

1) *Decomposition* of a complex object into more manageable parts is an old methodological principle. However, sheer decomposition is of no avail if the relations between the parts are too complex or ill-defined. Therefore, decomposition must be conducted in a systematic fashion and a number of guidelines have been proposed and illustrated : information hiding (Parnas 71), conceptual abstraction (Dijkstra 72), ease of modification and extension, measures of intermodule coupling (Myers 75) protection of sensitive information, decentralization of resource allocation decisions.

2) *Abstraction* is the intellectual operation whereby a representation, or abstract model, of the behaviour of a complex object is constructed, which only retains some relevant properties and omits irrelevant ones. An abstract model is nothing but the well-known mathematical concept of an equivalence class. The construction of an abstract model results from an explicit choice of the equivalence relation (the selection of the "relevant" properties). An *abstract machine* (Dijkstra 72) is one which exhibits a defined pattern of behaviour regarded as appropriate to the solution of a specific problem. A *abstract data-type* (Liskov 74) is a mechanism which allows the designer to construct information sets which may only be manipulated through a specified set of access functions, and whose behaviour is defined independently of their implementation. This point will be developped in a later section of this paper.

3) *Refinement* is the process by which abstract objects are eventually implemented. The elementary refinement step is to construct an object in terms of more primitive objects by the application of a set of composition rules. A "good" set of composition rules is therefore an essential tool.

Criteria of "goodness" are conceptual simplicity, ease of use and understanding, provability (in a more or less formal sense), efficient implementation. Some agreement has been reached on such elementary sets of composition rules : record structuring for data (Hoare 72), elementary conditional and iterative constructs for sequential programs, monitor structures for concurrent processes. In spite of the availability of such tools, the refinement process does not follow an automatic procedure and relies on the designer's insight and the application of a systematic method. The use of so-called "structured programming" primitives is by no means an insurance against the production of incorrect programs, as illustrated for instance, in (Henderson 72, Gerhart 76). The second reference contains an analysis of a number of errors found in "example" programs published in papers or texts about structured programming. However, the use of well-designed constructions has a positive influence on the process of refinement because it forces the designer to state his assumptions more explicitly. This in turn should eventually make the programs more amenable to an informal "proof".

4) Since design is not a purely deductive activity, the design process usually involves *iteration*. It is well-known that a good way to improve the quality of a design is, at a certain point, to start everything again from scratch, with the augmented knowledge and insight gained from the first attempt. Some caution should however be exercised against the overconfidence and tendency to oversophistication known as the "second system effect" (Brooks 75).

The application of systematic methods to all phases of the life cycle of an information system (from initial requirements to maintenance and modification) is greatly enhanced by the use of appropriate tools. The most widely known are programming languages. However, other kinds of tools have been developped in the recent years and it now appears that programming languages (or more precisely, their compilers) are only parts of more general systems for assistance to system development. It is now widely realized that source program texts, and more generally all sorts of texts such as specifications may be considered as data on which a number of processing operations may be made. The "standard" processing on a source program text is its translation into executable code; but other operations may be considered such as source program transformation, documentation retrieval, analysis of requirements.

In the following section of this survey, we shall review the evolution of the design process and of the tools which may assist the designer in his task. Then we shall give an account of the current trends and perspectives in the design of data and control structures.

2. - FROM SPECIFICATION TO IMPLEMENTATION

During the process of design and development, an information system takes a number of different forms : initial proposal (at a very high degree of generality), overall requirements, functional specifications, component specifications are examples of such forms. The ultimate form is a set of hardware, software and operating rules which together constitute the operational system.

A number of these forms are essentially descriptions. Several terms are currently used in relation to these descriptions.

1) *Requirements* usually refer to an overall description, expressed in the terms of the user, of what the system is intended to do, and of various external constraints.

2) *Specifications,* while having the same general meaning usually have a more precise and even formal connotation.

3) *Documentation* is a general term that applies to all the written material that is used in conjunction with a project description. A more specific meaning is frequently associated with a detailed description of the final form of a system. This description is often (if at all) produced a posteriori.

The designer's dream would be a formal (automated) procedure to obtain the system from its requirements. Although such a goal seems out of reach, the strive for the application of rigorous methods to the design process has led to a number of very significant efforts towards a more systematic treatment of the specifications and documentation.

The main trends of this evolution may be summed up as follows :

1) The specification and documentation process is carried out in a continuous fashion throughout the design. The main result is that the documentation applies not only to the final product, but to all the intermediate stages of its evolution, i.e. to the design process itself. Thus, the main design decisions are made explicit.

2) There is an attempt to introduce more formality in the specifications. The main investigation lines are the definition of specification languages and the use of set-theoretical and algebraic techniques.

3) A consequence of the formalization of the design and specification process is that the use of computerized aids becomes possible. Thus, a number of systems for computer-aided development of software are currently being experimented with.

These ideas have been actually with us for a long time. For instance, an overall scheme for system design by continuous refinement and partial simulation was proposed in (Zurcher 68); a systematic approach to program specification and construction was also investigated by the same time. However, it is not until the recent years that these ideas were applied to the actual design of sizable programs.

We shall review the recent evolution along three main directions : systematic program construction, computer aids to system design and development, and formal approaches to specifications.

2.1. Methods for systematic program construction

Since the pioneering work of (Dijkstra 72), a large amount of literature has been published on the subject of systematic program construction. We shall not attempt to review this work, but we shall make the following remark : systematic programming (the original expression "structured programming" has somehow degenerated into a buzzword) refers to a methodological approach, to a new attitude towards the act of program design, rather than to the strict application of some recipe. As a consequence, it may be very difficult (as experience has shown), to promote the use of systematic methods if adequate tools are not available.

This is especially true in a production (as opposed to academic) environment, where external constraints may impose the use of ill-suited languages. With regard to this remark, we shall restrict this review to a very limited aspect : the use of some sort of formalized methodology to assist in the development of programs. The methods that we shall examine are designed to be used manually (without computer assistance) and they often rely on a graphical language. All of these methods are based on some form of decomposition and stepwise refinement. As a consequence, various forms of tree-structured diagrams are basic ingredients of the methods.

SADT (Ross 77), developed by Softech, HIPO and Composite Design (Myers 75) developed by IBM, involve decomposition of a system into units (parts, modules,.). The relations and interfaces between these parts are formally described. Design criteria such as minimal coupling may be applied. The diagrams are used for documentation, for review of the design before implementation, and as a guide to implementation.

A more formal approach is proposed in (Warnier 72) and (Jackson 75). Both methods are mainly designed for the construction of data processing applications (as opposed to operating systems or real-time software). The main idea of (Jackson 75) is to set up a mapping between the structure of a file, or set of files, and the structure of the program that operates on these data. File structures are constructed from elementary components by the operations of concatenation, selection and iteration; this structure is reflected in the programs. Refinement may be applied, if necessary, to both program and data structures. A simple graphical language is used to document the design process.

Finally, a still more formal method has been developed in (Abrial 74, 77). The initial requirements are expressed in a specification language based on sets and relations (a similar approach is followed in SETL (Schwartz 72)). Specifications written in this language are then transformed by hand, using a set of semantics-preserving transformations, into programs written in another language, which assumes more specific implementation choices. This process is iterated until a working program is obtained. The validity of the design process relies on the correctness of the program transformation mechanisms. This method has been successfully experimented in an industrial environment and appears as very promising.

2.2. Computer aids to systems design and production

Most computer aids to system design and production may be roughly classified under two headings. In the first class, emphasis is on the early steps of design, specification and evaluation. In the second class, actual programs are manipulated and executable code is produced. Both types of systems have evaluation, testing and documentation editing facilities. Current research is under way to construct systems that would encompass all phases of the design and production process.

A general model for a computer-based system for assistance to system design is given on Figure 1.

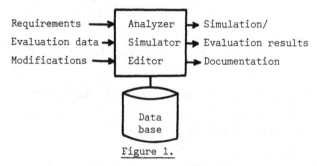

Figure 1.

All information relevant to the design is progressively entered into a data base which records every step of the development process. The data base is used for the production of documentation on the project and for evaluation of the design.

This type of system is exemplified by the PSL/PSA system (Teichroew 77) developed as a part of the ISDOS project at the University of Michigan. PSL (the "Problem Statement Language") allows the designer to describe a system design as a set of "objects" connected through "relationships". Specific types of objects and relationships are available for the description of a variety of aspects of information systems (input-output, hierarchical grouping, data structures, performance,...).

During the project, the result of every step in specification and development is
described as a PSL program. Descriptions written in PSL are processed by a Problem Statement Analyzer (PSA) which analyzes the information provided and enters
it into the data base. PSA also contains some evaluation features which may help
the analyst to evaluate the impact of potential improvements. The system is re-
portedly used in a variety of industrial environments.

Similar systems are CADES (CADES 73), DACC (Boehm 75), and TOPD (Henderson 73).
Although such systems offer no substitute to the design itself, they help the
designer by forcing him to formally express the requirements, by providing
checklists for relevant questions, by producing readable documentation in a
standard form and by evaluating the effect of design decisions.

Another class of computer aids may be represented by the general scheme of
Figure 2.

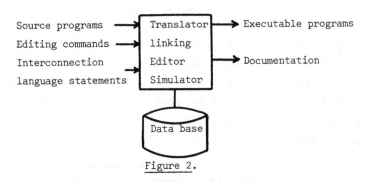

Figure 2.

These systems may be described as "software factories". The main capabilities that
they provide are as follows :

- creation, editing and modification of source programs
- program library management
- compilation and linking of programs
- debugging, testing and simulation
- documentation production

It should be noted that the production machine, on which these development tools
are implemented, may well be distinct from the target machine for which software
is produced.

Two important components of a software development system are the librarian in
charge of the program data base, and the language processor(s). Such an environment is well suited to the development of modular programs. The overall architecture of a system in terms of elementary components may be expressed in a
module interconnection language, while individual modules are developed using a

language processor that supports the concept of a module.

A number of production systems designed along these lines are currently being used or experimented with. The "workbench machines" described in (Ivie 77) are used as a network, connected to target machines by high-speed lines, and include development facilities for several target machines. Several recently developed systems are based on PASCAL or extended versions of this language (Donzeau-Gouge 75, Geschke 77, Krakowiak 76).

A natural extension would be a general system design and production facility that would integrate the capabilities described in figures 1 and 2, i.e. would encompass all stages of development from specifications to code production. This is actually a stated objective of some of the above mentioned systems (Boehm 75, Lucena 76, Teichroew 77). The essential steps of specifications writing and of building programs from specifications remains however the designer's task, but useful assistance may be provided (e.g. in the way of automatic consistency checks). Progress in this domain depends on advances in problem specification, a subject that will be reviewed in the next section.

2.3. Advances in specification techniques

The most widely used specification methods presently rely on natural language. Concern for software reliability has recently fostered the development of more formal methods. The purpose of such work is to allow the writing of specifications that could be amenable to formal verification and from which an implementation could be easily derived either by hand or by a formal procedure.

The first step is to define a unit for specification. Methods for decomposition and hierarchical structuring described for program design clearly extend to specifications. Therefore, most of the work on specifications has concentrated on specification techniques that apply to the basic building blocks that support simple abstractions (see section 3 of this paper), i.e. essentially multiple entry modules (Guttag 76, Liskov 75).

Specification methods fall into two classes : operational and definitional. In the operational approach, the specified operation is described in terms of some already defined "machine" (or set of operations). In the definitional approach, an operation is described by its effect, as a set of pre-and post-conditions for the state of the object upon which the operation is applied.

Current specification methods usually combine both types of definitions. An operational specification is often used as a guide for implementation, whereas a definitional specification is more readily usable as a guide for testing and verification.

Both methods are presently being investigated. The already mentioned work of (Abrial 74, 77) uses an operational specification, and the emphasis is on the stepwise transformation of this programmed specification into an implementation. On the other hand (Guttag 76) uses algebraic techniques for the specification of abstract data types and the emphasis is on completeness and consistency verification.

It seems that we are still a long way from using formal specification techniques in large scale projects. Meanwhile, the introduction of even primitive techniques to make specifications more formal will certainly provide an incentive towards a more systematic practice. Formal specifications per se cannot regarded as a panacea; after all, formality of the mathematical notation does not prevent mathematicians to occasionnaly write erroneous proofs! Instructive discussions of the relations between mathematical thinking and programming methodology may be found in (Dijkstra 76, Gerhart 76, Mills 75, Schwartz 72).

Finally, we should not leave the subject of specification without a world about the specification of the user interface, i.e. the language by which an information system and its human users achieve communication. This includes the design of the command language (including control and data description) by which the system is operated, as well as the design of the output language in which results are given. Little formal consideration seems to be given to these subjects, with the result that the above mentioned "languages" often hardly deserve this name at all. Notorious examples are the command languages used to instruct operating systems : the obscurity and lack of logical structure of most of these "languages" are well-known facts.

The interested reader should refer to part 4 of (Naur 74) which is devoted to a thorough survey of the design principles that apply to data interchange between man and computer. This is a difficult field of study where contributions are needed from ergonomists and psychologists. Advances in the technology of graphical information displays should open new directions for progress in this field.

3. - THE DESIGN OF CONTROL AND DATA STRUCTURES

An elementary information processing task may be described as the operation of
a *procedure* on some *data*. The execution of such a task is called a *process*. Concurrently executing (or logically independent) tasks are described by a system of
concurrent processes, which may interact through shared resources or informations.

Procedures, data and processes are thus the main components of any information
system. Much effort has been devoted to devise abstract (i.e. implementation
independent) models for these three classes of objects, and to develop structuring tools based on such models. In the following sections, we shall try to give
an account of the present state and future trends of this evolution.

3.1. Data structures

The use of abstraction for the design of data structures is more recent than for
programs. The underlying idea is that a data structure is more adequately defined
(for a user of this structure) in terms of the allowed access operations than in
terms of its implementation. The data abstraction operation consists in the definition of a model of behaviour (an *abstract data type*) according to which a class
of objects may be generated. The properties of such an object (an *instance* of the
type) are defined by the specifications of the abstract type. These properties do
not depend on the implementation of the object. Some important properties follow :

  1) The user of an object needs only to know the specifications of its
abstract type and should make no assumption on its internal structure.

  2) An object may be implemented in a number of ways. From a user's standpoint
all these implementations are equivalent (except perhaps as regards efficiency)
as long as they conform to the object's data type specifications.

  3) The user of an object may not access, retrieve or modify any part of this
object except through the specified access procedures.

A number of schemes have been proposed to implement the idea of data abstractions. This
variety is reflected in the number of different terms that were recently introduced : *abstract type* (Liskov 74), *abstract machine* (Dijkstra 72), *capsule*
(Horning 76) denote general abstraction mechanisms, while *class* (Dahl 72),
cluster (Liskov 74), *form* (Wulf 76) refer to specific implementations of such
mechanisms, and *module* is used in both (and other) contexts (e.g. Parnas 72,
Wirth 77).

The general pattern that appears to be common to these proposals is that an
object generated by an abstract data type may be described as follows :
  - the object is represented by a set of information ("state variables"),
together with a set of access procedures; the user interface is defined by these

access procedures,

 - the state of the object is defined at any time as the value of the state variables,

 - the state receives an initial value when the object is created;

 - the access procedures are the only way of access to the data part of the object; in most proposals, this restriction is enforced at compile time;

 - the effect of the access procedures may be specified in terms of an initial and final state. As a consequence, the state of the object, at any time, only depends on the sequence of operations that were executed since its creation.

On the other hand, some other issues are still controversial such as :

 - separate compilation of abstract types;

 - mechanisms for the construction of parameterized or generic abstract types;

 - efficient implementation;

 - mechanisms for parameter passing.

A number of experiments with the implementation and use of abstract types are currently under way. Experience with actual use of languages including this concept is still limited (Geschke 77) and seems to be restricted to the production of systems programs (see however (Hammer 76) for a discussion of the use of abstract types in data base design). Some tentative conclusions may be drawn from the first results :

 1) The use of new data structuring mechanisms does not automatically result in better (more reliable, understandable, efficient) programs. A good tool supplements the designer's skill but offers no substitute for it.

 2) A strict compile-time type checking system must tolerate some exceptions (for logical or efficiency reasons). Such exceptions should be made as explicit as possible to make the user aware of the potential dangers.

 As noted in (Geschke 77), early experience with this new data structuring concepts can be compared to experience with the use of "structured" control constructs. Such constructs help their user to acquire a good style of program design which may afterwards be put into practice with languages that do not support them. As a consequence, we would recommend early acquaintance with these mechanisms in computer science education.

3.2. Control structures for sequential programs

One of the main results of the recent advances in systematic programming (Dijkstra 72), (Knuth 74 b), (Mills 75), (Wirth 76) is that an adequate tool for the construction of sequential programs is the set of three elementary constructs: sequence, selection (*if-then-else*), and iteration (*while-do*), possibly supplemented by *case* and *repeat-until*, together with the basic abstraction device provided

by procedures. Even for widely used languages that do not include these constructs
(such as FORTRAN or COBOL), adherence to a programming discipline may be enforced
by the use of a preprocessor or by a set of standard rules of transcription.

The benefits of the systematic use of a small number of simple and well defined
constructs are presently recognized and largely illustrated by a number of published examples (e.g. in the references quoted at the beginning of this section).
However, an important feature that appears in the programs of many large scale
information systems is not easily captured by these constructs. The operation of
such systems may be described as a "normal case" algorithm together with a number
of "exceptions". Exceptions may include hardware failure, erroneous data, or any
condition specified by the designer. The exception-handling mechanisms often
account for a large fraction of the total size, cost and complexity of the system.

The problem of exception handling has been the subject of intense research since
its practical importance was realized. A number of methods have been proposed,
but it does not seem that a single solution to the problem has achieved pre-
eminence. A complete review of recent work, together with some new proposals,
may be found in (Levin 77).

Exceptions may be regarded as "special cases" and handled in the same way as
"normal cases" e.g. by means of return values that indicate abnormal return from
a procedure call. This way of doing, however, is detrimental to a good under-
standability of the programs. An acceptable exception handling mechanism should
be adapted to any abstraction-defining constructs used in the program : if an
object is defined by an abstraction mechanism that encapsulates its internal
structure, any exceptional conditions arising when the object is used should be
expressed in   terms of the abstraction by which it is defined. In other words,
for example, an exceptional condition detected when a programmer-defined data
structure is misused should not be expressed in terms of memory addresses, as is
too often experienced! An exceptional condition should be propagated through the
abstraction levels until enough information is available to allow its processing.

An adequate expression language for the definition of exception detection and
handling should allow to clearly separate what is considered a normal case and
what is considered an exception; it should also provide means for binding the
detection of an exception to its processing.

While the main issues in the design of exception handling mechanisms are now
being understood (at least for sequential programs), especially in the context of
abstraction- defining constructs, we are clearly lacking experience with the
actual use of such mechanisms. Some of the recent proposals are currently being
implemented under experimental conditions and user experience is eagerly awaited.

3.3. Parallel processes

Parallel processes provide a means for structuring systems in which a number of loosely coupled activities cooperate towards a common task. A great variety of methods have been devised to achieve interprocess cooperation. Semaphores provide a general tool which has been widely used in the design of operating systems, and which has been included as an elementary synchronizing operation in the hardware of a number of computers. However, some considerations have recently led to the development of more elaborate tools :

1) The trend towards the use of high-level languages for the design and implementation of systems programs : high-level synchronizing constructs were needed especially for inclusion in the data abstraction mechanism provided by these languages.

2) The growing concern for mechanisms amenable to precise specification and correctness proofs.

3) The advent of distributed systems, in which processes do not share a common store.

3.3.1. High-level synchronizing tools

Monitors (Hoare 74) were introduced to implement data structures which may be shared by several processes, and used through a set of access procedures. The synchronizing mechanism built into the monitor ensures mutual exclusion for the execution of access procedures, and allows to enforce a scheduling discipline among processes by means of a set of queues associated with activation conditions.

Monitors have been included in several programming languages (e.g. Concurrent Pascal (Brinch Hansen 77), Modula (Wirth 77)). Efficient implementations of monitors have been devised and some experience has been collected, which seems to demonstrate the usefulness of this construct. However, when programming with monitors, one must explicitly describe the scheduling operations in terms of waiting and activation primitives. In many cases, one would wish a more global and implicit expression of synchronizing conditions in terms of procedure executions, considered as elementary units of process activity. This has led to the development of more formal constructs.

Path expressions (Habermann 75) and various forms of event counters (e.g. Robert 77) were introduced in an attempt to express synchronizing conditions in a module in terms of procedure executions. These synchronizing conditions are described by regular expressions (path expressions) or by algebraic relations between the values of event counters. The formality of these expressions makes these mechanisms amenable to proofs. Experience with their actual use is still very limited. The main difficulty with their use seems to arise when synchronizing conditions in a module are execution-dependent, i.e. if they are expressed in terms of the value of internal variables of the module or of procedure parameters.

3.3.2. Process cooperation in distributed systems

A great deal of interest has arisen for distributed computing in the recent years. Three main reasons account for this interest :

1) The availability of low-cost computing power allows one to devise highly parallel computing systems constructed form a large number of interconnected processors.

2) The development of computer networks makes resource sharing possible netween geographically distant centers.

3) Increasing concern for reliability leads to the distribution of work between interconnected computers e.g. in industrial process management.

In spite of an intense activity, it does not seem that the ambitious goals set up several years ago have really been attained. A number of fundamental problems in distributed computing are still awaiting a solution. We shall try to analyse what appear to be the main issues in this fields.

We shall first set up a model of a distributed system as a set of entities connected by communication lines. We shall consider each of these entities as a self-contained module. Each of these modules is associated with a set of cooperating processes which share this module; communication between processes on different modules is achieved by asynchronous messages (this is the only possibility in the absence of a common store).

Besides the absence of a common store, à distributed system is characterized by the absence of a common clock. More precisely, the time scale which applies to message transmission is not negligible with respect to the local time scale in an individual process. Moreover, the transmission lines may usually not be regarded as reliable and message loss is not an exceptional event.

Some of the main problems in such a structure may be summed up as follows :
- how to achieve state consistency between data in different modules (this amounts to solve the mutual exclusion problem between two distant processes);
- how to ensure a sufficient overall reliability to the system in spite of the unreliable communication mechanism;
- how to express a computation distributed among several distinct modules (this may not be done by intermodule procedure calls because of the message mechanism, and new linguistic constructs are needed);

- how to survive a failure in one of the communicating modules.

Only partial solutions have been proposed so far to all of these problems. The model of a set of modules connected by asynchronous message lines seems to be the paradigm for a variety of situations : cooperating processes in the kernel of an operating system, multiprocessor systems, actor models in Artificial Intelligence, distributed data bases, loosely connected processors in industrial control applications. We think that a systematic investigation of this model (as initiated e.g. in (Feldman 77)) should contribute to give a sound framework to the design of distributed applications.

4. CONCLUSION

In this survey, we have tried to discuss a number of views pertaining to information systems design. Our conclusion will be very brief : design essentially remains a human activity, and no magic sophisticated device will ever replace thorough analysis, careful expression of requirements, clear separation of correctness and efficiency concerns, and strive for conceptual simplicity. The main achievement of the recent years' effort is that we are in the process of founding the designer's skill on an explicitly transferrable body of knowledge. In addition, we are learning to make a good use of computers to assist the designer as well as the implementor of information systems. In this respect, the importance of well-designed tools should not be underestimated, because the use of well chosen tools forces us to ask the "right questions", and because the applicability of a design method is greatly enhanced if the method is supported by a set of appropriate tools.

REFERENCES

ABRIAL J.R. : Data semantics, *Proc. IFIP Working Conf. on Data Base Management Systems* (Klimbie and Koffeman, eds.), North-Holland (1974).

ABRIAL J.R. : Méthode et langage de spécification. (Unpublished notes, 1977).

ALEXANDER C. : *Notes on the synthesis of form*, Harvard University Press, 1964.

BOEHM B.W., McCLEAN R.K., URFRIG D.B. : Some experience with automated aids to the design of large-scale software, Proc. Intern. Conf. on Reliable Software, SIGPLAN Notices 10,6 (juin 1975).

BRINCH HANSEN P. : *The architecture of concurrent programs*, Prentice Hall (1977).

BROOKS F.P. : *The mythical man-month*, Addison-Wesley, 1975.

CADES : Computer-Aided Design and Evaluation System (a series of articles in *Computer Weekly*, (July 1973).

DAHL O.J. : Hierarchical program structures, in *Structured Programming* (Dahl, Dijkstra, Hoare), APIC Studies in Automatic Programming n°8, Academic Press (1972).

DIJKSTRA E.W. : Notes on Structured Programming, in *Structured Programming* (Dahl, Dijkstra, Hoare), APIC Studies in Automatic Programming, n°8, Academic Press (1972).

DIJKSTRA E.W. : *A discipline of programming*, Prentice Hall (1976).

DONZEAU-GOUGE V., HUET G., LANG B., LEVY J.J. : A structure-oriented program editor : a first step towards computer-assisted programming, *Proc. ICS Conf.*, Antibes (May 1975).

FELDMAN J.A. : A programming methodology for distributed computing (among other things), *TR-9, Dept. of Computer Science, Univ. of Rochester* 1977.

GERHART S.L. and YELOWITZ L. : Observations of fallibility in applications of modern programming methodologies, *IEEE Trans. Software Engineering, SE-2, 3* (Sept. 1976).

GESCHKE C.M., MORRIS J.H., SATTERTHWAITE E.H. : Early experience with Mesa, *Comm. ACM,* 20, 8 (Aug. 1977).

GOULD I.H. : (Ed.) *IFIP Guide to concepts and terms in data processing,* North-Holland, 1971.

GUTTAG J. : Abstract data types and the development of data structures, Proc. SIGPLAN/SIGMOD Conf. on Data, *SIGPLAN Notices* 8,2 (march 1976). (To appear in Comm. ACM).

HABERMANN A.N. : Path expressions *Dept. of Computer Science, Carnegie Mellon University* (1975).

HAMMER M. : Data abstractions for data bases, Proc. SIGPLAN/SIGMOD Conf. on Data *SIGPLAN Notices,* 8,2 (March 1976).

HAMMING R.W. : One man's view of computer science, *Journal A.C.M.,* 16,1 (Jan.1969).

HENDERSON P., SNOWDON R. : An experiment in structured programming, *BIT* 12,1 (1972).

HENDERSON P., SNOWDON R. : A tool for structured program development; *Proc IFIP Congress 1974,* vol 2, North-Holland (1974).

HORNING J.J. : Some desirable properties of data abstraction facilities, *Proc. SIGPLAN/SIGMOD Conf. on Data, SIGPLAN Notices* 8,2 (march 1976).

HOARE C.A.R. : Notes on data structuring, in *Structured Programming* (Dahl, Dijkstra, Hoare), APIC Studies in Data Processing n°8 Academic Press (1972).

HOARE C.A.R. : Monitors : an operating systems structuring concept, *Comm. ACM,* 17, 10 (1974).

IVIE E.L. : The programmer's workbench - a machine for software development, *Comm. ACM* 20,10 (oct. 1977).

JACKSON M.A. : *Principles of program design,* APIC Studies in Data Processing n°12, Academic Press (1975).

KNUTH D.E. : Computer Programming as an art, *Comm. ACM* 17,12 (Dec. 1974 a).

KNUTH D.E. : Structured programming with goto statements, *Comp. Surveys,* 6,4 (Dec. 1974 b).

KRAKOWIAK S., LUCAS M., MONTUELLE J., MOSSIERE J. : A modular approach to the structured design of operating systems, *Proc. MRI Symp. on Computer Software Engineering,* Polytechnic Institute of New-York (1976).

LISKOV B.H., ZILLES S.N. : Programming with abstract data types, *Proc. SIGPLAN Symp. on Very High Level Languages,* SIGPLAN Notices, 9,5 (1974).

LISKOV B.H., ZILLES S.N. : Specification techniques for data abstractions, *IEEE Trans. Software Engineering,* SE-1 (March 1975).

LUCENA C.J., COWAIN D.D. : Toward a system's environment for computer assisted programming, *Inf. Proc. Letters,* 5,2 (June 1976).

MILLS H.D. : How to write correct programs and know it, Proc. Int. Conf. on reliable software, *SIGPLAN Notices* 10,6 (June 1975).

MYERS G.J. : *Reliable software through composite design*, Petrocelli/Charter (1975).

NAUR P. : *Concise survey of computer methods*, Studentlitteratur, Lund (1974).

PARNAS D.L. : Information distribution aspects of design methodology. *Proc. IFIP Congress* (1971).

PARNAS D.L. : On the criteria to be used in decomposing a system into modules, *Comm. ACM*, 15,12 (Dec. 1972).

POLYA G. : *How to solve it*, Princeton University Press (1971).

ROBERT P., VERJUS J.P. : Towards autonomous descriptions of synchronization modules, *Proc. IFIP Congress*, (1977).

ROSS D.T., SCHOMAN K.E. Jr : Structured analysis for requirements definition, *IEEE Trans. Software Engineering*, SE-3, 1 (Jan. 1977).

SCHNEIDER B.R. Jr : *Travels in computerland, or incompatibilities and interfaces*, Addison-Wesley (1974).

TEICHROEW D., HERSHEY E.A., III, PSL/PSA : A computer-aided technique for structured documentation and analysis of information processing systems, *IEEE Trans. Software Engineering*, SE-3,1 (Jan. 1977).

WIRTH N. : *Algorithms + data structures = Programs*, Prentice Hall (1976).

WIRTH N. : Modula, a language for modular multiprogramming, *Software Practice and experience* 7,1 (1977).

ZURCHER F.W., RANDELL B. : Iterative multi-level modelling : a methodology for computer system design, Proc. IFIP Congress (1968).

USING ASSERTIONS ABOUT TRACES TO WRITE ABSTRACT SPECIFICATIONS FOR SOFTWARE MODULES

Wolfram Bartussek

and

David L. Parnas

Department of Computer Science

University of North Carolina at Chapel Hill

Chapel Hill, North Carolina 27514, U.S.A.

ABSTRACT

A specification for a software module is a statement of the requirements that the final programs must meet. In this paper we concentrate on that portion of the specification that describes the interface between the module being specified and other programs (or persons) that will interact with that module. Because of the complexity of software products, it is advantageous to be able to evaluate the design of this interface without reference to any possible implementations. The first sections of this paper present an approach to the writing of black box specifications, that takes advantage of Guttag's work on abstract specification [9]. Then we illustrate it on a number of small examples, and discuss checking the completeness of a specification. Finally we describe a case history of a module design. Although the module is a simple one, the early specifications (written using an earlier notation) contained design flaws that were not detected in spite of the involvement of several persons in a series of discussions about the module. These errors are easily recognized using the method introduced in this paper.

I. <u>Introduction</u>

The Role of Specifications in Software Design

We are concerned with the building of software products that are so large
that we cannot manage the task unless we reduce it to a series of small
tasks. We further assume that each of the subtasks (which we call modules) will focus on one portion of the design and hide the details of
that aspect of the design from the rest of the system. This has become
known as the "information hiding principle," encapsulation, data abstraction, etc. [1,2,3]. The design process will only go smoothly if the intermodule interfaces are precisely defined. Ideally, the interface description states only the requirements that the component must satisfy
and does not suggest any other restrictions on the implementation. We
term such a description of the requirements a <u>specification</u> [19]. We
also note that any software product is but a module in a still larger
system; its requirements should be specified as precisely as each of
its components.

For a trouble-free development process it is also necessary that one be
able to verify the reasonableness of decisions before proceeding to make
further decisions. If we reverse one of our decisions later (or find
that it was inadequately described), we may have to discard all work
done subsequent to that decision. If we have written a formal specification for a module, we should be able to verify that the specification
has such basic properties as consistency and completeness. These aspects
will be discussed later in this paper.

What Are Specifications?

A fair amount of confusion has been caused by the fact that the word
"specification" is used with two distinct meanings in the computer literature. The dictionary definitions of the word "specification" cover any
communication which provides additional information about the object
being described - any communication that makes the description of the
object more specific. In engineering usage, the word has a narrower
meaning. A specification is a precise statement of the requirements
that a product must satisfy. A description of the number of ones in
the binary representation of a computer program is a specification in

the general sense but it is rarely a specification in the engineering sense.

In the remainder of this paper we will use the engineering sense of "specification."

Brief History of Work on Specifications

We distinguish two classes of specifications for software, which we shall denote as P/P (Precondition-Postcondition) and DA (Data Abstract). P/P specification techniques are based on the pioneering work of Floyd [4] and subsequent work by Hoare [5], Dijkstra [6], and others. P/P techniques describe the effect of a program in terms of predicates that describe acceptable states of data structures. The Precondition is a predicate that describes the states in which the program may be started. The Postcondition describes the states after program termination. Dijkstra's predicate transformers replace both of these predicates by a rule for transforming a postcondition into a precondition [6,7]. P/P specifications describe the change of state that the program must effect, but not how to effect it. Usually, the effect of each individual program is described separately and in terms of the data structure accessed by the program.

In DA specifications the specification of a module does not refer to the data structure used within a module. That data structure is not part of the requirement; it is part of the solution. It does not belong in a statement of requirements because it depends on implementation decisions. Early work on specifications that "hide" implementation data structures was done by Parnas [8]; more recent work by Guttag [9,10] put a sounder mathematical basis behind the work and suggested some notational improvements.

The DA specification work is motivated by a desire to give a "black-box" description of a software module. The user is told only of a set of programs that access the data structure within the module. Some of these (here termed V-functions) return values that give information about parts of the data structure. Others (here termed O-functions) change the internal data. In most cases, the execution of an O-function will eventually cause a change in the value of a V-function. The effects of the call of the O-function may not be visible in terms of V-function values until some other O-functions have been executed.

Parnas's early work was done on an ad hoc basis. The notation was developed to meet the needs of specific examples [8]. The early examples had the property that the effects of O-functions were immediately visible and could be described in terms of the new values of the V-functions. Only in later examples did Parnas and Handzel [20] seek to extend these techniques to cases where there were delayed effects.

The problem of delayed effects led Price and Parnas [21,11,12] to include "hidden" functions in their specifications. The "hidden" functions are not available outside the black box. They need not be implemented; their purpose is purely descriptive. The effects of O-functions are described in terms of the values of the hidden functions. These hidden functions are still in use at SRI [13] and elsewhere.

In spite of all disclaimers, the hidden functions do suggest data structures and possible implementations of the program. Liskov [14] and others have suggested writing specifications simply by giving possible implementations - i.e., by giving a program whose behaviour would be acceptable and asking that the programs produced be "equivalent."

The equivalent program approach and the hidden functions disturb us. They violate the basic motivation for DA specifications by providing information that is not a requirement. Some of the properties of an hypothetical implementation may not be required of the actual program. "One must be very careful not to read too much into such specifications" [14].

Guttag's method does not rely on hidden functions to describe delayed effects. His papers [9,10] describe a systematic way of writing the specification. However, there were cases that he could not handle without the introduction of hidden functions. One of those examples, the stack with overflow, will be used later in this paper [15].

In this paper, we propose yet another approach. It allows the specification of modules with delayed or hidden effects without any reference to internal data structures. The only statements made are about the effects of calls on user accessible O-functions or user accessible V-functions.

When Is a D/A Specification Complete?

For simplicity, we assume that our modules are always created in the same initial state and could be returned to that state (reinitialized). We further assume that for each access program (O-function or V-function) there is an applicability condition. If this condition holds, the program may be called. In states where the condition does not hold, the module will "trap" or refuse to return through the normal exit [16]. Values of V-functions after a trap occurs will not be discussed in this paper.

A trace of a module is a description of a sequence of calls on the functions starting with the module in the initial state. A trace is termed a legal trace if calling the functions in the sequence specified in the trace with the arguments given in the trace when the module is in its initial state will not result in a trap. A specification completely determines the externally visible behaviour of a module if for every legal trace ending with a call of a V-function, the value returned by that V-function can be derived from the specification. We term such a specification complete. A specification is consistent if only one value can be derived.

There are situations in which one may want a specification that is not complete in the above sense. In this paper, however, we will concern ourselves with the problem of recognizing complete and consistent specifications.

II. A Formal Notation for Specification Based on Traces

A specification will consist of two main parts. The first part, which we call syntax, gives the names of all of the access programs, and the type of each of the parameters. For O-functions we will indicate that it changes an object of the type being specified. For V-functions we will give the type of value that it delivers. This information is necessary for recognizing whether a program using the functions could be compiled by a typical compiler. The notation used is that used by Guttag [9,10].

The second part of the specification will be called the semantics. It consists of three types of assertions.

1. Assertions about trace legality. These assertions identify a subset of the set of legal traces, that is a set of traces such that calling the functions as described in the trace (starting with a module in its initial state) will not result in traps. Additional legal traces may be implied by the equivalence assertions (see below). Any traces that

cannot be shown to be legal using these assertions will be considered illegal traces.

2. Assertions about the equivalence of traces. These assertions specify an equivalence relation on traces, such that (1) equivalent traces have the same legality (either both are legal or both are not legal) and (2) that they have the same externally visible effect on the module or data item. These assertions of equivalence will often enable us to extend the class of traces known to be legal. Equivalence is usually weaker than equality. Two traces are equal if they are identical in every respect (the same sequence of function calls with the same parameters).

3. Assertions about the values returned by V-functions at the end of traces. These statements describe the values delivered by V-functions for a subset of the set of legal traces. The traces discussed directly in this section of a specification are called normal form traces. Using the equivalence statements, one can derive the values of V-functions at the end of other traces by finding an equivalent normal form trace.

Remark: In our examples, we have assumed that equality is defined for values of the types returned by the V-functions. In the unlikely event that we have no equality operator, V-function values would have to be described in terms of the operators that are available.

Since assertions about values of V-functions are made only using normal form traces, assertions about equivalence of traces will also be used to show that any legal trace can be transformed to a normal form trace.

The three classes of assertions together with the syntax definition form a specification or statement of requirements. An implementation will be considered correct if and only if the assertions are true of it. Any property that one can deduce from the assertions must be a property of any correct implementation.

A program that uses the module in such a way that the program's correctness depends only on properties of the module that can be deduced from the specification's assertions will be able to use any correct implementation of the module.

Notation

(1)   Notation for describing the syntax (taken from Guttag).

<Function Name>:   <type of parameter>X,...X<type of parameter>->
<type of result>

If the module maintains only one data item, that parameter need
not be explicitly named in each function call.

(2)   Notation for describing traces.

A trace will be represented as a string from the language described by the following syntax. The parsing of a trace into component subtraces is deliberately ambiguous. The trace denotes
execution of the functions named in a left to right sequence.

<subtrace> ::=   ⊔ |<syntactically correct function call>|
                 <subtrace>.<syntactically correct function call>

<trace> ::=   ⊔ |<subtrace>[.<subtrace>]*
[<T>]* denotes any number of occurrences of T .

"⊔"   denotes an empty trace. <u>Note that the symbol "⊔ " never</u>
         <u>occurs in a trace.</u>

We will sometimes use the following shorthand notation.

Let p_i, m≤i≤n, be a list of actual parameters and $X(p_i)$ a syntactically correct function call. Then $X_M^N(p_i)$ denotes the same as

$$X(p_M).X(p_{M+1}). \ldots .X(p_{N-1}).X(p_N)$$

If the list of parameters is empty, then X_M^N is simply X.X. .. .X
with n-m+1 repetitions of X. If M>N, then X_M^N denotes the empty
trace. For N ≥ 1 we write $X_1^N(p_i)$ as $X^N(p_i)$.

It is always assumed that a function call correctly adheres to
the rules of the syntax section.

(3)   Describing legality of sequences.

We introduce the predicate $\lambda(T)$ where T is a trace. $\lambda(T)$ is true
if T is a legal trace. The appearance of the assertion $\lambda(T)$ in a
specification is a requirement that calling the functions as described in T will not result in a trap.

Assuming that the module will not "trap" if it is not used, we
<u>always</u> assume $\lambda(⊔) =$ <u>true</u>. (The empty trace is always legal).
It follows from our discussion of traces that if T is a trace

and S is a subtrace, then

$$\lambda(T.S) \implies \lambda(T).$$

In other words, the prefix of any legal trace is a legal trace.

(4) Describing the values of V-functions at the end of traces.

If T is a legal trace, X is a syntactically correct call on a V-function, and $\lambda(T.X)$ is TRUE, then $V(T.X)$ describes the value delivered by X when called after an execution of T.

(5) Describing equivalence of two traces.

If T_1 and T_2 are traces then $T_1 \equiv T_2$ is an assertion that:

for any subtrace S (including the empty subtrace),
$$\lambda(T_1.S) <=> \lambda(T_2.S),$$
and

for any subtrace S (including the empty subtrace) and V-function X,
$$\lambda(T_1.S.X) \implies V(T_1.S.X) = V(T_2.S.X)$$

Then " $\equiv$ " is an equivalence relation. Note that the equivalence of two traces does not imply that they are the same in every respect, only in those respects specified above. For example, one may not conclude that two equivalent traces have the same length or that the prefixes of equivalent traces are equivalent. Note too that the above does not define a particular equivalence relation; that is done in each specification.

In the following specifications we have omitted universal quantifiers for variables representing traces (T) and values of specific types.

III.   Some Simple Examples (To be explained and discussed in the next
                            Section.)

Example 1.   A Stack for Integer Values

Syntax:

    PUSH: <integer> x <stack> -> <stack>
     POP: <stack> -> <stack>
     TOP: <stack> -> <integer>
   DEPTH: <stack> -> <integer>

Legality:

    (1) $\lambda(T)$ => $\lambda(T.PUSH(a))$
    (2) $\lambda(T.TOP)$ <=> $\lambda(T.POP)$

Equivalences:

    (3)   T.DEPTH $\equiv$ T
    (4)   T.PUSH(a).POP $\equiv$ T
    (5) $\lambda(T.TOP)$ => T.TOP $\equiv$ T

Values:

    (6) $\lambda(T)$ => V(T.PUSH(a).TOP) = a
    (7) $\lambda(T)$ => V(T.PUSH(a).DEPTH) = 1 + V(T.DEPTH)
    (8)   V(DEPTH) = O

Example 2. An Integer Queue

Syntax:

ADD:	<integer> x <queue> -> <queue>
REMOVE:	<queue> -> <queue>
FRONT:	<queue> -> <integer>

Legality:

(1) $\lambda(T) \Rightarrow \lambda(T.ADD(a))$

(2) $\lambda(T) \Rightarrow \lambda(T.ADD(a).REMOVE)$

(3) $\lambda(T.REMOVE) <=> \lambda(T.FRONT)$

Equivalences:

(4) $\lambda(T.FRONT) \Rightarrow T.FRONT \equiv T$

(5) $\lambda(T.REMOVE) \Rightarrow T.ADD(a).REMOVE \equiv T. REMOVE.ADD(a)$

(6) $ADD(a).REMOVE \equiv$ ⊔

Values:

(7) $V(ADD(a).FRONT) = a$

(8) $\lambda(T.FRONT) \Rightarrow V(T.ADD(a).FRONT) = V(T.FRONT)$

The above specification assumes that only one queue exists and omits the queue parameter in the calls on the access programs.

Example 3.   Sorting Queue = (SQUEUE)

Syntax:

INSERT:	<integer> x <squeue> -> <squeue>
REMOVE:	<squeue> -> <squeue>
FRONT:	<squeue> -> <integer>

Legality:

(1) $\lambda(T) \Rightarrow \lambda(T.INSERT(a))$

(2) $\lambda(T) \Rightarrow \lambda(T.INSERT(a).REMOVE)$

(3) $\lambda(T.FRONT) <=> \lambda(T.REMOVE)$

Equivalences:

(4) $\lambda(T.FRONT) \Rightarrow T.FRONT \equiv T$

(5) $T.INSERT(a).INSERT(b) \equiv T.INSERT(b).INSERT(a)$

(6) $INSERT(a).REMOVE \equiv \sqcup$

(7) $\lambda(T.FRONT)$ cand $(V(T.FRONT) \leq b) \Rightarrow$
   $T.INSERT(b).REMOVE \equiv T$

Values:

(8) $V(INSERT(a).FRONT) = a$

(9) $\lambda(T.FRONT)$ cand $V(T.FRONT) \leq b \Rightarrow$
   $V(T.INSERT(b).FRONT) = b$

Note the value of X cand Y is false if X is false, and the value of X cand Y is the value of Y if X is true. Y need not have a defined value if X is false.

Example 4.   Stack that Overflows (Stac)

Syntax:

    PUSH: <stac> x <integer> -> <stac>
     POP: <stac> -> <stac>
     VAL: <stac> -> <integer>

Legality:

    For all T, $\lambda(T)$

Equivalences:

$0 < N \leq 124 \Rightarrow PUSH^N(a_i).POP \equiv PUSH^{N-1}(a_i)$

$PUSH(a_o).PUSH\ {}_1^{124}.(a_i) \equiv PUSH\ {}_1^{124}(a_i)$

$T.VAL \equiv T$

$N \geq 0 \Rightarrow POP^N.PUSH(a) \equiv PUSH(a)$

Values:

$V(T.PUSH(a).VAL) = a \mod 255$

Example 5. Alternative Formal Specifications (Guttag Type) for STAC

This alternative includes two "hidden functions," which are marked in
the syntactic specifications with asterisk.

TYPE:

    stac

SYNTACTIC SPECIFICATION:

          NEWSTAC: -> <stac>
          PUSH(s,I):   <stac> X <integer> -> <stac>
          POP(s):   <stac> -> <stac>
          VAL(s):   <stac> -> <integer>
          SPSLFT(s):   <stac> -> <integer>
        *ADD(s,I):   <stac> X <integer> -> <stac>
        *DEQ(s):   <stac> -> <stac>

SEMANTIC SPECIFICATION:

          SPSLFT(NEWSTAC) = 124
          SPSLFT(ADD(s,I)) = SPSLFT(s) - 1
          POP(NEWSTAC) = NEWSTAC
          POP(ADD(s,I)) = s
          DEQ(NEWSTAC) = NEWSTAC
          DEQ (ADD(s,I)) = if SPSLFT(s) = 124
                                then s
                                else ADD(DEQ(s),I)
          PUSH(s,I) = if SPSLFT(s) > O
                                then ADD(s,I)
                                else ADD(DEQ(s),I)
          VAL(NEWSTAC) = undefined
          VAL(ADD(s,I)) = I mod 255

        *denotes a hidden function

IV.   Discussion of the Simple Examples

Example 1 is the classic example for abstract specifications. It is a stack with unlimited capacity. The legality section shows that any sequence of PUSH operations is a legal trace. The first statement in the value section shows the value of TOP after any trace that ends with a PUSH. (7) shows that PUSH always increments the value of DEPTH. (8) specifies the initial value of DEPTH to be zero. The equivalence section allows us to reduce any legal trace with PUSH, TOP, and POP to one that is equivalent but contains only PUSH operations. We will be able to determine the value of the V-functions for any legal trace by making such reductions.

In Example 2 (an integer queue) the "legality" section allows traces that consist of any number of ADDS but each occurrence of REMOVE or FRONT must be preceded directly by an ADD. However, the equivalence statements allow other traces because the sequence ADD.REMOVE may either be replaced by REMOVE.ADD or (at the start of a trace) deleted and the resulting trace will be equivalent to the original trace. The value section shows the value of FRONT after (a) an item is added to an empty queue and (b) an item is added to the queue that already has a value of FRONT (same as before). To find the value of FRONT after a trace that has REMOVES in it, one must apply (5) and (6) repeatedly until one has an equivalent trace that does not contain a REMOVE. Each application of (5) can move a REMOVE to the left one place. When REMOVE follows the first ADD directly, both can be deleted using (6).

In Example 3 we have a queue that always shows the largest item at the front. The largest object is also the one removed by REMOVE. The legal traces are the same as those in Example II (except for an obvious change of function names). The most important difference is (5) in which it is asserted that the order of two consecutive inserts is irrelevant. Assertion (7) shows the effect of a REMOVE after an INSERT that had a parameter larger than the value at the front of the SQUEUE. In that case it simply cancels the effect of the INSERT. However, because of (5), we can always rearrange the order of INSERTs so that the last one is the one that inserts the largest value. This allows us to use (7) for any REMOVE at the end of a trace with at least two inserts in it. (6) describes the effect of REMOVE in the case that it is preceded by only one INSERT. The value section shows us the value of FRONT after an

INSERT in an empty queue and after inserting a value that is greater
than the value of FRONT.

The discussion of the first three examples is intended to show that the
formal specifications do correspond to our intuitive notions of the way
that these modules perform. The correspondence with intuition must, of
necessity, remain informal. The demonstration of completeness can be
performed systematically. This will be discussed lateron.

The fourth example is the problem that John Guttag could not specify
without the use of hidden functions [15] (which follows from restrictions
of the mathematical model underlying his technique). His specification
is included as Example 5. We believe that the brevity of our specification shows the advantages of the trace method. This is a situation
in which the values of V-functions for some legal traces are deliberately
not defined. Any syntactically correct trace is legal. The module will
never "trap". However the value of VAL initially (or after a POP on an
"empty stack") is not defined. The implementation can deliver any value
in these situations without violating the specifications. If a value,
I, greater than 255, is inserted only I mod 255 will be stored.

The above examples show a number of advantages over previous methods of
DA specifications. There appears to be no need for hidden functions;
the specifications are quite compact and the individual statements are
simple. The derivations needed to demonstrate completeness are sometimes
quite involved but they need not be performed during the implementation
or during the verification that an implementation is correct.

The ideas are rather new and we are aware of a number of important un-
answered questions. Nonetheless, we believe that this report demonstrates
that the method is as good as any of the previously published ones and
can help to discover design errors early in the design process.

V. A Compressed History of the Development of an Abstract Specification

In this section we present the history of the development of an abstract
specification for a "table/list"-(T/L) module. The programs offered by
this module support the processing of linearly ordered data structures,

regardless of whether they are implemented as tables or lists. This module is currently implemented to help in generating address translation tables as we need them for a virtual memory mechanism within a family of operating systems [17]. It is also expected that this specification can be used for various other table or list handling purposes.

An Informal Picture of the T/L Module

Because it is the purpose of this report to introduce a method of describing such modules, we must begin with an intuitive description of our example. One physical implementation of this module would be by means of a set of children's blocks where it is possible to write one "entry" on the upper surface. The blocks are arranged in a single row and covered with an opaque lid with a single window. Through this window one may read the entry on a single block, insert and remove blocks, or change the entry written on the block that shows through the window. The entry on the block that shows through the window is referred to as the current entry. Because the cover is opaque it is not possible to tell how many blocks are currently under it, but the cover is fitted with signals that tell whether or not there is a block to the right of the current entry, whether or not there is a block to the left of the current entry, and whether there are any blocks under the cover at all.

The operations that we want to perform include reading the value of the current entry, moving the lid one place to the right, moving the lid one place to the left, moving the lid and all blocks at the right hand side of the current block to the right so that a new current block may be inserted through the window, and removing the current block (moving the lid and all blocks to the right of the deleted block one place to the left).

It was our goal that all operations that could be easily performed with the physical model described above be allowed by our specification.

In our specification we will have five operations (O-functions): INSERT, DELETE, ALTER, GOLEFT, and GORIGHT. ALTER will just be a shorthand for a sequence of DELETE and INSERT. The first two indicators mentioned above will be named EXLEFT (EXist entries to the LEFT), EXRIGHT, and the third is represented by EMPTY. The current entry will be available through the V-function CURRENT. The precise relationship among the V-functions and the way that their values are changed by the module's operations will be described in the specifications.

Lecture Notes in Artificial Intelligence 2637

Subseries of Lecture Notes in Computer Science
Edited by J. G. Carbonell and J. Siekmann

Lecture Notes in Computer Science
Edited by G. Goos, J. Hartmanis, and J. van Leeuwen

Springer
Berlin
Heidelberg
New York
Barcelona
Hong Kong
London
Milan
Paris
Tokyo

Example 6.   (Incorrect)   Version of a Specification for a Table/List
Module

Syntax of Functions

O-Functions: INSERT(e):   <entry> x <TL> -> <TL>
                  DELETE: <TL> -> <TL>
                  ALTER(e):   <entry> x <TL> -> <TL>
                  GOLEFT: <TL> -> <TL>
                  GORIGHT: <TL> -> <TL>

V-Functions: CURRENT: <TL> -> <entry>
                  EMPTY: <TL> -> <boolean>
                  EXLEFT: <TL> -> <boolean>
                  EXRIGHT: <TL> -> <boolean>

Legal Traces

(1) $\lambda(T)$ => $\lambda(T.INSERT(e))$
(2) $\lambda(T)$ => $\lambda(T.INSERT(e).CURRENT)$
(3) $\lambda(T.CURRENT)$<=> $\lambda(T.EXLEFT)$
(4) $\lambda(T.CURRENT)$<=> $\lambda(T.EXRIGHT)$
(5) $\lambda(T.CURRENT)$<=> $\lambda(T.ALTER(e))$
(6) $\lambda(T.CURRENT)$<=> $\lambda(T.INSERT(e).GOLEFT)$
(7) $\lambda(T.GOLEFT)$ <=> $\lambda(T.GOLEFT.GORIGHT)$

Equivalences

(8)   T.EMPTY ≡ T
(9)   T.INSERT(e).DELETE ≡ T
(10)   T.GOLEFT.GORIGHT ≡ T
(11)   T.ALTER(e) ≡ T.DELETE.INSERT(e)
(12) $\lambda(T.CURRENT)$ =>   T.CURRENT ≡ T
(13) $\lambda(T.EXLEFT)$ =>   T.EXLEFT ≡ T
(14) $\lambda(T.EXRIGHT)$ =>   T.EXRIGHT ≡ T

Values

(15)   V(EMPTY) = true
(16) $\lambda(T)$ =>   V(T.INSERT(e).CURRENT) = e
(17) $\lambda(T)$ =>   V(T.INSERT(e).EMPTY) = false
(18) $\lambda(T)$ cand (V(T.EMPTY) = true) =>   V(T.INSERT(e).EXLEFT) = false
(19) $\lambda(T)$ cand (V(T.EMPTY) = false) ∧ (V(T.EXLEFT) = false) =>
                  V(T.INSERT(e).EXLEFT) = true
(20) $\lambda(T)$ =>   V(T.INSERT(e).EXRIGHT) = V(T.EXRIGHT)
(21) $\lambda(T.GOLEFT)$ =>   V(T.GOLEFT.EXRIGHT) = true
(22) $\lambda(T.GORIGHT)$ =>   V(T.GORIGHT.EXLEFT) = true
(23) $\lambda(T.ALTER(e))$ =>   V(T.ALTER(e).CURRENT) = e
(24) $\lambda(T.ALTER(e))$ =>   V(T.ALTER(e).EMPTY) = V(T.EMPTY)
(25) $\lambda(T.ALTER(e))$ =>   V(T.ALTER(e).EXLEFT) = V(T.EXLEFT)
(26) $\lambda(T.ALTER(e))$ =>   V(T.ALTER(e).EXRIGHT) = V(T.EXRIGHT)
(27)   V(T.INSERT(e).GOLEFT.CURRENT) = V(T.CURRENT)
(28)   V(T.INSERT(e).GOLEFT.EXLEFT) = V(T.EXLEFT)

A. The First Version (Example 6)

We do not display the original specification but instead present a translation using traces. We were not using traces for specification purposes at the time that the original was written. The use of traces makes many deficiencies in the first version obvious. They were originally discovered after much hard labor. We show an abbreviated histroy of the development to provide evidence controverting the claim that abstract specifications state "only the obvious."

The "syntax" section is as in the earlier examples. We use elements of a type "entry" only to store them into the data structure of the T/L module, or to fetch them. We assume that the relation of equality over entries is defined elsewhere.

Statements (3) through (5) tell us that V-functions EXLEFT and EXRIGHT and O-function ALTER(e) have the same applicability condition as CURRENT.

The "equivalences" section should allow the reader to transform any legal trace to one shown to be legal by (1) through (7). The alert reader will notice that this section does not satisfy this requirement. This will be investigated in some detail later.

Statement (8) is unconditional because a call on EMPTY can always be added to or removed from any trace without making the module trap.

Statements (9) and (10) say that subtraces INSERT(e).DELETE and GOLEFT.GORIGHT have no effect. Statement (11) is supposed to tell us that a call on ALTER has the same effect as two consecutive calls on DELETE and INSERT, provided that INSERT has the same actual parameter as ALTER. Statements (12) through (14) tell us that V-functions CURRENT, EXLEFT, and EXRIGHT can be removed from a legal trace to get an equivalent trace.

Statement (15) gives the initialization of the module. Statements (16) through (20) describe the effects of INSERT at the end of a legal trace on the values of EMPTY, CURRENT, EXLEFT, and EXRIGHT.

Statements (23) through (26) define the effects of ALTER at the end of a trace on the four V-functions. Note that only CURRENT is changed.

Statements (27) and (28) say that two consecutive calls on INSERT and GOLEFT have no effect on the values of CURRENT and EXLEFT.

B. Discussion of Flaws in the First Version of the T/L Module
    Specification

The use of traces and the way in which the present specifications are
divided into sections allows us to discuss flaws in version 1 of the
T/L module in a straightforward way and to omit two or three intermediate stages of the original development. However, all errors below were
actually included in the original design of the T/L module (where a
different method of specification was used) and allowed to remain in
the design after formal discussions among the members of our group.

Incompleteness

In examining the first specification we first attempt to make certain
that the specification is complete. We will (by definition) consider
the specification to be incomplete if there are some traces ending in
calls on V-functions which can be shown to be legal but for which no
value can be derived.

One example of incompleteness concerns the value of the function EXRIGHT.
Only (20) and (26) make any statement about the value of EXRIGHT and
these make no statement about the initial value of EXRIGHT or
V(INSERT(e).EXRIGHT) which can be shown to be legal.

The specification is similarly incomplete with respect to EXLEFT.

Another form of incompleteness can be found by attempting to derive the
value of V(INSERT(a).INSERT(b).GOLEFT.EMPTY). There is no statement
about the value of EMPTY when immediately preceded by GOLEFT and no
equivalence assertion that would allow us to remove GOLEFT.

Specification Versus Intuitive Understanding

In addition to the instances of incompleteness that have been demonstrated, we can show that a number of statements in the "legal trace"
section and "equivalences" section do not meet our intuitive expectations. There is a problem with the legality of traces beginning with
a call on GOLEFT. For example, we would expect that a call on GOLEFT
before the first entry has been inserted into the data structure should
not be permitted. However, the value of λ(GOLEFT.GORIGHT) can by statement (10) always be calculated to be λ($\llcorner\lrcorner$), which is (by definition)
"true". Since by definition λ(T.X) => λ(T) we can conclude that (for
T $\equiv$ GOLEFT and X = GORIGHT) we have λ(GOLEFT) = true. A similar problem
exists concerning the legality of traces ending with a call on GOLEFT.

Statements (2) and (6) eliminate the possibility of insertion to the left of the leftmost entry. We can move the window in our cover over the leftmost entry but not further. An insert would then make EXLEFT true again (statement (19)) but we would have inserted to the <u>right</u> of the leftmost entry.

The mnemonic "EMPTY" was an obstacle to a straightforward solution. Imagine that one moves left from the left end. By statement (18), EMPTY would become true although there are entries in the data structure.

We will eliminate these problems by renaming "EMPTY" to "OUT" and allowing one move to the left beyond the left end. The value of CURRENT is then undefined, while OUT is true, EXLEFT is false, and EXRIGHT is true. This is in contrast to the new initial state (no entries in the data structure) where EXRIGHT is false.

A problem that initiated the development of the specification technique presented in this paper is best formulated by posing the following question.

How can the designer be sure that he specified the effects of all traces that he wants to be executable programs?

Or, put in other way and applied to our example, how do we determine the subset of

$$(INSERT(e),DELETE,ALTER(e),GOLEFT,GORIGHT,$$
$$CURRENT,OUT,EXLEFT,EXRIGHT)*,$$

(where "*" is the Kleene star) that comprises the set of executable, i.e. legal traces? (Rules for including V-functions are easy to find and are therefore not considered now.)

We now note some quantitative properties of such traces: Let $|X|$ denote the number of calls on X in a given trace. Then for all legal traces:

$$|GOLEFT| > |GORIGHT|$$
$$|INSERT| > |GOLEFT| - |GORIGHT|$$
$$|INSERT| > |DELETE| + |GOLEFT| - |GORIGHT|$$

These relations alone, however, help little. The obviously unreasonable trace

$$GORIGHT.GOLEFT.GOLEFT.INSERT(a).INSERT(b)$$

satisfies the above inequalities.

We therefore have to make some additional assertions to characterize the set of legal traces.

The specification of Example 6 did not capture the language of the module, as we intuitively understand it.

Example 7.   Table/List Module with Unlimited Capacity

Syntax

O-Functions:
INSERT:	<entry> x <TL> -> <TL>
ALTER:	<entry> x <TL> -> <TL>
DELETE:	<TL> -> <TL>
GOLEFT:	<TL> -> <TL>
GORIGHT:	<TL> -> <TL>

V-Functions:
CURRENT:	<TL> -> <entry>
OUT:	<TL> -> <boolean>
EXLEFT:	<TL> -> <boolean>
EXRIGHT:	<TL> -> <boolean>

Legal Traces

(1) λ(T) => λ(T.INSERT(a))
(2) λ(T) => λ(T.INSERT(a).GOLEFT)
(3) λ(T.GOLEFT)<=> λ(T.CURRENT)

Equivalences

(4)   T.OUT $\equiv$ T
(5)   T.EXLEFT $\equiv$ T
(6)   T.EXRIGHT $\equiv$ T
(7) λ(T.CURRENT) => T.CURRENT $\equiv$ T
(8) λ(T.GOLEFT) => T.GOLEFT.GORIGHT $\equiv$ T
(9)   T.INSERT(a).DELETE $\equiv$ T
(10) T.INSERT(a).GOLEFT.DELETE $\equiv$ T.DELETE.INSERT(a).GOLEFT
(11) λ(T) => T.INSERT(a).INSERT(b).GOLEFT $\equiv$
          T.INSERT(b).GOLEFT.INSERT(a)
(12) T.ALTER(a) $\equiv$ T.DELETE.INSERT(a)

Values

(13) V(OUT) = true
(14) V(EXLEFT) = V(EXRIGHT) = false

(15) λ(T) => V(T.INSERT(a).CURRENT) = a
(16) λ(T) => V(T.INSERT(a).OUT) = false
(17) λ(T) => V(T.INSERT(a).EXLEFT) = not V(T.OUT)
(18) λ(T) => V(T.INSERT(a).EXRIGHT) = V(T.EXRIGHT)

(19) λ(T.CURRENT) => V(T.INSERT(a).GOLEFT.CURRENT) = V(T.CURRENT)
(20) λ(T) => V(T.INSERT(a).GOLEFT.OUT) = V(T.OUT)
(21) λ(T) => V(T.INSERT(a).GOLEFT.EXLEFT) = V(T.EXLEFT)
(22) λ(T.GOLEFT) => V(T.GOLEFT.EXRIGHT) = true

For example:

$$\lambda(\text{INSERT}(a).\text{INSERT}(b).\text{GOLEFT}.\text{GOLEFT}) = \underline{\text{false}}$$

Other examples can easily be found.

C. The Current Specification for the T/L Module

After discovering the above errors (over a period of several months)
we made an observation that allowed us to write the specification given
in Example 7.

Any legal trace for the T/L module must be equivalent to a trace in
which there is a (possibly empty) sequence of INSERTs followed by any
number of repetitions of the sequence INSERT.GOLEFT. This observation
is based on our intuitive model of the object that we are trying to
specify. (We have no other possible basis). We could create the table
contents a_o, a_1 ... a_i ... a_N, where a_i is the current entry by suc-
cessivly inserting a_o,a_1 ... a_i and then executing INSERT(a_j).GOLEFT
for j = n, n-1......i+1. Each INSERT(a_j).GOLEFT sequence leaves CURRENT
unchanged but inserts a block to the right of current.

Traces in this form are the normal form traces of this module. We will
therefore have to provide a set of assertions that allow to transform
any legal trace to such a normal form trace.

The assertions labeled "legal traces" in Example 7 ((1) - (3)) state
that all traces in normal form (and some additional traces) are legal.
We also indicate that CURRENT may be called whenever a GOLEFT would be
allowed.

The assertions (4) - (7) state that the V-functions do not effect any
changes on the module. (8) and (9) give the obvious facts that GOLEFT
can be cancelled by a GORIGHT that follows it and that an INSERT can be
cancelled by a DELETE that follows it. Note that (8) only applies when
GOLEFT is legal.

If our specification is a good one, we should be able to show that every
legal trace is equivalent to a trace in normal form. The V-functions can
be trivially deleted. We are able to delete a DELETE if it immediately
follows an INSERT and a GORIGHT if it follows immediately after a GOLEFT.
Using statement (11) we can move a GOLEFT right or left through a sequence of INSERTs to get an equivalent trace. That will allow us to
remove instances of DELETE by bringing an INSERT up to them if only
GOLEFTs intervene. Using assertion (10) one may transform sequences
containing GOLEFT.DELETE and DELETE.GOLEFT into equivalent sequences

where either the DELETE has been moved to the left (bringing it closer
to the INSERT that it cancels) or the GOLEFT has been moved to the right
(bringing it close to any GORIGHT that would cancel it). Assertion (12)
allows the removal of all occurrences of ALTER. Repeated application of
these rules allows the removal of all functions except INSERT and GOLEFT.

Completeness of the Current Specification

To demonstrate completeness we examine primarily the value section (13) -
(22). (13) and (14) specify the initial values of all V-functions except
CURRENT. The failure to specify an initial value for CURRENT is not an
instance of incompleteness because CURRENT is not a legal trace. Using
(15) - (18) we have specified the values of all four V-functions for
traces containing only INSERT.

Using (19) - (22) we can determine the values of the V-functions for
any trace of the form T.INSERT(a).GOLEFT provided that we know the values
of those functions after T. It follows that we know the values for any
trace in the normal form. Since the equivalence statements allow any
legal trace to be reduced to an equivalent trace in that form, the specification is complete.

Consistency

Demonstration of consistency is more complex. It is quite clear that
the value section ((13) - (22)) is in itself consistent, but it is
necessary to show that the transformations allowed by the equivalence
section that produce a trace ending in a given V-function result in
traces with the same value. Such a proof is beyond the scope of this
paper.

VI. Conclusion

It is clear that when we entered into the design of the T/L module interface we did not expect the difficulties that we encountered. Each pro-

posal seemed intuitively obvious and the formal specifications that we
wrote appeared to correspond to our intuition. Several people examined
the specifications (which were written using weakest preconditions);
all thought that they were acceptable. The types of difficulties described in connection with the first version of the T/L module specification came as a complete surprise. We had expected that writing the
formal specifications was "only a formality" for so simple a module.

Our first conclusion then is simply that writing the formal specifications is useful _even_ for simple modules. Had we been forced to make
the change from the first version to the second version _after_ coding
was underway, it would have been expensive in terms of the amount of
code (both in the module and in programs that use the module) that
would have needed revision.

Once we became aware of the difficulties, we found attempts to convince
ourselves of the correctness of new versions to be extremely frustrat-
ing. The specifications that were written (using predicate transformers
for programs consisting of calls on the functions) did not lend themselves well to examination for completeness and consistency. The mathematical model underlying those specifications is complex and there were
difficulties instrinsic in the decision to talk about programs rather
than traces. Although we have not yet produced a complete formal proof
that this specification is complete and consistent, the intuitive justifications are far more convincing than our more formal arguments about
the old specifications. Our second conclusion therefore is that the
concept seems to be superior to other forms of data abstract specification known to us.

It is becoming popular among software specialists to speak of "front
end" investment. The proposal is that by investing time and intellectual
energy in the early design phase one can reduce the overall systems
costs because of time saved at the later stages. A weakness of the majority of such proposals is that they provide little in the way of
specific suggestions about what to do at those early stages. There is
little evidence that the effort invested in the early stages will actually pay off. There is lots of evidence that just writing vague statements of good intentions ("The system will have a user-oriented interface") will _not_ pay off. In this paper we have made a specific proposal
for the use of that "front end" energy. We have shown how to write such
specifications, and indicated how one may evaluate them for completeness
and consistency.

Further work on verifying properties of these specifications is clearly necessary. As Price has shown [21], there are clear advantages to doing as much verification as possible before implementation begins. Similar views are found in [18], but Price included some (machine assisted) proofs.

Acknowledgement

The authors are grateful to Professor D. Stanat for his advice while the research was being performed and on the writing of this paper. Dave Weiss, Lou Chmura, John Shore, and Janusz Zamorski also made helpful comments. This research was supported by the U.S. Army under contract #DAAG 29-76-G-O240. W. Bartussek was also supported by the German Academic Exchange Service (DAAD) under stipend #4-USA-CDN-AUS-NZ-3-EB.

REFERENCES

[1] Parnas, D.L. "Information Distribution Aspects of Design Methodology." Proc. IFIP Congress, 1971

[2] Parnas, D.L. "On the Criteria to be Used in Decomposing Systems into Modules." Communications of the ACM (Programming Techniques Department), December 1972.

[3] Parnas, D.L., Shore, J.E., and D. Weiss. "Abstract Types Defined as Classes of Variables." Proc. Conference on Data: Abstraction, Definition, and Structure, pp. 22-24, Salt Lake City, Utah, March 1976.

[4] Floyd, R.W. "Assigning Meanings to Programs." In "Mathematical Aspects of Computer Science" (J.T. Schwartz, ed.). Proc. Symp. of Applied Mathematics, Vol. 19, American Math. Society, Providence, 1967, 19-32.

[5] Hoare, C.A.R. "An Axiomatic Basis for Computer Programming." Comm. ACM 12, 10. October 1969, 576-583.

[6] Dijkstra, E.W. "Guarded Commands, Nondeterminancy, and the Formal Derivation of Programs." CACM 18, 8, August 1975.

[7] Dijkstra, E.W. A Discipline of Programming. Prentice Hall, 1976.

[8] Parnas, D.L. "A Technique for Software Module Specification with Examples." Comm. ACM, May 1972.

[9] Guttag, J. "The Specification and Application to Programming of Abstract Data Types." Ph. D. Thesis, CSRG TR 59, University of Toronto, September 1975.

[10] Guttag, J. "Abstract Data Types and the Development of Data Structures." SIGPLAN/SIGMOD Conference on DATA: Abstraction, Definition and Structure (to be published in CACM).

[11] Parnas, D.L. and W.R. Price. "The Design of the Virtual Memory Aspects of a Virtual Machine" Proceedings of the ACM SIGARCH-SIGOPS Workshop on Virtual Computer Systems, March 1973.

[12] Parnas, D.L. and W.R. Price. "Using Memory Access Control as the

Only Protection Mechanism." Proc. of International Workshop on Protection in Operating System, 13-14 August, IRIA.

[13] Roubine, O. and L. Robinson. "Special Reference Manual" (Second Edition), Technical Report CSG-45, Stanford Research Institute, Menlo Park, Calif.

[14] Liskov, B. and V. Berzins. "An Appraisal of Program Specifications." Research Direction in Software Technology (P. Wegner, ed.). To be published by MIT Press.

[15] J. Guttag. Private communication, 1976.

[16] Parnas, D.L. and H. Wuerges. "Response to Undesired Events in Software Systems." Proc. of the 2nd International Conference on Software Engineering, 13-15 October 1976, San Francisco, California.

[17] Parnas, D.L., Handzel, G. and H. Wuerges. "Design and Specification of the Minimal Subset of an Operating System Family." Presented at 2nd International Conference on Software Engineering, 13-15 October 1976; published in special issue of IEEE Transactions on Software Engineering, December 1976.

[18] Neumann, P.G., et.al. A Provably Secure Operating System: The System, Its Applications, and Proofs. Final Report, Stanford Research Institue, 11 February 1977, Menlo Park, California

[19] Parnas, D.L. "The Use of Precise Specifications in the Development of Software." Proc. IFIP Congress 1977, North Holland Publishing Company.

[20] Parnas, D.L. and G. Handzel. "More on Specification Techniques for Software Modules." Technical Report, Technische Hochschule Darmstadt, Darmstadt, West Germany, February 1975.

[21] Price, W.R. "Implications of a Virtual Memory Mechanism for Implementing Protection in a Family of Operating Systems." Technical Report (Ph. D. Thesis), Carnegie-Mellon University, June 1973, AD766292.

PROTECTION IN LANGUAGES FOR REAL TIME PROGRAMMING

P. Ancilotti
Istituto di Elaborazione della
Informazione,CNR, via S. Maria
Pisa - Italy

M. Boari
Istituto di Automatica, Facoltà
di Ingegneria, Viale Risorgimento
Bologna - Italy

N. Lijtmaer
Istituto di Elaborazione della
Informazione,CNR, via S. Maria
Pisa - Italy

ABSTRACT

A protection mechanism which may be embedded in an object oriented
language for real time programming permitting definition of abstract
data types, is proposed in this paper.

This mechanism provides support for designing highly reliable concur
rent programs; in fact it allows the detection at compile time of a
large class of time dependent errors. To verify the versatility of
the proposed mechanism it is firstly characterized abstractly in terms
of a protection model; then some linguistic features enforcing protec
tion are defined.

1. Introduction

The role of a protection mechanism in a programmed system is to prevent processes from acceding to the objects defined in the system in
an unauthorized or undesirable way; a protection mechanism must there
fore guarantee that each process may both accede only those objects
for which it has legitimate rights and perform only meaningful acces
ses to those objects.

Protection mechanisms are usually supported by operating systems. Recently, however, the opportunity to incorporate protection facilities
in programming languages has been recognized. In fact software reliability may be considerably enhanced if access control restrictions a-
re expressed directly in the language and enforced by the compiler of
that language. Thus, access control errors can be captured at compile
time so that programs may be written to be well-behaved with respect
to access control restrictions [Wul 74,76; Jon 76] .

A mechanism suitable for incorporation in object oriented languages
was presented in [Jon 76]. Such a proposal is based on capability
protection mechanisms provided by some operating systems.

Following this approach, a protection mechanism, which may be embedded in an object oriented language for real time programming providing abstract data type definition, is proposed in this paper. This
mechanism also provides support for designing highly reliable concur
rent programs; in fact it allows the detection at compile time of a
large class of time dependent errors.

Unlike some real time languages recently proposed [Bri 75, Wir 77] ,
the possibility that objects may be dynamically allocated to processes is considered. The protection mechanism must guarantee, also in
this case, a compile time checking of access control. To verify the
versatility of the proposed mechanism it is firstly characterized abstractly in terms of a protection model; then some linguistic features enforcing protection are defined.

2. PROTECTION: A MODEL

The versatility of protection mechanisms can be abstractly characterized in terms of a *protection model*. A protection model sees the *system* as a collection of components. System components may be partition
ed in two disjoint subsets, namely, *subjects*, that is·processes and
objects, that is resources. Furthermore, a protection model defines

the access rights of each subject to each object. Following Lampson's proposal on 1971, a protection model can be represented in the form of a *protection matrix*. Subjects are associated with rows of the matrix, while objects are associated with columns [Lam 71]. Given a pair subject-object, the corresponding entry of the matrix defines the set, possibly empty, of access rights that the subject has to the object. A *protection domain* defines the set of access rights that one subject has to the objects of the system.

A first aspect of the protection mechanism is related to the enforcement of protection. It means that , at any time, a subject can exercise only those access rights that belong to its domain. In terms of the protection model a subject can accede an object if and only if the matrix entry, corresponding to the pair subject-object, is not empty. Furthermore the subject can exercise on that object only those access rights specified in the matrix entry. This aspect of a protection mechanism is called the *enforcement rule* [Jon 73].

As far as system reliability is concerned it should be convenient to maintain protection domains as small as possible in order to restrict the number of objects that a subject can use to those significant to the performed activity. This concept has been called the "principle of least privilege" [Lin 76].

While a small program operates on a small number of objects, a larger one normally needs to operate on a great number of them. To conciliate this characteristic with the principle of least privilege, a large program must run in many different protection domains and then it must be able to switch protection domains during execution. In this way the protection domain in which a program runs, can be held time to time as small as possible.

Domain switching represents the second aspect of a protection mechanism and it is called the *domain binding rule* [Jon 73].

From the model point of view since a subject may run in more than one protection domain, it cannot be represented by only one row of the protection matrix. Thus, rows coincide with domain and a subject must be specified by a pair process-domain.

Fig. 2.1 shows that each process P running in the protection domain D_i can exercise access rights α and β on the object R_j while object R_k is not accessible.

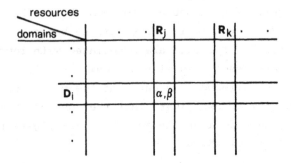

Fig. 2.1

When a process needs to change its current domain to acquire or release a limited number of rights, but one wants to avoid the complete
domain switching, the protection mechanism must be able to transfer
rights into and out of domains. This is the third and last aspect to
be covered by the protection mechanism and it is called the *transfer
rights rule* [Jon 73].

In terms of the model we must then allow changes in the protection
matrix. This aspect, related to the assignment of access rights to
domains, is strongly connected with the problem of resource allocation in the operating system sense.

Let us now characterize objects in relation to the way in which access rights are assigned to domains. If access rights to use an object are allocated to domains during the initialization phase and these
rights are no longer modifiable, then the object is statically allocated. Otherwise, if a domain may acquire or release rights to accede an
object, this object is dynamically allocated.

If access rights to accede an object belong to only one domain this object is said a dedicated resource. In particular a dedicated resource
statically allocated to a domain represents a private resource of that
domain. If access rights belong to more than one domain the object is
called a *shared resource*.

In order to avoid time dependent errors, concurrent accesses to the
same objects must be generally avoided. For this purpose we will suppose that at most one process at a time can run in a protection domain.
Furthermore shared resources will require exclusive access and synchronization among processes competing for their use.

3. LANGUAGE ENFORCED PROTECTION

The main goal of this paper is to show how a protection mechanism mo
deled as described can be embedded in an object oriented language for
real time applications. Concurrent Pascal has been chosen as the reference language [Bri 75].

From a linguistic point of view, an object or resource is defined as
a uniquely distinguishable structured encoding of information. An *ac
cess* to a resource is an algorithm to reference the resource in order
to transform it or to extract information encoded within it. The access algorithm depends upon the internal representation of those resources to which the algorithm can be applied. Thus the set of all re
sources is partitioned into equivalence classes, each called a *type*.
The type of a resource determines the accesses that can potentially
be applied to that resource.

Any description of a resource type is completely specified by the de
finitions of applicable accesses. Knowledge of the internal represen
tation of resources need not be available outside the access algorithms [Jon 73].

Then each resource type may be conceived as an *abstract data type* [Lis
75].

Following Concurrent Pascal notation, a dedicated resource type is de
fined as a *class:*

> *type* dedicated-resource-type = *class (.,.,.,.);*
> *begin .... end;*

Each instance of a class represents a dedicated resource to which only the accesses specified by *entry* procedures can be applied.

Analogously, a shared resource type may be defined as a *monitor:*

> *type* shared-resource-type = *monitor (.,.,.,.);*
> *begin .... end;*

In this case the entry procedures are mutually exclusive, so that at
most one process can be active in a monitor, at any time. Furthermore,
queue variables are allowed as part of the shared data base of a monitor in order to permit the specification of the coordination among
processes competing for this resource [Bri 75, Hoa 74].

Note that instances of abstract data types correspond to columns of
the protection matrix.

Scope rules of the language establish that local objects of abstract data types are accessible only by the procedures of those types. Then each instance of an abstract data type identifies a particular domain represented by a row in the protection matrix.

Each protection domain can be conceived as a capability list, where capability is a token that identifies an object and a set of access rights; possession of the token confers those access rights for that object. From the model point of view a capability is represented by an entry of the protection matrix.

The syntax of a capability declaration is:

$$var \ v: \ T \ \{\alpha, \beta, \ldots\}$$

The semantics of such a declaration specifies that a resource of type T is accessible through the capability v; α, β,.. are the only procedural accesses available to refer or alter that resource through v. They correspond to a subset of procedure entries of type T.

The concept of capability does not correspond to the traditional concept of variable as an object containing a value that can be modified by means of assignments. A capability can be conceived rather as a pointer that can be bound to a particular object containing the value [Jon 76].

Rules to bind capabilities to objects are different from assignment rules involving traditional variables. In fact, the first ones correspond to assignment rules involving variables of type pointer. A capability, bound to an object, can be conceived as an access path for the object. Both, the type of the object and the type of the capability must be the same, in order to guarantee a correct binding.

For simplicity we will use the notation resource or object x of type X instead of resource or object referred by the capability x of type X.

A process can refer and use a resource if and only if it executes in a domain owning an access right to that resource, that is a capability bound to the resource.

As stated in the previous paragraph assignment rules of access rights to domains may be either static or dynamic, and then static and dynamic binding of capabilities to resources must be provided.

Nevertheless while resources can be dynamically allocated they are created statically and will exist forever after the system initialization; that is no dynamic resource creation is considered in this paper.

First of all the static assignment of access rights to domains, that is the static capability binding, is considered. There are two diffe rent ways to statically bind a capability to a resource: the first one corresponds to the allocation of a dedicated resource, while the second one allows resource sharing.

A capability y of type Y is created in any domain containing the declaration:

$$var \ y: Y \ \{\alpha, \beta, \dots\}$$

that is, in any instance of an abstract data type containing the previous declaration.

A resource of type Y, that is an instance of an abstract data type Y, is created during the initialization of y:

$$init \ y \ .$$

The init statement allocates space for private variables of the resource, initializes them and binds the capability y to the resource [Bri 75].

Thus, given a domain associated with an instance x of type X, in order to statically allocate to x the access rights α, $\beta, \dots$ to a private resource of type Y, where Y must be a class, it is sufficient to declare the capability:

$$var \ y: Y \ \{\alpha, \beta, ..\}$$

into type X. During the initialization of x its local variables, included the variable y, are allocated in the name space of x and are initialized. That implies the initialization of y, that is the creation of a resource of type Y and the binding of y to this resource.

The static allocation of shared resources involves the definition of a resource for which more than one, statically declared, access path exists.

The problem now is to allow a set of domains associated with instances $x_1, x_2, \dots, x_n$ of types $X_1, X_2, \dots, X_n$ respectively, to share a resource of type Y, where Y must be a monitor. For this purpose the capability:

$$var \ y: Y \ \{\alpha, \beta, ..\}$$

is declared in the main program. As stated above, during the initialization of y a resource of type Y is created and y is bound to it.

Then, in order to statically assign to each domain $x_i (i = 1...n)$ the access rights $\alpha_{1_i}, \beta_{1_i} ...$ to the resource y, a formal parameter of type Y with access rights $\alpha_{1_i}, \beta_{1_i} ...$ is associated to the system type X_i, for instance:

$$type \quad X_i = \text{.......}(\text{.....},z : Y \{\alpha_{1_i}, \beta_{1_i}, ...\}...)$$

and y is declared as the actual parameter during the initialization of x_i:

$$init \; x_i \; (..., \; y \, , \, ...).$$

In this way, a capability $z : Y \{\alpha_{1_i}, \beta_{1_i}, ...\}$ is allocated in the name space of x_i, and, during the initialization of x_i, z is bound to the resource y.

Note that $\{\alpha_{1_i}, \beta_{1_i}, ...\}$ must be a subset of $\{\alpha, \beta, ...\}$ in order to guarantee the correct binding of z to the resource y.

In order to allow dynamic resource allocation, bindings between capabilities and resources must be delayed until execution time. If a resource of type Y is dynamically requested by an instance x of type X it means that x, initially, does not own any access path to accede to the resource. A path must be granted to x only when the resource is needed and then the path must be reclaimed. For this purpose the capability:

$$var \; y : Y \; \{\alpha, \beta, ..\} \; empty$$

is declared in the abstract data type X. In this way the capability y is created in any domain associated with an instance x of type X.

During the initialization of x all its capabilities, identified by the key word *empty*, are not bound to a particular resource. The domain x can acquire the access rights $\alpha, \beta, ...$ to use a resource of type Y. For this purpose the capability y must be bound to the resource.

Assignments between capabilities, that is new bindings of capabilities to resources, may appear only in instances of a particular type called *manager* with the following restrictions:

a) Each manager may modify only bindings between capabilities and resources of a particular type.

b) Only an *empty* variable may appear as the left member of an assignment involving capabilities. Then bindings statically declared cannot be modified.

c) The set of access rights associated with the left member of an assignment involving capabilities must be a subset of the access rights associated with the right member [Jon 76].

In this way a manager can both bind an empty capability $y : Y \{\alpha, \beta, \ldots\}$ to a specific resource of type Y - resource allocation - and assign the value *nil* to the *empty* capability - resource reclaim. In other words a manager handles the dynamic allocation of resources of a particular type. This type can be either a class or a monitor.

The syntax of a manager definition is similar to that of a monitor or a class definition [Sil 77, Anc 77a, Anc 77b]. The main difference resides in the form of the heading. In fact, together with the identifier of the defined type, the identifier of the resource type handled by the manager is also required.

> *type* <identifier> = *manager of* <resource type identifier>
>
> (formal parameters);
>
> <local declarations> ;
>
> ⋮
>
> <*entry* procedures> ;
>
> ⋮
>
> <local procedures> ;
>
> ⋮
>
> *begin* <initialization> *end*;

Let us resume the most important characteristics of the proposed method for dynamic binding of capabilities to resources:

i) A declared *manager* of resources of type Y is able to modify bindings among capabilities and resources of type Y. Then the manager must have access rights to a certain number n of resources of type Y. For this reason n capabilities of type Y are declared into the body of the manager. In other words any instance of a manager represents a domain with access rights to a certain number of resources of type Y. By means of capabilities assignments a manager can transfer its access rights to other domains. Following the previous restriction c) a manager is able to transfer only its own access rights.

ii) Two *entry* procedures, each of them with an *empty* capability

of type Y as a formal parameter, are declared into each manager handling resources of type Y. The first procedure allocates one of the resources handled by the manager and therefore in that procedure one of the capabilities of type Y is
assigned to the formal parameter. The second procedure deal-
locates the resource by assigning the value *nil* to the formal parameter.

iii) If the capability:

$$var\ y : Y\ \{\alpha, \beta, \ldots\}\ empty$$

is declared into a system type X, then X has a formal parameter whose type corresponds to a manager of resources of type
Y. Each instance x of type X requires the manager to allocate
a resource of type Y in order to operate on it. This request
is done by calling the allocation procedure of the manager.
When the resource is no longer needed, it must be released by
calling the deallocation procedure of the manager. The variable y is passed as an actual parameter to both procedures.
The set $\{\alpha, \beta, \ldots\}$ must be a subset of the access rights associated with the formal parameters of the allocation procedure
of the manager.

In order to guarantee completely controlled accesses to resources at
compilation time, the programmer must specify for each dynamically allocated resource R all the program regions in which R is referred. In
this way the compiler is able to establish, for each access to a resource R, whether the access rights will belong to the domain in which
the requesting process is running.

For this purpose we can use the notation of critical regions introduced by Brinch Hansen [Bri 73] in connection with the mutual exclusion
problem. In fact we can assume that references to a dynamically allocated resource R may appear only within structured statements called
allocation regions. An allocation region is represented by the notation:

$$region\ v\ do\ S;$$

where v is an *empty* capability. The allocation region associates a
statement S with the empty capability v. This notation enables the
compiler to check that empty capabilities are used only inside allocation regions and to place a call to the appropriate allocation procedure before the statement S and a call to the deallocation proce-

dure after S. Obviously these procedures must belong to a manager
handling resources of the same type of the *empty* capability v.

Let the *empty* capability:

$$var \quad v : V \; \{\alpha, \beta, \ldots\} \; empty$$

belong to a domain C. If more instances $m_1, m_2, \ldots, m_k$ of manager handling resources of type V are accessible to C, then for each allocation
region associated with the capability v, the name m_i of the manager
instance, to which requests will be directed, must be passed to the re
gion as a parameter.

$$region \; (m_i) \; v \; do \; S;$$

Since resources may be required in a nested way, nesting of allocation regions must be allowed.

To conclude let us show how the three protection aspects presented in
the previous paragraph are implemented by the proposed mechanism.

    i)   The enforcement rule is implemented by the scope rules of
            the language.

   ii)   The domain binding rule is implemented by the calling proce
            dure mechanism of the language; in fact a domain switching
            takes place every time a procedure entry of an abstract da-
            ta type is called.

  iii)   The transfer rights rule is implemented by introducing in
            the language new features, namely: empty capabilities, man-
            agers and allocation regions. In fact the manager allows the
            dynamic allocation of access rights to those domains in which
            an empty capability was declared. Furthermore, the alloca-
            tion region enforces the correct sequence of request use and
            release of access rights.

CONCLUSION

In this paper we have proposed a general protection mechanism to be
embedded in an object oriented language for real time programming.

Some linguistic features have been introduced to allow compile time
checking of access rights. Moreover dynamic allocation of access rights
to protection domains is permitted.

REFERENCES

[Anc 77a]   Ancilotti,P., Boari,M., Lijtmaer,N. - Dynamic resource man
agement in a language for real time programming - *AICA 77*
Pisa, Italy, October 1977.

[Anc 77b]   Ancilotti,P., Boari,M., Lijtmaer,N. - A mechanism for allocating resources and controlling accesses in languages
for real time programming - *Internal Report n.B77-23*; IEI
CNR, Pisa, Italy, December 1977.

[Bri 73]   Brinch Hansen,P. - *Operating System Principles* - Prentice
Hall, 1973.

[Bri 75]   Brinch Hansen,P. - The programming language Concurrent
Pascal - *IEEE Transac. on Software Engineering*, Vol.SE-1,
n. 2, June 1975.

[Hoa 74]   Hoare,C.A.R. - Monitors: an operating system structuring
concept - *Comm. ACM n. 10*, October 1974.

[Jon 73]   Jones,A.K. - Protection in programmed systems - *Dept. of
Computer Science, Carnegie-Mellon Univ., Pittsburgh* - June 73.

[Jon 76]   Jones,A.K., Liskov,B.H. - A language extension for control
ling access to shared data - *IEEE Transactions on Software
Engineering*, vol. SE-2, December 1976.

[Lam 71]   Lampson,B.W. - Protection - *Proc.Fifth Annual Princeton
Conf. on Information Sciences and Systems*, 1971.

[Lin 76]   Linden,T.A. - Operating System structures to support secu
rity and reliable software - *ACM Computing Surveys*, Dec. 76.

[Lis 75]   Liskov,B., Zilles,S. - Specifications techniques for data
abstractions - *IEEE Trans. on Software Engineering*, Vol. 1,
n. 1, March 1975.

[Sil 77]   Silberschatz,A., Kieburtz,R.B., Bernstein,A. - Extending
Concurrent Pascal to allow dynamic resource management.
IEEE Transactions on Software Engineering, Vol. SE-3, n.3,
May 1977.

[Wir 77]   Wirth,N. - Modula: a Language for modular multiprogramming -
Software-Practice and Experience, 1, 1977.

[Wul 74]   Wulf,W.A. - Toward a language to support structured programs - *Tech. Report Carnegie-Mellon Univ.,Pittsurgh, Pa.*
April 1974.

[Wul 76]   Wulf,W.A., London,R.L., Shaw,M. -.Abstraction and verification in Alphard: introduction to language and methodolo
gy - *Techn. Report, Carnegie-Mellon Univ., Pittsburgh* 1976.

AN EXPERIMENT IN COMPUTER AIDED INFORMATION SYSTEMS DEVELOPMENT

PER AANSTAD

TROND JOHANSEN

GEIR SKYLSTAD

COMPUTING CENTER OF THE UNIVERSITY

OF TRONDHEIM (RUNIT)

NORWAY

ARNE SØLVBERG

DIVISION OF COMPUTING SCIENCE

THE NORWEGIAN INSTITUTE OF TECHNOLOGY

THE UNIVERSITY OF TRONDHEIM

NORWAY

ABSTRACT

During the last decade there has been a growing interest in the use of computers to assist systems analysts and designers in the development of systems specifications and in the implementation of program systems consistent with these specifications. [1, 2].

An experimental software system, CASCADE/2, was developed at the Norwegian Institute of Technology in 1972-74 for this purpose. CASCADE/2 has been used in several software development projects. Experience with three cases is described. Further research and development are discussed both from a short term and long point of view.

1. THE SOFTWARE PROJECT PROBLEM

Development and operation of a software system proceed through phases, commonly called the system life cycle. Starting with a problem definition and analysis of user needs, crude requirement specifications are massaged into an information system design specification. Man/machine interfaces are agreed upon, equipment is ordered, software is produced, testing schemes are developed, manuals are written and users are trained. The resulting software product is put into operation, and the long, tedious process of changes, adjustment, tuning and user-correction proceeds.

The general charateristics of the software engineering project do not differ from those of engineering design projects in genereal. The project staff must be managed and the product specification reviwed and approved. The product must be tested for correctness and quality, costs must be controlled and the product must benefit the buyer.

Beneath the surface, however, there are distinct differences between software engineering and "engineering engineering". The skill and the knowledge of the software development staff are manifested, not in the shape of a physical structure like a bridge, an aeroplane or a telephone switchboard, but in a rather abstract set of rules governing the transfer and creation of data in a computer. This collection of rules, i.e. the computer program and its documentation, is the product.

The behaviour of physical structures is governed by the laws of nature. The behaviour of software products is governed by the laws of computer manufacturers (as manifested in operating systems, compilers, database systems, etc.) and by the behaviour of the persons who interact with the software.

The properties of physical structures can be specified in formal models which are consistens with the laws of nature. Formal calculations may be made to see if a proposed physical structure is feasible and to see if the proposed designs behave properly. If this was not so, our industrial societies would be based on craftmanship alone, not on technical science and craftmanship.

There is, so far, no generally accepted model which capture the relevant features of a software product, both with regard to the behaviour of computer resources, and with regard to the behaviour of the user community in which the software product is used.

There is an increasing volume of research concerned with the problem of information systems modelling. The modelling problem is attacked from different angles, such as data semantics, systems analysis, theorem proving, operating systems modelling, computer performance modelling, human engineering, etc.

In this paper, we shall describe one specific experiment in computer assisted information system development. The project has been carried out at the Norwegian Institute of Technology, Trondheim, Norway, in a joint venture between the univerity computing centre and the computing science division. The research effort started in 1969-1970.

We shall describe the system models which have been used and a systems documentation package (CASCADE/2) which has been developed to support these models [3,4,5]. We shall give an account of some real-life software development projects where CASCADE/2 has been used. Experiences with these experiments will be discussed.

2. MODELS FOR INFORMATION SYSTEMS SPECIFICATION

A complete specification model should provide concepts which enable the software engineer to capture and formally state the relevant features of any information system. The model should be so rich that all life cycle phases are covered, starting with the requirement definition and proceeding through design to implementation, operation and maintenance.

The information systems specifications are used for different purposes in different phases of the life-cycle. During requirement definition the most important use of the specifications is to enhance communication between the systems development staff and the user community so that the project group can get the requirements right before proceeding with implementation. The requirements specification is the information systems design at a crude level. The requirement specification therefore contains the basic design of the software system to be produced. During the subsequent phases of the lifecycle this basic software system specification is enriched by more and more detail. There is a shift en emphasis in the use of the systems specifications, from communicating with the user on the users terms, to communicating with the computer on the computers terms. This shift in emphasis is reflected by a change in the need for modelling concepts. The terminology and structure of programming languages, operating systems, communication networks etc., must be reflected in the systems model, if such a model is to enable the software engineer to specify the relevant properties of his software product.

One basic property of a high quality product is that is consistent with the specifications of that product. A complete information system specification model should provide opportunity for testing the consistency between the detailed software specification and the requirement specification. The modelling concepts of the software specification must be consistent with the modelling concepts of the requirement specification.

We have so far not managed to solve this modelling problem completely. Our models should therefore be regarded as a step in this direction, rather than as a porposal for a final solution.

Our basic modelling concepts are object classes, binary relations between object classes and attributes of object classes. A model is characterized by its object classes, relations and attributes. Three different kinds of models have been used: One software model, one information system model and one organisation model.

An information system is a part of a larger system called the total system. The information system model contains the specification of requirements to the software system. The organization model represents other relevant parts of the total system which might interact with the software system, i.e. the information system environment. The software system model represents an implementation which satisfies the requirement specification.

The organization model and the information system model are developed in collaboration with user representatives and must therefore reflect user terminology. The software model is developed by computing professionals and must reflect computer terminology.

2.1 The Information System Model

The basic object classes of our information system model are

INF — objects representing information which is produced and used in the total system, e.g. transactions, documents, archives.

IPS — objects representing information processing such as production, use, transmission, retrieval of information.

SIG — objects representing the flow of control in the information system, the sequencing of IPS-objects.

The basic model relations are

I — the input relation, $I \subset INF \times IPS$, relating information objects to those IPS-objects which use the information objects.

O — the output relation, $O \subset IPS \times INF$, relating IPS-objects to those INF-objects which are produced by the IPS-object.

C — the component relation,
$C \subset (IPS \times IPS) \cup (INF \times INF)$,
relating an object to its components. The C-relation is used to represent the hierarchical decomposition of processes and information respectively.

N — the entry relation, $N \subset SIG \times IPS$, relating the initiating control signal to the IPS-object to be activated.

X — the exit relation, $X \subset IPS \times SIG$, relating IPS-objects to those control signals which are produced when the IPS-object leave their active state.

The basic information system model is an 8-tuple

(INF, IPS, SIG, I, O, C, N, X)

The 4-tuple (IPS, INF, I, O) represents the flow of information between processes.

The 4-tuple (IPS, SIG, N, X) represents the flow of control between processes.

The tuple   (IPS, C) represents the hierarchical decomposition of processes.

The tuple   (INF, C) represents the hierarchical decomposition of information objects.

The leaf-nodes of an information tree are called TERMs (abbreviation: TR) analogously to the data item concept in database system terminology.

An auxiliary model concept is the information-type concept (abbreviation: INFTY). Information objects of the same INFTY appear several places in the information system description. The type-concepts is used to decrease the amount of writing associated with a systems description by permitting equivalent information structures to share the same structural definition, i.e. be declared to be of the same information type.

2.2 The Organization Model

The information system is part of a larger system, which we call the total system. To enhance the possibility of proper requirement definition the information system should be discussed in the total system context. The organization model is intended to describe those parts of the total system which do not belong to the information system but which are relevant to the requirements definition.

No general organization model, in terms of predefined object classes, relations and attributes, may be prescribed. This is so because the total system characteristics may be very different from case to case. The information system may in one case be a process control system for a chemical reactor, in another case it may be an accounting system in a retail business, or a project planning system in a shipyard.

The organization model is intended to describe the information system environment and must therefore be defined from case to case. The modelling concepts are object classes, binary relations and attributes.

One model that has been used to describe a civil service system <6>, consists of:

Object classes

    LEGISLATION   - laws and ruled that regulate the behaviour of public bureau-
                     cracy.

    ORGANIZATION   - bureaucratic units which are responsible for performing/
                     supervising certain civil service tasks.

    TASK   - functions which the civil service have according to laws
                     and regulations imposed by government/parliament.

Model relations

RESPONSIBILITY ⊂ ORGANIZATION X TASK

a certain organizational unit has responsibility for supervising, controlling, performing certain tasks.

LEGAL RIGHT ⊂ LEGISLATION X (ORGANIZATION U TASK)

an organizational unit exists because of some piece of legislation, a task is to be performed because of some piece of legislation.

COMPONENT ⊂ LEGISLATION X LEGISLATION
U ORGANIZATION X ORGANIZATION
U TASK X TASK

the component relation is used to represent the hierarchical decomposition of legislation, organization and tasks.

The organization model and the information model are interrelated by two binary relations:

SOLUTION ⊂ TASK X IPS

information processing systems represent the solution of tasks defined by legislation.

SUPERVISION ⊂ IPS X ORGANIZATION

an information system is supervised and controlled by an organizational unit.

The organization model and the information system model give the formal framework for requirement specifications development.

2.3 The Software Model

The basic object classes are

PROGRAM — objects representing computer programs.

SUBR — objects representing subroutine-type programs.

FILE — objects representing data files.

RCL — objects representing record classes.

EM — objects representing error messages from PROGRAM-objects.

The basic model relations are

REF ⊂ PROGRAM X SUBR U SUBR X SUBR

which is used to represent the reference structure (subroutine call-structure) in the software system.

MEM⊂ RCL X RCL, the membership relation,

which represent the owner/member relationship in a database network structure.

DBOP = FIND U GET U DELETE U STORE U MODIFY,

the database operation relations,

DBOP ⊂ SUBR X RCL, which represent the kind of database operations the SUBR-objects perform on RCL-objects.

The model contains additional facilities for representing how records consist of data items. The software model facilities also contain an object class called PROC, which represents chunks of declaration statements for subroutines, and one object class INLINE which represents chunks of active statements for subroutines. Object classes PROC and INLINE are related to SUBR-objects by REF-relations.

Several attributes, e.g. program size, number of records, record size, are defined in the software model and are used to represent properties of software objects.

The software model and the information system model are interrelated by the relation.

IMPL⊂ IPS X PROGRAM U INF X FILE which represents how IPS-objects are implemented by programs and how INF-objects are implemented by files.

3. THE SOFTWARE PACKAGE CASCADE/2

The CASCADE/2 program system has functions for systems description, system presentation and computer program generation. CASCADE/2 is designed for interactive use, but can also be operated in batch mode. It was developed for the UNIVAC 1100 series in Fortran IV. A Honeywell Bull 6000 Series version is now also available.

3.1 The most Important Design Criterion was Flexibility

The CASCADE/2 software package was developed to support research in the area of systems analysis and design. A major design criterion has been flexibility, to prevent rigidity in the software support making impossible experiments with new system model propositions.

A system is represented as a set of interrelated objects. Properties of objects and relations are described by attributes. CASCADE/2 has functions for storing, manipulating and presenting system descriptions based on this "object-relations-attribute" model. New kinds of objects, relations and attributed can be introduced ad lib. Consequently, CASCADE/2 is a very flexible tool for the investigations of systems models.

3.2 Free Format Input Language

CASCADE/2 has a free format input command language which reflects the object-relation-attribute concept.

In figure 3-1 is shown a very simple system where a process P (IPS) has A and B as
input related objects (INF) and Q as output related object (INF).

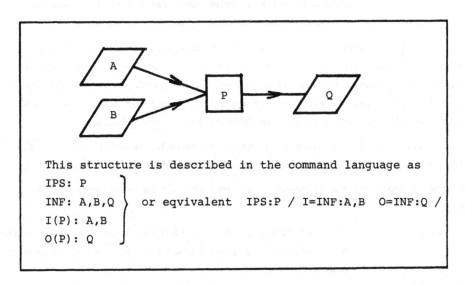

Figure 3-1. A SIMPLE SYSTEM.

In general, the user gives names to the system objects, and indicates to which
object class they belong. Two (or more) objects are related by specifying the name
of the (binary) relationship, and the name of the objects which participate in the
relation.

The standard repertoire of CASCADE/2 contains 49 different object classes, 36 types
of relationships and 10 types of attributes.

Relations are directed and the inverse relationships are automatically maintained
by CASCADE/2.

3.3 The User May Defind His Own Models

CASCADE/2 is designed such that the user is free to define his own models. This is
done by introducing the new object classes, relationships and attributes in the
CASCADE/2 database.

Objects are denoted by different naming systems: By unique (global) names or by
qualified names. Three different naming mechanisms give the user a wide choice in
selecting object names.

3.4 Simple Report Generator

A simple report generator has been developed to present the system description, which is stored in the CASCADE/2 database, in different ways.

The report specification language is closely related to the object-relation-attribute concept. The user specifies the contents of a report by referring to the relevant object classes, relations and attributes. The report formats are lists, cross reference lists, structure lists and tables.

There is also a facility for "navigating" through the stored description. The "route" is specified by a sequence of objects/relations/attributes selections. This navigation facility may be used to answer questions like "Which of the outputs from subsystem A have any influence on system B?".

3.5 Model-dependent Functions

The facilities which are mentioned so far are independent of the actual systems description models which are being used. Some features of CASCADE/2 are model-dependent:

3.5.1 Diagram Generator

An automatic digramming facility is available for system descriptions based on the information system model of chapter 2.1. The user specifies those parts of the description of which he wants a diagram. There is a choice of six different types of diagrams. The diagrams focus attention on different aspects of the structure by ignoring others. Examples of diagrams are systems flowcharts, process flowcharts and activity charts.

3.5.2 Consistency Test

A top-down decomposition means that the same system is described on different levels of detail. If the information system model (chapter 2.1) is used, CASCADE/2 contains facilities for testing the consistency of descriptions on two different levels of detail, as shown in figure 3-2.

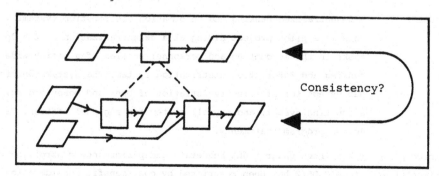

Figure 3-2. CONSISTENCY TEST.

3.5.3 Automatic Software Generation

By using the information system model, a system may be specified to a level of detail where processes are described either by Fortran code or decision tables. From this type of specifications, CASCADE/2 offers the possibility of generating Fortran source code from the detailed requirement specifications. The facility is called APG (= Automatic Program Generator).

4. EXPERIMENTS OF USE IN SOFTWARE PROJECTS

CASCADE/2 has been used in several projects of different character, ranging from projects concerned with organizational analysis and problem definition alone, to projects with the main emphasis on software implementation aspects. We shall in this paper concentrate on application projects at our own organization. This is partly because we know the projects very well, but also because a major emphasis in these projects has been on developing operational software, based on formal requirement specifications. Three software development projects have given especially valuable experience.

4.1 Three Development Projects

The experience presented in this paper are mainly based on the following three software development projects carried out by RUNIT.

BIB-SYS: Automation of university libraries.
This is a multi user, CRT-terminal oriented transaction system. It contains subsystems for document acquisition, cataloguing and borrowing and lending transactions. Ca. 60.000 program statements.

SAPO: An accounting and planning system.
This system was developed for a research organization with project-oriented activities. It contains subsystems for project accounting, invoicing, resource and financial accounting. Ca. 40.000 program statements.

DOLS: A desentralized banking and information distribution system.
This is a pilot project aiming at a decentralization of the dp tasks in the accounting and personnel section of a nation-wide banking and information distribution system. The system contains subsystems for off-line registration of bank and money orders, accounting system, and payroll system for regional offices. Ca. 40.000 program statements.

The three projects have been under tight budgetary and performance control, so the experimental use of CASCADE/2 has been restricted by cost/benefit considerations.

259

The first project was initiated in 1971, the last project in 1975. The CASCADE/2 software has been substantially developed over this period, so there has been different degrees of sophistication in the projects' use of CASCADE/2.

4.2 The Use of CASCADE/2 for Requirement Specification

The use of CASCADE/2 has been somewhat different in the three projects. In the library project CASCADE/2 was to be used for requirement specification purposes only. No software specification facilitites had been developed when the library project started. The Automatic Program Generator (APG) was developed in parallel with the library project. Because APG was not operational at the start of the proramming phase of that project, it was only used to a limited extent. The libraries, represented by librarians with little or no knowledge of data processing, played an active role in the development of the formal requirement specifications.

In the SAPO project, the use of CASCADE/2 was advanced another step towards automatic software production. The crude design specifications were decomposed to the program statement level. The APG facilities were used for software production as a matter of routine.

In the DOLS-project a new software specification model (chapter 2.3) was used. This model reflects computer programming terminology more accurately than the information systems model which was used in the two other projects. The requirement specification in the DOLS project was perceivable, and a manual documentation within the framework of the information systems model was used instead of computer assisted documentation.

4.3 Software Specification in the DOLS-project

The need for computer assisted documentation increased in the programming phase. Up to that point, the amount of "manual" documentation concerning the requirement specification was managable.

More persons were added to the project staff. The number of programs and subroutines increased rapidly. It became necessary to have up-to-date documentation to answer questions like:

- Is the name of the subroutine which is to be programmed used before?

- If I change the layout of a record in a file, which programs are then affected and who is to be informed of this?

- Which subroutine(s) writes a certain error message?

To store, maintain and present this kind of documentation, data about the software must be entered into the CASCADE/2 database: Descriptions of programs, subroutines, record-classes and so on. The project's programming secretary performed this operation.

The reports which were produced by CASCADE/2 supplementd the "manually" produced requirement specification and software specification, which consisted of text and flow diagrams.

Three main groups of report types was produced: overview lists, cross reference reports and deviation reports. A detailed specification of the three groups of reports is as follows:

- Overview lists

    transaction orders, error messages, transaction system tables, source programs, data fields, data field-types, file structure, record class descriptions.

- Cross reference reports between

    subroutines and transaction orders, PROCs and subroutines, record classes and subroutines, data field and record class, data field-type and data fields.

- Deviation reports

    not referenced subroutines, transaction orders not attached to a sub-system, record classes only read (not created/modified), record classes not referenced.

The project library contained an updated set of all CASCADE/2 documentation reports.

5. EXPERIENCES AND GENERAL CONCLUSIONS

Our objective in this research project has been to develop a basis for an integrated software development tool for all phases of the system lifecycle.

We have been reasonably successful in mastering the requirement definition part of the development project. We are not, however, satisfied with our results in the formal interfacing of the requirement specification to the programming and operation phases of the system lifecycle. Even if we can use our computerized documentation package in each separate phase we have not been able to integrate the formal specifications of the different project phases.

Because our research objectives have been ambitious, and because we have very thorougly tested our software support tools, we feel that our experiences can be of general value for further research in this area.

In all of the three reported projects of chapter 4, formal requirement specifications were developed. The requirement specifications were developed in several
steps:

- Systems knowledge was gained through analysis and description of the existing administrative systems in all three cases.

- Decisions were made regarding which parts of the exisiting system to reorganize and automate.

- An overall design of the new organization was made, including specifications
  of response times, man/machine interaction, implementation cost, computer
  system architecture.

So the requirement specification included both organizational design and baseline
software system design.

In the formal specification, main empasis was put on the use of the information
system model of chapter 2. The organizational environment was not formally specified in any of the projects.

5.1 Experiences in Automatic Programming

In two of the projects, basic software system design specifications were detailed
to a program statement level using the information system model. Suitable "chunks"
of code (on the average 15-30 statements for each "chunk") and decision tables were
linked together by the Automatic Program Generator (APG) of CASCADE/2, and program
modules were produced. Computer programs were produced from the design specification. During debugging we found, to nobody's surprise, that programmers want to do
their "firefighting" on the programs where the bugs are discovered. They resisted
going back to the design specification, correcting the bug, producing a program with
a slightly unfamiliar pattern, and starting all over again to get accustomed to the
pattern of the new program.

One basic problem is to keep track of the changes in the software product and the
software specifications at the same time, so that a change in the software product
(e.g. because of debugging), is reflected in the software specifications, and vice
versa. If the software specifications do not reflect the changes of the software
product (i.e. the program system), the value of the specifications are rapidly
so seriously degraded that the benefit of developing a detailed specification may
be seriously questioned when comparing the benefits with the costs involved.

5.2 Separation of Requirement Specification and Software Specification

Because we did not succeed in solving the automatic software generating problem in
our first trial we chose to separate the requirement specification and program
specification in our last project (DOLS). The basic reason for this decision was

that we at that time, did not have any facilities to ensure consistency between the program product and the software specification.

The software system model of chapter 2.3 was used for the program specification and the information system model of chapter 2.1 was used for requirement specification. For the reasons mentioned above, we did not try to relate the two specifications such that the software specification could be formally tested against the require- ment specification for completeness and consistency. The requirement specification therefore tended to serve as a body of basic systems knowledge, rather than a for- malization of system constraints and objectives against which the software could be formally tested.

The use of computer assistance during the requirement definition had consequently to be decided on different criteria than in the SAPO-project. Because the volume of the requirement specification was moderate, and the project staff experienced, the project staff chose to use manual methods for documentation purposes. Computer assistance was, however, used for software specification purposes, where the docu- mentatiom volume is larger.

5.3 Experiences from the Requirement Specification Phase

One of the most pleasant experiences has been the way that the projects have man- aged to get the users involved. Use of the information systems model of chapter 2.1, combined with functional decomposition to handle systems complexity, has had a pro- found diciplining consequence on the project staff and their user contacts and collaborators. Of special value was the hierarchical consistency test of CASCADE/2 (chapter 3.5.2).

The automatic flowcharting facilities of CASCADE/2 proved to be of considerable value in enhancing communication between project staff and users. The data diction- ary facilitites proved to be useful also from a project management point of view, both concerning integration aspects and control of project vocabulary.

Most of the positive effects can be obtained by using manual documentation methods. The decision to use a computer tool like CASCADE/2 is dependent on two points. The first point is the size and complexity of the information system. CASCADE-like tools tend to be more useful for large and complex systems than for small systems, even if computer assistance is used only for requirement definition as a separate task. The second point is the degree of formal integration between the requirement specification and the software product via the software specification. A high de- gree of formal integration means that formal requirement specifications can be used directly for checking the consistency and completeness of software specifi- cations and software products. Consequently the benefits of a thorough requirements specifications might easier outweight the cost of establishing the formal specification.

5.4 Experiences from the Software Specification Phase

We have earlier in this paper pointed out that we did not succeed in establishing a workable automatic programming environment from a cost/benefit point of view. In the DOLS-project, where the software specification was separated from the requirement specification, CASCADE/2 was used mainly as a simple data dictionary system. Our general experience is that the CASCADE/2 produced documentation is valuable for development and especially for maintenance of the software product. Cross reference reports have been very valuable. These give a satisfactory overview of the consequences of specification changes.

Automatic control of consistency between the documentation and the software system is impossible with the present system. It is left to the user to ensure that all alterations of the program code also involve the corresponding updata of the CASCADE/2 documentation. This is a weak point with our present system. To have the full benefit of the software specification, the consistency between product and specification must be maintained. A part of a solution might be to develop a "Data Division" generator such that any change in the datatype specifications would initiate a genereation of new Data Division and a subsequent recompilation of the program which used those Data Divisions. We are aware that some Data Dictionary systems, which are currently marketed, provide this kind of facility.

5.5 Conclusions

CASCADE/2 is a prototype. It was designed to support research primarily in the systems analysis and design phases of the life cycle. We had reasona   success in supporting the development of requirement specifications. Our appetite grew and we tried to use CASCADE/2 for software specification and software production by adding new facilities to our package.

CASCADE/2 is general in the way that it supports a wide variety of systems models. This generality has to be paid for in terms of computational efficiency. If we abandon the idea of integrating the documentation of the different life cycle phases, tailormade software support for each type if specification might bring down computer costs to a more pleasant level. The drawback of aiming at separate specifications is that what is documented in one phase will not be directly usable in subsequent phases except as a general body of knowledge about the system.

There is a considerable cost associated with a formalization of systems specifications. Without short term benefits from the formalized system specification which are comparable to the costs of formalizing the project staff will be reluctant accept the idea of developing formal specifications. Only if a development project is so large that controlling and maintaining the requirement specification becomes

a serious problem will the idea of computer aided systems development be appealing
to the project management.

Despite the difficulties we have mentioned in this paper, we want to point out that
the production rate in the projects which have used CASCADE/2, measured in lines
of code/man year, competes very favourably with production rates reported in the
literature [7]. What we do not know is if this should be attributed to the use
of CASCADE/2, to the use of formal systems models, to high competence in the project staff or to a combination of these factors.

We also want to make explicit, that incompetent project management can blow any
project, despite the quality of development techniques which are used.

We have in this research project shown that information systems specifications can
be formalized, that the formal specifications can be handled by a computer and that
there is some benefit associated with doing so.

There is room for substantial improvements in this technique. One of the major
long term problems will be to interface the different kinds of specifications to
the software product such that automatic programming is realized on a practical
scale and such that product changes are automatically reflected at the software-
and requirement-specification level.

The alternative of developing tailormade software tools for each phase seen in isolation has already been mentioned.

There is a lack of theoretical knowledge in the field of systems development. We
believe that a substantial improvement of the state-of-the-art is dependent on a
solution of some problems which still belong to basic research. We still lack a
workable definition of the concept of information. We also have problems concerning
concepts like consistency, completeness and flexibility, just to mention a few.
A theory of software design still seems to be quite far away.

The state-of-the-art of information systems development, based on formal specification technique, is dependent on the level of knowledge of these subproblems.
When comparing with the progress in the solution of important subproblems over
the last few years, we are convinced that our general framework for computer aided
information systems development will prove to be valid in the years to come.

265

REFERENCES

[1] 2nd International Conference on Software Engineering, 13-15 Oct. 1976.
San Fransisco, California, Proceedings.

[2] Bubenko, Langefors, Sølvberg (eds):
"Computer Aided Information Systems Analysis and Design",
Lund, Sweden, 1971.

[3] Aanstad, Skylstad, Sølvberg:
"CASCADE - A Computer Based Documentation System", In [2]

[4] Auglænd, Sølvberg:
"A Technique for Computerized Graphical Presentation of Information
Systems to be Used in Systems Design",
In: "Approaches to Systems Design", National Computing Center, England.

[5] Sølvberg:
"The Use of Models and Associated Software in the Design of Wicked Systems"
In: Grochla, Szyperski: "Information Systems and Organizational Structure",
de'Gruyter, Berlin, 1975.

[6] Fredriksen, K:
"Brukererfaringer CASCADE",
In: Proceedings, Nord Data 75, Oslo 23-26 juni 1975, (in Norwegian)

[7] C.E. Walston, C.P. Felix:
"A Method of Programming Measurement and Estimation",
IBM Systems Journal, No. 1, 1977, pp 54-72.

THE DESIGN OF RELATIONAL INFORMATION SYSTEM

ACCORDING TO DIFFERENT KINDS OF DEPENDENCIES

C. DELOBEL[1] & E. PICHAT[2]

ABSTRACT

The purpose of this paper is to present

(a) a survey of different approaches proposed for the conceptual design of logical schemas for relational data base systems,
    and
(b) a new approach based upon the study of the decomposition structure of a relation which ensures the complete joinability of data.

(1) – *Laboratoire I.M.A.G., Université Scientifique et Médicale de Grenoble, B.P. 53, 38041 GRENOBLE Cedex (France).*

(2) – *Institut d'Informatique d'Entreprise, C.N.A.M., 292 Rue Saint Martin, 75141 PARIS Cedex 03 (France).*

1. INTRODUCTION

The purpose of this paper is to present
- (a) a survey of different approaches proposed for the conceptual design of logical schemas for relational data base systems, and
- (b) a new approach based upon the study of the decomposition structure of a relation which ensures the complete joinability of data.

In the design of data base schema there are basically two approaches : the decomposition and the synthetic approach.

The decomposition approach has been proposed by Codd [COD 70] [COD 71] in the presentation of the relational data model. In this model, the concept of functional dependency is an important element when one is considering how to group attributes to form relations. The properties of functional dependencies, studied in [BOI 69], [COD 71], [DEL 71], [DC 73], [ARM 74], are the basic elements used to define the three CODD's normal forms and the normalization process. This normalization process can be viewed as the decomposition of a relation into two sub-relations (or more), so that the original one can be regenerated by the composition operation of the two sub-relations.

The synthetic approach attacked the same problem. The objective is that a designer should synthetize the data base schema algorithmically from a given set of semantic properties of data.

The first attempts based only on functional dependencies [DC 73], [WW 75], [SS 75] led to some imperfections and do not guarantee that the relations are in CODD's third normal form. The work of Bernstein [BER 75], [BER 77] develops an effective procedure for synthesizing relations satisfying CODD's third normal form. A similar approach is proposed in [LR 76], [LRP 77] with efficient algorithms to derive both the closure and a minimum covering set of functional dependencies by using the equivalence between the operations on functional relationships and the operations in boolean algebra.

If the functional relationship concept plays probably the most important role in the process of defining relational schema, it is not the only one. From an idea contained in [BOI 69] we have defined [DL 74] another concept : the first order hierarchical decomposition which allows to decompose relations independently of functional relationships. More recently, FAGIN [FAG 76] and ZANIOLO [ZAN 76], [ZAM 77] have introduced the notion of multivalued dependency which includes, as a special case, the functional relationship.

The following section, section 2 of the paper, reviews the definition of a relation and the condition for a relation to be decomposable.

Section 3 presents the different types of constraints describing a relation :

functional dependency, multivalued dependency, first order hierarchical decomposition. The properties of this type of constraints are reviewed and their role in the process design is illustrated through examples.

In section 4, the basic properties of the decompositions of a relation are studied and it is shown how it is possible, from these properties, to develop a new approach for the design of logical schemas. In this approach, the concept of maximal decompositions for a family of decompositions is an important element of the process.

2. THE RELATIONAL MODEL

The intent of this section is to review some definitions. Familiarity with the fundamental concepts of the relational model as presented in [COD 70] and [BER 77], and especially with the concept of functional dependency in [ARM 74] is assumed. We use throughout the section the terminology found in [BFH 77].

2.1. Relations and constraints

In the database terminology, the word relation is sometimes confusing because it is used to denote a set of tuples and a structural description of sets of tuples. It is important to make the distinction clear: when we talk about set of tuples we shall use the word relation and in the other case we shall use the word relational schema. With this approach a relation is an instance of a relational schema.

Attributes are identifiers taken from a finite set $A_1, A_2, \ldots, A_n$. Each attribute A_i has associated with it a domain, denoted by $DOM(A_i)$, which is the set of possible values for that attribute. We shall use the letters A, B, ..., for single attributes and the letters X, Y, ..., for sets of attributes.

An X-*value* is an assignment of values to the attributes of X from their domains.

Also, if X and Y are sets of attributes (not necessarily disjoint), then we write $X \cup Y$, $X \cap Y$ for the union and the intersection of X and Y, but if X and Y are disjoint, we shall use (X,Y) for the union.

A *relation* on the set of attributes $\{A_1, A_2, \ldots, A_n\}$ is a subset of the cartesian product $DOM(A_1) \times DOM(A_2) \times \ldots \times DOM(A_n)$. The elements of the relation are called *tuples*. A relation R on $\{A_1, A_2, \ldots, A_n\}$ will be denoted $R(A_1, A_2, \ldots, A_n)$. If one wants to distinguish among the attributes the sets of disjoint attributes. X and Y, we shall use the notation R(X,Y,...). In this case if $X = \{A_1, A_2\}$ and $Y = \{A_3, A_4, A_5\}$ the tuple (x,y,...) stands for $(a_1, a_2, a_3, a_4, a_5, \ldots)$.

A *constraint* involving the set of attributes $A_1, A_2, \ldots, A_n$ is a predicate on the collection of all relations on this set. A relation $R(A_1, A_2, \ldots, A_n)$ obeys the constraint if the value of the predicate for R is "true". A constraint is defined by giving a notation or a language for expressing it and the condition under which a relation obeys it.

A constraint can be seen as an intrinsic property of the data; for example, suppose that parts in an inventory are described by a relation R(PART-NUMBER, COLOR, PRICE, ...). A priori, any relation of this form can exist in the database. However if one specifies a constraint : the PRICE must range between $0 and $100 a piece, then only relations in which this constraint is valid can exist in the database. Similarly the specification that the knowledge of the PART-NUMBER-value implies the COLOR-value is also another type of constraint. In this case, this type of constraint is called a functional dependency and denoted PART-NUMBER → COLOR.

In this paper, we are investigating only some constraints : functional dependency, multivalued dependency, first order hierarchical decomposition which are presented in section 3.

2.2. Operations on relations

In his original presentation of the relation model [COD 70], Codd introduced the relational algebra as a data manipulation language. There are two basic operations that will be of some interest of us : projection and natural join.

The *projection* of a relation R(X,Y,Z) over a set of attributes X is the restriction of R over the attributes in X; this operation will be denoted R[X], and defined by R[X] = {x|∃y ∃z : (x,y,z) ∈ R}.

As a special case of projection we denote by R[x,Y] the projection of R over Y from an X-value x, R[x,Y] = {y|∃z : (x,y,z) ∈ R}.

The *natural join* operation is used to make a connection between attributes that appear in different relations. Let R(X,Y) and S(X,Z) be two relations: then the natural join R∗S is the set of {(x,y,z)|(x,y) ∈ R and (x,z) ∈ S}; R∗S is a relation defined over the attributes {X,Y,Z}.

2.3. Decomposition of a relation

Let R(X,Y,Z) be a relation ; we shall say that R is *decomposable* if there exists two relations S and T such that :
      (a) S and T are projections of R : S = R[X,Y], T = R[X,Z]
      (b) the natural join of S and T is R : R = S∗T.

In other way, we shall say that the pair ((X,Y), (X,Z)) constitutes a *decomposition* of R.

This concept of decomposition is very important in the process for designing a database schema, because in place of storing the relation R in the database, we can store only the projections of R. The purpose of this paper is to study under what conditions we can decompose a relation.

First, we shall recall basic characterizations for a relation to be decomposable ; the proofs have been given elsewhere.

Proposition 1 : R(X,Y,Z) is decomposable iff for all X-values x which are in R
$$R[x,Y,Z] = R[x,Y] \times R[x,Z]$$
where the operation × denotes the cartesian product.

<u>Proposition 2</u> : R(X,Y,Z) is decomposable iff for all X-value and Y-value which are in R :

$$R[x,y,Z] = R[x,Z].$$

<u>Example 1</u> : Let R(BOOK,CLASS,STUDENT,PROFESSOR) be a relation where the interpretation can be given intuitively as follows. Each CLASS has various STUDENTs, but a STUDENT is in one CLASS, a CLASS has one PROFESSOR. Each CLASS has a given set of BOOKS as reference's books. A sample of the relation is given by the table below.

R :	BOOK	CLASS	STUDENT	PROFESSOR
	b_1	math	peter	mike
	b_1	math	john	mike
	b_3	programming	jane	mike
	b_2	math	peter	mike
	b_3	programming	james	mike
	b_2	math	john	mike

One can notice that by application of proposition 1, the relation R is decomposable into two relations S = R[BOOK,CLASS] and T = R[CLASS,STUDENT,PROFESSOR].

S :	BOOK	CLASS
	b_1	math
	b_2	math
	b_3	programming

T :	CLASS	STUDENT	PROFESSOR
	math	peter	mike
	math	john	mike
	programming	jane	mike
	programming	james	mike

At the present time, we only make a constatation that relation R is decomposable ; we shall see in the next chapter that the decomposition property is related to the type of constraints.

3. THE TYPE OF CONSTRAINTS

3.1. Functional dependencies

3.1.1. Definition

Functional dependencies form a family of constraints. The properties of functional dependencies have been studied extensively in [DC 73], [ARM 74], [BER 77] ; therefore we recall here only the definition.

A *functional dependency* (abbr. FD) is a sentence denoted f : X → Y where f is the name of the functional dependency and X and Y are sets of attributes. A functional dependency f : X → Y holds in R(U) where X and Y are subsets of U, if for every tuple u and v of R u[X] = v[X]implies u[Y] = v[Y] (u[X] denotes the projection of the tuple u on X). According to the definition f can be seen as the unique application from R[X] to R[Y], therefore we can omit the name of the FD and write only X → Y.

A *full functional dependency* X → Y is an FD such that there exists no proper subset X' ⊂ X with X' → Y.

We can relate the concept of functional dependency to the concept of decomposition according to proposition 3 which is easy to prove.

Proposition 3 : If R(X,Y,Z) is a relation such that functional dependency X → Y holds then R is decomposable and we have R = R[X,Y] * R[X,Z].

The proposition 3 is the first proposition which establishes an association between a constraint and the decomposition property.

3.1.2. Properties of functional dependencies

The properties of FD's are important in the design of relational schemas and we shall use them later. These properties can be seen as inference rules for FDs, that is rules that deal with the implication of new FDs from a given set of FDs.

Let R(U) be a relation defined over a set U of attributes, and X,Y,Z,W be subsets of U.

F1 - Reflexivity : if X ⊇ Y then X → Y
F2 - Augmentation : if X → Y, then for all Z   X ∪ Z → Y ∪ Z
F3 - Transitivity : if X → Y and Y → Z then X → Z
F4 - Pseudo-transitivity : if X → Y and Y ∪ Z → W then X ∪ Z → W
F5 - Additivity : if X → Y and X → Z then X → Y ∪ Z.

Example 2 : We can define for the relation R given in example 1, the FDs according
to the relationships properties between attributes

STUDENT → CLASS

CLASS → PROFESSOR.

According to F3 we can derive STUDENT → PROFESSOR by transitivity.

3.1.3. Decomposition based upon functional dependencies

It is well known that if someone adopts for the relational schema the relation R only,
this relation suffers from a lot of anomalies. These anomalies occur when one wants
make an update, insertion or delete operation. The problem of designing relational
schemas is equivalent to replace the relation R by a set of relations which give the
same information without destroying the relationships between attributes. This
problem is equivalent to find for the relation R a decomposition such each relation
in the decomposition contains less anomalies than the original one. This problem
has been, first, identified by CODD and leads him to the definitions of normal's
form of a relation. Two main approaches have been recognized as valuable to produce
suitable normal forms : the decomposition approach and the synthetic approach related with the irredundant covering technique.

The decomposition approach consists to apply successively the proposition 3 to different functional dependencies. For example, the application of STUDENT → CLASS in
example 1 gives the decomposition of the relation R defined over the attributes
STUDENT, CLASS, PROFESSOR, BOOK into two relations R1 defined over the attributes
STUDENT, CLASS and R2 defined over STUDENT, PROFESSOR, BOOK. It is still again possible to decompose R2, because the FD STUDENT → PROFESSOR holds also in R2, into
two relations R21 defined over STUDENT, PROFESSOR and R22 defined over STUDENT,
BOOK.

The overall decomposition process can be represented by a tree as the one shown in
figure 1. The terminal nodes describe the attribute sets of each relation obtained
at the end of the decomposition process. To each non-terminal node we can associate
an attribute set equal to the union of the attribute sets of their successors. Each
non-terminal node corresponds a decomposition step which is labelled according to
a FD.

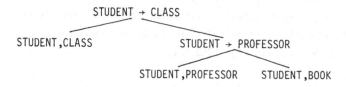

Fig. 1 - A decomposition of a relation

The difficulty of this approach is the possibility of multiple decompositions. It may be seen that the relation R has in addition the following decomposition as shown by figure 2.

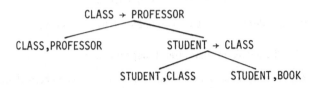

Fig. 2 - A decomposition of a relation

These two decompositions possess an anomaly because both they generate the relation R[STUDENT,BOOK]. The meaning of this relationship between the attributes is : s ∈ STUDENT and b ∈ BOOK are related if there exists a CLASS c such the STUDENT s is in the CLASS c and the BOOK b is used by the CLASS c. In other way one can say that STUDENT and BOOK are not directly related.

The irredundant cover technique is more systematic. It proceeds in different steps. The first step is to construct the closure and an irredundant cover.

Let F be a given set of FDs, the *closure* of F denoted F^+ is the set of FDs which can be derived from F through the inference rules of FDs.

A *cover* C is a set of FDs from which all others can be derived, that means that : $C^+ = F^+$. An *irredundant cover* F_m is a set of FD$_s$ such the closure of F_m is equal to closure of F, i.e $(F_m)^+ = F^+$, and there exists no proper subset F' of F such $(F')^+ = F^+$.

The problem of obtaining an efficient irredundant cover algorithm has received wide attention [DC 73], [LR 76], [LRP 77].

As we can see from our previous example 2,the irredundant cover of F = {STUDENT→CLASS, CLASS → PROFESSOR} is equal to F. So, the two relations built from the two FDs in the irredundant cover, say R1 defined over STUDENT, CLASS and R2 defined over CLASS, PROFESSOR,are not enough to regenerate R.

The second step is to synthetize valuable relational schemas from an irredundant cover set of FDs regarded as semantic primitives. Different algorithms have been proposed to obtain optimal set of relations in third normal form from an irredundant cover [BER 75], [BER 77], [FLO 77], [OSB 77].

3.2. Multivalued dependencies

3.2.1. Definition

In a functional dependency X → Y, the knowledge of the X-value determines a unique

Y-value. In a multivalued dependency the X-value determines a set of Y-values. According to the definition given by Fagin [FAG 76] and Zaniolo [ZAN 76], a *multivalued dependency* (abbr. MVD) is a sentence denoted $g : X \twoheadrightarrow Y$ where g is the name of the multivalued dependency and X and Y are sets of attributes. A multivalued dependency $g : X \twoheadrightarrow Y$ holds in R(X,Y,Z), where X,Y,Z are disjoint sets of attributes, if for every X,Z-value, (x,z) that appears in R, we have

$$R[x,z,Y] = R[x,Y].$$

As we do for FDs, here also we usually omit the name g and write only $X \twoheadrightarrow Y$.

One can notice that the definition of an MVD uses the same condition as proposition 2.

Example 3 : For the relation R defined in example 1, one can notice that the MVD CLASS $\twoheadrightarrow$ BOOK holds in R, because for every CLASS, STUDENT, PROFESSOR-value (c, t, p) that appears in R we have :

$$R[c,t,p,BOOK] = R[c,BOOK].$$

Proposition 4 : The relation R(X,Y,Z) obeys the MVD $X \twoheadrightarrow Y$ iff R is decomposable into two parts R[X,Y] and R[X,Z].

This proposition is a direct consequence of the definition and proposition 2.

$\square$

The decomposition given in example 1 is based upon proposition 4 applied to the MVD CLASS $\twoheadrightarrow$ BOOK.

In the definition of MVD given here, as in Fagin's paper [FAG 76], we require the left and right sides of an MVD be disjoint. There are two reasons for that : first, a transitivity property does not always hold if the restriction is lifted, and second, in all practical situations a database designer will define this type of constraint with disjoint sets of attributes. In [BFH 77] the reader can find a complete study of MVD properties; nevertheless we shall recall the basic properties without proof.

3.2.2. Properties of MVDs

The properties of MVDs are very similar to FDs. In this subsection we discuss inference rules for MVDs, that is rules that deal with the implication of new MVDs from a given set of MVDs.

Let R(U) be a relation defined over a set U of attributes, where X,Y,Z are disjoint sets of attributes contained in U.

MO reflexivity : if X ⊇ Y then X ↠ Y

M1 complementation : if X ↠ Y then X ↠ U-(X ∪ Y)

We have to note that this complementation rule has no equivalent for the FDs.

M2 augmentation : if X ↠ Y then for all Z then X ∪ Z ↠ Y ∪ Z

M3 transitivity : if X ↠ Y and Y ↠ Z then X ↠ Z

M4 pseudo-transitivity : if X ↠ Y and Y,Z ↠ W then X,Z ↠ W

M5 additivity : if X ↠ Y and X ↠ Z then X ↠ Y ∪ Z

M6 decomposition : if X ↠ Y and X ↠ Z then X ↠ Y ∩ Z.

3.2.3. Decomposition of relation with MVDs

We have seen in section 3.1.3 that the decomposition of a relation R structured only by FDs can lead to some anomalies. Now we can repeat the decomposition process of the relation.

R(BOOK, STUDENT, CLASS, PROFESSOR) structured by the FDs

STUDENT → CLASS, CLASS → PROFESSOR and the MVD CLASS ↠ BOOK.

It is easy to obtain the decomposition illustrated by figure 3.

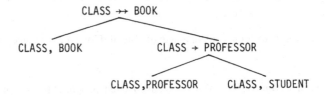

Figure 3

We can see that in this decomposition (figure 3), we obtain a relation where the attributes CLASS and BOOK are directly related.

3.2.4. MVDs defined on a subset of attributes

Sometimes it is possible to interpret the concept of MVDs as an extension of the concept of FDs by saying that the MVD X ↠ Y holds in R means that the X-value determines a unique set of Y-value. This interpretation can be misleading. This fact is more easily clarified by an example.

Example 4 : Consider the following relation R over the attributes BOOK, CLASS, STUDENT, PROFESSOR with a different semantic from example 1. Each CLASS has different STUDENTs and PROFESSORs but a STUDENT is only in one CLASS. The BOOK is determined by a PROFESSOR and a CLASS. Then we have the FDs : STUDENT → CLASS and CLASS, PROFESSOR → BOOK.

An instance of the relational schema is given by the table below.

R :	BOOK	CLASS	STUDENT	PROFESSOR
b_1	math	peter	mike
b_1	math	john	mike
b_2	programming	jane	mike
b_2	math	peter	ronald
b_2	math	john	ronald
b_2	programming	james	mike

In the relation R given in example 4, the knowledge of CLASS-value determines the
set of PROFESSOR-value, but is not true that CLASS $\twoheadrightarrow$ PROFESSOR because :

$$R[math, PROFESSOR] = \{mike, ronald\}$$
$$R[b1, math, peter, PROFESSOR] = \{mike\}.$$

This is due to the fact that the validity of the MVD, $X \twoheadrightarrow Y$ in $R(X,Y,Z)$ depends not
only on the values of X and Y, but also on the values of Z. It is possible that
$X \twoheadrightarrow Y$ is not valid in $R(X,Y,Z)$, but that $X \twoheadrightarrow Y$ is valid in the projection
$R[X,Y,Z']$ where Z' is a subset of Z.

For example, the constraint CLASS $\twoheadrightarrow$ PROFESSOR is only valid on R[CLASS, STUDENT,
PROFESSOR]. Furthermore we can interpret this constraint as "all the students
attending a class are taught by all instructors".

One can check that the MVD CLASS $\twoheadrightarrow$ STUDENT is valid in R ; then the relation R
is structured by the FDs : STUDENT $\rightarrow$ CLASS, CLASS,PROFESSOR $\rightarrow$ BOOK and the MVDs
CLASS $\twoheadrightarrow$ STUDENT and CLASS $\twoheadrightarrow$ PROFESSOR which are only valid on the set of attributes CLASS, PROFESSOR, STUDENT.

To capture this new type of constraints, Fagin calls this case *embedded multivalued
dependencies*. To denote an embedded MVD we need to specify precisely the sets of attributes which occur.

This concept of embedded multivalued dependencies is similar to the concept of first
order hierarchical decomposition which has been previously defined in [DL 74] and
which is more general than the concept of MVDs.

3.3. First order hierarchical decomposition

3.3.1. Definition

A *first order hierarchical decomposition* (abbr. FOHD) [DEL 77] is a sentence denoted
$X : Y_1|Y_2|\ldots|Y_k$ where $X,Y_1,Y_2,\ldots,Y_k$ are disjoint sets of attributes.

A relation $R(X,Y_1,Y_2,\ldots,Y_k,W)$ where $X,Y_1,Y_2,\ldots,Y_k,W$ are disjoint sets of attributes, obeys the FOHD if for every X-value we have
$R[x,Y_1,Y_2,\ldots,Y_k] = R[x,Y_1] \times R[x,Y_2] \times \ldots \times R[x,Y_k].$

According to proposition 1, it is easy to prove that this condition is equivalent to the condition :

$$R[X,Y_1,Y_2,\ldots,Y_k] = R[X,Y_1] \star R[X,Y_2] \star \ldots \star R[X,Y_k]$$

which expresses the decomposability of the projection R over $\{X,Y_1,\ldots,Y_k\}$ into k projections.

The definition of FOHD given here is more general than the definition of decomposition that we have introduced in section 2.3.

In the FOHD, $X : Y_1|Y_2|\ldots|Y_k$, X is called the *root segment*, $Y_1,Y_2,\ldots,Y_k$ are called *segments* and the pair (X,Y_i) is called a *branch*. We shall say that a FOHD is a *full first order hierarchical decomposition* or a *generalized multivalued decomposition* (abbr. GMVD) if W is the empty set. We shall denote a GMVD as $X \twoheadrightarrow Y_1|Y_2|\ldots|Y_k$. In the case where the root X is empty, this means that the projection of R over the set of attributes $\{Y_1,Y_2,\ldots,Y_k\}$ is the cartesian product of the projections of R respectively on $Y_1,Y_2,\ldots,$ and Y_k.

For the relation R given in example 4, the FOHD

CLASS : STUDENT|PROFESSOR

holds. We have also seen that the MVD CLASS $\twoheadrightarrow$ STUDENT holds in R, thus proposition 4 can be applied to give the FOHD :

CLASS : PROFESSOR,BOOK|STUDENT.

As one can see, the definitions of FOHDs and MVDs are two facets of the same problem : under what conditions is a relation decomposable ? Effectively, each time that a constraint on a relation can be paraphrased by a sentence like "all the students attending a seminar are taught by all instructors", we have a constraint of the family of FOHDs, and this sentence refers to the condition of proposition 1. On the other hand, a sentence like "the type of seminar determines uniquely the set of books" can generate a constraint of the family of MVDs, and this sentence refers to the condition of proposition 2.

From the definition of FOHD and MVD, it is clear that we have the following relationships : a GMVD represents a set of MVDs, for example if $X \twoheadrightarrow Y|Z|U$ holds in $R(X,Y,Z,U)$ then we have $X \twoheadrightarrow Y$, $X \twoheadrightarrow Z$ and $X \twoheadrightarrow U$ and the converse. Nevertheless if $X \twoheadrightarrow Y$ holds in $R(X,Y,Z,U)$ we have $X \twoheadrightarrow Y|Z,U$. And if $X : Y|Z$ holds in $R(X,Y,Z,U)$ we cannot deduce any MVD.

3.3.2. Properties of FOHDs

In this section, we discuss inference properties for FOHDs from a given set of FOHDs. The properties are only reviewed here since they have been presented in [DEL 77]. Throughout this section, we assume that the relation R is defined over a given disjoint set of attributes $X,Y_1,Y_2,\ldots,Y_k,W$.

H0 permutation : if $X : Y_1|Y_2|...|Y_k$ holds in R and π a permutation of $1,2,...,k$, then $X : Y_{\pi(1)}|Y_{\pi(2)}|...|Y_{\pi(k)}$ holds in R.

H1 addition of branches : if $X : Y_1|Y_2|...|Y_k$ holds in R then $X : Y_1 \cup Y_2|...|Y_k$ holds also in R.

H2 deletion of a branch : if $X : Y_1|Y_2|...|Y_k$ holds in R then $X : Y_2|...|Y_k$ holds in R.

H3 projection of a branch : if $X : Y_1|Y_2|...|Y_k$ holds in R then $X : Y_1'|Y_2|...|Y_k$ holds in R where Y_1' is any subset of Y_1.

H4 root augmentation : if $X : Y_1|Y_2|...|Y_k$ holds in R then $X \cup Y_1' : Y_1-Y_1'|Y_2|...|Y_k$ holds in R, where Y_1' is any subset of Y_1.

H5 projection of the root and decomposition of a branch :
if $X : Y_1|Y_2 \cup Y_3|...|Y_k$ and $X \cup Y_1 : Y_2|Y_3$ holds in R then
$X : Y_1|Y_2|Y_3|...|Y_k$ holds in R.

H6 decomposition : if $X : Y_1|Y_2|...|Y_k$ and $X : Z_1|Z_2|...|Z_r$ hold in R with
$Y_1 \cap Z_1 \neq \emptyset$ and $Y_1 \cap Z_2 \neq \emptyset$ then $X : Y_1 \cap Z_1|Y_1 \cap Z_2|Y_2|...|Y_k$ holds in R.

3.4. Axiomatization for FDs and MVDs

Armstrong [ARM 74] proved that the properties of reflexivity, augmentation and transitivity constitute a complete set of inference rules for FDs. The completeness theorem can be stated as follows : for every set F of FDs there exists a relation R such that the FDs which are valid in R is exactly F^+.

This property is only valid for the FDs ; we now turn our attention to a more general problem. What could be a set of inference rules for a set F of FDs and a set M of MVDs ? In the previous section, we have seen properties for MVDs but these properties are not enough to define a complete axiomatization since certain combinations of FDs and MVDs imply additional FDs that cannot be derived by the use of the above rules. We need to introduce three additional properties :

FM1 : if $X \rightarrow Y$ then $X \twoheadrightarrow Y$

FM2 : if $X \twoheadrightarrow Y$ and $W \rightarrow Z$ where $Y \supseteq W$ with $Y \cap Z = \emptyset$ then $X \rightarrow Z$.

Berri and al [BFH 77] proved that the rules $\{F1,F2,F3,M1,M2,M3,FM1,FM2\}$ are a complete set of inference rules for the family of functional and multivalued dependencies. The completeness theorem can be stated as follows : for every set F of FDs and M of MVDs there exists a relation R such that the set of FDs and MVDs valid in R is exactly $(F \cup M)^+$.

The problem of determining a complete set of inference rules for FOHDs is an open research problem. It has been solved only in the special case of full FOHD.

4. DECOMPOSITION PROCESS USING DECOMPOSITION STRUCTURE

4.1. General analysis of the decomposition approach

The different approaches for the design of logical schemas presented in section 3 are based on the properties of FDs and MVDs. In all of these approaches, the process starts with a list of dependencies given by the designer. From a designer view point, it is not necessary to consider all the given dependencies.An irredundant cover is adequate for this purpose, since the remaining dependencies are derivable by the inference rules. In the previous section (section 3.4), we have remembered that we have to consider 8 basic inference rules.

Instead to look at the dependencies, there exists an alternative approach where the family of decompositions of a relation is directly studied. This approach presents the following advantages :
- to study the decomposition properties of a relation we need only 4 basic inference rules
- the family of decompositions for the relation R can be ordered by an order relation in order to simplify the designer's task for finding a minimal cover. To achieve this goal, first we shall define a *-operation that generates from two decompositions a new one, and second to avoid the multiplicity of decompositions we introduce an order relation denoted $\subseteq$ and according to which the *-operation is isotonous. This property is very important when one is looking for a minimal cover
- to consider only decompositions which ensures the complete joinability of data. A similar idea has been introduced by Rissanen [RIS 77] with the notion of independent component and by Zaniolo and Melkanoff [ZAM 77] with the concept of complete relatability.

4.2. The decomposition of a relation

Let R be a relation over U. A *decomposition* of R is a pair (X,Y) of subsets of U with $X \cup Y = U$ such that

$$R = R[X] \star R[Y]$$

A decomposition can be considered as a GMVD : $X \cap Y \twoheadrightarrow X-Y|Y-X$ with two branches. By similarities the sets X-Y and Y-X will be called the segments of the decomposition and $X \cap Y$ the root segment.

Using the definition, Armstrong and Delobel [AD 77] proved that the following inference rules (or axioms) for decompositions are complete.

D1 empty decomposition : $(\emptyset,U)$ holds in R
D2 symetry : if (X,Y) holds in R then (Y,X)
D3 augmentation : if (X,Y) holds in R and $Z \subseteq U$ then $(X \cup Z, Y \cup Z)$

D4 intersection : if (X,Y) and (Z,W) hold in R with Z ∩ W = Y then (X ∩ Z, W).

The completeness theorem can be stated as follows : for every set D of decompositions there exists a relation R such that the decompositions which are valid in R are exactly D^+. The very important point with this axiomatization is that it captures directly the decomposition structure of the relation R.

We introduced now some additional rules that we have found to be useful for the manipulation of decomposition :

D5 absorption : if X ⊆ U then (X,U) holds in R
D6   : if (X,Y) and (Z,W) hold in R with Z ∩ W ⊆ Y then
          (X ∩ (Y ∪ Z), Y ∪ W) holds also in R.

The proofs are given in [AD 77] and they are direct consequence of properties D1-D4.

4.3. Analysis of an initial set of decompositions

To study the decomposition structure of the relation R we need at the initial stage of the process to have a set of decompositions. This can be done only by the designer in interpreting, from his understanding of the case at hand, the dependencies in terms of decompositions. The following example will explain the initial stage.

Example 5 : Let R be a relation over the attributes Student, Class, Professor, Book and Hour. On an intuitive background the relation R expresses a time table. A student is only in one class. A class has different students and professors. A professor at one hour is teaching to only one class. A class uses different books. Then we can consider that the relation R is structured by the following set F of dependencies (FDs and MVDs), F = {f1,f2,f3,f4}

f1   S → C
f2   C ↠ S
f3   C ↠ B
f4   H,P → C

By application of proposition 3 successively to f1-f4, we derive the following decompositions d1-d4 :

d1   S ↠ C|B,P,H
d2   C ↠ S|B,P,H
d3   C ↠ B|S,H,P
d4   H,P ↠ C|S,B

It is from d1, d2, d3 and d4 that the designer will study the various decompositions of the relation R.

We now present an effective methodology for the design of relational schema using the decomposition approach. The different stages of this methodology are :
- starting from an initial set of decompositions, derive all the possible decompositions
- construct a minimal cover for the initial set of decompositions
- select from the minimal cover a decomposition that ensures complete joinability of data.

4.4. An algorithm for finding the maximal decompositions

4.4.1. The ∗-operation

The reference rule D4 for decomposition can be generalized into two directions by deleting the condition and by considering a decomposition in more than two branches. This can be done by considering the proposition 5 applied to GMVDs.

Proposition 5 : If $Y \twoheadrightarrow B_1|B_2|\ldots|B_J$ and $Z \twoheadrightarrow A_1|A_2|\ldots|A_I$ hold in R then :

(1) $Y \cup (B_k \cap Z) \twoheadrightarrow B_1|B_2|\ldots|B_{k-1}|B_k \cap A_1|B_k \cap A_2|\ldots|B_k \cap A_I|B_{k+1}|\ldots|B_J$,
with $k \leq J$, holds also in R.

Proof :
By setting $B_k \cap Z = C$, $B_k \cap A_1 = C_1$, ..., $B_k \cap A_I = C_I$ then we have to prove that :

(2) $Y \cup C \twoheadrightarrow B_1|B_2|\ldots|B_{k-1}|C_1|C_2|\ldots|C_I|B_{k+1}|\ldots|B_J$

Let

(3) $(y,c,b_1,b_2,\ldots,b_{k-1},c_1,c_2,\ldots,c_I,b_{k+1},\ldots,b_J)$

(4) $(y,c,b_1',b_2',\ldots,b_{k-1}',c_1',c_2',\ldots,c_I',b_{k+1}',\ldots,b_J')$

be two tuples into R.

$Y \twoheadrightarrow B_1|B_2|\ldots|B_J$ implies $Y \twoheadrightarrow B_k|B_1,\ldots,B_{k-1},B_{k+1},\ldots,B_J$ according to H1 ; then we can infer the existence of the tuples :

(5) $(y,c,b_1',b_2',\ldots,b_{k-1}',c_1,c_2,\ldots,c_I,b_{k+1}',\ldots,b_J')$

(6) $(y,c,b_1,b_2,\ldots,b_{k-1},c_1',c_2',\ldots,c_I',b_{k+1},\ldots,b_J)$

$Z \twoheadrightarrow A_1|A_2|\ldots|A_I$ implies $Z \twoheadrightarrow A_1|A_2,\ldots,A_I$ according to H_1 ; then we can infer from (3) and (6) the tuple :

(7) $(y,c,b_1,b_2,\ldots,b_{k-1},c_1',c_2,\ldots,c_I,b_{k+1},\ldots,b_J)$

and from (4) and (5) the tuple :

(8) $(y,c,b_1',b_2',\ldots,b_{k-1}',c_1,c_2',\ldots,c_I',b_{k+1}',\ldots,b_J')$;

the tuples (3), (4), (7) and (8) imply the existence of the GMVD

(9) $Y \cup C \twoheadrightarrow C_1 | B_1, B_2, \ldots, B_{k-1}, C_2, \ldots, C_I, B_{k+1}, \ldots, B_J.$

It is easy to prove that we have also by replacing successively C_1 by C_2, until C_1 by C_I :

$$Y \cup C \twoheadrightarrow C_2 | B_1, B_2, \ldots, B_{k-1}, C_1, C_3, \ldots, C_I, B_{k+1}, \ldots, B_J$$

(10) $\vdots$

$$Y \cup C \twoheadrightarrow C_I | B_1, B_2, \ldots, B_{k-1}, C_1, C_2, \ldots, C_{I-1}, B_{k+1}, \ldots, B_J.$$

From GMVDs (9) and (10) we can derive :

(11) $Y \cup C \twoheadrightarrow B_1, B_2, \ldots, B_{k-1} | C_1 | C_2 | \ldots | C_I | B_{k+1}, \ldots, B_J$

according to inference rule H6. But the two branches $B_1, B_2, \ldots, B_{k-1}$ and $B_{k+1}, \ldots, B_J$ in (11) can be decomposed since $Y \twoheadrightarrow B_1 | B_2 | \ldots | B_J$; then we get finally

$$Y \cup C \twoheadrightarrow B_1 | B_2 | \ldots | B_{k-1} | C_1 | C_2 | \ldots | C_I | B_{k+1} | \ldots | B_J$$

which is equal to (2).

□

We shall denote by $\underset{k}{*}$ the operation :

$(Y \twoheadrightarrow B_1 | B_2 | \ldots | B_J) \underset{k}{*} (Z \twoheadrightarrow A_1 | A_2 | \ldots | A_I) =$

$(Y \cup (B_k \cap Z) \twoheadrightarrow B_k \cap A_1 | \ldots | B_k \cap A_I | B_1 | B_2 | \ldots | B_{k-1} | B_{k+1} | \ldots | B_J).$

The *-operation can be considered as an application from $G \times G$ to G, such the image by the *-operation of two GMVDs g1 and g2 given with respectively i and j branches is a GMVD with i+j-1 branches. This operation looks like the consensus operation in boolean algebra defined in [KP 77].

<u>Remark</u> : Let g_1 and g_2 be GMVDs and $g = g_1 \underset{k}{*} g_2$; then the root of g contains at least one of the roots of g_1 or g_2.

4.4.2. Definition of an order relation

We shall define an order relation, denoted $\subseteq$, on the decompositions of a relation as follows :

$(Y \twoheadrightarrow B_1 | B_2 | \ldots | B_J) \supseteq (Z \twoheadrightarrow A_1 | A_2 | \ldots | A_I) \Longleftrightarrow$

$(Y \supseteq Z) \wedge (\forall i = 1, 2, \ldots, I, \exists j \in \{1, 2, \ldots, J\} : B_j \cup Y \supseteq A_I)$

<u>Proposition 6</u> : The relation $\subseteq$ is a partial order.
The proof is given in appendix.

□

<u>Examples</u> : $A,D \twoheadrightarrow B|C,E \supseteq A \twoheadrightarrow B,D|C|E$

$\qquad\qquad A,D \twoheadrightarrow B|C,E \supseteq A \twoheadrightarrow B|D|C,E.$

<u>Remarks</u> :

(1) if an GMVD g' is derivable from g by addition of branches (inference rule H1) then $g' \supseteq g$.

(2) if an GMVD g' is derivable from g by augmentation (inference rule D3 or H4) then $g' \supseteq g$.

These two remarks show that to obtain the closure of a given set of decompositions it is not necessary to consider the addition of branches and the augmentation inference rules. We shall prove in the next proposition an important property that relates the *-operation and the order relation $\subseteq$.

<u>Proposition 7</u> : The *-operation is isotonous according to $\subseteq$.

The proof is given in appendix.

$\qquad\qquad\qquad\qquad\qquad\qquad\qquad\qquad\qquad\qquad\qquad\qquad\qquad\qquad\qquad\quad$ □

This property means that if $(Y \twoheadrightarrow B_1|B_2|...|B_J) \supseteq (Z \twoheadrightarrow A_1|A_2|...|A_I)$ then

$(X \twoheadrightarrow C_1|C_2|...|C_K) * (Y \twoheadrightarrow B_1|B_2|...|B_J) \supseteq (X \twoheadrightarrow C_1|C_2|...|C_K) * (Z \twoheadrightarrow A_1|A_2|...|A_I)$.

4.4.3. Maximal decompositions algorithm

The determination of a maximal covering of decompositions is greatly simplified according to proposition 7 because we have only to consider the maximal decompositions according to the order relation $\subseteq$. This will be illustrated on an example.

Let G be an initial set of decompositions as in our example 5, G = {d1,d2,d3,d4}

$$\begin{array}{ll} d1 & S \twoheadrightarrow C|B,P,H \\ d2 & C \twoheadrightarrow S|B,H,P \\ d3 & C \twoheadrightarrow B|S,H,P \\ d4 & H,P \twoheadrightarrow C|S,B \end{array}$$

From d2 and d3, we derive according to proposition 5

$$d5 \quad C \twoheadrightarrow S|B|HP.$$

As $d2 \supseteq d5$ and $d3 \supseteq d5$ we can delete d2 and d3 from G and we introduce d5 in G. At the end of the process, we have only the maximal decompositions MAX(G^*). In the present example, we have MAX(G^*) = {d5,d6,d7}

$$\begin{array}{ll} d5 & C \twoheadrightarrow S|B|H,P \\ d6 & S \twoheadrightarrow C|B|H,P \\ d7 & H,P \twoheadrightarrow C|S|B. \end{array}$$

Now we have only to search an irredundant covering among the maximal decompositions. In the present case it is MAX(G^*).

The designer has the choice between three different decompositions d5, d6 and d7 ; we have to choose the ones which satisfy a special condition called "complete joinability of data".

4.5. Complete joinability of data

The notion of normal forms introduced by CODD provide a guideline to study the decomposition of a relation.

Many authors have modified the definition of the third normal form by looking at the type of anomalies in a relation, so we have Boyce, Codd, Kent, Irreducible and fourth normal forms. We propose a more general criterion for a decomposition, the *complete joinability of data*. We shall say that a decomposition has the complete joinability of data, intuitively, if this decomposition preserves all the joins.

Definition : Let F be a given set of dependencies (FDs and MVDs) for a relation R and D the initial set of decompositions derived from F as explained in subsection 4.4. Let $Y \twoheadrightarrow A_1|A_2|\ldots|A_I$ be a generalized decomposition of R belonging to D ; then we consider

$F_{Y \cup A_k}$ $k = 1,2,\ldots,I$ as the set of dependencies which hold over the attribute set $Y \cup A_k$. To each $F_{Y \cup A_k}$ $k = 1,2,\ldots,I$, we associate $D_{Y \cup A_k}$ which is the set of decompositions derived from $F_{Y \cup A_k}$.

We shall say that the generalized decomposition $Y \twoheadrightarrow A_1|A_2|\ldots|A_I$ has the complete joinability property iff :

$$(\bigcup_{k=1}^{I} D_{Y \cup A_k})^+ = D^+$$

Example : If we consider decomposition d6 $S \twoheadrightarrow C|B|H,P$, we have

$$F_{S,C} = \{S \rightarrow C, C \twoheadrightarrow S\}$$
$$F_{S,B} = \emptyset$$
$$F_{S,H,P} = \emptyset$$

and

$$D_{S,C} = \{S \twoheadrightarrow C|B,P,H , C \twoheadrightarrow S|B,H,P\}$$
$$D_{S,B} = \emptyset$$
$$D_{S,H,P} = \emptyset.$$

One can check $(D_{S,C} \cup D_{S,B} \cup D_{S,H,P})^+ \subset D^+$, then d6 has not the complete joinability property of data.

Now, if we consider d5 $C \twoheadrightarrow S|B|H,P$, we have

$$F_{S,C} = \{S \rightarrow C, C \twoheadrightarrow S\}$$

$$F_{C,B} = \{C \twoheadrightarrow B\}$$

$$F_{P,H,C} = \{P,H \rightarrow C\}$$

and

$$D_{S,C} = \{d1,d2\}$$

$$D_{C,B} = \{d3\}$$

$$D_{P,H,C} = \{d4\}$$

then as $D_{S,C} \cup D_{C,B} \cup D_{P,H,C} = D$, the decomposition d5 has the complete joinability of data.

5. CONCLUSIONS

The main objective of this paper was to establish a rigorous foundation for the design of conceptual schemas for a database system. Because of its formal nature the relational data model was used as a framework.

In a first part, we introduced an analysis of the state of the art, by considering the different kind of dependencies : functional dependencies, multivalued dependencies and hierarchical dependencies. We also recognized that dependencies and irredundant covers supply a power tool for the analysis and design of relational schemas.

In a second part, we proposed the decomposition approach where the initial relations are broken into subcomponents. This decomposition approach is based upon the concept of generalized multivalued dependency. The formal properties of these dependencies yields a number of new concepts, such the *-operation which offers the possibility of deriving new decompositions from an initial set of decompositions ; a technique is proposed for finding the maximal decompositions among the whole set of decompositions.

APPENDIX

<u>Proposition 6</u> : The relation $\subseteq$ is a partial order.

1. $\subseteq$ is reflexive.
2. Let us show that $\supseteq$ is transitive, that is :

$(X \twoheadrightarrow C_1|C_2|\ldots|C_K) \supseteq (Y \twoheadrightarrow B_1|B_2|\ldots|B_J)$ and

$(Y \twoheadrightarrow B_1|B_2|\ldots|B_J) \supseteq (Z \twoheadrightarrow A_1|A_2|\ldots|A_I)$

$\Longrightarrow (X \twoheadrightarrow C_1|C_2|\ldots|C_K) \supseteq (Z \twoheadrightarrow A_1|A_2|\ldots|A_I).$

Indeed $X \supseteq Y$ and $Y \supseteq Z \Longrightarrow X \supseteq Z$

$\forall i = 1,2,\ldots,I, \exists j \in \{1,2,\ldots,J\} : B_j \cup Y \supseteq A_i$; $\exists k \in \{1,2,\ldots,K\}$:

$$C_k \cup X \supseteq B_j.$$

Hence : $\forall i = 1,2,\ldots,I, \exists k \in \{1,2,\ldots,K\} : C_k \cup X \supseteq B_j \cup Y \supseteq A_i.$

3. $\subseteq$ is antisymetric because

$Y \twoheadrightarrow B_1|B_2|\ldots|B_J \supseteq Z \twoheadrightarrow A_1|A_2|\ldots|A_I$ and $Z \twoheadrightarrow A_1|A_2|\ldots|A_I \supseteq Y \twoheadrightarrow B_1|B_2|\ldots B_J$

give, thanks to the antisymetry of $\supseteq$: $Y = Z$ and

$\forall i = 1,2,\ldots,I, \exists j \in \{1,2,\ldots,J\} : B_j \cup Y \supseteq A_i$

or $\quad B_j \supseteq A_i$ because A_i and Y are disjoint,

that is by symetry : $\forall i = 1,2,\ldots,I, \exists j \in \{1,2,\ldots,J\} : B_j = A_i.$

$\square$

<u>Proposition 7</u> : The $*$-operation is isotonous according to $\subseteq$.

Let us show that

$(Y \twoheadrightarrow B_1|B_2|\ldots|B_J) \supseteq (Z \twoheadrightarrow A_1|A_2|\ldots|A_I) \Longrightarrow (X \twoheadrightarrow C_1|C_2|\ldots|C_K) \underset{k}{*}$

$(Y \twoheadrightarrow B_1|B_2|\ldots|B_J)$

$(X \twoheadrightarrow C_1|C_2|\ldots|C_K) \underset{k}{*}$

$(Z \twoheadrightarrow A_1|A_2|\ldots|A_I).$

At first $\forall k = 1,2,\ldots,K$, isotonous $\cap$ and $\cup$ relatively to $\subseteq$ give :

$$Y \supseteq Z \Longrightarrow X \cup (C_k \cap Y) \supseteq X \cup (C_k \cap Z).$$

Secondly $\forall i = 1,2,\ldots,I, \exists j \in \{1,2,\ldots,J\} : B_j \cup Y \supseteq A_i,$

hence $\forall k = 1,2,\ldots,K : C_k \cap (B_j \cup Y) \supseteq C_k \cap A_i,$

a fortiori : $\qquad (C_k \cap B_j) \cup X \cup (C_k \cap Y) \supseteq C_k \cap A_i.$

289

REFERENCES

[BOI 69]   J. BOITTIAUX, Etude mathématique d'un ensemble de notions. Contrat DGRST
            67.01.015 (1969).

[COD 70]   E.F. CODD, A relational model of data for large shared data banks,
            CACM, June 1970, pp. 377-387.

[COD 71]   E.F. CODD, Further normalization of the database relational model,
            Courant Computer Science Symposium 6, Database Systems, Prentice
            Hall, N.Y. (May 1971), pp. 65-98.

[DEL 71]   C. DELOBEL, Aspects théoriques sur la structure de l'information dans
            une base de données, RIRO, B.3, 1971, pp. 37-64.

[DC 73]]   C. DELOBEL & R.G. CASEY, Decomposition of a data base and the theory
            of boolean switching functions. IBM Journal of Research and
            Development, Vol. 17, n° 5, 1973.

[DL 74]   C. DELOBEL & M. LEONARD, The decomposition process in a relational
            model. International Workshop on Data Structures, IRIA, Namur
            (Belgique), May 1974.

[ARM 74]   W.W. ARMSTRONG, Dependency structures of database relationships.
            Proc. IFIP 1974, pp. 580-583, North Holland, 1974.

[WW 75]   C.P. WANG & H.H. WEDEKIND, Segment synthesis in logical data base design.
            IBM Journal of Research and Development, January 1975.

[SS 75]   H.A. SCHMID & J.R. SWENSON, On the semantics of the relational data
            model. Proceedings ACM SIGMOD, W.K. King (Ed.), San José, Calif.,
            May 1975, pp. 211-223.

[BER 75]   P.A. BERNSTEIN, Normalization and functional dependencies in the rela-
            tional data base model. Ph.D Thesis, Departement of Computer
            Science, University of Toronto, 1975.

[FAG 76]   R. FAGIN, Multivalued dependencies and a new normal form for relational
            data bases. TODS 23 (Sept. 1977), pp. 262-278.

[RIS 76]   J. RISSANEN, Independent components of relation. IBM Research Laborato-
            ry, San José, October 1976.

[LR 76]   M. LEONARD & F. REYNAUD, Existence du consensus et caractérisation des
           couvertures et bases irredondantes d'une fonction $\sum_i \mu_i\ A_i^j$.
           Rapport de recherche, September 1976, Computing Laboratory, Uni-
           versity of Grenoble (submitted to "Discrete Mathematics").

[ZAN 76]   C. ZANIOLO, Analysis and design of relational schemata for data base
           systems. Ph.D Dissertation, UCLA, 1976.

[BER 77]   P.A. BERNSTEIN, Synthesizing third normal form relations from functional
           dependencies. ACM TODS, January 1977.

[BFH 77]   C. BEERI, R. FAGIN & J. HOWARD, A complete axiomatization for functional
           and multivalued dependencies in database relations, Proc. 1977
           ACM SIGMOD, D.C.P. Smith (ed.), Toronto, pp. 47-61.

[RLP 77]   F. REYNAUD, M. LEONARD & E. PICHAT, Calculation of the prime implications
           and their irredundant covers in data bases. Research Report #3,
           June 1977, CNAM.

[KP 77]   A. KAUFMANN & E. PICHAT, Méthodes mathématiques non numériques et leurs
           algorithmes. Tome I : algorithmes de recherche des éléments maxi-
           maux. Masson, Paris, 1977.

[OSB 77]   S. OSBORN, Normal forms for relational data bases. Ph.D Thesis, University
           of Waterloo, November 1977.

[FLO 77]   A. FLORY, Un modèle et une méthode pour la conception logique d'une base
           de données. Thèse d'Etat, Université de Lyon, France, Novembre
           1977.

[AD 77]   W. ARMSTRONG, C. DELOBEL, Decompositions and Functional dependencies
           in Relations. Département d'Informatique et Recherche Opération-
           nelle, Université de Montréal, October 1977 (submit to ACM-TODS).

[DEL 78]   C. DELOBEL, Normalization and hierarchical dependencies in the relational
           data model (to appear in ACM-TODS).

[ZAM 77]   C. ZANIOLO, M. MELKANOFF, Relational schemas for data base systems.
           Technical Report, UCLA, November 1977.

ASPECTS OF DATA SEMANTICS: NAMES, SPECIES

AND COMPLEX PHYSICAL OBJECTS

Ronald Stamper
London School of Economics
Houghton Street,
London WC2

ABSTRACT

Investigating the definition of administrative information systems in the LEGOL Project using legislation as experimental material, a new approach to data semantics is found necessary. The emphasis is upon the operations linking data to reality. The operations are performed in the Discourse System by people according to language norms, thus the social reality of natural language links the structures in the Formal System to the Object System. Data representing things in the real world may be regarded as subroutines in programs which people can interpret. This principle is being used to explore the possibility of a canonical data model based on an operational semantics. This is contrasted with other data models. To illustrate the principle, it is used to examine names and species, an aspect of the problem of universals and particulars. The results enable structures to be defined for the complex physical objects: collectives and systems. The use of an 'alias' function to express the operational identity of entities with different representations is introduced.

(The work is supported by UK Science Research Council with additional support from IBM)

ASPECTS OF DATA SEMANTICS: NAMES, SPECIES AND COMPLEX PHYSICAL OBJECTS

Introduction

This paper is another in the series explaining the concepts underlying the semantic
model of the LEGOL language*. It is a sequel to a paper on the identifiers of
physical objects (ref 1) which skated over the problem of universals and particulars.
The treatment here is made self-contained by treating some conclusions of earlier
papers without further argument.

The LEGOL Project is exploring the relationships between formal information systems
and the world they represent and endeavour to control. As experimental material,
the project uses legislation because, for example, a body of tax statutes in effect
defines a large and complex formal information system. By trying to devise a formal
language which can express whatever is in the statute, we are forced to explore
the fundamental problems of systems definition at the highest level: i.e. saying
what should be done without saying much about the procedures of how to do it. By
testing each version of the language, as it evolves, against samples of legislation,
well-founded progress is being made. These samples provide tests far more severe
than do the synthetic examples usually conjured up to test theories of data modelling.

Program, data and human performance

The LEGOL language is interpreted by a computer**. Superficially it resembles a
computer language but this is misleading. Only a subset of a body of legislative
rules is sufficiently cut-and-dried to be interpreted without human judgement;
LEGOL has to encompass both the mechanical rules and the exercise of judgement.
LEGOL is therefore a language for specifying a formal organisation, not merely the
computer programs used within it. It is a language for the systems designer from
which programs (very inefficient ones) are derived automatically.

---

* The LEGOL Project is supported by the UK Science Research Council with additional
 support from IBM UKSC, Peterlee. Reports are obtainable from the author at the
 London School of Economics, Houghton Street, London WC2.

** Development of the first prototype system is complete and work has begun on the
 second.

Judgements are exercised by people whenever they report observations or perform specified actions. Whereas input to a computer system is achieved by the electrical or mechanical transduction of characters from one representation to another, the prime input to an organisational information system depends upon a person choosing a symbolic representation for what he observes or feels. Similarly, output from a computer system in the form of character strings only becomes an output from an information system when a command is obeyed, and this entails human judgement. Therefore a formal information system to help run an organisation depends both on the computer and upon human performance.

It is useful to think of the data which the computer holds as a means of 'programming' human behaviour. An entry on the list called DISTANCE is like a subroutine call with a precise operational meaning which will be modified by the context in which it is used. The entry may be in a typical LEGOL example:

20 miles, Lightship X, Portsmouth Point, 1952-

If this forms a part of the data base which is assumed to contain our relevant knowledge for a task, then it can supply an assertion to an enquirer in response to his request for information. This he may employ instead of making an independent observation. If he is sceptical he may treat it as an hypothesis to be tested by navigating between and making observations upon 'Lightship X' 'Portsmouth Point' and possibly some third location. In the LEGOL semantic analysis we construe the meaning of the data in terms of these operational procedures. The data make use of calls to various 'subroutines' in the sense of programs directing human behaviour. Thus 'distance' indicates a procedure for assigning a value jointly to 'Lightship X' and 'Portsmouth Point'. These two identifiers of the objects mentioned are also subroutines which permit a person to find two particular objects which are distinct from each other but which can be relocated repeatedly. The enquirer, to be able to make a precise operational interpretation of the data, has to be given the necessary 'code' for the subroutines and other bases for his actions. These include:

knowledge of the language
understanding of the problem context and
motivation to respond

each of them being provided by the society in which he forms a part. Living in a community and solving the practical problems of day-to-day living, he acquires a set of norms of perception, evaluation, cognition and behaviour. Language does not enter into all these social norms but into a high proportion of them. Such are the language norms to which we shall refer in the rest of this paper.

Language norms should be a primary concern of any information system designer. If he creates a system which does not make appropriate use of the ways in which people interacting with the system have been 'programmed' linguistically, then that system is likely to function incorrectly. Unfortunately, the literature of computer science almost totally ignores the social foundation of language (e.g. refs 2-5). This is true even in the field of data modelling where the current search for a well-founded theory of conceptual schemas could profit most from this simple observation.

Canonical forms for data models

To understand the significance of the kind of semantic model being evolved in the LEGOL Project, it is worth considering briefly the quest for canonical data models. We should not expect there to be a single canonical form like a holy grail for a mathematical crusade: there should be several to serve different purposes. Each contains the minimal information for some purpose, subject to relevant constraints that make its expression irreducibly simple in that context. Some purposes are:

  (a) to specify access paths among data elements
  (b) to specify an information retrieval interface
  (c) to describe data movements
  (d) to characterise operational meanings.

The last of these is the purpose of a canonical data model for the LEGOL language. Let us examine each of them in slightly more detail.

Access structures will be equivalent in a fundamental way if they refer to the same data elements and if they embody the same pathways among pairs of elements. A binary relational structure is natural for this task and authors such as Senko (q.v.) have adopted this approach. A totally formal definition of meaning can be based upon transformations which preserve the structure. It may be possible to find certain constraints under which these transformations will result in a standard representation. It does not matter that this binary model requires the decomposition by artificial devices of relations that are 'irreducible' by criteria relevant to another purpose.

295

A retrieval interface should impose more structure than is needed to characterise access pathways. Codd's third normal form aims to give the user a view of his data which is irreducibly simple in respect of functional dependencies among attributes. This model moves away from the totally formal definition of meaning to one which is implicit and dependent upon intuition unless a purely empirical treatment of the functional dependencies is adopted. For example, that an employee has only one manager may be observed or it may be thought to be likely or it may even be prescribed; the model does not distinguish. There is an arbitrariness about the translation from real world properties to data representations but once chosen the representation can be transformed into this canonical form.

Data movements relate to messages and Langefors has developed a notion of 'elementary messages' to show how complex messages can be decomposed (refs 9, 10 and 11). The notion of meaning employed in this analysis is also intuitive. Langefors introduces 'elementary concepts' to construct his model. This explicit treatment of meaning (e.g. refs 9 p.229, 11 p.50), however, is conceptualist. The result is a data model appropriate if one is concentrating upon the flows of data in a particular organisation. The associated canonical form preserves the essential features of these data flows.

Operational meanings which underlie the LEGOL language are the patterns of behaviour which enable the data to 'program' the human user in an organisation. They embody distinctions which are organisationally significant (e.g. is 'to each man, one manager' a descriptive or a prescriptive constraint?) but computationally or procedurally irrelevant (see ref 12 for details). Given enough constraints of this kind, it should be possible to eliminate most of the arbitrariness of the information analysts' views of the data permitted by the approaches of Codd and Langefors. The goal we refer to as a 'semantic normal form'. The data in this model do not correspond to messages but to hypotheses and there is no conceptual treatment of meaning but an explicit operational one. This paper applies these principles to the cases of some complex objects.

Operational ontology

To understand the problem of the meanings of the data in a formal information system we need far more than intuition. Being concerned with organisational behaviour rather than data manipulation, the LEGOL Project forces upon the researchers a regard for the ways in which operationally (rather than intuitively) the data are linked to the real world. The use of legal material, by itself, makes one conscious of problems of evidence and judgement which are the key aspects of an operational semantics. The data model which underlies the LEGOL language must express the implications of the data in terms of human performance. Ideally, given any datum in the system and the question 'what does it mean?' then there should be a route, through rules and decisions, back to the observations and value judgements and another route forwards to the actions that anchor that datum's meaning in reality. A standard way of doing this which enables one to check equivalence of meaning or to characterise precisely differences of meaning should lead to a semantic normal form. This is not a substitute for any of the other three types of data model described above. Each one entails concepts or constraints that are irrelevant to the others although they may overlap in other respects.

Universals and particulars

For the other data models described above, this ancient problem of universals (e.g. woman, beauty) and particulars (e.g. Grace, Jennifer) is not important. It may have some relevance for the message-orientated model which may use syntactic categories bearing these names, as a feature of its intuitive treatment of semantics. In the LEGOL data model, and any other concerned with operational semantics, the problem is fundamental. Obviously the operational processes of referring are different for universals and particulars. We have to say precisely how. This is easier if we confine our attention to what we say about physical objects. Abstract objects can be dealt with at a later stage.

As a background to the issue, let the main philosophical views on the problem be presented in an outline so brief it may seem like a parody*. The theories are generally about the relationships between mental concepts and external reality. The Realists assert that, just as concepts of particulars correspond to physical objects (Grace and Jennifer), so do universals (woman and beauty) correspond to ideal objects. For Platonic Realists these are the transcendental ideals (Woman and Beauty) which physical reality dimly reflects. For Aristotolean Realists the form of woman and the form of beauty are only immanent in the matter (Grace and Jennifer) with which they co-exist. Nominalists, on the other hand, reject the existence of real universals saying that these are only names which refer to particulars in a general manner in propositions. Objective Nominalists accept external particulars which cause the corresponding internal concepts. Subjective nominalists take only the concepts as real, the external world being a projection of them.

---

* For a still brief but more substantial summary see Lyons' useful book on semantics (ref 13).

These traditional views are deeply embedded in Western thought and one or other of them tends to inform any intuitive view of meaning (e.g. Platonic Realism in mathematics). They are all <u>conceptualist</u> in the sense that they treat <u>concepts</u> as one of the kinds of things being studied. Concepts, however, are not much use to an empirical scientist, neither are they much use to anyone designing an information system for running an organisation. As designers of systems to handle and share data in the public domain <u>we should avoid a theory of data which needs access to the processes inside someone's head</u>. This objection leads the LEGOL model towards what might be called an <u>Operationalist</u> theory of semantics which differs in its ontology from all those described above. This is a significant departure from the conceptualism which is implicit in most data base work today.

Operational view of universals and particulars

As pointed out earlier, the character strings in our data bases are linked to the external reality by people who use these data items according to linguistic norms. These norms depend upon the purpose for which the data are being used. (This was argued at length with copious illustrations in ref 1.) Linguistic behaviour is not mechanically uniform so that meanings, in the sense of signs signifying objects or states of affairs, will be rough and ready. If these variations hinder the performance of some practical task, or the resolution of a definite problem, they are too rough and will be adjusted by those involved, otherwise further precision is superfluous. The majority of data processing specialists seem to treat all data as though they had some definite meaning in terms of an external reality. (In this naive confidence in words and numbers lies the most serious social threat posed by the computer and its technologists!)

To illustrate the operational view of universals and particulars, attention is confined, at first, to physical objects. This is relatively safe ground where we can feel some confidence when asked to explain the meanings of our data*. The problem in this context we may narrow down to the use of names for individuals and for species. For the sake of brevity, let these be called 'names' and 'species'.

---

* However, even in the case of individual physical objects, the reality we know is partly a product of our use of language. As was argued in ref 1, the partitioning of the world into components bearing different labels is partly under the control of those labels and their use in a given problem context. (e.g. my son's new bicycle is a parcel as I carry it home but a vehicle when he rides to a friend and therefore subject to different laws on its two journeys).

Names and species

The words that we are talking about are used by people according to language norms
which they acquire in a natural way whilst solving problems and performing tasks.
The Formal System that we are interested in designing employs these words (or codes
for them or other equivalent signs) to represent things in the external Object
System. The linkage between the Formal and Object Systems is dependent upon the
informal use of natural language in the Discourse System. To be effective, the
signs in the Formal System must imply precisely (to the degree warranted by the
practical problem) how a person in the Discourse System should locate an object
in the Object System.

This is straight-forward in the case of individual physical objects. If the problem
we are dealing with requires the use of an object with certain properties but
does not require us to relocate precisely the same object, we can ask for it by
using a species or common name:

    a prawn   a pipe section

Though generally a gourmet may not wish to know that the prawn he is eating was
called 'Fred' nor the engineer that the pipe section being laid is '1234', these
particulars are important to others. The prawn was called 'Fred' by the keeper
of an aquarium from which it was stolen; his keeper was enamoured of Fred's
quite distinct personality among the other prawns and fishes. The inspector, who
originally failed the pipe section for use under high pressure after an X-ray
check, was careful to report its exact code name. The way a species name is used
will be learned, and there will be pressures both to retain and to alter the
language norm: a chef who confuses 'prawn' and 'crayfish' will offend his clien-
tell but a frozen food manufacturer who ennobles the shrimp into a prawn with
enough breadcrumbs may make himself a great deal of money.

A species name enables a person 'programmed' by the appropriate language norm, to
locate an individual which meets criteria appropriate for some practical task.
Provided that the individual remains literally in the grasp of the finder, and it
is treated in a way that prevents confusion with others of the same species, it
may be given a local name, perhaps merely

    the pipe section

and if there are several of the same kind, they may be distinguished by adjectives
or pronouns:

    my pipe section   your pipe section
        mine   yours

all of which serve as local names of individual pipe sections.

A name enables a person to 'navigate' within the Object System and repeatedly
return to the same individual object. For completeness, the notion of sameness
can be defined operationally in terms of the continuity of a person's grasp upon
the object, starting with simple cases (e.g. a book) and gradually generalising
to more difficult ones (e.g. airport, star), whilst individuality is operationally
defined in terms of the separation of the grasped object from others of the same
kind. These operational definitions are learned and generalised during the
acquisition of language. We may revert to the same type of instruction whilst
training people in specialised skills relevant to an information system which
we are designing*. A local name is easily constructed and it serves symbolically
the same function of control as the grasping of the object. The problem of naming
becomes severe when we put down the object among many of its kind. The local
name is then useless. A name must be given in such a way that it is equivalent
to a description of the individual, sufficiently precise to distinguish it from
all others. In this process of name-giving, the local name will be employed.
Legislation governing the registration of vehicles or of births includes rules
for assigning names. A local name 'the vehicle' can be given during the registration procedure when an alias 'UNP 313F' is uniquely associated with that machine,
at least in the wider context of the UK. When several individuals are distinguished
locally, as in multiparous births, the physical 'grasp' on the individuals must
ensure that individuality is established and maintained: in the delivery room and
the nursery, careful management of the babies and their labels is necessary to
maintain even the local names of Stamper 1, Stamper 2, Stamper 3.

The above account is exaggerated if it seems to suggest that species names are
always established informally by norms and that proper names are assigned by
formal procedures for which local names are the natural starting point. A few
particular objects of sufficient importance may be named by natural norm formation
among a group of people. Conversely, species names established by informal norms
are not precise enough for many tasks and refinements have to be introduced by
formal procedures. In all cases where formal procedures are used to establish
proper or species names, recourse must ultimately be made to unsupported language
norms.

The method of treating universals and particulars adopted as a semantic principle
in LEGOL avoids the usual philosophical approaches sketched earlier. There is no
need to talk of concepts if a strictly operational treatment of the question is

---

* A pertinent example is the design of an information system for computer-aided
  design of D.P. systems. The analyst must learn what is meant by the individual
  D.P. system and its sameness.

used, in keeping with the engineering approach of the project. It is possible to
account for the paradox that language creates, to some extent, the reality that
it represents, by fixing attention upon the ostensible phenomena of language as
a social instrument for problem solving.

There are three ostensible phenomena. It is convenient to distinguish between
them by using three types of brackets, if there is a danger of confusion. The two
obvious ones are recognised by all authors dealing with the conceptual schema:

> the sign used in the Formal System: e.g. ⟨book⟩
> and the physical book in the Object System: [book]

No mental object or concept need detain us. The link between these two is established by the phenomenon neglected in computer science:

> the language norm sustained in the Discourse System: (book)

which is a pattern of behaviour and is itself ostensible. If we are pressed hard
to explain the meaning of ⟨book⟩ we can avoid any meta-language other than
demonstration by pointing at ⟨book⟩ , [book] and (book). Generally we shall
find a common-sense use of natural language an aid to our explanation. However
we must take care to recognise that the explanations have not a philosopher's
purpose (some ultimate unravelling of metaphysical mysteries?) but the pedestrian
object of building better information systems. Not until our explanations
palpably fail in that problem context have we any need to revise them. This
point has to be made lest it be thought that this is the beginning of a vicious
infinite regress through layers of semantic analyses.

Elaborations upon the idea of physical objects

A semantic model to represent individual physical objects and their species must
be elaborated to deal with complex physical objects such as collectives and systems
and to extend to abstract objects. The same kind of semantic analysis must be
applied. That is, each new data-construct should be treated as a 'program' inter-
pretable by members of the social group upon whose language norms our meanings
depend.

To direct a person, who has learnt the appropriate language subroutines, towards
any individual of a certain kind we need give him only a value for
> ⟨species⟩
where the value is from a domain limited to the species names of objects. To direct
him towards an individual he needs
> ⟨species⟩ , ⟨name⟩

where the name is a value domain defined within the context of the named species.
A more specific species included within the species governing the name domain
can be employed:

⟨Swedish citizen⟩ , ⟨national ref. no.⟩
⟨citizen of Lund⟩ , ⟨national ref. no.⟩

Also, we should note that the individual has associated with it, as a matter of
ontological necessity, a period during which it is said to exist. That is,
during which the 'programme' given to the person to locate the object is
operationally effective. Thus

⟨species⟩ , ⟨name⟩ , ⟨period⟩

constitute the parts of a data-construct which can represent a physical object.
This structure may be subdivided into

⟨identifier⟩ ∷ = ⟨species⟩ , ⟨name⟩
⟨necessary attribute⟩ ∷ = ⟨period⟩

where the identifier provides a set of data values which correspond to the unique
individual. When applied to the task of searching the real world, we refer more
appropriately to the identifier of the individual whilst 'key' is reserved for
locating a record about it. The period of existence is not required to find an
extant individual but it is an attribute inseparable from any such entity and
tells us when it would be fruitless to search for anything but reports of the
individual.

Collectives

Often we wish to talk about rather amorphous objects such as a box of index cards,
a stock of nails, a consignment of sugar or a roll of paper. These have many of
the attributes associated with physical individuals but may lack coherent
spatial features (e.g. a flock of sheep).

A collective is not merely a set of individuals nor is it so much as that. A set
is represented in LEGOL by the values of the individuals:

(⟨species⟩ , ⟨name⟩ , ⟨period⟩)
PERSON   TOM   1934-72
PERSON   DICK   1918-69
PERSON   HARRY   1940-78

If the members of this set change then the set changes. A collective, however,
e.g. The Dairylea Herd of Pedigree Ayreshires, continues its existence whilst the
membership changes. The important features of a collective are that its membership, though varying, remains of the same kind and that it has a focus, a collector
or 'shepherd', as it were. The collective as a whole has a species which is not

the kind of its members. Any collector may have associated with it more than one collective of a given kind of individual, in which case it will need a name local to the collector and kind. Thus a representation for a collective may be:

CATTLE

($\langle$species$\rangle$, $\langle$name$\rangle$, $\langle$collector$\rangle$, $\langle$kind$\rangle$, $\langle$period$\rangle$)

HERD	A	DAIRYLEA	AYRESHIRE	JAN 56 -
HERD	B	DAIRYLEA	AYRESHIRE	MAY 62 -
HERD	-	DAIRYLEA	FRESIAN	APR 48 -
DELIVERY	123	DAIRYLEA	FRESIAN	JAN 78 -
SHIPMENT	99	DAIRYLEA	AYRESHIRE	JAN 78 -

where 'herd' is the collective species word for a stock of cattle; we may imagine two herds of Ayreshires are run on different lines of management for experimental reasons whilst there is a single herd of Fresians (for which no local name is required); the delivery and shipment are other species of collectives which have distinct roles within the management of the live-stock; the existence of a delivery will cease once it has been merged into a herd, and of a shipment when it has been accepted by the customer, these rules being explicitly treated in the LEGOL formulation.

The identifier of the collective is given by:-

COLLECTIVE

($\langle$species$\rangle$, $\langle$name$\rangle$, $\langle$collector$\rangle$, $\langle$kind$\rangle$)

in which the collector has a value drawn from a domain of identifiers of individuals. This identifier enables the collective to be referred to as though it were a simple individual object. The attributes $\langle$collector$\rangle$ and $\langle$kind$\rangle$ are also <u>identifiers</u>. As an aside, it may be of interest that, in the LEGOL language, the identifier attributes may be referenced by symbols in brackets following the entity name, e.g. HERD(FARM,ANIMAL), the $\langle$name$\rangle$and $\langle$period$\rangle$ attributes remaining implicit. An unusual feature of the collective is that it draws one identifier, $\langle$kind$\rangle$, from a set of universals not particulars. No other entity yet encountered has required this device.

One problem remains. What happens if the Dairylea Fresians are sold to the Archers' Farm? Does not the collector change and the herd cease to exist only to become another one? In practice the Archers would probably want the herd to continue to be regarded as the same as the old Dairylea herd because if its reputation and accompanying good-will in the dairy industry. The way this is handled is to continue using the value 'Dairylea' as the name of the herd. (Note that this has no implication for ownership, in a legal sense. Ownership would have to be represented as a relation quite independently.) Operationally, the collector is the link through which access to the members of the herd can be established. This

is now "Archers'" To record the continuity of the herd we can express the
equivalence of names which have different periods of use:

    IDENTIFIER. APR 49-MAY 78: (HERD, ———,DAIRYLEA FRESIAN)

        ALIAS. May 78 -   : (HERD,DAIRYLEA,ARCHERS' FRESIAN)

Thus may continuity of an entity's existence and individuality be traced through
changes of name or during the use of a multiplicity of names.

Systems

Systems are like collectives but they have additional structure: as a minimum, components with designated roles and probably individually named components. The components may be simple individual objects or other, complex ones.

The construction of a particular system may be arrived at, either _analytically_ if
one starts with an individual and then decomposes it into components, or _syn-
thetically_ if the individual components are identified before the system is
constructed.

We also need to talk about systems _generically_ especially in such applications as
design or product specification. To define a system as a universal we must at
least state the components' roles. For this we need

    ROLE NAMES

   ⟨role⟩ , ⟨system⟩

    FATHER   FAMILY

    MOTHER   FAMILY

    CHILDREN FAMILY

Whether the role were associated with an individual, a collective or another
system would have to be indicated as additional structural information about the
system. Using collectives or systems as components we are able to make it a
requirement of a system definition that at any one time there should be assigned
to each role no more than a single object. This will ensure that names can be
assigned unambiguously via systems and their role names, as will now be illustrated.

To specify a particular family, the names of the individual members must be given.
In the Family Allowances Act 1968 the family system is synthesised from specified
individuals who satisfy certain conditions and relationships. The legislation
does not deal with the vital administrative questions of how to name a family and
when a family begins and ends its existence. To resolve disputes the legislation
can rely upon the use of local names for dealing with each specific case. A
semantic model for the administration of Family Allowances over extended time
periods must handle names less informal. This naming problem concerns a synthesised

system which cannot be localised and perceived independently of its components
(e.g. each member of a family may be in a different town yet still constitute
a family). To solve the naming problem we must refer to the system via some
integrally perceivable object. Normally this will be one of the components; for
a machine assembly it would probably be the chassis but for a family in the
context of the Family Allowances Act 1968 it is the mother, unless the family
has only a male parent. The data structure for a specific system such as a
family will include the values:

ROLE ALLOCATION

⟨species⟩	,	⟨type⟩	,	⟨compenent⟩	,	⟨whole⟩	,	⟨period⟩
FAMILY	,	FATHER	,	JOHN SMITH	,	MARY SMITH*,		1960 -
FAMILY	,	MOTHER	,	MARY SMITH	,	MARY SMITH*,		1960 - 1978
FAMILY	,	CHILDREN	,	CH(MARY SMITH*),		MARY SMITH*,		1960 -

where probably the name of the family will eventually be a code number but it may
initially be given as MARY SMITH*, using the name of a key number to refer to the
whole system. The naming of the children as a collective (if we are only interested
in the number of them and not individually) or as a system (if their individual
identities are relevant) is also based on the system called MARY SMITH* as the
collector or reference individual. Similarly for an analytically defined system
the components would be named by taking the whole system as the reference individual.

In the table for the Smith family, Mary Smith ceased playing her role as mother
in 1978 when in fact she died. The ontological problem posed by this change is
whether the family continues or ceases to exist, becoming a new family based on
John Smith as the reference object. The family, whilst having many physical
features is also to some extent a legal fiction, being dependent on the rules
defining it. Operationally there is no incontrovertible answer to questions of its
continued existence. Rules must therefore prescribe when the family begins and
ends. The rules in the Family Allowances Act establish that there is one family
existing until 1978 called MARY SMITH* and another from that date called JOHN SMITH*.
They might be treated administratively as distinct so that a new Family Allowances
Pass Book would be issued to mark the change. However, the continuity of a group
may be regarded as sufficient to preserve the family identity as a system despite
the change of reference object. This, as in the case of a change of collector for
a collective, could be established by a definition of an alias relation between the
names:

IDENTIFIER.1960-1978:(FAMILY,MARY SMITH*)

ALIAS   .1978   :(FAMILY,JOHN SMITH*)

This again illustrates the interplay between reality and our use of language.

305

Conclusions

A radically new approach to data modelling has been introduced. It was used to show how the signification of a data value or a suitable group of values can be explained precisely in terms of the operations a person would follow to link the sign for an object, ⟨object⟩ and the real, tangible object, [object] . The linkage depends upon the relevant language norms which we may call (object). The method of analysis was extended from simple individual objects to complex objects such as collectives and systems.

Throughout the analysis it was necessary to draw upon the distinction between universals, such as species of objects, and particulars, such as individual objects. It was shown that the conceptualist, philosophical problems of universals and particulars are avoided by an operationalist view of meaning. In the LEGOL semantic model, particulars are data-structures linked to physical reality within certain operational or formal constraints. The same formal constraints are obeyed by abstract particulars. Universals are data-structures that establish domains of values and their formal interrelations linked to the social reality of language norms.

The practical applications of this analysis will become evident as computer-based systems in organisations become larger and more complex. The explicit and precise analysis of operational meanings will then be necessary to prevent automation from wreaking organisational havoc. The work dovetails with analyses of semantics at a different level which is exemplified by a paper of Biller and Neuholdt (ref 14). They examine the problem of demonstrating formal equivalence of data schemata but do so on the basis of a definition of meaning that they merely indicate. We feel that the treatment of meaning which is evolving from the LEGOL Project will provide the underpinning which data-base management scientists require.

Acknowledgements

I wish to thank my colleagues on the LEGOL Project - Susan Jones, Peter Mason and Paddy Mudarth for their help. In particular, Susan Jones deserves my thanks for many sessions of critical discussion on collectives!

References

1. Stamper, R.K. 'Physical Objects, Human Discourse and Formal Systems' in reference 5.

2. Klimbie, J.W. and Koffeman, K.L. (eds) Data Base Management, North-Holland, Amsterdam, 1974.

3. Nijssen, G.M. (ed) Data Base Description, North-Holland, 1975

4. Nijssen, G.M. (ed) Modelling in Data Base Management Systems, North-Holland, Amsterdam, 1976

5. Nijssen, G.M. (ed) Architecture and Models in Data Base Management, North-Holland, Amsterdam, 1977

6. Senko, M.E. 'The DDL in the Context of a Multilevel Structured Description: DIAM II with FORAL' in ref. 3.

7. Senko, M.E. et al 'Data Structures and Accessing in Data Base Systems', IBM Systems Journal 1973, 12, 30-93

8. Codd, E.F. Normalized Data Base Structure: a Brief Tutorial, IBM RJ935, November 1971.

9. Langefors, B. Theoretical Analysis of Information Systems, Auerback 1973 (4th Edition), Studentlitteratur, Sweden.

10. Langefors, B. & Sundgren, B. Information Systems Architecture, Petrocelli/ Charter, New York, 1975.

11. Langefors, B. & Samuelson, K. Information and Data in Systems, Petrocelli/ Charter, New York, 1976

12. Stamper, R.K. 'Towards a Semantic Model for the Analysis of Legislation', Informatica e Diritto (in press)

13. Lyons, J. Semantics Vol. I, Cambridge University Press, 1977

14. Biller, H. & Neuhold, E.J. 'Semantics of Data Bases: the Semantics of Data Models' Information Systems Vol. 3, No.1 pp11-30, 1978

A MODEL FOR THE DESCRIPTION OF THE INFORMATION

SYSTEM DYNAMICS

A. FLORY, J. KOULOUMDJIAN
Laboratoire d'informatique
Université Claude Bernard - LYON

This paper deals with the description of the information system dynamics at the conceptual level. A model is put forward to take into account both the structural and dynamic aspects of data base. The data model (an entity-relation model) helps to build the data into a structured set of relations that correspond to a specific third normal form. These relations do not play the same role in data base evolution : we propose a classification of *"event"* relations - the tuples of which are the starting point of treatments - and of *"permanent"* relations. The description of processings deriving from a primary event may be divided into elementary chronological steps with the help of a hierarchy graph. Each step can be analyzed in term of *"states"* which express the stage of treatment of the triggering event.

1 - INTRODUCTION :

Nowadays, it is agreed by most people that the design of a data base must begin with the definition of a conceptual schema which *"represents the enterprise's view of the structure it is attempting to model in the data base"* (ANS 75).

This conceptual schema is derived from a model which defines the correspondence between the data stored in the base and the facts of the real world which are of interest for the organization. The symbolic representation of these facts can be considered from both a static point of view (description of the real world objects and of their mutual relationship) and a dynamic one (description of the data base modifications induced from the insertion of a new fact). Thus some authors (HUI 76) speak of a *"completude criterion"* of the model, which indicates its ability to take into account the characteristics of the real world.

A lot of models have been put forward mainly in the field of the relational models. One advantage of this type of models is to keep the conceptual schema free from implementation aspects. Since their first description (COD 70), these models have been studied into 2 directions : first the analysis of their mathematical properties (for example : research of the minimal cover of a set of functional relations, decomposition and normalization processes), second, their semantic purport because fact representation is not very easy with such mathematical models. For this reason, entity-relation models were created (CHE 76).

The introduction of the dynamic components of information systems at the conceptual schema level can take several shapes in scientific literature : Association of integrity constraints to the data model (MAC 76) (HAM 76) (ESW 75) (WEB 76), use of high level procedural languages (HUI 76) or definition of a set of "evolution rules" triggered by "events" (BEN 76).

However it seems to us that none of these models clearly shows the link which exists between the stored data and its evolution deriving from the storage of a new fact. The aim of this paper is to describe a model which includes both structural and dynamic aspects. In our model which belongs to the entity-relation type :

a) every real world entity of interest for the organization, appears as a third normal form relation tuple in the data base.

b) therefore when a new event (entity) is taken into account in the organization, modifications of the data base can be expressed very easily as tuple updatings or creations.

The next paragraph concerns a short description of the part of the model which deals with the structure of data. A more complete presentation is given in (FLO 78). In paragraph 3, a classification of the relations which represent the entity types of the real world is proposed according to the type of the domains composing these relations. In this typology, the relations are splitted into two classes : first the relations representing events, called "event" relations. The tuples of such relations can trigger procedures which event relations refer but which are not the starting point of treatment. Finally, in paragraph 4, all the concepts needed for the description of the dynamics of the information system are introduced.

2 - STRUCTURAL COMPONENTS OF THE MODEL :

In our model the data structure is expressed as a set of relations each of which is composed of a set of elementary attributes. In each relation, a subset of attributes (possibly one attribute) plays a key role, that is to say that the other attributes of the set are functionally dependant of this subset. The attributes that do not belong to the key are called characteristics of the relation.

Each relation is in third normal form (COD 72) and is called an entity-type (denoted E.T. later on).

Two kinds of E.T.'s can be distinguished :

- first order E.T.'s the key of which includes a single attribute. These E.T.'s are images of object types of the real world. 1-order E.T. characteristics express the properties of these objects. For example the E.T. "ITEM" (the key is underlined)

(ITEM # , PRICE, NAME, QTY-ON-HAND)

corresponds to the products manufactured in the enterprise.

- n-order E.T.'s the key of which is composed of more than one attribute. These E.T.'s depict the relation existing between two or more 1-order E.T.'s. For instance the n-order E.T. "ORDER-LINE" :

(ORDER # , ITEM # , ORDERED-QTY)

expresses a relation between the 1-order E.T. "ORDER" and "ITEM" which have respectively ORDER # and ITEM # as key.

Let C be the set of all the attributes which occur in the information system and S the set of the E.T.'s formed on C.

S is called an entity type system (noted E.T.S. later on) if the following conditions are satisfied :

1) Each attribute of C belongs at least to one E.T.,

2) If an attribute of C belongs to several E.T.'s,

- either it plays a key role in one of the E.T.'s

- or it cannot appear in the righ side of a functional relations.

An example of an E.T.S. is given below (FIG 1). It has five E.T.'s called

"DELIV-SLIP", "ORD", "CUST", "ITEM" and "REPR" (corresponding respectively to the object type DELIVERY-SLIP, ORDER, CUSTOMER, ITEM and REPRESENTATIVE of the real world).

FIG. 1

REPR   (REPR # , NAME)

CUST   (CUST # , ADRESS, BALANCE, REPR ≠)

ORD   (ORD # , CUST # , DATE, ITEM # , ORD-QTY)

DELIV-SLIP   (SLIP # , ORD ≠ , DELIV-QTY, AMOUNT)

ITEM   (ITEM # , QTY-ON-HAND, PRICE)

(we suppose there is one order line for each order).

Using an arrow to represent a functional relation (noted F.R. later on) between two attributes, this data structure can be visualized as follows (FIG. 2) (each rectangle represents a E.T.).

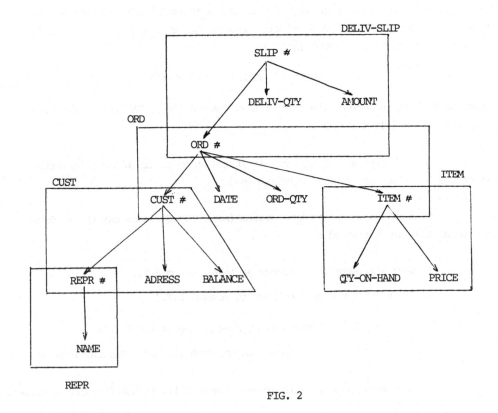

FIG. 2

On this graph we can see that the E.T.'s are linked together through the F.R. existing between their keys (An attribute which occurs in more than one E.T. must necessarily be the key in one of the E.T.'s). So the F.R.'s are defined either inter E.T.'s (e.g. ORD # $\longrightarrow$ CUST #) or intra an E.T. (e.g. CUST # $\rightarrow$ ADRESS).

Algorithms leading to the E.T.S. are given in (FLO 77). In a conceptual schema, the E.T.S. is shown to be unique.

In order to define the role played by the E.T.'s in the dynamics of the information system, we must classify the attributes according to two criteria : first, the way they appear in the system (calculated by the information system, or given by the environment) ; second, the stability of their values in the time. By combining the first criterion (calculated/not calculated) and the second one (stable/unstable), we can define four classes of attributes. Thus, for the afore mentioned example, we have the following classification.

	stable	unstable
calculated	SLIP # DELIV-QTY AMOUNT	BALANCE QTY-ON-HAND
not calculated (given)	CUST # DATE ORD-QTY ORD # ITEM # REPR #	ADRESS PRICE NAME

The definition of the data structure can be viewed at two different levels : A "*detail*" level in wich the description includes all the attributes (as in FIG. 1), and a "*macroscopic*" level which handles only the E.T. keys and inter E.T. functional relations. At this level, to a set of F.R.'s there corresponds a graph which defines a partial order on the keys, which we call key hierarchy. In our exemple, the hierarchy will be (FIG. 3)

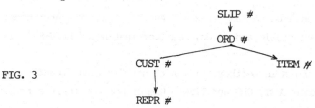

FIG. 3

This hierarchy is quite valuable for the description of the information system dynamics.

3 - LOGICAL AND CHRONOLOGICAL F.R.'s :

3.1. - Classification of the entity types :

The informations to be memorized belong to two classes : they refer either to the "*living*" objects of the real world or to the manifestations of these objects. The properties of the first class of objects can change as time elapses : for instance the address of a customer or the surface of a room. On the contrary, the manifestations of "*living*" objects do not change and are (explicitly or implicitly) dated. They correspond to the notion of events : for instance, a hotel reservation is an event referring to the "*living*" objects "*person*" and "*hotel*". A date is associated to a reservation and any modification of the reservation characteristics defines a new event which deletes the previous reservation.

We think that in the information system the most important part of the dynamics is inferred from the events as defined above. It is therefore important to be able to distinguish them in the data base. This can be done by analyzing the type of the E.T. attributes.

Calculated-unstable attributes are state variables, which are introduced to facilitate the date base management but they are not essential for the description of the data base. Indeed, the present state of a data base can be deduced from its initial state and from the set of events which occured since the beginning. For that reason, we call these attributes "*situational*".

Thus that kind of attributes is not helpful if one wants to differentiate the E.T.'s we will rather use the three other kinds of attributes to define a typology of the E.T.'s which is useful in the description of the dynamics :

An E.T. will be called "*external event*" if all its attributes (other than situational) are stable and not calculated (for example the E.T. "ORDER").

An E.T. will be called "*internal event*" if all its attributes (other than situational) are stable and calculated (for instance "DELIV-SLIP").

An E.T. which is neither "*external*" nor "*internal event*" is called permanent (in our example CUST, ORD and ITEM). One or more of their attributes must be unstable- not calculated.

Thus we define an "$event$" as a tuple of a relation whose value does not change as time alapses and which may have several different "$states$", as will be seen later on. We suppose that its life time, before it is archived, is short. Its insertion, in a relation may start a procedure of the information system. On the other hand, the permanent E.T. tuples have a longer life time and are referred to by the events.

3.2. - F.R. semantics :

The semantics of an "$inter$" E.T. functional relation are different according to the types of the E.T.'s involved in F.R.. We will perhaps give some definitions to begin with :

A F.R., A $\longrightarrow$ B, will be called $strong$ if it is defined for each value of A. It will be called $stable$ if it is always the same values of B which is associated to a given value of A. Finally, a F.R. will be $total$ if it is both $strong$ and $stable$.

Given two E.T.'s X and Y, a $total$ relationship between their keys may have two different meanings : first it may mean that an entity of type X is an $emanation$ of an entity of type Y . For instance : class-room $\#$ $\longrightarrow$ school $\#$. In this case, the F.R. will be called $logical$.

Second, it may indicate that there exists a $chronology$ in the creation of the two entities. For example the total R.F.

$$\text{INVOICE } \# \longrightarrow \text{ ORDER } \#$$

implies that the INVOICE is created $after$ the order.

If a chronological functional dependency exists between two E.T. keys, then at least one of the E.T.'s belongs to the event type. On the contrary, a F.R. between two permanent E.T. keys expresses a logical relationship.

For example, let us consider the key hierarchy given in figure 4.

Logical and chronological relationship is noted respectively L and c. From this very simple example, we can note that the $primary\ event$ is the order and that its treatment involves $two\ steps$: first, the creation of a delivery slip which then leads to an invoice.

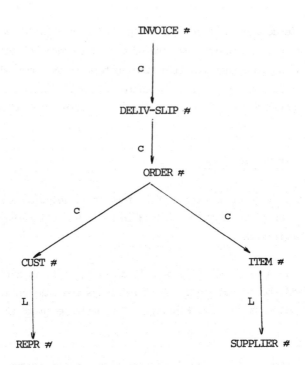

Figure 4.

In some way, and with the help of the key hierarchy we will be able to define the "*gross architecture*" of the treatments associated to a primary event. Though a few precautions are needed when the hierarchy is handled in order to get the processing sequence, the notion of chronology seems a useful tool in the analysis of an information system.

The next paragraph will deal with the concepts used in the description of the treatment of an elementary step (e.g. order ⟶ delivery slip or delivery slip ⟶ invoice).

4 - DESCRIPTION OF INFORMATION SYSTEM DYNAMICS :

Similarly to several authors (HAM 76) (ESW 75) (BEN 76) we will assume that for the description of information system dynamics, one must define the valid states of the data base and the allowed transitions between these states. Before we delineate what a data base state is, let us notice that in this paper we only consider *individual* processing, that is to say the processing resulting from a new fact (event) collected in the data base. We do not take into account set processings (such as statistical treatments) starting from an external decision.

4.1. - Tuple state in an E.T. :

On way of indicating at which stage of processing a given tuple is in an E.T., is to define the concept of tuple state. This notion must be define previous to the definition of data base state.

For instance let us consider a tuple of the E.T. "ORD" : we are not able to say if it concerns an order which has already been processed or a new order. So we will define a tuple state as an information associated to each tuple of a E.T. and which may have a finite number of values. For instance, in the management of an order, one can define four valid states :

State 0 : registered order

State 1 : refused order

State 2 : waiting order

State 3 : fulfilled order

Processing an order is equivalent to realizing a transition between two of the above described states. Only certain transitions are allowed. A graph can be used for the representation of these transitions (figure 5). In this graph the number between parenthesis indicates the state.

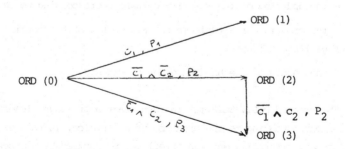

Figure 5.

The existence of an arc between the nodes i and j of this graph means that a given tuple of the E.T. "ORD", in state i, may be processed to state j by the procedure P_{ij} provided that the predicate C_{ij} is fulfilled.

In our example, we consider two predicates which can be expressed under the following form (we assume that X, Y, Z, T are tuples respectively of CUST, ORD(0), ORD(0) U ORD(2) and ITEM).

$$C_1 \; : \; X.CUST \# = Y.CUST \# \wedge X. BALANCE > \text{given value}$$

$$C_2 \; : \; Z.ITEM \# = T. ITEM = \wedge X. ORD\text{-}QTY \leqslant T.QTY\text{-}ON\text{-}HAND.$$

So the set of the E.T. "ORD" is partitioned into four equivalence classes the population of which will have the same processing.

In figure 5, states 1 and 3 are final states and refer to completely processed information (information to archive or delete).

4.2. - Description of the information system dynamics :

Because of the correspondence between the objects of the real world and the entity types of our data model, the dynamics can be described in a simple way, for in this case the tuple is a *"natural"* manipulation unit :

Once more, let us take our example and suppose that a new order is inserted (in state 0), which satisfies the predicate $\bar{c}_1 \wedge c_2$ (figure 5). The procedure P_3 involves :

- the updating of the quantity-on-hand corresponding to the ordered item
- the creation of a tuple in the E.T. DELIV-SLIP according to the calculation rules of its attributes.
- the change of the handled order to state 3.

The created DELIV-SLIP may in turn start a procedure leading to the tuple insertion corresponding to the creation of an invoice. In this example we can see that a primary external event (an order) may be handled by a procedure which created a tuple (in DELIV-SLIP) : this tuple appears then as a (secondary) internal event.

Thus the dynamics are expressed in term of process synchronization, each process been associated to an *"event"* which is a tuple creation in our model.

Though the purpose of this paper is only to give some concepts which seem to us, rather useful, it appears that a high level set oriented language would be very convenient in this approach of the information system dynamics.

4.3. - Data base state :

The data base state is defined by the set of the tuple states in the various E.T.'s ; the data base state changes when a transition between two states of a given tuple takes place. As seen above the state changement may involve situationel attribute updating, tuple creation or deletion.

The data base is in a stable state if all the tuples of *"event"* E.T.'s are either in final states or in states for which no predicate is satisfied. A change in the data base can then occur only if an external event appears.

5 - CONCLUSION :

Though up to now, very few papers have linked data structure and the description of dynamics at a conceptual level, it seems to us that these two components are strongly dependent on each other. It is according to the data model that a more or less convenient description of the evolution of the data base will be reached. Our model is an entity-relation one. It introduces in a simple way the notion of event commonly used in literature but seldom clearly defined. Thus the resulting data structure is in an *"optimal third normal form"* as called by SCHMIDT (SCH75). In the data base, relations play either *"event"*or *"permanent"* role. In most cases, we think that the event role can be deduced from the type of relation attributes. The processing inferred from a primary external event can be broken down into a succession of elementary steps each of which is characterized by a state graph.

It appears that the concept of state is a common one with data users and thus it may be a useful tool in the analysis of an information system.

B I B L I O G R A P H Y

(ANS 75) ANSI/X3/SPARC interim report (1975)

(BEN 76) BENCI G., BODART F., BOGAERT H, CABANES A. : "Concept for the design of a conceptual schema". Modelling in D.B.M.S. North Holl. PUBL. (1976) pp. 181-200.

(CHE 76) CHEN P. : "the entity-relationship model : toward a unified view of data" A.C.M. Transactions on data base systems. Vol.1 n° 1 (march 1976) pp 9-36.

(COD 70) CODD E.F. : "A relational model of data for large shared data banks" Com. A.C.M. 13 (June 1970) pp. 337-387.

(COD 72) CODD E.F. : "Further normalization of the data base relational model" Current Computer Schience Symposia, Data Base System Vol. 6 Prentice Hall (1971)

(ESW 75) ESWARAN P., CHAMBERLIN P. : "Functional specifications of a subsystem for data base integrity" Proc. Very Large Data Base Conf. (Sept. 1975) pp. 48-68

(FLO 77) FLORY A. : "Un modèle et une méthode pour la conception logique d'une base de données" Ph. Thesis (nov. 1977).

(FLO 78) FLORY A., KOULOUMDJIAN J. : "A model and a method for logical data base design" Very Large Data base conf. West Berlin Sept 1978 pp. 315-326.

(HAM 75) HAMMER M., Mc LEOD J. : "Semantic Integrity in a relational data base system" Proc. Very Large Data Base Conf. (Sept. 1975) pp. 25-47.

(HUI 76) HUITS M.H.M. : "Requirements for languages in data base systems" DATA Base Description - North-Holland pub. (1975) pp. 85-110.

(MAC 76) MAC GHEELS C. : "A procedural language for expressing integrity constraints in the coexistence model" Modelling in D.B.M.S. North-Holland Pub. (1976) pp. 293-301.

(SCH 75) SCHMID H.A., SWENSON J. : "On the semantics of the relational data model" A.C.M. SIGMOD Conf. on managt of data (mai 1975).

(WEB 76) WEBER H. : "A semantic model of integrity constraints on a relational data base". Modelling in D.B.M.S. North-Holland Pub. (1976) pp. 269-292.

MODELLING APPROACH FOR DISTRIBUTED DATA BASES

Michel ADIBA

Laboratoire d'Informatique

Université de Grenoble

B.P. 53 / 38041 GRENOBLE Cedex (France)

ABSTRACT

This paper describes a general model for distributed data (MOGADOR) together with a language for describing and manipulating dispersed data.

In providing an homogeneous level for the description and the behaviour of distributed data bases, MOGADOR can be viewed also as a logical tool for designing heterogeneous distributed data bases management systems.

1. INTRODUCTION : DIFFERENT APPROACHES TO DISTRIBUTED DATA BASES MANAGEMENT SYSTEMS

The advent of computer networks and the increasing development of data base technology brought a great potential for sharing data among heterogeneous computing facilities.

This area of research is currently refered to as distributed data bases, one of the main problems being the design and the implementation of a distributed data base management system (DDBMS) [12][23].

In France, a national project ("SIRIUS") projet [24]) sponsored by IRIA, coordinates several research projets on this area such as the one described here which is in process at the Grenoble University.

There is a common agreement to recognize two kinds of DDBMS [11] :
1) Homogeneous or standardized DDBMS where the description and the manipulation of the distributed data base components are made by the same kind of DBMS which is implemented on each sites [26][12].

2) Heterogeneous or integrated DDBMS where these description and manipulation functions are assumed by heterogeneous DBMS such as I.M.S, I.D.S, SOCRATE, etc ... [25][14].

The second approach seems to be more realistic in the way that a great variety of DBMS are to day commercially available. Often in some big enterprises or administrations data processing has been made by sectorization then creating several data bases

This research is supported by IRIA SIRIUS Project (contract 77 076).

with heterogeneous implementations but having however semantic links. It is this common semantic which allows the gathering of different data bases in order to implement new applications.

It is not conceivable to come back to a centralized approach which goes against the natural entreprise structure, but rather to have a distributed data base approach in which each component data base keeps a part of autonomy, while being able to share data with other data bases.

Making the assumption that several data bases are currently in existence and that we want to use them without modification, leads to a cooperation approach where the distributed data base corresponds to the gathering of data stored in these existing data bases [25].

On the other hand, the implementation of a new data base with a distributed DBMS is rather a distributed approach which is easier because of the freedom we have to define the component data bases [14][12].

These two approaches are possible either with homogeneous or heterogeneous DDBMS.

The goals of the POLYPHEME project developed at the Grenoble University are the study and the design of an heterogeneous DDBMS in a cooperation approach [25].

The system architecture, a prototype of which is currently being implemented, stands upon a relational data model (MOGADOR General Model for distributed data [5]) which provides an homogeneous level for :
1) the description and the manipulation of the cooperating data bases and of the distributed data base.
2) the behaviour of the cooperating data bases in order to be able to share data. Each data base is considered as a standard abstract machine.

The goal of this paper is to present this particular data model MOGADOR.

At section 2, we define basic concepts of MOGADOR, i.e level and spaces, object, category, functions.

In section 3, we describe LDDM, i.e a language for distributed data description and manipulation based upon MOGADOR concepts.

With this language, it is possible to ensure, at the local level the homogeneisation of the cooperating data bases by describing them with local views. Through the global view concept this language is used also to describe and manipulate the distributed data base.

2. MOGADOR : BASIC CONCEPTS

A distributed data base is first a data base and we make the assumption that it is
described by a kind of conceptual schema [7] and that the users access it through
external schemas (Figure 2.1).

Let us ignore for the moment this distinction between conceptual and external schemas in order to define the nature of the model and of the corresponding tools
(languages) we need, to implement new applications involving the cooperation of
different and heterogeneous, data bases.

It is a well-known fact that relational models can be used to describe data which
are structured in a hierarchical or network way [18][19].

In a previous paper [4], using Abrial's Data Semantics [1] formalism we give a first
methodological approach for distributed data bases.

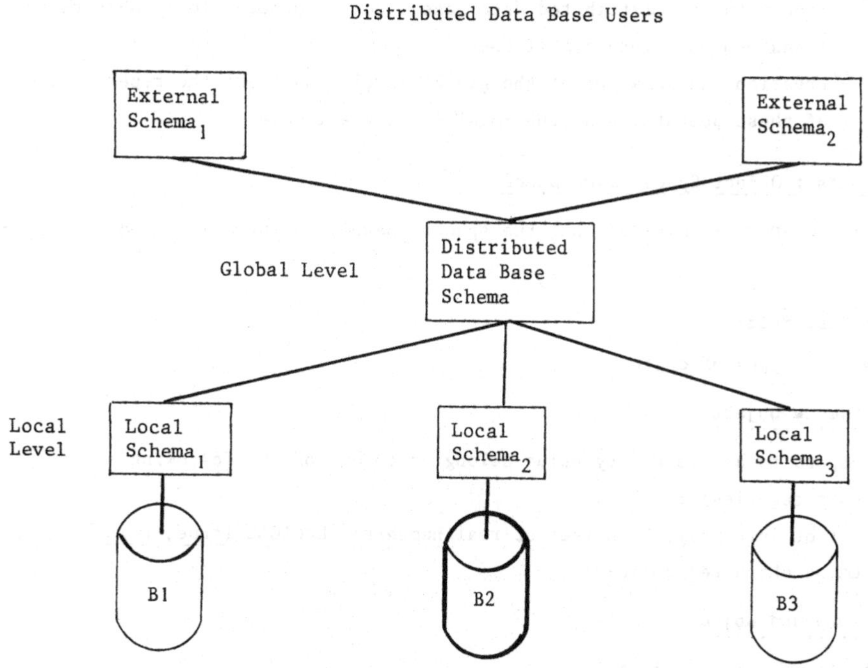

Distributed Data Base Users

Fig. 2.1 - Local Data Bases and Distributed Data Base

In going further in this direction, we define a general model for distributed data :
(MOGADOR) which, in the framework of POLYPHEME project, allows us to implement tools
for describing, retrieving, updating distributed data, tools which are available at
different system levels by interfaces and languages.

Besides set theory, MOGADOR is based on three fundamental concepts :

    1) <u>element</u> concept : objects and names

    2) <u>category</u> concept : set of elements

    3) <u>function</u> concept to express relationships between categories.

These concepts are used at three levels :

    1) At the <u>level of the definition</u> of MOGADOR itself with pre-defined categories and functions.

    2) At the <u>local level</u> of the cooperating data bases to describe the behaviour of these data bases by making homogeneous the data description (local names and objects) and the operations they can execute.

    3) At the <u>global level</u> which concerns the local data bases cooperation, in order to describe the distributed data base schema and its manipulation (global names). Particularly we have to define the mapping between this level and the local ones for the following two types of operations (global rules) :

        - access to the distributed data base : how to process local objects to transform them into global ones ?

        - creation and updating at the global level : what are the repercussions of these operations at the local data base levels ?

2.1. Elements : Object Space, Name Space

Elements in MOGADOR are divided into two spaces, namely object spaces and name spaces.

2.1.1. Objects Space

We define four types of objects.

2.1.1.1. Simple object

They correspond to an elementary value belonging to one of the following sets (predefined categories) :
INTEGER (set of integers), REAL (set of real numbers), LOGICAL (<u>true</u>, <u>false</u>), STRING (set of character strings).

2.1.1.2. Compound object

It is a tuple of simple objects.
For example <F56, NEW-YORK, 525, 10000>.

2.1.1.3. Program

A "program object" corresponds to the set of instructions executable by a given machine.

2.1.1.4. Process

Execution of a given program by a given machine.

2.1.2. Names Space

It concerns the description of an object space as we are going to see in section 3.

A name is a character string. It is used to give names to categories and functions.

By convention, we use upper case letters for categories and lower case letters for functions and we make the distinction between functions which send back one element (monovalued functions) and those which send back a set of elements (multivalued functions [2]).

We consider also that there exists a special name space constituted by predefined categories and functions (3.1).

2.2. Categories

2.2.1. Definition of Categories

In MOGADOR we suppose the existence of predefined categories like INTEGER, REAL, STRING, LOGICAL but also those which correspond to the description of data bases. We shall find in section 3.4.1. a table giving these main predefined categories.
It is possible to define a category using already defined ones.
For this, we use the following operators

1) Assignment ":="

A := B define category A as the set B.

2) Cartesian Product "×"

A×B = {(a,b) | a ∈ A, b ∈ B}.

3) Restriction "(predicate)" after a category is used to define a subset of this category :

A (predicate) = {a ∈ A | predicate (a) = true}.

Examples

i) AGE := INTEGER (18..65) define AGE as a set of integers which are between 18 and 65.

ii) DAY := INTEGER (1..31)
    MONTH := INTEGER (1..12)
    YEAR := INTEGER (0..99)
    DATE := DAY × MONTH × YEAR.

iii) LCC := STRING (length ≤ 8) defines LCC (Local Concrete Category) as a set of strings (names set).

2.2.2. Operations on categories

Four basic operations are defined on categories :

1) Creation of an element of a category.

2) Deletion of an element.

3) Test of the existence of an element in a category.

4) Enumeration of all the elements of a category.

For each space we define operators to realize these operations :

- for object space we have manipulation operators and

- for name space description operators.

These operators will be used in the language for describing and manipulating distributed data (LDDM), see section 3.2.

2.2.3. Abstract Categories

They correspond to sets of simple or compound objects upon which we cannot apply creation and deletion operators. This means that abstract objects already exist in our universe and that we can use them directly.

For example AGE is an abstract category. The character string "AGE" is the name of the category and AGE is the name of a set of integers.

If AC is the name of the abstract categories set we have

$$AGE \in AC \quad (Name)$$
$$\text{and for instance } 26 \in AGE \quad (Object).$$

This notion of abstract category can be viewed as the domain notion in the relational data model [18][19].

At the local level we consider local abstract categories (LAC) and at the global level, global abstract categories (GAC).

2.2.4. Concrete Categories

They correspond to sets of objects upon which the category operations are defined (section 2.2.2.).

This notion is analogous to the relation concept in Codd's relational model but as it was pointed out by J.M. Smith and D.C.M. Smith in [10], this notion supports two distinct forms of abstraction : aggregation, i.e materialization of a relationship into a set of objects, and generalization, where similar objects are regarded as a generic object.

In MOGADOR, to make explicit the difference between these two forms, we consider the function concept as it is described in section 2.3.

To define a concrete category we need at least two elements :

1) the name of the category, for example PERSON, RESERVATION

2) the cartesian product of abstract categories which can be used to identify the concrete object (key). We call it the identifier name of the concrete category.

For instance, if a set of persons are being identified by a social security number, we have :

- SSN := STRING (length = 13)
- Concrete Category PERSON identified by SSN.

If RESERVATION is a set of couples SSN and H♯ (Hotel number) we have Concrete Category RESERVATION identified by SSN × H♯.

We consider local concrete categories (LCC) and global concrete categories (GCC) together with local identifier name (LIN) and global identifier name (GIN).

2.3. Functions

The function concept is a well known mathematical notion [13] which has been applied by Abrial in [1] to data models.

This concept presents a double aspect :

- static aspect, namely the existence of a named relationship f, for example, between two sets A and B
- a dynamic aspect, namely given one object $a \in A$ and a function f how the related object f(a) can be obtained. If function f is completely determined by the existence of its graph (i.e by the set of couples (a, f(a)), then from a given a, we can obtain f(a) by accessing objects in the graph.

The second possibility is to have the set of operations (the equation) to apply on a to obtain f(a).

Applying these mathematical notions to distributed data bases, provides a very flexible way for :

1) expressing the existing relationships between local objects
2) taking into account logical access paths between categories of objects
3) making a given data base execute some data access programs
4) expressing new relationships between distributed objects.

2.3.1. Names and objects

A function in MOGADOR is defined by the following elements :

- the name of the functions (written in lower case letters)
- the type of the function, namely if it is mono or multi-valued
- the source and target, i.e if f goes from A to B, A is the source of f and B its target. Note that A and B can be cartesian product, for instance : birthdate is a monovalued function from PERSON to DATE (DAY × MONTH × YEAR)
- if a relationship between A and B is completely determined by the graph of a function f, this mean that there exists, for example in a local data base, a

set of objects belonging to a concrete category C :
$$C = \{(a, f(a)) \mid \forall a \in A\}.$$

Note however that we are not concerned by the physical representation of concrete categories.

- To denote the inverse function of a function f we use the notation <u>inv</u> f
- Functions can be composed to form new functions
- There is an identify function, named "id".

2.3.2. Operations on functions

We define four basic operations on functions :

1) <u>Access</u> i.e given a to obtain f(a) which can be an element if f is monovalued or a set if f is multivalued.

   By extension if f applies to a set this means that it has to be applied successively to each element of the set :
   $$X = \{x_1, x_2, \ldots, x_n\} \subseteq A$$
   $$f(X) \equiv \{f(x_1), f(x_2), f(x_3), \ldots, f(x_n)\}.$$

2) <u>Link</u> a set of objects to a given object, for example :
   $$f(a) := \{b\}$$
   $$\text{or } g(x) := \{y_1, y_2, \ldots, y_n\}.$$

3) <u>Erase</u> the link between an object and its related objects
   $$f(a) :\neq \{b\}.$$

4) <u>Graph</u> : to obtain the graph of a function $\{(a, f(a))\}$.

3. LDDM : A LANGUAGE FOR DISTRIBUTED DATA DESCRIPTION AND MANIPULATION

3.1. Predefined Categories and Functions

As we have said at the beginning of section 2 the basic concepts of MOGADOR are used at three levels, the first one concerning MOGADOR definition, the second and third ones concerning respectively local and global levels.

Predefined categories, functions and corresponding operations are basic elements of the LDDM language. This language is intended to provide an homogeneous way to describe and use both the components of the distributed data base and the distributed data base itself. Our purpose is not to provide a complete and new data base language like SEQUEL [21] or an equivalent language, but rather to define a minimum set of primitives for describing and manipulating dispersed data, primitives available in a high level host language like PL/1.

It is beyond the scope of this paper to give a complete list of all the predefined categories and functions [5].

The predefined functions are divided into descriptive functions to express relationships between names in a given space or between two different spaces (local and global), and manipulating functions which are in fact operators.

The following table T shows some of the main predefined categories and functions and explanations on its content are given in the following sections.

3.2. Operators

Operators in the LDDM are divided into description and manipulation operators.

3.2.1. Description Operators

They are used to describe both local and global views. These descriptions are given to the cooperation system which stores them in an internal format into local and global machines (Figure 3.1).

These operators apply to predefined categories and functions in order to define name spaces. To simplify the description these operators are combinations of elementary operations seen at section 2.2.2. and 2.3.2.

For example to create a name of a local concrete category (LCC) and to link it with its identifier name (LIN), we use two operators lcc and lin in the following manner :

lcc PERSON lin NAME × FIRST-NAME.

To define a global monovalued function (GOF), together with its source and target we write :

gof age from PERSON to AGE.

328

3.2.2. <u>Manipulation operators</u> (see table TOP)

They are used to manipulate local data bases through the local views and the distributed data base through the global view. Software systems which manipulate the distributed data are viewed as standard automata or abstract data base machines.

We assimilate the name of each local data base with the name of the machine which
permits its utilization (see figure 3.1) and we say that the global machine (named
"g") is the one accessed by distributed data base users.

Each machine is able to execute two kinds of operations :
- <u>primitive operations</u> on categories and functions according to the corresponding local or global view ;
- <u>operations on objects or set of objects</u> : these operations can be applied to
  the result of "enumerate", "access" and "graph" primitives. They are used to
  derive new sets of objects upon which other operations can be applied and so
  on.

Sets of compound objects are in facts n-ary relations so we find here a complete set
of relational operations [18].

MOGADOR Global Machine (g)

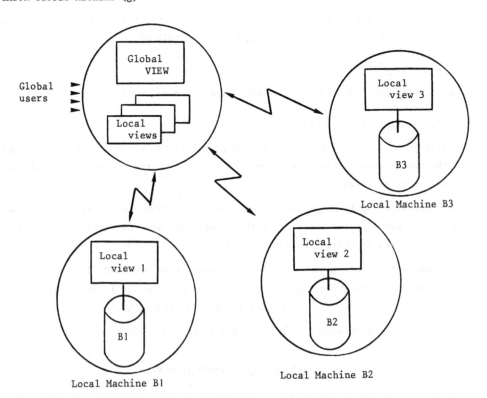

<u>Fig. 3.1</u> - Logical Structure of the Distributed Data Base System

SPACES / LeVELS	1 OBJECTS	2 NAMES-OBJECTS MAPPING	3 NAMES
1 LOCAL	. BASE local data bases . LSO local simple objects . LCO local compound object . LID local identifiers (keys) . LPC local programs codes . LPS local processes	Manipulation operators on local categories and functions (table TOP for m ∈ BASE) (local machines)	. LAC abstract categories . LCC concrete categories . LIN local id. name . LAF (LOF, LMF) mono and multi-valued functions . LPD local programs description
2 LOCAL/GLOBAL MAPPING	GR : global rules on categories and functions (objects localization) GEN, GCR, GDEL, GEX, GFA, GFL, GFE	Decomposition Process (section 3.5.1)	Name's localization GAC → Gacloc → BASE×LAC GCC → Gccloc → BASE×LCC GAF → Gafloc → BASE×LAF BASE GAC → locnamegac → LAC BASE GCC → locnamegcc → LCC BASE GAF → locnamegaf → LAF
3 GLOBAL	GSO global simple object GCO global compound object GID global identifier GPC global program codes GPS global processes	Manipulation operators on global categories and functions (table TOP for m = g) (global machine)	. GAC global abstract categories . GCC global concrete categories . GAF (GOF, GMF) global functions . GIN global identifier name . GPD global program description

Table T - Predefined Categories in MOGADOR

Table TOP : operators

Primitive Operations

Operation	Syntax	Output
creation	create (m, c, i)	$\ell \in$ LOGICAL
deletion	delete (m, c, i)	ℓ
existence test	test (m, c, i)	ℓ
enumeration	enumerate (m, c)	X(Set of i)
access	access (m, f, X) or f(X)	Z
link	link (m, f, x, Y)	ℓ
erase	erase (m, f, x, Y)	ℓ
graph	graph (m, f)	Z

Operations on Simple objects

Operation	Syntax	Output
addition	add (s1, s2)	s3 = s1 + s2
subtraction	sub (s1, s2)	s3 = s1 - s2
multiplication	mult (s1, s2)	s3 = s1 * s2
division	div (s1, s2)	s3 = s1 / s2

Operations on Compound objects

Operation	Syntax	Output
union (∪)	union (X, Y)	Z
intersection (∩)	inter (X, Y)	Z
difference (-)	diff (X, Y)	Z
cartesian product (×)	cart prod (X, Y)	Z
projection	project (X, filter)	Z
restriction	select (X, filter)	Z
join	join (X, Y, condition)	Z
division	divide (X, Y, condition)	Z
cardinality	card (X)	s $\in$ INTEGER
sum, product	sum (X), prod (X)	s $\in$ INTEGER ∪ REAL
eliminate redundancy	unique (X)	Y
concatenation	concat (X, Y)	Z

Notations

- machine name m m $\in$ BASE ∪ {g} for categories and functions operations (g stands for "global")
- category c $\in$ LCC ∪ GCC
- function f $\in$ GAF ∪ LAF (N.B. Composition of function is a function)

- objects ℓ ∈ LOGICAL

        x, y, z objects, X, Y, Z set of objects

        i identifier i ∈ LID ∪ GID

        s simple object.

3.3. Local Level (T11, T12, T13 of table T)

3.3.1. Local view description

Given one local data base in use under a given DBMS, we consider that this data base corresponds to the following elements :

- a local object space constituted by the data stored in the data base : we assume that it is always possible to see this data as simple and compound objects (T11) [3][5]
- a local name space constituted by the data base schema which is re-interpreted in MOGADOR terms to form what we call the local view of the data base (T13).

This means that the data is seen as a collection of n-ary relations but explained in terms of local abstract categories (LAC), local concrete categories (LCC) with their identifier names (LIN) and local access functions (LAF).

Furthermore, associated to this view, we consider a serie of local programs which are pre-compiled in the data base and which realize at least the elementary operations of creation, updating and accessing.

Let us consider a simple example of distributed data base. We have a big enterprise managing several factories making several products. These factories are distributed over the country but data processing is done in 3 computing centers C1, C2 and C3. In C1 we consider data base B1 implemented under a codasyl-like system with the following schema :

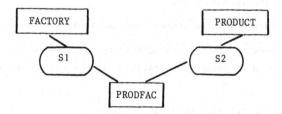

S1 and S2 are codasyl-sets.

Each factory is described by a number F#, a town, the number of employees (NBEMP), the total of all salaries (TOTSAL) and the functioning budget (FBUD).

Each product has a number (P#), a name (PNAME).

Set S1 links a factory to all the products made in this factory and set S2 links a product to all the factories which made it.

In one record PRODFAC we find a factory number (F#), a product number (P#) and the number of products made per day (NBPD).

In relational terms, we have :

B1
$\begin{cases} \text{FACTORY } (\underline{F\#}, \text{ TOWN, EMPNB, TOTSAL, FBUD)} \\ \text{PRODUCT } (\underline{P\#}, \text{ PNAME)} \\ \text{PRODFAC } (\underline{P\#}, \underline{F\#}, \text{ NBPD)} \end{cases}$

With MOGADOR concepts, the local view of this data base B1 is :

LAC = {F#, TOWN, EMPNB, TOTSAL, FBUD, P#, PNAME, NBPD}

LIN = {F#, F# × P#, P#}

LCC = {FACTORY, PRODUCT, PRODFAC}.

Local Function (LAF) :

Function's name	Type	Source	Target	Graph
town	mono	FACTORY	TOWN	in FACTORY
empnb	mono	FACTORY	EMPNB	in FACTORY
totsal	mono	FACTORY	TOTSAL	in FACTORY
fbud	mono	FACTORY	FBUD	in FACTORY
pname	mono	PRODUCT	PNAME	in PRODUCT
nbpd	mono	PRODFAC	NBPD	in PRODFAC
factory	multi	PRODUCT	FACTORY	in PRODFAC
production	multi	FACTORY	PRODFAC	in PRODFAC
fabrication	multi	PRODUCT	PRODFAC	in PRODFAC

N.B. To be complete, this local view example must contain local program descriptions involving :

- the program name
- the type of operation (access, update)
- the nature of inputs and outputs.

The program code is supposed to be stored into the local data base.

N.B. Programs are linked to concrete categories rather than to functions. In fact an access program linked to category C realizes all the monovalued functions having C as source.

In data base B2, we consider also works and products :

B2
MANUFACTORY (M#, TOWN, EMPNB, MBUDGET)

PRODUCT (P#, PNAME)

PROMAN (P#, M#, NBPD).

Note that in B2 the budget is not split into two components as in B1 (TOTSAL and FBUD).

In MOGADOR, we have :

LAC = {M#, TOWN, EMPNB, MBUDGET, P#, PNAME, NBPD}

LCC = {MANUFACTORY, PRODUCT, PROMAN}

LIN = {P#, F#, P# × F#}

Local functions of B2

Name	Type	Source	Target	Graph
town	mono	MANUFACTORY	TOWN	in MANUFACTORY
empnb	mono	MANUFACTORY	EMPNB	in MANUFACTORY
nbudget	mono	MANUFACTORY	MBUDGET	in MANUFACTORY
pname	mono	PRODUCT	PNAME	in PRODUCT
nbpd	mono	PROMAN	NBPD	in PROMAN
manufactory	multi	PRODUCT	MANUFACTORY	in PROMAN
production	multi	FACTORY	PROMAN	in PROMAN
fabrication	multi	PRODUCT	PROMAN	in PROMAN

Finally, in B3 we have :

B3   PRODUCT (P#, PNAME, PDESCR, SELLPRICE, COSTPRICE)

For each product made by the enterprise we have here a complete description with cost and selling prices.

In MOGADOR :

LAC = {P#, PNAME, PDESCR, SELLPRICE, COSTPRICE}

LCC = {PRODUCT}

LIN = {P#}

Local functions of B3 :

Name	Type	Source	Target	Graph
pname	mono	PRODUCT	PNAME	in PRODUCT
pdescr	mono	PRODUCT	PDESCR	in PRODUCT
sellprice	mono	PRODUCT	SELLPRICE	in PRODUCT
costprice	mono	PRODUCT	COSTPRICE	in PRODUCT

3.3.2. Local data base manipulation

The local view is stored by a MOGADOR local machine (see figure 3.1) which is an abstract machine whose physical components can be distributed. This machine can at least execute basic operations on local categories and functions (see T12). These executions involve in fact calls to local programs which are executed by the local DBMS. This MOGADOR local machine provides a standardized behaviour for heterogeneous data bases. It is used by a MOGADOR global machine form which global users manipulate the distributed data base (Figure 3.1).

3.4. Global level and global view

The description of the distributed data base can be logically divided into three parts :
- definition of global names
- localization or mapping between global and local names
- global rules on categories and functions.

3.4.1. Global names space (T31)

It is composed with names of global abstract categories (GAC), global concrete categories (GCC) with their global identifier name (GIN) and global functions GAF (GOF, GMF). Global categories and functions are of two kinds :
- distributed where the global name has some synonym into several local views. This mean that the global objects are in fact local objects dispersed over several object spaces.
- calculated where the global name has no equivalent in the local views. This means that the corresponding global objects are going to be elaborated at the global level by mean of a calculation expressed by a global rule (see section 3.4.3).

N.B. From this global name space, we consider that it is possible to derive external schemas given to users of the distributed data base and for whom the distribution of objects will be transparent.

Example of global view for B1, B2 and B3 :

At the global level we want to see data bases B1, B2 and B3 in the following manner : we consider two global concrete categories (GCC) namely :

- FACTORY which corresponds to distributed but not duplicated objects on B1 and B2. The global identifier name (GIN) is F#. We consider that the criterion for distributing factories depends on the value of the town attribute. For example factories located in New York, Boston or Washington are managed by data base B1 and factories located in Denver, Los Angelès, San Francisco are managed by data base B2. We shall come back to this point in section 3.4.3.1.

- PRODUCT which corresponds to distributed and duplicated objects over B1, B2 and B3 (GIN is P#). We make the assumption that each product is described at least in B3 i.e B3 contains the general catalogue of all the products.

We consider the following global abstract categories (GAC) :
- distributed : F#, TOWN, NBEMP, P#, PDESCR, PNAME, SELLPRICE, COSTPRICE and BUDGET.
The last one is not a strictly distributed category because it exists in B2 (MBUDGET) and not directly in B1 (TOTSAL+FBUD) ; this will be expressed together with the global function "budget" (see 3.4.3).
- calculated : let TOTALP be, for a given product the total number of this product made per day, over all the factories (B1 and B2).

We consider also the following global functions (GAF) :

GAF name	Type	Source	Target	Graph
town	mono	FACTORY	TOWN	in B1 or B2
nbofemp	mono	FACTORY	NBEMP	in B1 or B2
budget	mono	FACTORY	BUDGET	calculated in B1, exists in B2
pname	mono	PRODUCT	PNAME	in B1, B2, B3
pdescr	mono	PRODUCT	PDESCR	in B3
costprice	mono	PRODUCT	COSTPRICE	in B3
sellprice	mono	PRODUCT	SELLPRICE	in B3
production	multi	FACTORY	PRODUCT	in B1 or B2
inv production	multi	PRODUCT	FACTORY	in B1 and B2
totalp	mono	PRODUCT	TOTALP	calculated

3.4.2. Localization on names (T23)

For each distributed GAC, GCC and GAF we have to give the corresponding LAC, LCC and LAF using predefined functions gacloc, gccloc and gafloc (see table T : T23).

For instance

- gacloc (F♯) := ((B1,F♯), (B2,M♯))
- gccloc (PRODUCT) := ((B1,PRODUCT), (B2,PRODUCT), (B3,PRODUCT))
- gccloc (FACTORY) := ((B1,FACTORY), (B2,MANUFACTORY))
- gafloc (nbofemp) := ((B1,empnb), (B2,empnb))
- gafloc (production) := ((B1, inv factory), (B2, inv manufactory))
- gafloc (inv production) := ((B1,factory), (B2,manufactory))
- gafloc (budget) := ((B1, calc), (B2, budget))
- gafloc (totalp) := ∅.

3.4.3. Global rules (T21)

Global rules are a very important notion relating to distributed data bases. We have defined two kinds of global rules, i.e on global concrete categories (GCC) and on global functions (GAF). These global rules express what are the repercussions of a global operation concerning GCC or GAF at the local levels.

For example :

- how to execute the creation of a factory ?
- how to enumerate all the products ?
- how to calculate the TOTALP of a given product ?

Obviously to express all these semantics we need manipulation operators. In the following two sections, we give some examples of global rules. We want to stress, that global rules can express semantic properties of the distributed data base and in this way, they have to be written by a global administrator. However in some simple cases these rules can be deduced automatically by the cooperation system given, for example, only global names and name localizations.

3.4.3.1. Global rules on concrete categories (GCC)

They concern the four operations creation, deletion, enumeration, existence test.
For example to express the following semantic :

→ GR1 global enumeration of factories is realized by enumeration of local components, we write :

gen (FACTORY) := enumeration (gccloc (FACTORY)).

Note that gccloc (FACTORY) gives the set :

((B1, FACTORY), (B2, MANUFACTORY))

so the global rule will be interpreted as (see section 2.3.2)

(enumeration (B1,FACTORY), enumeration (B2,MANUFACTORY)).

At name's level this corresponds to the creation of two independant processes which can be executed in parallel one on the local machine B1, the other on B2.

→ GR2. To express that the creation of a factory depends on the value of attribute TOWN, we suppose the existence of a special global function named "locfactory" from TOWN to BASE and whose graph is :

TOWN	NEW-YORK	BOSTON	WASHINGTON	DENVER	LOS ANGELES	SAN FRAN-CISCO
BASE	B1	B1	B1	B2	B2	B2

Then the global rule for the creation of a factory is :

gcr (FACTORY, F#, TOWN) := create (locfactory (TOWN), locaname
                                   locnamegcc (locfactory (TOWN), FACTORY), F#)

Creation of factory F15 located in DENVER will be :

create (B2, MANUFACTORY, F15).

(locnamegcc is defined in T23).

N.B. Note that if the graph of the localization function (here "locfactory") is not available, this function will be calculated. This allows more complex localization criteria.

→ GR3. The enumeration of all the products is to be made only on B3

gen (PRODUCT) := enumeration (B3, PRODUCT)

→ GR4. The deletion of a product is not allowed :

gdel (PRODUCT) := not allowed.

3.4.3.2. Global rules on functions (GAF)

They concern the three basic operations : access, link, erase.

For example :

→ GR5 : access to the number of employees of a given factory (GAF is nbofemp)

gfa (F♯) := (access (gafloc (nbofemp), F♯))

Two accesses will be generated, one on B1, one on B2 since we dont know where the
factory is located (unless we define a localization function on factories).

GR6 : access to a factory budget (GAF is budget)

Let us give all the description of this function :

guf budget from FACTORY to BUDGET

locgaf (budget) := ((B1, calc), (B2, mbudget))

gfa (F♯) := (add (access (B1, totsal, F♯), access (B1, fbnd, F♯)),

access (B2, mbudget, F♯))

gfl ( ) := impossible the link operation is impossible at the global level

gfe ( ) := impossible idem.

GR7 : Access to all the products made by a given factory :

global multivalued function "product"

gfa (F♯) := (access (gafloc (production), F♯))

N.B. The result set will come only from B1 or B2.

GR8 : Access to all factories which made a given product : global and multivalued
function "inv production".

gfa (P♯) := concat (access (B1, factory, P♯), access (B2, manufactory, P♯)).

The result set is the concatenation of the two sets coming respectively from B1 and
B2 because of duplication of product.

GR9 : Obtain the total number of product per day : global monovalued function
totalp

gfa (P♯) := add (sum (access (B1, nbpd o fabrication, P♯)),

sum (access (B2, nbpd o fabrication, P♯)))

gfl ( ) := impossible

gfe ( ) := impossible.

The operator "o" denotes the composition of functions.

3.5. Manipulation of the distributed data base

We have seen the main elements of LDDM language but these elements cannot constitute
the external form of this language given to an end-user. It is beyond the scope of
this paper to give the syntactic form of this external LDDM but we shall discuss
briefly two points, i.e decomposition of a global transaction into local operations
and execution of these local transactions.

338

3.5.1. Decomposition Process

Let us consider a sample of global transaction :

Q1 give number and town of all the factories which made products whose costprice
is > p.

Q1 can be expressed in LDDM as :

X ← F♯ : id × TOWN : town [(inv product (inv.costprice (> p)))]

This means that from "p" we apply the inv costprice function to find all products
whose costprice is greater than p. On the result (set of P♯) we apply inv-product
which gives all the factories (set of F♯) making those products. On this set of F♯
we apply two functions, "town" and "id" (the identity function) to form a set of
tuple :

$$(F♯, TOWN).$$

Since inv.costprice is not defined in the global view, this expression is in fact :

X ← F♯ : id × TOWN : [town (inv.production (project (select

(P♯ : id × PRICE : costprice (enumeration (PRODUCT)), PRICE > p), P♯)))].

Which can be transformed into the following graph, showing the macrosynchronization
of operations :

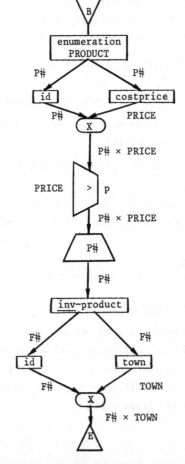

All operations in square boxes are going to be decomposed into local operations using global rules. This will give another graph where some parts are to be executed by local machines. From this graph we have to generate a distributed program and to execute it [29].

3.5.2. Execution process

The distributed program is composed of several procedures which are distributed over several sites. On each site mechanisms are provided to execute these procedures some of them involving calls to procedures which are located in another site [15][16].

Therefore a global transaction is transformed into several global procedures which call local procedures in order to initialize local program execution and which are called themselves by local procedures when local objects are available.

CONCLUSIONS

We have presented here the basic concepts of our distributed data model MOGADOR together with the elements of a language for describing and manipulating distributed data.

In providing an homogeneous level for the description and the behabiour of distributed data bases MOGADOR is not only a data model but also a logical tool for the design of heterogeneous distributed data base management systems.

ACKNOWLEDGMENTS

We are grateful for the comments of J.R. Abrial, C. Delobel and M. Léonard and of all SIRIUS people. We also acknowledge the contribution of all the POLYPHEME team : J.M. Andrade, E. André, J.Y. Caleca, P. Decitre, C. Euzet, Nguyen Gia Toan and A. Stiers. We also acknowledge Professor M. Shave for correcting our english.

REFERENCES

[1] J.R. ABRIAL, Data Semantics, IFIP-TC2 Working Conference, Cargèse, Avril 1974.

[2] J.R. ABRIAL, Langage de spécification Z. Paris, Mai 1977.

[3] M. ADIBA, C. DELOBEL, M. LEONARD, A unified approach for modelling data in
     logical data base design. IFIP-TC2, Freudenstadt, January 1976.

[4] M. ADIBA, C. DELOBEL, The cooperation problem between different Data Base
     Management Systems. IFIP-TC2 Working Conference, Nice, January 1977.

[5] M. ADIBA, Projet POLYPHEME : MOGADOR : Un MOdèle GénérAl de DOnnées Réparties.
     Laboratoire d'Informatique. Research Report 81, July 1977, Grenoble.

[6] ASTRAMAN et al., System R. Relational Approach to Data Base Management.
     ACM-TODS, Vol.1 n°2, June 1976.

[7] ANSI/SPARC, Interim Report ACM Sigmod FDT.7, December 1975.

[8] G. BRACCHI, G. PELAGATTI, P. PAOLINI, Models views and Mappings in multilevel
     Data Base representation. Politechnico di Milano. 1976.

[9] BROOKS, CARDENAS, NAHOURAII, An approach to data communication between diffe-
     rent GDBMS. Very Large Data Base Conference. Brussels, September 1976.

[10] J.M. SMITH and D.C.P. SMITH, Data base abstractions. Aggregations and Genera-
     lization ACM-TODS. Vol.2 Nb.2, June 1977.

[11] M.E. DEPPE, J.P. FRY, Distributed Data Bases : a summary of research.
     Computer Networks 1.1976.

[12] J.B. ROTHNIE, N. GOODMAN, An overview of the preliminary design of SDD-1 :
     a system for distributed data bases C.C.A. Cambridge. 1977.

[13] C.C. PINTER, Set Theory. Addison Wesley Publishing Company. 1971.

[14] G. GARDARIN, M. JOUVE, C. PARENT, S. SPACCAPIETRA, Designing a distributed
     data base management system. AICA. October 1977.

[15] E. ANDRE, P. DECITRE, On providing Distributed Applications Programmers with
     control over synchronizations. Accepted for publication in computer
     network protocols symposium, Liège, February 1978.

[16] DANG, G. SERGEANT, System and Portable Language intended for distributed and
     heterogeneous network applications. ENSIMAG, December 1976.

341

[17]   R. DEMOLOMBE, M. LEMAITRE, Rôles d'un modèle commun dans la conception d'un
        SGBD réparti : analyse des principaux modèles CERT. Research Report.
        March 1977.

[18]   E.F. CODD, A relational model of data for large shared data banks.
        CACM 13, 6 (June 1970).

[19]   P. PIN-SHAN CHEN, The Entity-Relationship model. Toward a unified view of
        data ACM TODS, March 1976.

[20]   J.Y. CALECA, J.P. FORESTIER, L'interrogation simultanée de plusieurs bases de
        données. Rapport de D.E.A., Université de Grenoble. Juin 1976.

[21]   D. CHAMBERLIN et al., Sequel 2. A unified approach to Data Definition,
        Manipulation and Control. IBM Journal of Research and Development,
        Vol.20 n°6, November 1976.

[22]   D. CHAMBERLIN et al., Views Authorization and locking in a relational Data
        Base System. Proc. 1975 National Computer Conference Anaheim Ca.,
        May 1975.

[23]   Canning Publications, Distributed data Systems. EDP Analyser, June 1976,
        Vol.14 n°6.

[24]   J. LE BIHAN, SIRIUS Project, IRIA, Domaine de Voluceau, 78150 LE CHESNAY,
        France.

[25]   POLYPHEME, Propositions pour un modèle de répartition et de coopération de
        Bases de Données dans un réseau d'ordinateurs. Laboratoire Informatiques
        CII/ENSIMAG/USMG, Université de Grenoble. Rapport de Recherche n° 29,
        Décembre 1975.

[26]   E. NEUHOLD, M. STONEBRAKER, A distributed data base version of INGRES.
        Memorandum n° ERL-M612, Septembre 1976, Université de Californie,
        Berkeley.

[27]   M.E. SENKO, DIAM as a detailled example of the ANSI/SPARC architecture.
        IFIP-TC2 Working Conference, Freudenstadt, Germany, January 1976.

[28]   M. STONEBRAKER et al., The Design and Implementation of INGRES. ACM TODS
        Vol.1 n°3, September 1976.

[29]   M. ADIBA, J.Y. CALECA, Modèle relationnel de données réparties, problème de
        décomposition. Journées sur le modèle relationnel. Paris. Avril 1978.

PORTABILITY OF THE PROGRAMS
USING A DATA BASE MANAGEMENT SYSTEM

Michel DEMUYNCK   -   ELECTRICITE DE FRANCE - DER/IMA.
                        Centre de Documentation. 1 Avenue du Général de Gaulle 92140
                        CLAMART.

Patrick MOULIN   -   ELECTRICITE DE FRANCE - DSFJ/STI/DEMA
                        21, rue Joseph Bara 92132 ISSY-LES-MOULINEAUX

Serge VINSON   -   ELECTRICITE DE FRANCE - DSFJ/STI/DEMA
                        21, rue Joseph Bara 92132 ISSY-LES-MOULINEAUX.

ABSTRACT :

This paper deals with the problem raised by the portability of programs using a data
base management system (DBMS).

The study of this problem led us to introduce a number of data access primitives.

Finally we study the translation of these primitives into the data manipulation
language for our DBMS.

INTRODUCTION.

Whether one has to execute an application under various data processing system (software and hardware) or to transfer an application from one system to another, data management raises the rather complex problem of portability.

Already quite apparent in file management systems (FMS), the problem becomes even more crucial in DBMS. This paper is concerned with the solution to the latter problem.

We shall first discuss the design and programming of applications. This will suggest a solution to the above-mentioned problem. To put such an approach into practice, we need to define a set of general purpose primitives. In a second section, the primitives will be studied in detail and equal attention will be paid to the syntax and semantic of the primitives.

We shall then proceed to examine the problem raised by the transformation of primitives into statements which can be understood by the system.

Finally, it may prove interesting to analyze the scope of such a method and its possible applications.

1 - PORTABILITY.

1.1. Outline of the operation of a program using a DBMS.

The major functions of a DBMS are either one of the following :

   a) - Definition of the data and file structures.

   b) - Manipulation of data stored in the bases. This may be performed either by
        using a DBMS embedded in the "host language" (2) or through a command lan-
        guage, or else by the management of transactions between terminals and Data
        bases.

This paper will deal only with the problems raised by the use of a DBMS, in the
"host language" approach. Thus, the pattern of operations of a program using a DBMS
will be more closely examined.

To work with a data base in an application, the user has at his disposal a Data Manipulation Language (DML) which he invokes by means of a CALL, Macro-call to the
DBMS, or simply by means of the commands (eg, READ, WRITE) already implemented in
the language.

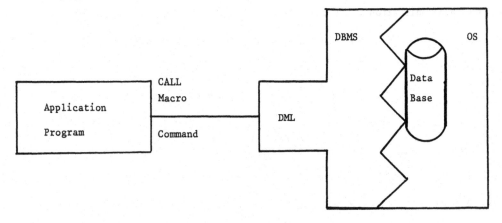

In general, this language allows those three kinds of actions :

   a) - Control commands (open, close,...)

   b) - Search commands (locate, reach, obtain, ...)

   c) - Modification commands (add, delete, replace, reorganize ...).

Due to the physical organization of data, and to the operating system environment,
manipulation languages vary from one DBMS to another.

While the programs are made less dependent upon data organization by the use of a DBMS, they remain quite dependent upon the DML of the DBMS used.

1.2. The problem raised by portability.

The purpose of our study of portability is to reduce or eliminate the aforementioned dependance.

The issue is transforming a program using a DBMS 1 into a program using a DBMS 2. This raises the following problems :

a) - The transformation of the data description in terms of DBMS 1's DDL into a description in terms of DBMS 2's DDL.

b) - The transformation of data managed by DBMS 1 into data manageable by DBMS 2.

c) - The transformation of the program operating with DML 1 into a program operating with DML 2.

The above must obviously be carried out without altering the semantics of program procedures, and data managed by the DBMS.

Problem (a) is outlined in (3), (4), (5) ; for our purposes, we shall consider that transformation (a) is performed manually by the data manager.

The solution to problem (b) depends on whether a solution exists or not to problem (c).

Finally, assuming, solutions to a) and b), the problem of portability comes down to transforming a program using a DML 1 into an equivalent program using a DML 2.

1.3. Application Design.

The model underlying the ANSI (1) report is a three-level model of Data Base Management systems : an internal schema, an external schema and a conceptual schema.

The major quality of the conceptual schema is that it is a semantically stable point. The external schemas are particular views of the conceptual schema and may evolve as long as the semantics are not altered. The impact of such modifications will be felt at the level of the interfaces between the conceptual schema and the external schemas The same is true for the modifications of the internal schema which again will affect only the interface between the internal schema and the conceptual schema.

One may thus measure both the value of the semantically stable point and the importance of the interfaces between the schemas.

As to the portability of programs using a DBMS, the problem results from the differences between the various DBMS, with respect to data description as well as data
manipulation.

For the purposes of our problem, it would be interesting to isolate a stable point
in relation to the internal data description and the internal program description,
in the same manner as in the ANSI report, in which the conceptual schema is independent from the internal schema.

This stable point would thus be a description of data and programs at the conceptual
level.

The translation of the data structure description into the internal level is then
accomplished by interpreting this description. For programs this process must take
into account the physical interpretation of the conceptual structure of the data.

This method can be represented as follows :

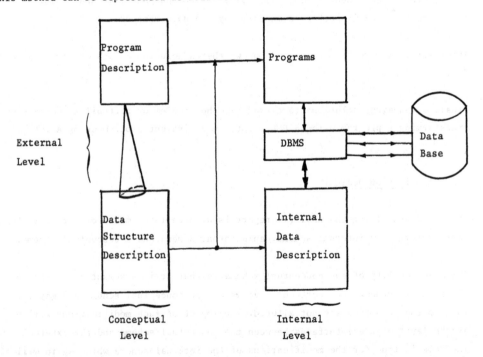

1.3.1. Conceptual model

For our purposes, the conceptual model must be able to define both the data structure
and the processing of these data.

Moreover, the model must be completely free of physical implementation constraints.
The "Entity-Relationship" models proposed in (13) and (14) have the advantage of being
easily understandable. But the drawback is not taking into account the description of
processing. (15) suggests an approach to represent processing through evolution rules.
However, these concepts do not seem currently directly applicable.

As we started to work on the issue of portability, J.R. ABRIAL was developing a tool
for rigorous description of data processing applications : the Z specification lan-
guage(7). After experimental use demonstrated that it was best suited to our purposes, this tool was selected for the conceptual model. In order to make the remainder
of this paper more easily understandable, we shall briefly describe the basic concepts of the language Z.

1.3.1.1. Elements allowing the static description of application.

- Objects and sets of objects.
- Functions between sets of objects.
- Conditions pertaining to objects and functions.

Objects.
There exist several kinds of object denotations :
    . name of object, e.g. : *dupont, 3*
    . function evaluation result, e.g. : *ageofcar (c)*
    . object variable, e.g. : *p*
    . "tuple" (i.e., ordered collection) of objects, e.g. : *(march, 10, 1977)*
    . conditional objects.

A distinction is made between concrete objects (e.g. : person *p*) and abstract objects (e.g. *:dupont, 3*   etc.) (10).

Object identifiers begin with a small letter.

Sets of objects.
There are various ways to designate sets of objects, such as :
    . name of set e.g. : *Name, Person,* etc...

. function evaluation result e.g. : *Personnamed (dupont)*

. set variable

. cartesian product of sets e.g. : *Day* x *Month* x *Year*

. union, intersection, difference of sets.

Set identifiers begin with a capital letter.

Functions :

- single-valued functions. The identifiers of single-valued functions begin with a small letter.

The value of the function is an object of the target set.

e.g.   *Person* $\dfrac{name\ of\ person\ (1)}{}$ *Name*

A single-valued function is total if $\forall a.\exists b$ such that $f(a) = b$ (notation : $f(1)$)

A single-valued function is partial if it is not total.   (notation : $f(0)$)

- multi-valued functions. The identifiers of multi-valued functions begin with a capital letter.

The value of the function is a subset of the target set.

e.g.   *Name* $\dfrac{Personnamed\ (n,m)}{}$ *Person*

The numbers in parentheses (n and m) are the minimum and maximum numbers of elements in the image of an element of the source set.

Conditions :

The static description of application is complemented by stating the conditions related to the objects and functions. These conditions are expressed by means of mathematical symbols.

e.g. $\forall c \in Car.ageofcar\ (c) < 10$

1.3.1.2. Processing_description_language.

This language manipulates the preceding elements and is made up of the following types of phrases :

- assignment phrases for objects or sets of objects :

e.g. *name of person (p)* ← *dupont*
car of person (p) : ∋*c*

- conditional phrases :

e.g. *if condition then action else action end*

- iterative phrases :

e.g. *for c* ∈ *Car ... action ... end*
while condition ... end

1.3.2. Internal model

Data are implemented and accessed by using the possibilities offered by the existing DBMS.

The DBMS are either of "hierarchical - network" type, (CODASYL, SOCRATE (9), IMS (8),...) or of relational type (system R, PRTV (11) , SYNTEX (12)).

The choice of an implementation involves two aspects :

- implementing sets of objects,
- implementing relations.

In order to examine what happens at the internal level, to object's sets of conceptual level, the notion of object's representation should be introduced.

This representation may portray the objects in two ways :

Either, the objects are given a unique representation, in other words, a one-to-one correspondence between the conceptual model set of objects and the internal model set of objects.

These sets may be implemented in, for example, the form of root segment in IMS or in the form of entity in SOCRATE.

Or, the objects are given a multiple-potential representation ; in other words, for one conceptual model object, there will be at least one corresponding internal model object.

These sets may be implemented, for example, as dependent segments or fields in IMS, or, in the case of SOCRATE, as interleaved entities or entity characteristics.

Let us examine the sets of objects of the conceptual model. It has been seen that there are two types of objects : abstract objects which have a value of, for example: 25, toto ... and concrete objects which are compositions of abstract objects, able to represent a physical reality as for example : the person p. The sets of abstract objects will be translated at the internal level by a multiple-potential representation. On the contrary the two possibilities of representation are applicable to the sets of concrete objects.

1.3.3. <u>External model</u>

The external model represents the view of what is real for a given user. This model is made up of a sound and consistent subset of objects, sets and functions of the conceptual model.

1.4. <u>Method of resolution.</u>

The conceptual schema of a data base consists partly of an object description and the relations linking these objects, and partly of the non-algorithmic specification of the processing to be performed on the data.

The processing description uses the objects and relations described in the conceptual data structure. It should be said that, at this stage of design, no other physical contingency occurs, as the problem of the physical implementation of data is dealt with at the level of the internal schema (ref. § 1.3).

From this step, it is necessary to go on to convert this processing formulation into procedures. These procedures are expressed by means of a pseudo-code type of language. (Level Z1 of the Z language).

This phase leads, therefore, to a processing description in algorithmic form, which, as it is important to point out, manipulates the objects and relations of the conceptual structure.

The transition to the following phase will now depend on the options for implementing

351

data and programs. We then enter into the phase : internal schema description :

. description of internal data structure,
. conversion of the pseudo-code into programming language.

This transition to the internal model is the most common one. It corresponds to the well-known approach : systems analysis-programming. All the same, one may wonder if it would be possible to modify this approach, and rather attempt to convert the programs, leaving the data in their conceptual form. The programs would then be clearly independent of the internal data structure. One solution consists in converting the pseudo-code phrases used to manipulate the data, into primitives.

The primitives act as an interface between the concepts contained in the conceptual model and the internal zones of the procedure.

The other procedure phrases may then be directly translated into program phrases by the use of a standard programming language. These phrases will use only strictly portable verbs or variables : as the definition of primitives is equally standard, the program obtained will be portable by construction.

The diagram of 1.3 would then become :

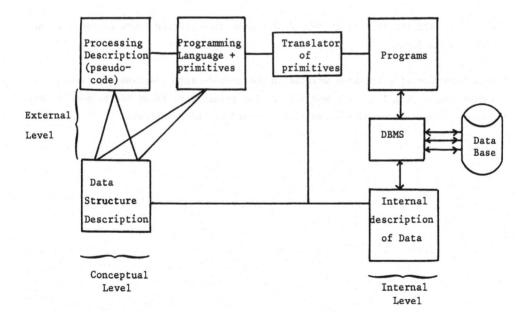

In order to analyze the process in greater detail, let us take the example of a management program using a data base. "Data base" meaning a set of files, we could as well be concerned with conventional files as with a data base managed by a DBMS.

Three types of instructions appear in this program :

    a) – Computing and transaction instructions, e.g. : statement of algebraic expression, assignment, processing of character chains...

    b) – The control structures for the program, e.g. : if –then –else, do while,...

    c) – CALLs to the DBMS or to access methods and, when describing internal data to the program, declarations of structure, e.g. : *CALL DBMS (selection filter)* or *READ FILE*.

In analyzing the three types of instructions, it appears that the statements of type (a) are independent of the Data Management system. The use of a standardized programming language would suffice to make this part portable.

For the code of type (b), only those statements which express the logic associated with computing and processing areas (type a), are portable.

For the other instructions are associated with the use of a data management system, they are not necessarily portable (this is the case for diagnosis tests, access description, etc...).

The code of type (c) is totally dependent on the manipulation language and the physical implementation.

A transformation of procedures written in pseudo-code into programs expressed in a standard programming language, and data access primitives, leads to the non-portable procedure statements being systematically masked by the primitives.

2 - DEFINITION AND USE OF PRIMITIVES.

The methodology selected leads to design programs using primitives which manipulate the elements defined in the conceptual schema.

The procedure will be programmed by using a standard programming language.

For this study, the COBOL ANS (6) language will be used.

The different functions that the primitives must fulfill are :

- creation of objects or links,

- deletion of objects or links,

- modification of links,

- acquisition of objects.

The creation of objects will be defined by the primitive *CREATE* ; the creation and modification of links, by the primitive *LINK* ; the deletion of objects, by the primitive *KILL*; the deletion of links, by the primitive *UNLINK*, and the acquisition of objects, by the primitive *GET*.

Set traversal primitive will be defined as *OPEN*, *NEXT* and *CLOSE* and an object variable primitive as : *DEFINE*.

These primitives essentially manipulate two types of parameters : ∗ COBOL variable :

- CR : return code composed of a diagnostic and/or of an anomaly-processing address.

- Zc : communications area which can be understood by the rest of the procedure and interface with the definition at the conceptual level of data.

- Adress : disconnecting address at the end of set traversel.

∗ Z variable :

- objects. Note : Z variables are marked $ object in order to distinguish them from COBOL variables.

- sets.

- relations.

Let us examine the syntax and semantics of the primitives in greater detail.

Note :   The Z_o specifications of the problem relating to the different examples
          are given in the Appendix.

2.1. Creation of objects or links.

The creation of objects or links refers either to the creation of an object a in a
set A or to the creation or a link between two objects a and b , where b is defined
as the result of the multi-valued function, $F(a)$.

2.1.1. Creation of an object.

Suppose that a person, p, is to be created in the set *Person*.

The Z_1 notation is :

p — *create (Person)*

The syntax of the corresponding primitive will be :

CREATE (cr, $\not{p}$, Person)

Where : CR is the return code,

       $\not{p}$ the Z variable designating the new element of the set,

    Person   is the set defined in Z_0.

2.1.2. Creation of a link.

✱ Suppose that the date of birth of a person is to be created. The phrase which will
be found in the Z_1 description is :

dateofbirth (p) —   *(january, 31, 1977)*

The syntax of the primitive will be :

LINK (cr, dateofbirth ($\not{p}$) = $\mathcal{B}c$)

where : cr is the return code,

      $\not{p}$ the Z variable designating the person,

$⊄c$, the communications area belonging to the host programming language.

<u>Note</u> :   In this case, $⊄c$ will contain a tuple composed of :

(january, 31, 1977).

✱ Suppose that a link is to be created between a person, p , and a car, c .

The corresponding Z_1 expression is :

Carofperson (p) : ∍ c

The primitive will be written :

LINK (cr, Carofperson (⊄p) = ⊄c)

Now, assume that the creation of the link between the person p and the car, defined by its number N, was desired.

The following would have been written :

LINK (cr, Carofperson (⊄p) = carofnumber (N)).

2.2. <u>Deletion of objects or links</u>.

Just as for the creation of objects or links, the following can be considered in the deletion of objects or links :

- either, an object belonging to a set,
- or, a link between two objects, each belonging to a different set.

If an object a of a set A is deleted, all the links stemming from or leading to this object will consequently be broken.

It is therefore appropriate to examine the type of implementation of the resulting sets of objects.

If the set of objects has a unique representation, the resulting object is not destroyed. If the set has a multiple-potential representation, the internal representation of this object is destroyed. It is evident that the links will be destroyed, if they stem from or lead to this object. This process may therefore start off a chain reaction of deletions of links and object representations.

2.2.1. Deletion of an object.

★ Suppose that a car c in the set, *Car*, is to be deleted. The corresponding Z notation is :

kill (c)

The syntax of the primitive will be :

KILL (cr, ∮c, Car)

This primitive will have the effect of deleting the object c as well as all its links with the sets *Person, Make, Numbers* and *Number*.

★ Suppose that a person p of the set *Person* is to be deleted. The Z notation is :

kill (p)

The corresponding primitive is :

KILL (cr, ∮p, Person)

This primitive will have the effect of deleting the object p as well as all its links with the sets *First-name, Name, Sex, Year, Month, Department, City, Day, University, Branch and Distinction.*

Moreover, if all these sets have a multiple-potential representation, this primitive will thus also delete the representations concerning the objects.

2.2.2. Deletion of a link.

★ Suppose that a person p, pays a rent n at date $d1$. If at date $d2$ this person becomes the owner of the house, he (or she) will no longer pay rent. The corresponding expression in Z will be :

rent (p, d2) — nil

Which will be translated into the following primitive :

UNLINK (cr, rent (⊄p ⊄d2) = nil)

✱ Suppose that the link between a car c and the person p is to be deleted. The corresponding Ƶ expression will be :

Carofperson (p) : ⊄c

The translation of this expression in primitives will be :

UNLINK (cr, Carofperson (⊄p) = ⊄c)

This primitive will have the only effect of deleting the link between *Car* and *Person* if the set *Car* has a unique representation.

The car could have been designated by its number, using the relationship *numberof - car* between the sets *Car* and *Numbers*.

The primitive would have thus become :

UNLINK (cr, Carofperson (⊄p) = carofnumber (N))

2.3. Modification of links.

Two cases are involved in link modification :

   a) - If a link is to be modified between two objects connected by a single-valued function, the related primitive will be LINK and its use will be identical to that of a link creation. (§ 2.1).

For example, if the age of the car designated by c , moves from 3 years to 4 years, the Z expression of this link modification will be :

ageofcar (c) — 4

The corresponding primitive will be :

LINK (cr, ageofcar (c) = 4)

    b) - Link modification between two objects connected by a multi-valued function does not involve actual modification. What it does is in fact a link deletion followed by a link creation.

Indeed, suppose that a person p no longer possesses the car c , but now possesses the car c' : the Z expression for this action will be :

Carofperson (p) : ∌ c
Carofperson (p) : ∋ c'

and the natural translation into primitives will yield :

UNLINK (cr, Carofperson (p) = c)
LINK (cr, Carofperson (p) = c')

2.4. Obtaining objects.

Objects are obtained through the primitive GET. The purpose of this primitive is to fetch certain objects belonging to sets defined in Zo.

Just as for the preceding primitives, several objects can be obtained by a single call of the primitive.

Example : One wishes to know the make of the car designated by c . This information can be obtained by simply writing : *makeofcar (c)* . There is therefore no corresponding Z_1 expression.

The syntax of the primitive will be :

GET (cr, makeofcar (c), Bc)

2.5. Set traversal.

It is quite often necessary in programs, to perform an action not only on an object, but also on a set of objects. This set of objects can be a set in the Z sense of the term, or the result of a multi-valued function.

In order to obtain the successive elements of a set, it is necessary to define a set traversal variable. This variable can be used by the previously examined primitives.

a) - The primitive OPEN permits the definition of a common $\check{z}$ variable.

b) - The primitive NEXT is used to initiate the evolution of the common $\check{z}$ variable, so as to point to a new object of the set.

c) - The primitive CLOSE is useful for indicating the end of traversal, liberating the $\check{z}$ variable defined at the time of OPEN.

Example : Suppose that the age of each car is to be increased by one year. The corresponding Z_1 pseudo-code is :

c — openget (Car)
while c ≠ nil
ageofcar (c) — ageofcar (c) + 1
c —get (Car)
end

When translated into primitives, the procedure becomes :

OPEN (cr, ¢c, Car)
DEB. NEXT (cr, ¢c, Fin)
GET (cr, ageofcar (¢c), ฿c)
ADD 1 TO ฿c
LINK (cr, ageofcar (¢c) = ฿c)
GOTO DEB.
FIN. CLOSE (cr, ¢c)

Note : If the only cars whose ages must be increased are those belonging to the person p, then the OPEN primitive will be :

OPEN (cr, ¢c, Carofperson (¢p))

(No other change would occur in the rest of the procedure).

2.6. Variable definition.

Just as in the preceding paragraph a set traversal variable was defined, it may be useful to define an object variable designating a particular object. This is the case where, for example, reference is made several times in a program to the same object.

This is the role of the primitive *DEFINE* . This primitive allows both a simplified writing and the possibility to gain access (logical, then physical for the internal model).

For example, if several references are to be made (by using *GET* ,*LINK, UNLINK...*) to the name of the person p, we shall write :

DEFINE (cr, ɸn, nameofperson (ɸp))

Thus, each time that a reference is made to the name of the person p, ɸn will be included in the primitives.

3 - TRANSLATION OF PRIMITIVES

3.1. General principles.

We've already investigated the static aspects of the application in Z_0, and its dynamic aspects in Z_1. The next step will be to translate the resulting Z_1 pseudo-code into COBOL programs which operate on the Z_0 conceptual model. These programs are made up of primitives representing the access to data and of COBOL instructions representing data manipulation.

These programs are perfectly independent of the internal model chosen to represent the data ; in other words, independent of the internal structure of the data in the DBMS and independent of the DBMS itself.

The next step will consist in describing an internal model and translating the access primitives in accordance with the internal model.

Describing the internal model involves converting the Z_0 conceptual structure, into an internal schema, dependent on the DBMS selected and on its own specifications as well as on the system analyst's ideas about implementation.

This schema will be described by means of the data description language (DDL) of the DBMS selected. Discussion of this problem is not included in this study.

In order to be translated, the primitives (*GET, CREATE, LINK...*) are converted into a language which is understandable by the DBMS. Such a translation requires knowledge of the choices made for the internal structure.

Once the primitives have been translated, they will be entirely dependent on this internal structure. A change in DBMS or in structure will call for retranslation of primitives.

We shall now turn our attention to the translation of a primitive.

A primitive consists of a name and a list of parameters. These parameters are :

- the status returned after completion of the operation,

- the elements of the conceptual model invoked by the primitive,

- the communication area.

The name of the primitive indicates the kind of action which has to be performed.

This action, except for the definition of variables and set traversal(s), will always be ponctual and its effect will be limited to those objects and links of the conceptual model which are designated in the primitive.

It is thus necessary to evaluate the Z-expression of the primitive.

The result of this evaluation will be the access path to the internal representation of an object, or a link of the conceptual model. The result will be called the *context* of the Z-expression.

This evaluation will be performed by a dedicated automaton . This automaton obviously requires some information ; in particular, it must know how sets and functions of the conceptual model are represented in the internal model.

This information, needed by the translator, is included in some tables called *translation tables*.

These tables are attached to an internal representation of the conceptual model. Any change in the internal structure of the data involves changes in the translation tables. Other informations will as well be necessary to evaluate Z-expressions. These are informations related to Z-variables. For instance, if we want to evaluate the expression

nameofperson ($p) ,

then we need information on the function *nameofperson* which is contained in the translation tables and we have to identify which person is designated by the variable *$p. This information will be found in a particular table, associated with the variable. We call it the *context of variable*.

3.2. Description of translation tables.

These tables are made of different kinds of information :

- informations about functions :
    - range and domain
    - allowable access paths
    - translation of the function
    - type of the function (single - or multivalued)

- <u>informations about sets</u> :
    - . internal representation (field, segment, record,...)
    - . mapping of name
    - . designation of I/∅ buffers
    - . type of representation (unique or multiple)
    - . coding of data.

- <u>informations about buffers</u> :
    - . description of I/∅ areas
    - . description of access paths.

3.3. <u>Variable context</u>.

Ƶ-variables occurring in Ƶ-expressions, written as "$ variable", are introduced by means of the primitives for variable definition : e.g. CREATE,DEFINE,OPEN.

We associate a variable context with each Ƶ-variable. This context contains :

- a definition of access to the Ƶ-object internal representation
- a pointer to the active access path
- an indicator of the validity of the Ƶ-object internal representation
- an indicator of the existence of the Ƶ-object internal representation.

3.4. <u>Evaluating Ƶ-expressions</u>.

The purpose of evaluating a Ƶ-expression is to find a context for the expression. This context will permit physical access to the Ƶ-object internal representation. To this aim, we must build a block, the functions being taken one by one from the most internal Ƶ-object of the expression to the left most function. This block will be the definition of the access path. The evaluator builds this block by visiting the translation tables for each function.

3.5. <u>Code generation</u>.

Using the action designated by the name of the primitive, the result of the Ƶ-expression evaluation, and the informations contained in translation tables, the translator will generate Cobol and DML code corresponding to the primitive.

CONCLUSION

We have suggested in this paper a system of primitives which allows programs using a Data Base Management System or a File Management System to be portable. Our approach is limited to programs and does not at all deal with the portability of data structures (it is the hypothesis of this paper that this is dealt with manually).

Our system is founded on the conceptual model chosen (used as the fixed point of the application) : the Z specification language, designed by J.R. ABRIAL.

The primitives have been tested on a few programs.

Studies are now under way to examine the problem of translating these primitives into executable code.

This system, which solves the problem of program portability, may as well be of some interest in the field of distributed data bases, at the level of the global view of local data bases.

o

o   o

ACKNOWLEDGMENTS

We would like to thank both, J.R.ABRIAL for all the advice given throughout the development of this system and for the interest he always manifested during our many discussions and J.RANDON for his fruitful remarks and for his constant optimism.

REFERENCES.

(1) ANSI/X3/SPARC - Study group on DBMS. Interim Report (75.02.08).

(2) CODASYL - Feature Analysis of generalized DBMS - May 1971.

(3) M. ADIBA, C. DELOBEL, M. LEONARD - A unified approach for modelling data in logical data base design - Proceedings IFIP TC 2 Working conference - Freudenstadt - January 1976.

(4) M. ADIBA, C. DELOBEL - The problem of the cooperation between different DBMS. Proceedings IFIP TC 2·Working conference - Nice, January 1977.

(5) H. TARDIEU, H. HECKENROTH, D. NANCI - Etude d'une méthodologie d'analyse et de conception d'une base de données - Mai 1975 - Research contract IRIA N°74180.

(6) ANSI/X3 - American National Standard Programming Language COBOL (23-1974).

(7) J.R. ABRIAL - Z specification language manual - Unpublished.

(8) IMS/V2 - General Information Manual IBM (SH20 - 0912 - 4).

(9) SOCRATE - Manuel d'utilisation et Manuel d'opération. ERIA ECA Automation.

(10) J.R. ABRIAL - Data semantics IFIP TC 2. Working Conference - Cargese 1974.

(11) S.J.P. TODD - The Peterlee Relational test vehicle - A system overview - IBM system journal n° 4 1976.

(12) R. DEMOLOMBE - Syntex - Data structure models for information systems. Institut d'informatique de Namur 1975.

(13) P.P. CHEN - The Entity - Relationship Model - Toward a Unified view of Data. ACM TODS Vol. 1 N°1 (March 1976).

(14) P. MOULIN, J. RANDON, S. SPACCAPIETRA, S. SAVOYSKI, H. TARDIEU, M. TEBOUL - A conceptual model as a data base design tool. Proceedings IFIP TC 2. Working Conference - Freudenstadt. January 1976.

(15) G. BENCI, F. BODART, H. BOGAERT, A. CABANES - Concepts for the design of a Conceptual Schema - Proceedings IFIP TC 2. Working Conference - Freudenstadt - January 1976.

APPENDIX : <u>CONCEPTUAL SCHEMA OF EXAMPLES.</u>

The examples found throughout this paper refer to the following conceptual model.

1 - <u>THE SETS OF OBJECTS ARE.</u>

    Person

    Name

    Firstname

    Car

    Make

    Numbers

    University

    Branch

    Distinction

    Date

    Number

    City

    Department

    Month

    Year

    Sex

    Day

2 - <u>THE LINKS BETWEEN OBJECTS ARE EXPRESSED BY THE RELATIONSHIPS.</u>

Person $\dfrac{\text{ss number (1)}}{\text{person of ss number (0)}}$ Sex x Year x Month x Department x City x Nb

Person $\dfrac{\text{Diploma (0, -)}}{}$ University x Branch x Distinction

Person x Date $\dfrac{\text{rent(0)}}{}$ Number

Person $\dfrac{\text{Carofperson (0, -)}}{}$ Car

Car $\dfrac{\text{ageofcar (1)}}{}$ Number

Car $\dfrac{\text{makeofcar (1)}}{}$ Make

Car $\dfrac{\text{numberofcar (1)}}{\text{carofnumber}}$ Numbers

Person $\dfrac{\text{dateofbirth (1)}}{}$ Day x Month x Year

Person $\dfrac{\text{nameofperson (1)}}{\text{Personnamed (0, -)}}$ Name

Person $\dfrac{\text{firstnameofperson (1)}}{}$ Firstname

A Multi-Level Approach for Data Description and Management
of a Large Hierarchical Database Supporting
a Hospital Patient Information System

K. SAUTER, W. WEINGARTEN,
J. KLONK and P.L. REICHERTZ
Department of Biometrics and Medical Informatics
Medical School Hannover

D-3000 Hannover, Fed. Rep. of Germany
P.O.B. 610180

ABSTRACT

The paper describes a systematic approach to satisfy the various - and partly conflicting - data processing requirements occurring in a large operational hospital database.

The central patient data bank of the Medical System Hannover contains at the present time the data of about 124000 patients with a large number of medical and administrative data for accounting and reporting. This data is stored in several databases under the hierarchical database management system IMS and in standard OS-files. A number of parametric update and retrieval programs is in operation on this data bank and the main design criteria for the data structures had to be the support of the efficiency of these routine programs in addition to database administrator considerations such as storage space economy, security and stability.

In the recent years, the need for more flexible tools to describe and process this large amount of data has become apparent. Therefore, a data description system handling the various views of data has been designed and implemented in a first version.

The choice of the appropriate type of data processing language for a specific task depends on a number of criteria, such as efficiency, input/output requirements and the complexity of the database problem. Therefore, the following hierarchy of user-languages had to be developed:

(1) The conventional PL/1-programming using
    the DBMS sublanguage DL/1
(2) A procedural database language
(3) A descriptive query language.

The main features of the languages (2) and (3) are described in the paper.

1. INTRODUCTION

1.1 System Environment

An integral part of the Medical System Hannover - an operational computer-supported hospital information and control system for the Medical School Hannover (MSH) (11) - is the patient information system handling medical and administrative data of 124000 patients with 178000 stays (May 1978). These data are stored in several hierarchically structured databases using IMS (6) as database management system (DBMS) in conjunction with a number of files of various organization using OS for data management (13). This combination is referred to in the following as the data bank.

Parametric program systems for updating of the data bank, for conversational enquiries, for various periodical reports and for analyses of data are in routine use, along with utilities for database administration, with particular attention being paid to the maintenance of database integrity (11, 13).

1.2 Problems and Objectives

The main layout criteria necessitate that the routine requirements of such an application-oriented data bank be concerned with system reliability and performance.

However, the ever changing information demands of a dynamic hospital environment require that information processes be modified as new needs arise. The appropriate solution of the related database processing requirements has to cope with the following problems:

- that the creation of new application programs is expensive with regard to the qualified manpower involved and should be restricted to certain program classes (see section 3)
- that there is a need to answer 'ad-hoc' questions
- that the data involved may be embedded in various subsystems of the

MSH using different data management systems
- that the maintenance of documentation consistency and actuality become more difficult.

These considerations led to the following objectives:
- Standardization and formalization of the documentation by centralized data and program description (see section 2)
- Standardized interface to presently used data management systems
- Integration of distributed data sets
- Simplification of application programming with a path-oriented procedural data manipulation language (see section 3.1)
- Provision of a descriptive query language (see section 3.2)
- Provision of interfaces to data analysis and presentation systems
- Simple establishment of permanent and temporary data aggregates.

2. DATA DESCRIPTION CONCEPT

A prerequisite for the implementation of powerful data processing tools such as data manipulation languages is a formal description of the data bank. The basic features of the 'Data and Program Description System (DAPRO)' have already been outlined (14), especially the underlying data model consisting of the data element, the record and the file/database as structural levels in addition to relations describing structural objects and their inter-relationships.

To handle the various views of data as they are seen by the data management system, by an application programmer or by a non-EDP expert at a terminal, the main ideas of the 3-schema-model proposed in the literature (1), are used for DAPRO. This approach comprises essentially:
a) The "Internal Schema" (IS) describing the data as it is stored and handled by the data management system, i.e. not on the elementary machine-dependent storage structure level. In the MSH, the IS consists essentially of several hierarchical structures.
b) The "External Schema" (ES) describing the data as it is viewed by the programmer/user and restricting the access to the application-specific subset.
c) The "Conceptual Schema" (CS) describing, in a way free of redundancy and independent of the implementation as well as of specific applications, those aspects of real world entities (i.e. patients) on which data is stored in the data bank.

The data description facility contains the description of:

- the entities within each schema level
- the relationships between these entities within each schema level
- the mappings between the entities and relationships of the different schema levels.

It is evident that the data description of the IS depends on the data(base) management systems implemented. So, the input data required to generate an IMS database description, are a subset of the IS of the respective database. With the design objective of DBMS-independent CS and ES the differences between the various DBMS are restricted to the IS, a major step towards the strategical goal of general system applicability.

Several classes of ES have been identified: the routine application programs are related to ES which are subsets of the IS whereas a query system for the database layman is based on his view of the data and therefore uses an ES which is closer to the CS.

The least consolidated level is the CS, representing the "natural" view of data, including semantic information on the substructures and the relationships between them, constraints to maintain data integrity, etc. A number of recent publications are devoted to this subject: in (9) the major candidate concepts are evaluated. But no realization of a CS is so far known to us, as has quite recently been stated during the IFIP CONGRESS 77 (16).

At the present time, the following objects and relationships are described in DAPRO:

- Internal Schema
  The IS describes "internal records", such as IMS-segments or file-records and the owner-member relationship between records. Internal records are composed of "internal items".

- Conceptual Schema
  As data model for the CS a single hierarchy of conceptual records (entities) was found to be sufficient within the MSH-environment. Conceptual records are composed of "conceptual items". Conceptual items are mapped into the IS by the following parameters:

-- identification of internal record
-- length and position within the internal record
-- format
-- simple mapping condition.

One conceptual item can be mapped onto several internal items, since there is redundancy at the internal level. Also several conceptual items can be mapped onto the same internal item, because the meaning of some internal items depends on the contents of other items from the same internal records.

- External Schema
  At present no separate ES-descriptions are stored. The IS is taken to be the ES for routine programming, whereas the ES for the query-system is a subset of the CS.

3. PROCESSING LANGUAGES

The choice of the appropriate type of data processing language for a specific application depends on a number of criteria. These are among others:
- performance
- input/output-requirements
- complexity of the database problem
Therefore the following hierarchy of languages has been developed:

LEV1) Conventional PL/1 - call DBMS:

This type of programming (host language with database sublanguage) will remain to be useful where:
- optimum performance is essential and/or
- specially formated interfaces for input/output are needed.
An example of this type of problem is the routine data-acquisition program. The support of this type of programming by the data description system and its possible replacement by the integration of general carrier systems (10, 18) into complex database processing remain under investigation.

LEV2) Procedural Data Processing Language:
Certain types of processing problems are not suitable for

formulation in a descriptive language (5). A procedural database language was implemented for solving such problems, this also serving as an intermediate language for the descriptive query processor (see 3.1).

LEV3) Descriptive Query Language:
Within a hospital environment, many of the database processing tasks consist of more or less complex retrieval requests. For such purposes a language was designed that contains features which:
- can be used without extensive training and
- can be implemented with reasonable effort and efficiency (see 3.2).

3.1 The Procedural Data Processing Language (PDPL)

An analysis of data processing requests leads to the identification of general processing functions such as: retrieve data items from data bases, compute data item values, perform logical condition testing and perform output operations. The different processing functions are placed before, between or after physical data accesses and are, in general, defined and checked out by PL/1-IMS programs.

The existing high-level database processing languages do not cope with all of these general problems, for example:
- They only support special data item types and not the whole spectrum of PL/1 - declares or more general data item types built up by sophisticated data transformation routines.
- They only support record accesses by use of special data management routines, with restrictions even on the manner of retrieving or modifying data segments.

The system DAPRO, mentioned above, defines as records all existing data item aggregates, these being records from files, tuples of relations, communication areas of subroutines, or segments of hierarchical data structures from databases. Many data processing requests involve data spread over different hierarchies or, at the very least, different records from a single hierarchy. Thus it is necessary to perform structure navigation by provision of sequences of accesses to data segments.

In the procedural language PDPL the first part of the so-called access path definition comprises the description of the navigational request to a certain record of a particular hierarchy. Data item processing is related to a segment access and therefore the second part of an access path definition contains the related data processing functions.

DAPRO provides support for data item processing, e.g. plausibility checking, extraction from or insertion into records, or automatic conversions between the internal and external formats of data items.

Similar efforts to implement generalized data processing by defining high level languages have been published, e.g.: A relational language of Frasson (4) supports hierarchical databases with the corresponding data type restrictions on its database description. The Lisp Data Manager LIDAM (12) uses a structure base to describe data items and hierarchies, the query being translated into COBOL-code. The University of Toronto (3) uses formal descriptions of data items and data structures for the TOTAL data management system, but has not yet implemented a data processing language. The General Information System (GIS) (7) of IBM enables search definitions to the same extent as DL/1 does for IMS databases, while Operating System files are included in addition. GIS uses a special data description facility (Data Definition Tables) and its language compiler creates modules for the execution of the therewith defined tasks. GIS is, however, not suited for the relational interactive query facilities required within the MSH.

The essential constructs of the PDPL are as follows:
    The data processing task is divided into different access paths. The data items involved in the task are defined once only. Any access path is defined by a sequence of statements having the format:
        <keyword> (<statement>)
    Eight different keywords can be used to describe the data processing tasks required. The format of the statements differs according to the keywords used and their intended function.

The PDPL thus provides a single solution to the problems arising when data items are
    - entered from an external medium, e.g. terminal
    - used for record access control (navigation)
    - inserted into records
    - extracted from records
    - computed by use of arithmetic or string operations
    - involved in logical condition testing
    - transmitted to an external medium.

The execution system of the PDPL is the General Update and Retrieval System (GURS) (17).
Conditional or unconditional actions of this execution system can be specified in the data processing language by use of commands, which complement the PDPL statement types.

These commands may be classified into three groups:
- modify GURS execution parameters such as modes of output
- branch to execution points within an access path or to any other access path
- execute service functions such as: Inspect data description contents, list data definitions, list the access result code, output special messages.

The navigation within hierarchical data structures is record-oriented. The language allows hierarchical record accesses as does DL/1, but it is easier to handle than DL/1.

GURS has been implemented both for batch- and teleprocessing applications.

3.2 The Descriptive Query Language

3.2.1 Language design
The general form of the language follows the usual pattern:
    SELECT < output list > WHERE <condition>.

Implemented or proposed hierarchical languages of this type were found to be not powerful enough. The semantics of the language are therefore based on an interpretation of hierarchies as a collection of relations, where each tuple contains the key of its hierarchical 'father' - tuple. This interpretation is one of the external schemata mentioned in section 2. The <condition> is interpreted as a predicate calculus expression as in ALPHA or QUEL (15). An analysis of typical and frequent retrieval requests showed that the complexity of the <condition> should not be restricted provided that the tuple variables (in the predicate calculus sense) are existentially quantified. The constructs, however, which replace the universal quantifier in other relational languages ('GROUP BY', aggregate functions) are both too complicated for the non-specialized user and not general enough where output analysis is concerned. These considerations led to two decisions:

1.) Grouping and functions are restricted to the defined structural
dependencies.

2.) The result of query execution is not a set or relation but a
file, because it may contain duplicates. Grouping (counting of
frequencies, cross-tabulation) or more sophisticated analysis are
performed in subsequent steps by a general statistical package
(SPSS (8)). Part of the input specifications (e.g. data
description) for this package is also generated by the query
system. In different environments other means of output analysis
may be appropriate (a report generator e.g. in a business
environment).

In most relational languages the structure of the database has to be
specified in each query (e.g. which field is in which relation, which
field of one relation is a primary key of other relations).
In the MSH query language the queries operate on a structured
external schema. By use of the data description system the following
structural information is added to the query:

- Names of data items are unique and predefined in the data description system (section 2), which allows the generation of the
usual RANGE or FROM expressions.
- Tuple variables and quantifiers do not appear explicitly in the
query but are generated through a set of defaults.
- Using the structural dependencies stored in the data description
system, the corresponding 'join' - constructs are added to the
query. The user is, however, free to specify additional
dependencies between the relations or to suppress the inclusion
of the 'join' - terms.

The following example illustrates these points. The request:

"Find the names and admission dates of all patients where the
diagnoses with the codes 5423 and 4711 are stored for this
admission".

is to be executed on the External Schema (ES):

        PAT: NAME, PATKEY,...
          ↓
        ADM: ADMDATE, ADMKEY,...
          ↓
        DIA: DIAG

and takes the form:

SELECT NAME, ADMDATE WHERE DIAG(1) = 5423 ∧ DIAG(2) = 4711.

On the corresponding set of relations:

    PAT: NAME, PATKEY,...
    ADM: PATKEY, ADMDATE, ADMKEY,...
    DIA: DIAG, ADMKEY

the equivalent query in the relational language QUEL (14) is:

RANGE OF X IS PAT
RANGE OF Y IS ADM   } RANGE definition
RANGE OF Z1 IS DIA   for the variables
RANGE OF Z2 IS DIA
RETRIEVE X.NAME, Y.ADMDATE
WHERE Z1.DIAG = 5423
    ∧ Z2.DIAG = 4711
    ∧ Z1.ADMKEY = Y.ADMKEY
    ∧ Z2.ADMKEY = Y.ADMKEY   } 'join' terms
    ∧ Y.PATKEY = X.PATKEY

The equivalent query in the relational language SEQUEL(2) is even longer.

For implementation reasons there are certain restrictions on the use of the language:

- The external structure must consist of a single hierarchy
- The query does not contain references between different occurrences of this structure. (In the example: Data from different patients cannot be compared in the <condition> or combined in one output tuple.)

3.2.2 Query Processing

In a relational file handling system there is no restriction on the sequence of tuple-retrieval calls. Query - translation can be fully oriented towards the structure of the query. This is a complicated process which is as far as we know still not fully understood (15). With a hierarchical data management system the situation is still more complex because here the sequence of DBMS calls has to follow the data structure.

For this reason the following solution was adopted:
Query-processing is divided into two parts: "retrieval" and "evaluation".

- Retrieval: All field names which appear in the query are collected and with the help of the data description system a path through the data bank is constructed which retrieves all values of these fields. This retrieval is carried out using the procedural data processing language PDPL (see 3.1).
- Evaluation: The output of the retrieval process is organized into relations and kept in core. The query is subsequently executed on these in-core relations without the need for complicated optimization.

Because of the language restriction mentioned at the end of the previous section, retrieval and evaluation can be repeated in a cycle for each occurrence of the hierarchical structure (i.e. for each patient in the example). Therefore only a moderate amount of core storage is needed for the intermediate relations.

For certain fields inverted files are maintained within the data bank, mapping these field values to the top-level key. These inversions are used whenever possible.

4. CONCLUSION

The concepts of formal data description and creation of powerful procedural as well as descriptive data manipulation languages based on a central data dictionary and representing a hierarchy of user languages has proven both applicable and useful.

A major objective of future work is the consolidation of the central data description system, especially of the conceptual schema, the incorporation of semantic information and more general mechanics for schema mapping.

REFERENCES:

(1) ANSI/X3/SPARC, Study Group on Data Base Management System, Interim Report, Washington DC: CBEMA, 1975

(2) Chamberlin, D., Boyce, R.: SEQUEL - A Structured English Query Language. IBM Technical Report RJ 1394, IBM Research Lab., San Jose, California, 1974

(3) Dubien, R.J., Corvey, H.D., Sevcik, K.C., Wigle, E.D.: A Data Base System Implementation Providing Data Independence for Medical Applications, Proceedings of the Second World Conference on Medical Informatics, Toronto, 1977

(4) Frasson, C.: A System to Increase Data Independence in a Hierarchical Structure. In: G. Goos and J. Hartmanis (Eds.): Lecture Notes in Computer Science, Vol. 34, Springer-Verlag, Berlin, 235-246, 1975

(5) Huits, M.H.H.: Requirements for Languages in Data Base Systems. In: Douque, B.C.M. and Nijssen, G.M. (Eds.): Data Base Description, North-Holland Publ. Co., Amsterdam, 1975

(6) International Business Machines Corporation (IBM): Information Management System /360, V.2 - General Information Manual, Form No. GH20-0765-1, 1975

(7) International Busines s Machines Corporation (IBM): Generalized Information System, V.2 (GIS/2) - Application Description, Form No. GN20-0360, 1970

(8) Nie, N.H., et al.: Statistical Package for the Social Sciences (SPSS), Second Edition, McGRAW-HILL Book Co., ISBN No. 0-07-046531-2, 1975

(9) Nijssen, G.M.: On the Gross Architecture for the Next Genera-tion Database Management System, Information Proceedings 77. In: Gilchrist, B. (Ed.): IFIP Congress 77, North-Holland Publ. Co., Amsterdam, 1977, 327-335

(10) Pocklington, P.R.: The Necessity for Requirements of and Basic Design of a General Data Interpretation and Evaluation System (DIES). In: Anderson, J. and Forsythe, J.M. (Eds.): MEDINFO'74, North-Holland Publ. Co., Amsterdam, 1974, 411-418

(11) Reichertz, P.L.: The Medical System Hannover (MSH). In: Collen, M.F. (Ed.): Hospital Computer Systems, Wiley & Sons, New York, 1974, 598-661

(12) Risch,T.: LIDAM - LISP Data Manager, Datalogilaboratoriet Report DLU 77/2, Uppsala University, 1977

(13) Sauter, K.: Structure and Functions of the Patient Data Bank in the Medical System Hannover. In: Guenther, A. et al. (Eds.): International Computing Symposium 1973, Davos, North-Holland Publ. Co., Amsterdam, 1974, 585-589.

(14) Sauter, K., Reichertz, P.L., Weingarten, W., Schwarz, B.: A System to Support High Level Data Description and Manipulation of an Operational Data Base System, Medical Informatics, Vol. 1, No. 1, 1976, 15-26

(15) Stonebraker, M., Wong, E., Kreps, P., Held, G.: The Design and Implementation of INGRES, ACM Transactions on Database Systems, Vol. 1, No. 3, 1976, 189-222

(16) Tsichritzis, D. (Chairman): Data Base Organization (Panel). IFIP Congress 77, Toronto, August 7-13, 1977

(17) Weingarten, W., Klonk, J., Sauter, K., Reichertz, P.L.: Individual Data Retrieval by the Non-Programmer and System Supported Data Manipulation in a Complex Hierarchically Organized Data Base System, Proceedings of the Second World Conference on Medical Informatics, Toronto, 1977, 83-86

(18) Wolters, E., Reichertz, P.L.: Problem-Directed Interactive Transaction Management in Medical Systems, Meth. Inform. Med. 15, 1976, 135-140

A COMPLEMENTARY APPROACH TO PROGRAM ANALYSIS AND TESTING

A. Endres and W. Glatthaar
IBM Corporation
Boeblingen/Germany

ABSTRACT

The paper is a contribution to the theory of program testing. We
first discuss why program testing is superior in some respects to
program proving. Then current strategies and criteria for test case selection are reviewed and their shortcomings identified. A methodology is proposed to facilitate the selection of a set of test cases
which is adequate for functional verification. It is necessary,
however, that certain complementary information be used which can
be derived from the program text through formal analysis.

KEYWORDS

Software engineering, software reliability, program verification,
program testing, program semantics.

INTRODUCTION

During the last ten years there has been an abundance of proposals on
how to improve the software development process. One school of thought,
as represented by *Dijkstra* (1969), *Mills* (1971) and *Wirth* (1971), puts
the emphasis on so-called *constructive* methods. In essence, these are
rules and guidelines on how one should proceed when developing software.
By definition, these methods are neither suited nor intended for communicating the end results of the development process. This function, however, is indispensable in practical environments, and has to be provided
by a complementary set of methods. We will refer to these methods as
analytical methods.

The most effective analytical methods are both objective and repeatable.
By *objective* we mean that the result should be as independent of the
person applying the method as possible and, most important, the method
should be applicable with minimal involvement of the program originator.
A *repeatable* method has the advantage that it can be applied by

various persons, either concurrently or at different times and locations, thus reducing the risk of failure and increasing the credibility.
This paper is concerned with program analysis and verification, two
analytical methods that are relevant with respect to program validity and
correctness. They fulfil the above postulates by being applicable to
the program text itself, and also by the fact that they are formal
in nature. A formal method has as a key property the facility of being
mechanized.
The approach described in this paper is influenced by the work of
Miller/Paige (1974), *Bauer* (1975), *Goodenough/Gerhart* (1975) and others.
Easily accessible tutorials on the subject are given by *Huang* (1975) and
Miller (1977). For most of the concepts of this paper, the corres-
ponding formalism has been developed elsewhere and can be found in
Endres (1977).

PROGRAM VALIDITY AND ANALYSIS

To define specifically the role of testing, we introduce first some
concepts and definitions that allow us to restrict as much as possible
the scope of information to be derived from a program through testing.
The basic idea that enables us to do this is reflected in the distinction between program validity and program correctness.

A certain string g is a *valid* program of some language L if it is a
member of the set G of all syntactically and semantically valid programs of that language. The following relations apply:

$$g \in G \quad \text{and} \quad G \subseteq L$$

It is the purpose of *program analysis* to show that these relations
hold. While the term syntactically valid may not need any explanation,
there is hardly any agreement in the literature on the term semantic
validity. We would like to adhere to the most stringent definition
possible. Stringent, in this connection, implies reducing as far as
possible the number of programs that are accepted as valid.

Semantically, programs are function definitions. Hence, if a string g
is a program, it specifies some mapping between a set of values,
called the *input domain* X, and another set of values, called the

output range Z.

$$g: \quad X \to Z$$

The sets X and Z are usually sets of tuples, i.e. Cartesian products
of simple sets.

We say a program is *semantically valid* if both X and Z are specified,
and if, for every tuple in X, the program terminates and the result is
contained in Z. Mathematically speaking, a valid program always defines a total function.
This concept can be illustrated by the following brief examples (in
PASCAL notation):

```
        var x,y: integer;
g₁:     y:= x div 0
g₂:     y:= x div (x - 5)
g₃:     while x > 0 do x := 2 * x - 20
g₄:     if x ≠ 5 then y:= x div (x - 5) else y:= x
g₅:     if x ≥ 0 then y:= x  else y:= -x
```

In most semantic language definitions the above examples would be classified as follows:

g_1:   invalid
g_2,g_3: partially valid
g_4,g_5: (totally) valid

In practical programming environments these classifications may be
incorrect. For some machine architectures, g_1 may give a valid result;
for others, g_5 may only be partially defined (i.e. if abs(Minint) $\neq$
Maxint).

As these examples indicate, the validity of a program is determined
by the operational semantics of the programming language, a semantics
which is dependent both on machine architecture and language implementation. In some cases the effort to do a thorough analysis may be quite
significant, in particular if the question of loop termination is involved.
The concept of validity in the sense of total function may appear
unusually restrictive to some readers. It is really the extension

of the major common idea inherent in several recent language proposals addressing either the type concept, as *Liskov* (1974) and *Wegbreit* (1974), or exception handling, as *Goodenough* (1975), or predicate analysis, as *Dijkstra* (1975).

In our opinion, it is simply irresponsible to consider program validity a run-time, rather than a compile-time, property. Compile-time, of course, is only meant in a generic sense, i.e. it stands for development time. Neither the lack of appropriate tools, nor the effort required, is a reason for not requesting a complete analysis. In other words, programs for which such basic aspects as arithmetic representation, subscript range or loop termination are not handled, should not be allowed to enter the testing phase.

How difficult the analysis of a program may be, it can either be performed, based on the program text alone, or not at all. Testing can and should not be used to contribute to the determination of program validity.

FUNCTIONAL CORRECTNESS AND VERIFICATION

After validity, the next important property of a program is its functional correctness. A program g is said to be *functionally correct* if the function F(g) realized by g corresponds to the *intended* function. The function realized by a program is the set of all pairs of input and output values.

An example:

$$\underline{var} \quad x,y: -2..3;$$
$$g_6: \quad \underline{if} \quad x < 0 \quad \underline{then} \quad y:= x \quad \underline{else} \quad y:= 0$$
$$g_7: \quad \underline{if} \quad x > 0 \quad \underline{then} \quad y:= 2 * (x - x) \quad \underline{else} \quad y:= 2 * x - x$$
$$g_8: \quad y:= x; \quad \underline{while} \quad y > 0 \quad \underline{do} \quad y:= y - 1$$

The function calculated by the above three programs is
$$F(g_6) = F(g_7) = F(g_8) =$$
$$\{ <-2,-2>,<-1,-1>,<0,0>,<1,0>,<2,0>,<3,0> \}$$

Hence, these three programs are functionally equivalent. They differ, however, with respect to storage and time requirements, with respect to

the type of operations used, etc. These are algorithmic properties
that are normally very important, but are irrelevant in this context.

Sometimes F(g) is referred to as the *graph* of the function defined
by g. The program g, on the other hand, constitutes the *rule* to
calculate the right hand elements of F for given left hand elements.
If we assume for a moment that the intended function of a program,
denoted by F*, can be specified in some formal way, then g is correct
with respect to F* if the following relation holds:

$$F(g) \supseteq F^*$$

The relation 'equal to or superset of' includes the possibility that
the realized function may be defined for a larger domain and hence
contains more elements than the intended function. Functional *verification* is any process which shows that the above relation holds.

It follows from the above definitions that program correctness can
only be determined relative to some other information. The availability of this information is the prerequisite for any form of verification.

It is a common problem for all formal verification methods that the
intended function F* is usually only an abstract or vague concept.
Apart from a few simple (and frequently quoted) examples, no concrete
and formal definition of this alternate function specification is
readily available.

A program is usually the result of a mental process, performed by
one or more persons, who assembled, abstracted, and generalized
various pieces of information that they may have obtained from multiple
sources. For a programmer who is well-versed in his trade, the program
text may indeed be the most precise and most concise means of function
description available to him. However, it is typical of the human mind
that at any given point it can concentrate on only a limited subset
or on certain aspects of a multi-element object. A realistic verification method, therefore, has to cope with the situation that the
intended function F* may itself be the product of a development
effort, if it is formally specified, or which is more frequent, that
it cannot be specified and remains a vague concept which takes on a
certain form while time is elapsing or while different people's view

of it are being combined.

PROGRAM PROVING AND PROGRAM TESTING

The two principal methods of functional verification are *program
proving* and *testing*. They differ mainly in the way that information
on the intended function F* is introduced. In the case of proving,
the information introduced is a pair of assertions, called *pre-con-
dition* and *post-condition*, defining conditions that the input values
and the output values have to fulfil. To be meaningful and adequate
as a means of verification, these predicates have to meet two important criteria.

First, the assertions have to be strong enough to describe precisely
the sets of input and output values and the correct relation between
them. One may easily find predicates that describe a certain property
of the output values, but are too weak to provide an alternate or dual
specification of the intended function. The literature on program verification abounds with examples of this. As a simple illustration, we
cite example g_5. This function is usually referred to as abs(x). The
following three formulas are possible postconditions

$$c_1: \quad y \geq 0$$
$$c_2: \quad y \geq x$$
$$c_3: \quad y = \underline{if}\ x \geq 0\ \underline{then}\ x\ \underline{else}\ -x$$

Obviously, formulas c_1 and c_2 are too weak; they describe properties
of $F(g_5)$ which may apply to many other functions as well.

Secondly, the assertions (the *verificator*) must have a higher degree
of authority as to the intent of the program than the program text
itself (the *verificand*). This normally cannot be expected if they
are developed at the same time as the program and by the same person.
As investigations on programming errors have shown(see e.g. *Boehm*
(1975), *Endres* (1975)), the most prevalent errors are so-called design errors. These errors occurred when either the problem to be
solved was misunderstood, or the knowledge of applicable algorithmic
solutions was inadequate. For this reason, it is mandatory that the
development of the dual specification be the result of an independent
thought process. In practice, the possibilities for achieving the
desired independence between object and means of verification, i.e.
between program text and assertions, are rather limited. Possible

ways include development in different languages, at different times
or by different people. None of these alternatives is very practical.
This may explain why the application of this method has been limited
to the few demonstration cases published in the literature.

In the case of program testing, the information on the intended
function F* is introduced by means of pairs of input and output
values, so-called *test cases*. As we shall see later, it is extremely
difficult to devise a set of test cases, or test set, which can serve
as a complete specification of the intended function.

The great advantage of testing, however, is the degree of independence that can be achieved very easily. Testing exposes the graph
of the function, which is a representation considerably different
from the rule. In many cases, testing makes visible the element of
time, which was only implied in the program text, if we think for example
of programs that control mechanical devices. The key point, however,
is that test cases can be developed and understood by persons who
are not programmers. We can thus introduce into the verification
process information that may not have been communicated in other forms
before. A final argument is that programs are written to be executed.
Only testing brings them into this important mode where they interact with their environment, and where errors or misinterpretations of
the input data may be exposed.

Testing has remained, therefore, in spite of many critical comments,
the most widely used and the most successful method of program verification. What is primarily lacking, however, is the theoretical understanding needed to develop effective strategies and meaningful criteria for test case selection and application.

PROGRAM TESTING STRATEGIES

The strategies in use today to select a set of test cases for verifying a given program fall essentially into two categories. The one
group considers the program to be tested as a 'black box'. In this case
two different test criteria are worth mentioning. Most frequently one
performs a test with an *arbitrary sample* of input values, probably
with special preference for some limit values. It is impossible in
this case to say which tests were significant (unless errors were found)
or which degree of verification was achieved. The only meaning-

ful test criterion under the black-box strategy is an *exhaustive* test. This means that every point in the input domain X is verified by a test case. Due to the typical number ranges of many input domains, an exhaustive test leads very frequently to astronomical time requirements and is hence not feasible.

All serious efforts to devise systematic test approaches which are also practical and feasible start out from the conviction that the black-box view is too limited. It is also our view that it is legitimate to combine in the verification process different types of information, provided it can be obtained in an objective and reproduceable way.

All methods that fall in this category make use of additional information obtained through a formal analysis of the program text. Since usually only the control structure of the program is taken into account, some authors refer to their approach as *structured* testing.

The weakest of the criteria in this group is the one requiring that every *statement* in the program should be executed at least once. For a conditional statement, this means that only one of the branches is taken. The tool needed for this level of testing is a tag or counter that is added to every statement. A much more stringent criterion, that includes the previous one as a subset, requires that every path in the program be executed at least once.

Although there appears to be a consensus that only a test covering each path at least once can be considered 'minimally thorough', nothing is said as to the degree of verification such a test has achieved. That a single execution of a path does not contribute significantly to the functional verification of the path will become clear from the following small example. For this we assume that along the respective path the function z := x*y is to be computed, where z is the output and x and y are input variables. If the test case used is the pair (x=2, y=2), the program will be considered correct even if the function actually evaluated was z := x+y.

A NORMAL FORM OF PROGRAMS

A key element for any testing strategy is the concept of a *path*. The precise definition of what constitutes a path varies among authors. We

want to use a definition which is based on the static text of a program and does not depend on the number of iterations for a loop. To do
this, we imagine that the given program g is transformed into a form
such that it does not contain any intermediate variables (see *Bauer*
(1975)). For every output parameter, we obtain a piece of code which
expresses the type of computation leading to it. It will have the form
of a generalized assignment statement:

$$f := p_1 \rightarrow e_1, \; p_2 \rightarrow e_2, \; \ldots, \; p_n \rightarrow e_n$$

In this scheme, f, p and e are metavariables; f designates the output
parameter, p is a Boolean expression and e is any expression. This form
can be derived from a given program text via backward substitution and
predicate transformations. Expression simplification, applied to both
p and e, may exhibit certain redundant or identical computations.

The above form resembles *McCarthy's* (1962) notation for recursive
function definitions. It will, therefore, be called *functional normal
form* or, simply, *functional notation*.

Each condition form $P_i \rightarrow e_i$ will be considered as a *path*, p_i will
be called *path predicate* of path i, and e_i is the corresponding *path expression*. All expressions will be in normal form as well.

Path expressions can be either elementary, i.e. depending on input
parameters and/or constants only, or recursive. In the second case, they
are also dependent on the output parameters. These output parameters
are invocations of the expression e_i itself for some other input value.
They are designated by using the name f of the output parameter as
function designator. Note that recursive expressions occur in practically all cases where loop programs are considered.

The above representation of programs differs from McCarthy's notation,
in that it requires that all p_i are mutually disjoint, hence the sequence of evaluation is irrelevant. We can assume that the functional
analysis has ascertained that the union of all subdomains X_1 through
X_n indeed covers the entire input domain X, that each expression e_i is
totally defined for the subdomain X_i as described by p_i, and that termination is assured for every point in X. The following equations
describe this situation:

$$X = \overset{n}{\underset{i=1}{\cup}} X_i$$

$$X_i = \left\{ \bar{v} \mid p_i(\bar{v}) \right\}$$

$$i \neq j \supset X_i \cap X_j = \{\} \qquad\qquad i,j = 1..n$$

Here $\bar{v}$ designates a particular tuple of values from the range of the input parameters v_1 through v_m, and n is the number of paths in a program. We do not require that n be minimal, that means, that any two path predicates p_i and p_j where e_i is identical to e_j be combined into a single path with the path predicate $p_i \vee p_j$.

The functional normal form described above is a very inefficient representation of programs by any convential measure, e.g. storage or time requirement. It is advantageous, however, in that it exhibits most clearly the functional dependency between input and output parameters.

ERROR AND TEST NEIGHBOURHOODS

The semantic analysis and the transformation into normal form will have eliminated an entire set of typical programming errors that have to do with the overall functional characteristics of the program, like its input domain, the number of independent paths, the number of input and output parameters, etc.

For the next step in the verification process, we take those properties as given. The type of errors that may be left have to do with the detailed functional specification of path predicates and path expressions. Similar to *Howden* (1976), the relevant properties can be classified as follows

    a)   correct path expression

    b)   correct path predicate

    c)   correct association between path predicate and path expression

These properties can, in fact, be verified through testing. Even assuming the normal form as described before, there are still many different valid programs that are functionally equivalent, e.g. all permutations and modifications of parameter names. They are indistinguishable to any test procedure.

We therefore introduce as our universe of discourse the space DF(G), which is the space of all functions F definable by programs in G. This space can be substructured in many subspaces, depending on some pro-

perties of elements in G. As an example, all single parameter programs
form such a subspace, all two parameter programs, etc. If we have to
verify a program g, we only have to verify it with respect to the subspace of functions in which it falls due to the properties derived by
the associated analysis of g. We call such a subspace a *neighbourhood*
of F(g). A neighbourhood of a function always includes the respective
function.

The most interesting neighbourhood of F(g), and the one that would be
most useful for any functional verification, will be called *error
neighbourhood* EN(g). We can imagine it as being the smallest subspace of
definable functions in DF(G) containing F(g) and all other functions
which can result from errors made while writing g. If we would have
sufficient and reliable empirical data as to the type of errors most
likely to occur when using certain construction elements in g, we could
define such a tolerance range for g and consequently for F(g). Testing
of a program would then mean testing the equivalence of two functions
F(g) and F*, both of which are elements of the subspace EN(g).

Unfortunately we are unable to define precisely the error neighbourhood of F(g). Therefore, we shall address the problem from the opposite direction by asking: What are function spaces in which we can
determine the equivalence of two functions through a point-by-point
comparison of a subset of the function's elements? We call such a space
the *test neighbourhood* TN(g) of F(g) (see Figure 1). Note that EN(g) and
TN(g) are really abbreviations for EN(F(g)) and TN(F(g)) respectively.

If such a function space exists, relative to a given g, it must be a
parameterization of g, i.e. we must be able to introduce additional
degrees of freedom in g and eliminate these again by point-to-point
comparison with the intended function F*. Of course, we must also first
ascertain that F* belongs to the same space as well.

A well-known class of functions, where individual member functions
can be uniquely determined by a finite number of points, are the algebraic polynominals.

Before applying this concept to any particular class of functions,
we have to define more stringently some previously used terms.
We refer to a *test case* t as any tuple

$$t = \quad \langle \bar{v}, f \rangle$$

of corresponding input and output values. A test case set, or *test set*, T(TN), is a minimal set of test cases, for a given test neighbourhood, which is adequate in a sense to be defined.

$$T(TN) \quad = \left\{ \; t_i \; \mid \; i=1..s \; \& \; \text{Cond}_{TN}(\bar{t}) \; \right\}$$

Here s stands for the number of test cases required to make up the set T(TN) and $\text{Cond}_{TN}(\bar{t})$ is a predicate to be fulfilled by the test cases t_1 through t_s belonging to the set. The predicate $\text{Cond}_{TN}(\bar{t})$ usually does not select a specific set T as the only adequate set of test cases, but rather an entire *class* TS of *test sets* of which T is an element; hence

$$T(TN) \; \in \; TS(TN)$$

If the path expression under consideration is limited by a path predicate p_i, then the applicable test cases have to be elements of the appropriate subdomain

$$TS_i(TN) \; = \; \left\{ \; T_j \; \mid \; j \; = \; 1..s \; \& \; \text{Cond}_{TN}(\bar{t}) \; \& \; p_i(\bar{v}) \; \right\}$$

Mathematically, each TS_i is a set of sets. There are many possible sets T that satisfy TS. Any individual member T may be selected at random.

TEST SELECTION FOR ELEMENTARY EXPRESSIONS

The type of path expressions considered first are elementary (i.e. non-recursive) arithmetic expressions without division. Their normal form are polynominals.

If for a given path expression e_i, we know that it depends on a single input variable only and the degree of this variable (in the normal form of the expression) is less than or equal to k, then the equivalence of $F_i(g)$ and F_i^* can be shown by k+1 test cases. This follows from the central law of algebra, saying that a polynominal of degree k is uniquely determined if k+1 disjoint points are given.
If v is the independent variable, the polynominals of degree k can be expressed by the following expression scheme:

$$P_1{}^k : a_k * v^k + a_{k-1} * v^{k-1} + \ldots + a_1 * v + a_0 \qquad \text{or}$$

$$\sum_{i=0}^{k} a_i * v^i$$

An individual member of this class is determined if the coefficients a_0 through a_k are determined. By using k+1 value pairs <v,f>, we obtain k+1 linear equations which can be solved for the k+1 'unknowns' a_0 through a_k. Hence the number s of test cases is

$$s = k+1$$

The same principle applies to polynominals with m variables ($m \geq 1$). The corresponding scheme for this class of functions is

$$P_m{}^k : \sum_{0 \leq i_1 \leq k} a_{i_1 \ldots i_m} * v_1{}^{i_1} * \ldots * v_m{}^{i_m} \qquad 1 = 1..m$$

In this scheme, different independent variables are designated by v_1 through v_m. While m is the number of variables, k is the highest degree in one of the variables. The summation above occurs over the set of m-tuples $(i_1, \ldots, i_m)$, such that $0 \leq i_1 \leq k$ and $1 = 1..m$. The number s of coefficients in this case is

$$s = (k+1)^m$$

In order that the respective system of linear equations becomes solvable, the points (test cases) chosen have to be linearly independent of each other. This results in the condition that for any chosen set of test cases, the determinant $D(\bar{v})$ formed from the row vectors has to be different from zero, or

$$\text{Cond}_{TN}(\bar{t}) : \quad D(\bar{v}) \neq 0$$

For examples including division, the normal form is (sometimes) a quotient of two polynominals. Here additional conditions arise, e.g. that the value of the polynominal in the denominator has to be different from zero. Since the properties of the division operator result in many other restrictions, we shall not treat this class any further.

What we have done above is the following. For a given path expression e_i, we have chosen as test neighbourhood the class $p_m{}^k$ of expressions. In addition, we have to pick upper bounds for both m and k. The number m of variables occuring on a given path is certainly limited by the total number of input variables to the program. The highest degree k of any variable is probably often a judgemental decision that should be made jointly by implementer and tester. In large classes of applications, degrees higher than 2 are unlikely to occur. Any erroneous expressions may in this case be caught via the semantical analysis.

If an error is made as to the functional class chosen as test neighbourhood, the result can be misleading. The approach is, therefore, only valid relative to a correctly chosen test neighbourhood.

If the program consists of multiple paths, the correct association between path condition p_i and path expression e_i can be verified if all test cases for e_i are chosen such that the output values are different from all output values of the test cases for other paths. This will give rise to additional conditions for each test set.

TEST SELECTION FOR RECURSIVE EXPRESSIONS

Since loops within a program will most likely result in recursive path expressions, it is important to treat this class as well. We make the restriction again that all expressions and subexpressions are polynominal (i.e. excluding division).
The functional scheme for this class of expressions can be given as follows:

$$R_{m,\ r}^{k}\ :\ \sum_{0\leq i_1\leq k}\ a_{i_1\ldots i_{m+r}}\ *v_1^{i_1}*\ldots*v_m^{i_m}*u_1^{i_{m+1}}*\ldots*u_r^{i_{m+r}}$$

$$\text{with}\ u_j = f(h_j(\bar{v})),\quad j = 1..r,\ l = 1..m+r$$

In this scheme the parameters to be picked are
   m: the number of different independent variables
   k: the highest degree in one of these variables
   r: the number of different recursive invocations of the function $f(\bar{v})$.

We consider two recursive invocations of a function as different if their respective argument expressions $h_j(\bar{v})$ are different. The degree k applies to variables occuring either in the auxiliary functions $h_j(\bar{v})$ or in the main function.

When calculating the number of coefficients to be determined (and hence

the number of test cases required), we treat each different invocation
of f like an additional independent variable $u_j (j=1..r)$. The corresponding formula is therefore

$$s = (k+1)^{m+r}$$

By doing this we pretend that the function to be verified is from a
higher function space, i.e. it has more degrees of freedom, than is
actually the case.

In addition to the function $f(\bar{v})$ itself, we have to be able to determine
the coefficients of the auxiliary functions $h_j(\bar{v})$ as well. This increases the number of test cases further. On the other hand, any independent test case t_i may trigger the execution of several dependent
test cases t_j, t_{j+1}, etc. The proper relation between values of h_j and
f is established by keeping track which test case t_i may trigger the
execution of several dependent test cases t_j, t_{j+1}, etc. The proper
relation between values of h_j and f is established by keeping track
which test case t_i relies on which other test case t_j. For a successful
verification,not only the values of all independent and dependent
test cases have to match with the expected values, but t_j must be a
correct predecessor of t_i in the history of the function.

This point can be explained in terms of recursive function theory as
follows. A recursive function f computes its nth value based on the
history up to the $n-1^{th}$ value. What has to be verified is how the nth
value is being calculated, based both on the actual input parameters
$\bar{v}$ and on that subset of values from the history of f that are being
used for this step.

AN EXAMPLE

As an illustration of the previously developed concepts we select a
well-known loop program, namely the determination of the greatest
common divisor (gcd). We first give the program in loop notation.

```
function z(x,y: integer): integer;
begin while x≠y do
    if x>y then x := x-y else y : = y-x;
    z := x end
```

Without giving the details, we perform a transformation into our functional notation where this program takes on the form:

$$f := p_1 \rightarrow e_1, \; p_2 \rightarrow e_2, \; p_3 \rightarrow e_3$$

with p_1: x>y & x>0 & y>0   and e_1: z(x-y,y)
      p_2: x<y & x>0 & y>0 e_2: z(x,y-x)
      p_3: x=y & x>0 & y>0 e_3: x

We are dealing with three path expressions e_1 through e_3, one of which is elementary and two are recursive. The dependent variable is called z, the independent variables are x and y. The domain X of this function is the set

$$X = X_1 \cup X_2 \cup X_3$$
$$= \{ <x,y> \mid <x,y> \in Int \; \& \; x>0 \; \& \; y>0 \}$$

This information on the actual domain is being derived from the semantic analysis of the program and is independently verified.

We now look at the individual paths. We start with the third one, since its path expression e_3 is non-recursive. This gives us at the same time an illustration for any loop-free program. The decision that must be made in order to proceed is to pick the proper test neighbourhood. We choose the class $p_2{}^1$ which means that no occurence of a variable of degree higher than 1 is anticipated. Here is where analysis and testing complement each other. The property assumed when testing has to be consistent with what the analysis has yielded. In terms of the variable names x and y, the above class is represented by the scheme:

$$P_2{}^1 (x,y): \; a_3*x*y + a_2*x + a_1* y + a_0$$

The test set class TS, which is adequate to verify any function relative to this function scheme, can be characterized by the following set equation

$$TS(P_2{}^1 (x,y)) = \{ <x_i,y_i,z_i> \mid i = 1..4 \; \& \; D(x,y) \neq 0 \}$$

Here suffixes are used to designate individual values from the range of the variables x, y and z. The notation $D(x,y)$ is an abbreviation for the following determinant

$$D(x,y) \;=\; \begin{vmatrix} x_1{}^*y_1 & x_1 & y_1 & 1 \\ x_2{}^*y_2 & x_2 & y_2 & 1 \\ x_3{}^*y_3 & x_3 & y_3 & 1 \\ x_4{}^*y_4 & x_4 & y_4 & 1 \end{vmatrix}$$

It should be noted that this determinant is not particular to the program in question, but rather to the function class $P_2{}^1$. TS is a set of 4-element sets.

If we pick three of the four test cases at random, then the determinant $D(x,y)$ results in a condition that the fourth test case has to fulfil. As an example, we choose the following values for the three first test cases:

$$\{ <1,1,1>,<2,2,2>,<3,3,3> \}$$

Then the condition resulting from the determinant is

$$\begin{vmatrix} 1 & 1 & 1 & 1 \\ 4 & 2 & 2 & 1 \\ 9 & 3 & 3 & 1 \\ x_4{}^*y_4 & x_4 & y_4 & 1 \end{vmatrix} \neq 0$$

Solving the determinant leads to the inequation

$$y_4 \;\neq\; x_4$$

Hence

$$\{ <1,1,1>,<2,2,2>,<3,3,3>,<1,2,1> \} \; \in \; TS(P_2{}^1)$$

In our particular case, the condition for the fourth test case is in conflict with the path condition

$$P_3 : \; x = y$$

This reflects the fact that the same function would be computed if in the program x were replaced by y. In other words, the class to be verified is really $P_1{}^1$, and not $P_2{}^1$; hence two test cases are already sufficient.

We now consider the two recursive path expressions e_1 and e_2. Making the same decisions as before, the test neighbourhood to be picked is the class $R_{2,1}^1$. In terms of the variables x and y, the corresponding scheme is

$$R_{2,1}^1(x,y) : a_7*u*x*y + a_6*u*x + a_5*u*y + a_4*x*y + a_3*u + a_2*x$$
$$+ a_1*y + a_0$$

In this case u stands for

$$u = z(h_1(x,y), h_2(x,y)),$$

where h_1 and h_2 are considered to be elementary functions from the class P_2^1. In order to also verify these auxiliary functions, we have to determine the coefficients in the following two schemes

$$h_1(x,y): b_3*x*y + b_2*x + b_1*y + b_0$$
$$h_2(x,y): c_3*x*y + c_2*x + c_1*y + c_0$$

A test set for this function class has to be a member of the following set of sets

$$TS(R_{2,1}^1(x,y)) = \left\{ <x_i,y_i,z_i> \mid i = 1..8 \,\&\, D(u,x,y) \neq 0 \right.$$

$$\left. \&\, D_{h_1}(x,y) \neq 0 \,\&\, D_{h_2}(x,y) \neq 0 \right\}$$

Here three different determinants occur, corresponding to the three different functional schemes to be tested. They are

$$D(u,x,y) = \begin{vmatrix} u_1*x_1*y_1 & u_1*x_1 & u_1*y_1 & x_1*y_1 & u_1 & x_1 & y_1 & 1 \\ \cdot & \cdot & \cdot & \cdot & \cdot & \cdot & \cdot & \cdot \\ \cdot & \cdot & \cdot & \cdot & \cdot & \cdot & \cdot & \cdot \\ \cdot & \cdot & \cdot & \cdot & \cdot & \cdot & \cdot & \cdot \\ u_8*x_8*y_8 & u_8*x_8 & u_8*y_8 & x_8*y_8 & u_8 & x_8 & y_8 & 1 \end{vmatrix}$$

$$D_{h_1}(x,y) = D_{h_2}(x,y) = \begin{vmatrix} x_1*y_1 & x_1 & y_1 & 1 \\ \cdot & \cdot & \cdot & \cdot \\ \cdot & \cdot & \cdot & \cdot \\ \cdot & \cdot & \cdot & \cdot \\ x_4*y_4 & x_4 & y_4 & 1 \end{vmatrix}$$

In this case it is not practical to first pick 7 test cases at random and then derive a condition for the eighth. Rather, a software tool could be conceived to keep a record of which dependent test cases are triggered by a randomly chosen independent test case.

Figure 2 gives an example of such a test record. It comprises three test sets T_1 through T_3 corresponding to the three paths 1 through 3. Note that both T_1 and T_2 show nine test cases each, the ones in parenthesis, however, are insignificant and hence not part of the set of relevant tests. All independent and arbitrary chosen test cases are numbered, with circled numbers ① through ⑧ . These eight test cases trigger the execution of 14 other different test cases. Also, the predecessor relation between test cases is given by an appropriate arrow. Note that test cases for one path may trigger a test case for another path. The imagined test tool should, for each new test case, check its significance for the appropriate path and gradually fill out and evaluate the respective determinants.

While the determinant $D(u,x,y)$ is the real defining criterion for the corresponding test set, the determinants $D_{h_1}(x,y)$ and $D_{h_2}(x,y)$ may be fulfilled by any 4 out of the 8 test cases for a path. For path 1, the test cases ① , ② plus its dependent test case, and ③ already result in the appropriate condition.

That the set T_1 in Figure 2 is an adequate set to verify the function computed along path 1 relative to the class $R_{2,1}^1$ follows from the fact that the two determinants $D(u,x,y)$ and $D_h(x,y)$, as given below, are both different from zero.

$$
D(u,x,y) \; = \;
\begin{vmatrix}
20 & 5 & 4 & 20 & 1 & 5 & 4 & 1 \\
3 & 3 & 1 & 3 & 1 & 3 & 1 & 1 \\
2 & 2 & 1 & 2 & 1 & 2 & 1 & 1 \\
96 & 16 & 12 & 48 & 2 & 8 & 6 & 1 \\
108 & 36 & 9 & 36 & 3 & 12 & 3 & 1 \\
54 & 18 & 9 & 18 & 3 & 6 & 3 & 1 \\
384 & 48 & 32 & 96 & 4 & 12 & 8 & 1 \\
128 & 32 & 16 & 32 & 4 & 8 & 4 & 1
\end{vmatrix}
\; = \; 155\ 520
$$

$$
D_h(x,y) \; = \;
\begin{vmatrix}
20 & 5 & 4 & 1 \\
3 & 3 & 1 & 1 \\
2 & 2 & 1 & 1 \\
48 & 8 & 6 & 1
\end{vmatrix}
\; = \; 45
$$

A test of the correct association between path predicates and path expressions is not performed in this example.

SUMMARY

In this paper we have presented a first step towards a practical and systematic approach to program verification by combining program analysis and testing. For certain classes of programs, we were able to clarify the question of what information can be obtained through testing and what constitutes an adequate test case set. It will be very interesting to extend this concept to other classes of programs, in particular to programs operating on data structures, lists or non-numerical data.

The testing strategy resulting from the proposed approach offers the great advantage that the information gained and the degree of verification performed increases in proportion to the effort invested. This makes it superior to most other testing strategies currently in use. The major disadvantage of the approach is the size of effort needed to establish and evaluate the determinants for each test set. This cost may be compensated by the savings achieved by the elimination of redundant test runs.

As other formal methods continue to fall short of their expectations, it becomes increasingly important that we better understand the role that testing can and should play. We hope that this paper may spur on more work in this important area.

LITERATURE

Bauer, F.L. : Variables considered harmful,Tech. Univ. Munich, Report 7513 (1975)

Boehm, B.W. et al: Some experience with automated aids to the design of large-scale reliable software, Proceedings Intern. Conf. on Reliable Software, Los Angeles (1975).

Dijkstra, E.W.: Notes on structured programming, Eindhoven Technical University, Report EWD 249 (1969).

Dijkstra, E.W.: Guarded commands, nondeterminacy and formal derivation of programs, Comm. ACM 18,8 (1975)

Endres, A.: An analysis of errors and their causes in system programs, IEEE Transactions on Software Engineering, 1,2 (1975).

Endres, A.: Analyse und Verifikation von Programmen,
    R. Oldenbourg, Munich(1977)

Goodenough, J.B. and Gerhart, S.L.: Toward a theory of test data se-
    lection, IEEE Transactions on Software Engineering, 1,2 (1975).

Goodenough, J.B.: Exeption handling: Issues and a proposed notation,
    Comm. ACM 18,12 (1975)

Howden, W.E.: Reliability of the path analysis testing strategy, IEEE
    Trans. in Software Engg. 2,3 (1976)

Huang, J.C.: An approach to program testing, ACM Comp. Surveys 7,3
    (1975).

Liskov, B.H.: A note on CLU, MIT Project MAC, Memo 112 (1974).

McCarthy, J.: Towards a mathematical science of computation, Proceedings
    IFIP Congress, Munich (1962).

Miller, E.F. and Paige, M.R.: Automatic generation of software test
    cases, Proceedings Eurocomp Conference, Uxbridge (1974).

Miller, E.F.: Program testing: art meets theory,IEEE Computer 9,5 (1977)

Mills, H.D.: Top down programming in large systems, In Debugging Tech-
    niques in Large Systems, R.Rustin (ed), New York University (1971).

Wegbreit, B.: The treatment of data types in EL1, Comm ACM 17,5 (1974).

Wirth, N.: Program development by stepwise refinement, Comm ACM 14,4
    (1971).

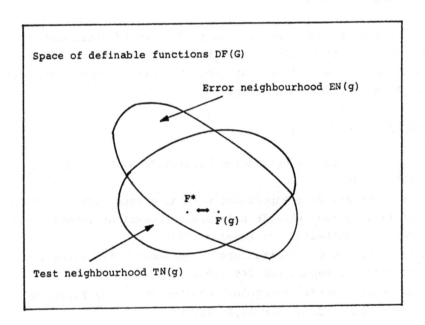

Fig. 1: Error and test neighbourhoods of a function F(g)

401

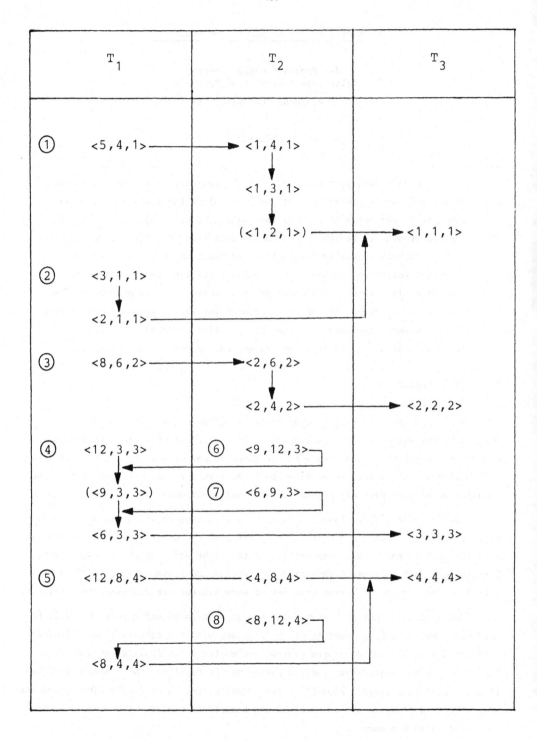

Fig. 2: Test record for the sample program

<u>*ORGANIZING THE SEQUENCING OF PROCESSES*</u>

Fred Lesh
Jet Propulsion Laboratory
California Institute of Technology
Pasadena, CA 91103/USA

ABSTRACT

The difficult part of many computer applications is the design and
implementation of control mechanisms which allow the necessary calculations to be performed at exactly the right time or in exactly the right
logical sequence. Designing systems of this kind involves the designer
in a variety of unusually interesting problems, and requires blending
classical design approaches with techniques peculiar to computational
control applications. This paper presents three examples of such applications and describes the approach adopted for each. The classical design
steps are then discussed, and elements peculiar to computational control
are highlighted by illustrations drawn from the three sample systems.

I.   INTRODUCTION

Many kinds of complexity plague designers of software systems. They are over-
whelmed by the number and sophistication of the equations of motion for an interplanetary spacecraft, innundated with the sheer volume of options and special cases
of a business application, or puzzled how to achieve a convergent algorithm in the
solution of coupled sets of partial differential equations.

But there is a wide class of computer applications in which the primary complexity stems not from the difficulty or variety of the underlying computations but
from the need to make those computations in the right order or at the right time.
Systems developed to achieve this kind of ordering and timing are called "computational control systems." Three examples of such systems are discussed in Section II.

Computational control systems are usually written assuming that they will be
used more than once for a variety of applications within a certain class. Programmers who use an already developed control system for a new application are called
"users." But a computational control system may be developed for a single application in order to achieve a flexible system architecture which avoids large reprogram-
ming efforts during the system checkout and acceptance period. The example of Section II-A is such a case.

The architecture chosen for a computation control system can make the difference
between a chaotic, inefficient program — difficult to understand and maintain — and
an orderly, even elegant program which pleases the maintenance staff as well as the

designer. The task of arriving at an orderly architecture for computational control
systems blends standard system design concepts with other concepts peculiar to problems
of computational control. Section III describes a set of standard steps in system
design and some novel features of the standard steps which arise from peculiarities of
computational control problems. The novel features and their impact on the standard
steps are presented in terms of illustrations drawn from the examples of Section II.

II.   EXAMPLES OF COMPUTATIONAL CONTROL SYSTEMS

Sections A, B, and C below contain examples of computer applications leading to
interesting and difficult computational control problems. A brief description of the
problem is given in each case, followed by an indication of the kind of system developed to solve the problem.

A.   Mosaic Tiling

Figure 1 shows a test frame produced by a program written to process TV picture
data from the camera aboard the Viking Lander spacecraft, which in late 1975 transmitted to earth the first pictures of the rock-strewn plains of Mars. This picture
is a mosaic of tiny tiles — so small that they can be seen in the 20 x 20 centimeter
original glossy only under a strong magnifying lens. The total frame is 1024 tiles
across and 1024 tiles deep. The tiles range from black to white in 64 shades of gray.
Unlike real ceramic tiles, the tiles originate aboard the spacecraft as 6-bit integers
which represent gray levels, reside for a while in computers as 8-bit bytes of data,
and end up as tiny spots on a photographic film. The tiles are called "pixels."

Processing of images, a complicated task that has been the subject of many
papers, is not the concern here. Instead, this paper concentrates on the problem of
producing the total frame — image plus borders, annotation, histograms, gray-level
calibration scale, etc. In this process, the image is assumed to be stored as data
bytes in bulk storage where it can easily be retained and inserted into the frame as
required.

The complications in compiling the frame arise from the fact that the number of
pixels in the image is too large (10^6) to store conveniently in random access memory.
Instead, it is necessary to generate the picture one line at a time and write the lines
on magnetic tape in the format needed by the film recorder, which produces the final
hard copy. The Viking Lander camera produced vertical data lines that sweep from bottom to top instead of the horizontal lines familiar from commercial TV that sweep
from left to right, so it is natural to produce vertical lines for the entire frame
also.

Figure 2 is an approximate expansion of the bottom right-hand corner of one of
the histograms which appears along the right edge of Figure 1. The area of Figure 2

404

Figure 1. Test Frame From a Mosaic Tiling Program

is shown broken into six subareas. Each of these subareas is associated with a program which generates the pixels for that area. These programs are called "pixel generators." The breakdown of the total area of Figure 2 into six subareas is necessary to make each of the pixel generators a simple, logical entity. Even with this much breakdown, the pixel generators for subareas 1 and 6 are far from trivial since the histogram and annotation both vary from one picture to the next.

The difficulty in building a single line of the frame in Figure 1 is that any single vertical line consists of pixels from many different areas. To produce the necessary pixels for a given vertical line, many different pixel generators must run in the right order, and that order is different for some vertical lines than for

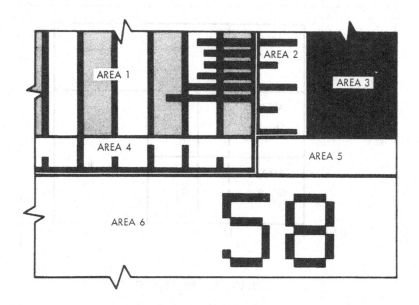

Figure 2. Lower Right Corner of a Histogram

others. Each pixel generator must produce only a small subset of its total pixels
(those for one line through the area associated with the pixel generator) and must
then somehow remember where it was until its turn comes to produce pixels for the next
line. The problem of coordinating the activities of all the pixel generators is an
exercise in computational control.

The control scheme adopted for this problem is table-driven, as illustrated in
Figure 3, which, for explanatory purposes, shows a simple pattern of areas unrelated
to those of Figures 1 and 2. But areas 3 and 5 require the same pixel generator (G_3),
and this is exactly the situation which arises with the histograms that appear along
the middle of the right-hand side of Figure 1. Even though areas 3 and 5 require the
same pixel generator, they need different data.

Each pixel generator is written as a pure procedure operating on a state vector
which can be pointed to by a data pointer P. The pixel generator is written as though
it begins at line zero, and each time it is called, it outputs a single vertical line,
increments the internal line count in its state vector, and returns. The state vector
for the pixel generator may contain just the internal line number, but it may also
contain data such as that used to produce the histograms at the right of Figure 1.

The first row of the control table of Figure 3 tells the control program to
produce lines 0 through 12 of the frame by calling pixel generator G_3 with data pointer
P_5, then calling pixel generator G_3 again with data pointer P_3, and finally calling

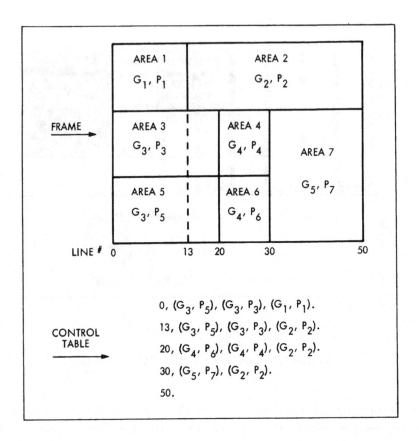

Figure 3. Sample Frame and Control Table for Mosaic Tiling

pixel generator G_1 with data pointer P_1. The control program simply produces lines according to the specifications of the control table, and writes the lines one at a time on magnetic tape.

B.   Spacecraft Control

An interplanetary spacecraft typically consists of as many as 20 subsystems such as the power subsystem, the propulsion subsystem, the attitude control subsystem, and the television subsystem. From a computer software point of view, operating a spacecraft is similar to operating an oil refinery or an automotive assembly plant. The operations to be performed are usually such simple ones as closing a switch. The complexity of the problem derives entirely from the necessity to close the right switch at exactly the right time.

At the Jet Propulsion Laboratory, the natural and historical partitioning of spacecraft into subsystems led easily to consideration of distributed systems of

microprocessors for spacecraft control and data handling, and a breadboard system was
constructed to test the concept [1, 2]. In the breadboard, each subsystem is assigned
a microprocessor with just enough capacity to control that subsystem and to acquire
and format its data. All the processors are tied together with a common data bus.

Even though the spacecraft control job is distributed, each microprocessor still
has several programs which have to run concurrently. The TV subsystem, for example,
has one program to drive a TV camera through a complex cycle of erasures needed to
eliminate the ghost of the old picture from the vidicon, another program to simultaneously drive data lines out of the alternate camera, and a third program to gather
miscellaneous engineering data needed by the system controller to monitor the health
of the TV subsystem.

Most real-time systems are completely interrupt-driven in the sense that all
activity initiations and completions are signaled by a processor interrupt. Interrupt-
handling programs turn on flags requesting that computations be performed, and processing is then done on a highest-priority-first basis. Because it is important to
be able to easily diagnose malfunctions on a spacecraft operating millions of miles
from earth, JPL's breadboard operates in a completely different manner. The only
interrupt is one which occurs simultaneously in all processors every 2.5 milliseconds.
There is nothing special about 2.5 milliseconds — it is simply an interval chosen to
give time resolution more accurately than any which would be needed for spacecraft
control, yet long enough to allow completion of calculations which need to be done
at a given point in time.

This 2.5-millisecond interrupt serves two related purposes. The first purpose
is to act as the escapement for a software clock. The 2.5-millisecond interval
between interrupts is called a "tick." There are 400 ticks to a second, and a software clock in each processor increments time (hours, minutes, seconds, ticks) at each
interrupt. The second purpose is to strobe out all input or output signals which have
been set up by the operation of software during the tick.

The interface circuitry between processors and their instruments is organized
in such a way that electrical control signals produced by software operation during
a given 2.5-millisecond interval are buffered until the end of the interval before
being sent to the external hardware. If programs A and B both run and produce output
signals during a 2.5-millisecond interval, the output signals are identical whether
program A or B runs first. The order of program execution within a 2.5-millisecond
interval is therefore irrelevant, and the programs are thought of as operating simultaneously and instantaneously.

Each program running in this system is required to do specific (usually very
simple) things at exactly the right times as kept by the software clock. The computation control problem in this application centers around the questions: How can
several programs — each of which must time its operations exactly — be run concurrently?

Can application programs be written in such a way that each is transparent to the others during development and operation? Can they be designed and written using standard DO, IF, and CALL structures?

It was originally proposed to organize the software system around time-event tables. But it is difficult to achieve the effect of DO loops in a time-event system. Also, it was desired to do structured design and coding.

The adopted solution to this problem involves two constructs, called WAIT and WHEN, which allow for program timing. If, for example, a programmer wishes to delay the execution of a sequence code defining some action A until the first tick for which his software clock reads SEC = 40, he writes:

WHEN SEC = 40

DO ACTION A

If the programmer wishes to delay the execution of a sequence of code defining action B for exactly 75 ticks, he writes:

WAIT 75 TICKS

DO ACTION B

These constructs can be written anywhere inside or outside of the ranges of DO, IF, or CALL statements. Consider, for example, the case in which it is desired to drive data from a 600-line TV camera. Assume that the camera must start taking the picture at the exact beginning of each new minute (when SEC = 0 and TICK = 0), that the camera takes exactly 10 lines of data each second, and that the line sweep requires 90 milliseconds followed by a 10-millisecond period during which the vidicon beam "flies back" to the beginning of the next line. The control loop for this operation is illustrated in Figure 4.

```
WHEN SEC = 0, TICK = 0
DO FOR LINE = 0, 1, 2,. . . , 799
        START LINE READOUT
        WAIT 36 TICKS
        STOP LINE READOUT
        START FLYBACK
        WAIT 4 TICKS
ENDDO
```

Figure 4. Design Language for Spacecraft Control Applications

All application programs written in this fashion are run under an executive. The WHEN and WAIT constructs are CALLs to WHEN and WAIT subroutines which operate as part of the executive. Each time an application program comes to a WAIT or WHEN, the executive takes over, remembers where control came from, and, on each subsequent TICK, tests whether the conditions stated on the WAIT or WHEN are satisfied. Only when they are satisfied does the executive return control to the application program at the location immediately following the stated test conditions.

Using this scheme, the executive runs, in any tick, only the short segment of a given user program between two logically consecutive WAIT or WHEN statements. The programmer of an application program never needs to be aware that this is going on, but it allows his programs to time their operations and simultaneously enables the executive to maintain control so it can run several application programs concurrently [3].

The only control table needed by the executive is shown in Figure 5 and consists of a set of eight pointers. Each application is assigned to one pointer. If the application program is not running, its pointer is zero. If it is running, its pointer points to the test which must be satisfied before control is returned to the program.

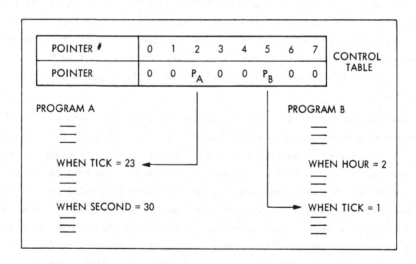

Figure 5. Control Table for Concurrent Program Operation

C. Spacecraft Simulation

The primary product of a spacecraft simulation is a real-time simulated telemetry stream. Months before a spacecraft is launched, teams of engineers and scientists spend hundreds of hours rehearsing their roles in monitoring and controlling its operation. The effectiveness of these rehearsals depends on realistic simulation of the spacecraft and its telemetry stream.

Spacecraft simulation is mostly a discrete problem. Commands arrive at the spacecraft and cause state changes; instruments turn on and off; the telemetry mode changes. All these are discrete events which can be queued and executed in a predetermined order. But if it is to be realistic, spacecraft simulation must also involve things like the three angles of spacecraft attitude, the battery charge, and the instrument temperatures, all of which change continuously. Continuous variables may be calculable directly as functions of time, but more often they are obtainable only as solutions to a simple set of differential equations. So the spacecraft simulation problem is essentially a continuous/discrete problem [4].

One complication in the computational control of simulation arises from the magnitude of the programming task. Programs which simulate given spacecraft subsystems are called "models," and usually 10 to 20 models running simultaneously are required to produce an effective telemetry simulation. A given programmer/analyst is often assigned to write and check out one to five models. Generally, no one programmer/analyst understands all the ramifications of certain spacecraft events.

For example, a program simulating subsystem A may set a logical variable with a given name. That logical variable may affect operation of subsystems B, C, D, and E. The programmer of the subsystem A model may know about some of these instructions, but not all. If he neglects to notify any subsystem model that he has changed a logical variable which affects it, the ramifications of the change will not be propagated correctly.

A second complication in this kind of problem is that there are often a variety of simple actions which must be performed at specified values of the continued variables. In simulating a spacecraft attitude control system, for example, there is a continuous variable — usually a combined position sensor and rate sensor output — which determines whether or not the tiny cold-nitrogen jets which control spacecraft attitude should be on or off. The jet turns on when this variable reaches one value and then shuts off when it reaches another. Simulating the turn-on or shut-off of the jet is a simple matter of changing the value of a parameter used in the differential equations which describe the spacecraft's rotatory motion. But that action must be taken at exactly the right point, and it is not possible to predict in advance the time associated with that point.

A third complication peculiar to spacecraft simulation is that simulated time must be kept close to real time, though not as close as might at first be thought necessary. The reason for this is illustrated in Figure 6. At distances where interplanetary spacecraft spend most of their time (lunar distances or greater), it takes many seconds for a command to pass from the transmitter on earth to the receiver on

411

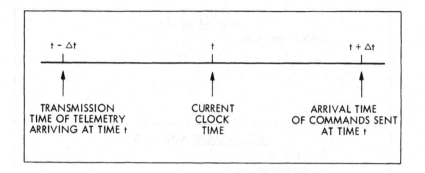

Figure 6. The Real-Time Window

the spacecraft. Telemetry changes resulting from any command require a similar time
to return to earth. Therefore, it is necessary only that simulated time τ remain
between $t - \Delta t$ and $t + \Delta t$, where t is real time and ΔT is the light travel time.
This scheme requires that commands sent at a time t_o be buffered until $\tau = t_o + \Delta t$
and that telemetry data produced at τ_d be buffered until $\tau_d = t - \Delta t$. But the rate
of command transmission is very low, and telemetry data buffers can be kept small by
maintaining τ near $t - \Delta t$.

Effective solution of the computational control problem for spacecraft simulation requires three major elements — one for each of the complications described
above.

To handle the complexity and interaction problem, models are organized as shown
in Figure 7. The first part of each model handles initialization of all continuous
and discrete variables associated with the subsystem. The second part handles discrete
variable calculations. Using a method reminiscent of the General Simulation Language
(GSL), this part of the model has many entries [5]. The code at each entry normally
executes one simple action (event). Figure 7 shows only three entries (10, 20, and
30) for ease of illustration. The third part of the model calculates the continuous
variables associated with the subsystem.

Propagation of effects between models is allowed for by providing a class of
logical variables which must always be set or reset with the statement

CALL LSET (LV,X)

where LV is the logical variable and X gives the value (0 or 1) to which LV is to be
set. The subroutine LSET changes LV, then checks a list associated with the variable
LV to see how to propagate the effects of the variable, and, in effect, enters the

```
┌─────────────────────────────────────────────────────────┐
│                    INITIALIZATION                         │
│      SUBROUTINE ACSI                                      │
│                                                           │
│          ─────                                            │
│          ─────                                            │
│          ─────                                            │
│                                                           │
│      RETURN                                               │
│ ─ ─ ─ ─ ─ ─ ─ ─ ─ ─ ─ ─ ─ ─ ─ ─ ─ ─ ─ ─ ─ ─ ─ ─ ─ ─ ─   │
│                DISCRETE EVENT HANDLING                    │
│      SUBROUTINE ACS                                       │
│      GO TO (10, 20, 30) K                                 │
│   10 ─────                                                │
│          ─────                                            │
│          ─────                                            │
│                                                           │
│      RETURN                                               │
│   20 ─────                                                │
│          ─────                                            │
│          ─────                                            │
│                                                           │
│      RETURN                                               │
│   30 ─────                                                │
│          ─────                                            │
│          ─────                                            │
│                                                           │
│      RETURN                                               │
│ ─ ─ ─ ─ ─ ─ ─ ─ ─ ─ ─ ─ ─ ─ ─ ─ ─ ─ ─ ─ ─ ─ ─ ─ ─ ─ ─   │
│             CONTINUOUS VARIABLE CALCULATION               │
│      SUBROUTINE ACSX                                      │
│                                                           │
│          ─────                                            │
│          ─────                                            │
│          ─────                                            │
│                                                           │
│      RETURN                                               │
└─────────────────────────────────────────────────────────┘
```

Figure 7. Structure of a Simulation Model

appropriate discrete event handlers with the correct control parameter k. The list is constructed automatically as the result of calls of the type

CALL LTAB (LV,T/F/B,M,K)

where the cases T, F, and B are defined as:

Case	Meaning
T	call model M with parameter K when logical variable LV goes from F to T
F	call model M with parameter K when logical variable LV goes from T to F
B	call model M with parameter K with logical variable LV changes

Most calls to LTAB are made at initialization time, but they can be and often are made from the discrete event subroutines.

The second major element is required to allow programmers to easily specify and achieve the execution of events at given values of continuous variables. To specify that an action be taken at a given time, the programmer writes:

CALL TSET (T,M,K)

Here T is the value of time (t) at which the action is taken, and the code which performs the action is at entry K of the discrete event handling portion of model M. To specify that an action be taken when a continuous variable V becomes equal to C, the programmer writes:

CALL YSET (V,C,M,K)

The set of items (V,C,M,K) is called a "Z-trigger," and the point at which V = C is called the "Z-trigger 0-point."

The third major element (required to keep simulated time close to real time) is the computational control algorithm itself. This algorithm is shown in Figure 8. The 0-points of Z-triggers are not easy to find, and the search requires iteration. The complexity of this search is completely represented by step 4 in the algorithm of Figure 8.

III. STEPS IN DESIGNING COMPUTATIONAL CONTROL SYSTEMS

System requirements are elusive, and gathering and documenting requirements is a major part of any system development. But the problem of requirements gathering is a subject in itself and too complex to be covered here.

So this paper assumes that a reasonably good statement of the requirements is available and discusses the standard system design steps that remain, emphasizing peculiarities and regularities which arise from restricting the problem class to that of computational control problems.

A.   Analysis

The first task of the system designer is to establish the elements from which the system will be constructed. In the case of the mosaic tiling system, for example, the elements are the individual pixel generators in the form of closed subroutines, the control table, and the system control program which uses the pixel generators and the table to construct the picture lines. In the case of simulation, the elements are more complex. There is still a system control program, but it is far more complicated than that for mosaic tiling. Instead of pixel generators, there is a set of models, but each model has a complicated internal structure which needs to be understood in

414

```
LET H BE THE MAXIMUM STEP SIZE FOR CONTINUOUS VARIABLE CALCULATIONS

LET δ BE THE ONE-WAY LIGHT-TRAVEL TIME

LET t BE REAL TIME

LET τ BE SIMULATED TIME

LET τ_L AND τ_R BE THE LEFT AND RIGHT ENDS OF A SIMULATED TIME INTERVAL

LET T_NEX BE THE SMALLEST T FOR ANY UNSATISFIED TSET CALLS

LET ε < δ/2
```

0. INITIALIZE ALL MODELS

1. SET $\tau_L = \tau$, $\tau_R = MIN(\tau + \delta/2, \tau + H, T_{NEX})$

2. WAIT UNTIL $\tau < t - \delta + \epsilon$

3. CALCULATE CONTINUOUS VARIABLES AT τ_R FOR ALL MODELS

4. IF ANY Z-TRIGGER 0-POINTS LIE BETWEEN τ_L AND τ_R ADJUST τ_R UNTIL IT IS THE FIRST Z-TRIGGER 0-POINT

5. INTERPOLATE AND BUFFER TELEMETRY FOR INTERVAL (τ_L, τ_R)

6. PERFORM ACTIONS ASSOCIATED WITH ANY TRIGGERS AT τ_R

7. REPEAT FROM STEP 1.

Figure 8. Computational Control Algorithm for Spacecraft Simulation

detail. Instead of a single control table, there are T-triggers, Y-triggers, and lists associated with logical variables.

It is important to note that in spite of the diversity of applications in Sections II-A through C, the systems have a certain regularity of structure. In each case, there is a control program which runs the system, one or more control tables used by the control program to determine what to do next, and a set of slave programs (pixel generators, application programs, or models) which do the desired calculations under the direction of the control program. These elements can be expected to appear in almost any computational control system, and recognition of this fact gives the system designer a quick start in drawing his system diagrams.

In the mosaic tiling example, the control table completely defines the sequence of calculations from start to finish. Sequencing does not depend on calculated results. In the spacecraft control and simulation examples, however, sequencing by the control program depends on the results of calculations or real-time interactions and cannot be predicted in advance. As a result, each of these last cases requires the development of mechanisms by which the user can communicate new sequencing requirements through the slave programs to the control program. In each case, these mechanisms involve language elements. In the spacecraft control case, the language elements used are the WHEN and WAIT. In the simulation case, the language elements used are CALL TSET, CALL YSET, CALL LSET and CALL LTAB.

These language elements can be thought of as statements which pass necessary sequencing information to the control program. In each case, they translate as simple CALLs to subroutines. Only in the case of the WAIT and WHEN is the control more complicated. These cases, too, are CALLs to subroutines, but the subroutines do not return control immediately as subroutines normally would.

So the designer can expect that if the sequence of control depends on calculated results, he will need some simple language elements to pass sequencing information from the slave programs to the control program. He can also expect the language elements to translate to subroutine CALLs.

B.   Synthesis

Synthesis is the process of putting the diverse elements of the system back together again, and consists primarily of designing the algorithms used by the control program. Even though analysis and synthesis seem like entirely different activities, they are so mixed together in early design stages that it would be almost impossible to say where one began and the other left off. They are like two sides of a coin of design that the designer flips from time to time to decide which to do next. Consider, for example, developing an algorithm to control operation of application programs in the spacecraft control system. The question immediately arises: How does the control program know when the application programs need control again and to what location control should go? Obviously, some kind of control table is needed. But there are countless table structures which can be used, and each different table structure requires a different algorithm. Deciding which tables to use and what structures they should have must be done concurrently with deciding on the algorithm to be used in the control program. One effective procedure involves writing down a clear picture of a control table, sketching an algorithm which uses it, and then looking for ways to simplify and speed up the algorithm by changing the control table. Several iterations of this procedure with constantly increasing precision of definition usually suffice to produce something good enough to serve as a starting point for the system architecture.

There are many ways to depict algorithms. The use of flow charts is an old, well understood technique and is perfectly acceptable. A better system involves the use of structured languages such as "Program Design Language" and the processor developed by Caine, Farber, and Gordon, Inc., of Los Angeles [6]. Restricting programming structures to DO, IF, and CALL generally leads to better program organization than allowing unrestricted branching. Structured English language statements are far more compact and efficient than flow charts, and — given a processor — far easier to change.

Users of structured languages should always be aware that the rules of structured usage are intended to simplify programming structures and make them easier to understand. Whenever adherence to structured usage clearly complicates programs or makes them more difficult to understand, the rules should be violated. A good example of this violation occurs in the WAIT and WHEN statements of the spacecraft control example. These subroutines do not return control to the user in the normal way, but instead return control to the control program.

C. Experiment

The system designer eventually completes his analysis and synthesis, has his control tables structured, has a specification for the language elements (if any) used by the slave programs to communicate control information back to the control program, and has some form of representation of the control program algorithm. It would seem that the next step should be coding. It probably is not. The next step should usually be for the designer to put himself in the shoes of a user unfamiliar with the system and try to develop some sample slave programs for the system. This may turn out to be just as easy as the designer thinks it will be. If so, it will not take long. But the designer may run into some surprises — particularly if he is lucky enough to be able to talk a friend into trying to develop slave programs. Because of his unfamiliarity with the underlying assumptions concerning the problem, the friend may try to do things which never even occurred to the designer.

An example of the kind of surprises which can hit the system designer at this stage occurred during the design of the spacecraft control system. Originally, the WAIT statement had been designed to specify the wait duration in terms of the number of minutes, seconds, and ticks. These are the elements of system time and are used in the WHEN statement. Using them in the WAIT statement seemed to give the system a pleasing uniformity. But as soon as slave programs were written using the WAIT system, it became apparent that the natural unit for the user to specify was the total number of 2.5-millisecond ticks he wanted to wait, even if that number exceeded 400 (1 second).

D. Articulation

The importance of letting others know the details of a proposed architecture varies greatly with the circumstances of the development. In the mosaic tiling system, there were no anticipated users of the system beyond the designer himself, and the primary concern was simply to get the job done fast. The system was already running before any significant amount of documentation was done, and the documentation, even when published, was ignored.

The other extreme is illustrated by the simulation example. Publication of the first draft of the system design in that case raised a storm of technical and managerial controversy which did not settle down for months. A second draft of the system design document with a considerable number of fundamental changes was required to settle the dust of the technical arguments, but formal management meetings and full-scale slide presentations of the system design were required to finally resolve the managerial disagreements.

So the system designer must try to anticipate the need for communication even as he designs the system. There is almost always a need for some communication of design ideas to some audience, and if the designer keeps that in mind as he draws his system diagrams, he may be able to make them usable for future presentations to his peers.

E. Iteration

It is not enough to iterate the analysis and synthesis during the early design phase, or even to iterate a system design based on inputs from a design review. It should always be planned, if time permits, to iterate a system at least once after it has been completely designed, implemented, and used. The importance of iteration cannot be overstressed — it is the essence of design. Systems of significant complexity can seldom be designed right the first time. A little experience with a system almost always makes glaringly obvious things which could be seen, if at all, only with peripheral vision during design.

Sometimes, because of schedule pressures, it may not be possible to iterate, and the system is left with glaring flaws. The mosaic tiling system contains a good example. That system was implemented so that any single picture line which deviated by even one pixel from the known line length caused a diagnostic message to print and processing to stop. That sounded like a good idea when thought of, but it meant that normally only one error per computer pass could be caught. If the entire picture had been allowed to complete, the finished copy could have pointed out errors in many pixel generators simultaneously.

Although the system does work, system redesign is usually valuable just because several benefits are gained by the redesign. The spacecraft control system affords a good example. This system was originally implemented with a WHEN statement which assumed that each item of system time was normally to be tested and used control bits to tell which items to ignore. Once the system was being used, it became obvious that the WHEN statement should simply name each item to be tested. This made the system independent of the time base used and even allowed programmers to say, in effect,

WHEN   FLAG F IS SET

CONTINUE

where flag F may be set by another program or by a data transmission from another machine. The resulting system was easier to use, was far more flexible, and ran faster than the old one.

IV.   CONCLUSION

Systems for computational control almost always consist of a single control program, a number of slave programs, and one or more control tables. The problem of analysis is to identify these elements and their structure. The problem of synthesis is to determine a simple way to tie the elements together with a foolproof algorithm for the control program. Usually, the algorithm needs to be flexible enough to handle an entire class of similar problems. Analysis and synthesis must be tackled together and constitute the early design stage.

But a first design, untested by the fires of criticism and use, seldom represents the designer's best work. Even before his design is cast in code, the designer needs to put himself in the user's place by writing slave programs and testing alternative architectures. He needs to explain his work to an audience of peers and solicit their helpful and honest criticism. And even when a first version of the system is running, the designer should, if time allows, plan to do a final iteration, for experience is an exacting referee, and iteration is the name of the game.

ACKNOWLEDGEMENT

This paper presents the results of one phase of research carried out at the Jet Propulsion Laboratory, California Institute of Technology, under Contract No. NAS7-100, sponsored by the National Aeronautics and Space Administration.

REFERENCES

1.   David A. Rennels, Borge Riis-Vestergaard, and Lance C. Tyree, "The Unified Data System: A Distributed Processing Network for Control and Data Handling on a Spacecraft," *NASCON Conference Proceedings*, May 1976.

2.   David A. Rennels, "A Distributed Microprocessor System for Spacecraft Control and Data Handling," *MIDCON/77 Conference Proceedings*, 1977.

3.   Hansen Per Brinch, "Concurrent Programming Concepts," *Computing Survey*, Vol. 5, No. 4, December 1973.

4. Fred Lesh, "A Continuous/Discrete Simulation for Interplanetary Spacecraft," *Proceedings of the Seventh Annual Pittsburgh Conference on Modeling and Simulation,* April 21-22, 1977, pp. 243-247.

5. Donald G. Golden and James D. Schoefflen, "GSL - A Combined Continuous and Discrete Simulation Language," *Simulation,* Vol. 20, No. 1, January 6, 1973, pp. 1-8.

6. Steven H. Caine and E. Kent Gordon, "PDL - A Tool for Software Design," *AFIPS Conference Proceedings,* Vol. 44, pp. 271-276, National Computer Conference, 1975.

A MACRO FACILITY FOR COBOL

J. M. Triance,
Computation Department,
University of Manchester Institute
  of Science and Technology,
Sackville Street,
Manchester, England.

Abstract

The British Computer Society's COBOL Specialist Group has designed a Macro Facility
for COBOL for the purpose of permitting users to obtain a more up-to-date portable
version of COBOL. This paper describes the facility and explores its implications.

INTRODUCTION

In terms of its wide usage COBOL must be regarded as the most successful of all the
programming languages. This continued success can be largely attributed to two factors. Firstly, it is kept up-to-date: the current version of CODASYL COBOL [1] contains extensive facilities for Database Handling, Telecommunications and Structured
Programming. Secondly, COBOL is, "all things to all men". For example, the addition
of Structured Programming facilities will not prevent (or even discourage) those users, who so desire, from writing programs in the same unstructured way as has always
been possible in COBOL.

But any advantages which CODASYL COBOL offers exist only in a diluted form in most
compilers. It is intended as the ideal form of COBOL which will, in due course, be
implemented on all relevant computers. But most users must wait until new features
are standardised and then until their compiler is modified or rewritten. Typically
this involves a delay of eight or nine years from when a facility is incorporated into CODASYL COBOL. Perhaps even more serious is the unbounded variation in this delay.
At one extreme CODASYL can adopt a facility which is already provided as an extension
in a particular implementation of COBOL and at the other extreme the same facility
might never be supported in another implementation.

This has led to vast variations amongst the COBOL compilers currently in use. Some
are based on the 1968 Standard [2] and some on the 1974 Standard [3]. Some are small
subsets of the Standard, others full implementations with many extensions. But this
vast selection of compilers is not available to the individual user since on many ranges of computer there is only one COBOL compiler available.

Problems arise when the compiler available is not suitable for the programming techniques used or is incompatible with other compilers which are also to be used. In
fact, a measure of users' dissatisfaction with their compilers is provided by the
large number of COBOL pre-processors currently available. The dissatisfaction is
with features in the compiler which are regarded as undesirable as well as with the
omission of features which are considered desirable. The undesirable features can,
of course, be avoided by installation standards and in some cases by compiler options.
On the other hand, extensions to the compiler can be handled by the pre-processors.
But pre-processors can only be regarded as a stop-gap measure, not as the ideal solution.

A better solution would be for the user to configure his compiler so that it provides
the version of COBOL he needs. Currently there is no mechanism for doing this, but
the British Computer Society's COBOL Specialist Group has designed a suitable mechanism based on Macros. This differs from many other macro facilities [4] in that the
macro calls are COBOL-like and the Macro Processor is an integral part of the compiler.

The Macro Facility was designed to meet the following objectives:

(1) It should not be necessary for the programmer to know which parts of the programming language are macro calls and which parts are pure COBOL.

(2) Macro definitions should be portable between different compilers and different computer systems.

(3) Macro definitions will be no more difficult to write, test and maintain than any COBOL program of comparable complexity.

THE MAIN FEATURES OF THE MACRO FACILITY

Macro Calls are COBOL-like

One of the requirements of the Macro Facility is that it can implement features of CODASYL COBOL which are not yet generally available. In order to satisfy this requirement the facility must accept any macro call which can be expressed in the COBOL Metalanguage, an example of which is shown in Figure 1.

```
SET index-name-1 [index-name-2] ...
 { UP   BY }  { identifier }
 { DOWN BY }  { literal    }
```

Figure 1.   Example of COBOL Metalanguage

The implication of this for the applications programmer is that the macro calls will look like normal COBOL. For example, CODASYL COBOL's in-line PERFORM (Figure 2) could be a macro call with the condition (in this case IND > LIMIT) and the imperative-statement (SET IND UP BY INCR) as arguments.

```
PERFORM  UNTIL  IND  >  LIMIT
SET IND UP  BY INCR
END-PERFORM
```

Figure 2.   Example of a Macro Call

As with any macro processor the macro call is replaced by equivalent coding referred to in this paper as Substitute Coding. The possible substitute coding for this example is shown in Figure 3.

```
    PERFORM  P1   UNTIL IND > LIMIT
    GO TO P2.
P1. SET IND UP BY INCR.
P2.
```

Figure 3.   Substitute Coding for in-line PERFORM

Substitute Coding is Lower Level COBOL

The Substitute Coding is processed by the compiler as if it, rather than the macro call, had appeared in the source program. This takes place within the compiler unseen by the applications programmers who need not know any more about the Substitute Coding than they do about the object code.

In order to achieve portability of macros the Substitute Coding is always a lower level of COBOL. If this rule is interpreted strictly it is estimated that it would still be possible to represent more than half of CODASYL COBOL by means of macros including the Structured Programming facilities, the Report Writer, the String Processing verbs, SORT, MERGE and the more involved formats of many other verbs.

Macros may be Nested

When the Substitute Coding is processed by the compiler a search is made for further macro calls. For example, if SET IND UP BY INCR is a macro call with Substitute Coding ADD INCR TO IND, then after expanding the nested macro call the Substitute Coding in Figure 3 becomes :

```
        PERFORM P1 UNTIL IND > LIMIT
            GO TO P2.
    P1. ADD INCR TO IND.
    P2.
```

Figure 4. Expansion of a nested macro call

This process is repeated until all levels of nested macro calls have been expanded, at which stage the Substitute Coding is compiled in the normal manner. Recursive macro calls are also permitted.

The Macro Call Format is Expressed in COBOL

Each macro contains a description of the format of the macro call (known in some macro processors as a prototype or template). The obvious representation in this case would appear to be the COBOL Metalanguage (Figure 1) but this cannot readily be produced on a line-printer. The solution adopted is to represent the macro call format by means of a COBOL record as shown in Figure 5.

The record definition tells us that, to satisfy this format, a macro call must consist of four parts (corresponding to the four level 3's). The first part must be the words PERFORM UNTIL, the second part must be a condition, the third part must be an imperative statement and the final part must be the word END-PERFORM. The USAGE ARGUMENT and CLASS clauses are extensions to COBOL which will be explained later.

424

```
PERFORM UNTIL condition
imperative-statement
END-PERFORM
1       IN-LINE-PERFORM.
  3    PERF-UNTIL PIC X(14).
   88 P-U                    VALUE "PERFORM UNTIL ".
  3    COND       USAGE ARGUMENT.
   88 CD                     CLASS "CONDITION".
  3    IMP-ST     USAGE ARGUMENT.
   88 I-S                    CLASS "IMPERATIVE-STATEMENT".
  3    END-PERF   PIC X(12).
   88 E-P                    VALUE " END-PERFORM".
```

Figure 5.   **Macro Call Format in Metalanguage and Equivalent COBOL**

Figure 6 shows how all the metalanguage constructs used in ANS 74 can be represented. A new clause OPTIONAL is introduced for optional parts of macro calls. There is also a need for an UNORDERED clause for the case when the sequence of components in a macro call is immaterial.

COBOL Meta-language	Meaning	Representation
CAPS	Key words	VALUE clause in level 88
CAPS	optional words	VALUE clause in level 88 following OPTIONAL clause
lower-case	semantic-class	CLASS clause in level 88
[ ]	optional clause	OPTIONAL clause
{ }	alternatives	REDEFINES clause for each alternative after the first or (when all alternatives are semantic classes) consecutive level 88's with CLASS clauses
...	repetitions	OCCURS clause with DEPENDING option

Figure 6.   **Method of Representation Formats of Macro Calls**

The macro call formats are stored in a file and used by the macro processing phase of the compiler to identify macro calls.

Each Macro is a Cobol Subprogram

Most macro processors employ a special language for writing the macro definitions. However, given the language extensions highlighted in the preceding section COBOL seems to be just as well suited for writing macros as it is for writing applications programs. It has the distinct advantages over existing special purpose languages of

(1) avoiding the need for extra training for the macro writers;

(2) being a complete and self documenting language; and

(3) being portable.

The fact that it is verbose and will lead to unnecessarily long macro definitions would seem unimportant when we consider that most COBOL users are willing to sacrifice brevity in return for readability.

The macro definition is written as a sub-program to achieve the greatest possible independence from other macro definitions. An example of the macro definition for the in-line PERFORM is shown in Figure 7.

```
 1 IDENTIFICATION DIVISION.
 2 PROGRAM-ID. PERFORM-FORMAT-5
 3            CLASS IS "IMPERATIVE-STATEMENT".
 4 ENVIRONMENT DIVISION.
 5 INPUT-OUTPUT SECTION.
 6 FILE-CONTROL.
 7     SELECT SUBSTITUTE-CODE ASSIGN SUBCODE.
 8 DATA DIVISION.
 9 FILE SECTION.
10 FD    SUBSTITUTE-CODE.
11 1     SUBST-CODE-REC.
12    3  SUBST-ST   PIC X(17) VALUE     "PERFORM P1 UNTIL ".
13    3  SUBST-COND USAGE ARGUMENT.
14    3  SUBST-MID  PIC X(15) VALUE     " GO TO P2.
15-                          "P1. ".
16    3  SUBST-STMT USAGE ARGUMENT.
17    3  SUBST-END  PIC X(6)  VALUE     ".
18-                          "P2. ".
19 MACRO-LINKAGE SECTION.
20 1     IN-LINE-PERFORM.
21    3  PERF-UNTIL PIC X(14).
22       88 P-U              VALUE "PERFORM UNTIL ".
23    3  COND      USAGE ARGUMENT.
24       88 CD               CLASS "CONDITION".
25    3  IMP-ST    USAGE ARGUMENT.
26       88 I-S              CLASS "IMPERATIVE-STATEMENT".
27    3  END-PERF  PIC X(12).
28       88 E-P              VALUE "END-PERFORM ".
29 PROCEDURE DIVISION USING IN-LINE-PERFORM.
30 SET-UP-SUBST-CODE.
31     MOVE COND TO SUBST-COND.
32     MOVE IMP-ST TO SUBST-STMT.
33     WRITE SUBST-CODE-REC.
34 END-MACRO.
35     EXIT PROGRAM.
```

Figure 7. A Macro Definition

This macro definition is compiled and placed in a library. When, subsequently, a call for the macro is encountered in an applications program the appropriate macro definition is executed after the details of the call have been placed in the Macro Linkage Section. (It will be noted that this Section contains the macro call format previously shown in Figure 5.)

Thus for the Macro Call shown in Figure 2 the contents of the Macro Linkage Section will be as follows :

field	contents
PERF-UNTIL	PERFORM UNTIL
COND	IND > LIMIT
IMP-ST	SET IND UP BY INCR
END-PERF	END-PERFORM

The Procedure Division coding uses these data items to set up the Substitute Coding and write it to the Substitute Coding File before returning control to the Macro Processor. The contents of this file are then processed as if they had been part of the original source program.

The non-standard use of VALUE in the File Section is explained later.

The Class of the Arguments is Specified

Most macro processors process arguments as strings of text and leave any error checking to the macro writer. This Macro Facility, however, requires the macro writer to specify explicitly the class of each argument. In Figure 5 the classes were CONDITION and IMPERATIVE-STATEMENT but in general could be any syntactic type in COBOL. This scheme which could be regarded as a development of Levenworth's Syntax Macros [5] offers many advantages:

(1) Greater precision in the identification of macro calls is possible. For example,

> PERFORM UNTIL STOP RUN
>
> SET IND UP BY INCR
>
> END-PERFORM

would not be accepted as a valid call of our sample macro since STOP RUN is not a condition.

This precision permits the early detection of errors and the generation of error diagnostics relating to the original macro call rather than to the Substitute Coding. This is vital if the error messages are to be meaningful to the applications programmer.

(2) Greater precision is possible in the generation of Substitute Coding. Imagine that, for a compiler which does not support the option, a macro is being defined for PERFORM with the VARYING option:

$$\underline{\text{PERFORM}} \text{ procedure-name-1 } \left[\underline{\text{THRU}} \text{ procedure-name-2}\right]$$
$$\underline{\text{VARYING}} \begin{Bmatrix} \text{identifier-1} \\ \text{index-name-1} \end{Bmatrix} \underline{\text{FROM}} \begin{Bmatrix} \text{identifier-2} \\ \text{literal-1} \\ \text{index-name-2} \end{Bmatrix}$$
$$\underline{\text{BY}} \begin{Bmatrix} \text{identifier-3} \\ \text{literal-2} \end{Bmatrix} \underline{\text{UNTIL}} \text{ condition.}$$

In the Macro Linkage Section the argument following VARYING will be defined as

3	COUNTER-ITEM	USAGE ARGUMENT.
88	ID	CLASS "IDENTIFIER".
88	IND	CLASS "INDEX-NAME".

to indicate that the argument is either an identifier or an index-name. The condition-names ID and IND can then be used in the normal way to discover whether the Substitute Coding should contain MOVE's and ADD's or SET verbs for initialising and incrementing the counter.

(3) Macro calls may be identified more efficiently. Since an argument of a macro call may itself contain macro calls it is necessary for each macro to have a class associated with it to permit full validation of the argument. The class of the macro is specified by the CLASS clause in the PROGRAM-ID of each macro definition. Having been specified this information can then be used by the Macro Processor to ensure that the source text is scanned only for Macro Calls of a permissible class in each context.

All Macros are Stored in a Library in a Separate Run

Before macros can be used they must be submitted to a library run. This extracts the macro call format from the Macro Linkage Section and stores it in a Macro Call Formats File. At the same time a Reserved Word File is updated to include any additional reserved words. Finally the macro definition is compiled and stored in a macro library file. These three files, when used in conjunction with the compiler's inbuilt features, determine the language which is available for applications programs.

IMPLEMENTATION CONSIDERATIONS

Handling Variable Lengthed Arguments

Most arguments are variable in length without any moderate maximum size. Even an identifier in CODASYL COBOL can exceed 72,000 characters without counting the embedded spaces. Since COBOL isn't well equipped to handle such fields the solution adopted is to store a pointer to the argument in the fields described as USAGE ARGUMENT rather than the argument itself. When this pointer is transferred to the Substitute Coding the Macro Processor will replace it by the original argument. The Macro Writer will not be able to access the actual argument.

The length of fields described as USAGE ARGUMENT would be implementor defined with the result that the macro writer should make no assumptions about them. But the macro writer must define the correctly lengthed Substitute Coding File record. The preferred solution to this problem is to set up the Substitute Coding in the File Section rather than in the Working-Storage Section and to enhance COBOL to treat VALUE's in the File Section in the same way as in the Report Section. In other words the field is automatically initialised immediately before writing each record. If this approach is unacceptable to CODASYL various other solutions are possible including using MOVE statements to achieve the same effect.

Distribution of Substitute Coding

Not all Substitute Coding will belong in the same physical location as the macro call. The actual location (e.g. Working Storage, End of Procedure Division) is indicated at the start of each Substitute Coding record. (This indicator was omitted from the example in Figure 7 to avoid confusion.) Error diagnostics are also output to the Sub-

stitute Coding File with an indicator to signify that they should appear with the
compilation error list.

Accessing Data Definitions

In some macro definitions it will be necessary to access the data definition of identifiers which are arguments. For example a macro for the CODASYL COBOL verb

INITIALISE identifier

would need to establish whether identifier was numeric or not to determine whether
the Substitute Coding should move zeros or spaces to the identifier.

The data descriptions are all stored in a compiler generated indexed file and the
data definition of any item can be accessed by using its identifier as the key.

Generating Unique Names

The Macro Definition in Figure 7 would not work if it was called twice in the same
program because two sets of identical paragraph-names P1 & P2 would be produced. This
is obviously unacceptable.

The problem is overcome by appending the desired combination of the following special
registers to such names :

- MACRO-ID   -   the unique identifier of the macro definition currently being
               executed
- CALLS-OF-CURRENT   -   the number of times the current macro has been called in
               the current source program
- CALLS-OF-ALL   -   the total number of calls of all macros in the current source
               program.

Each of the special registers also contains a symbol which will make any of these
generated names distinct from names defined in the original source text.   The
Macro Writer can also obtain a unique name by terminating any name in the Substitute
Coding with a hyphen, whereupon CALLS-OF-ALL will automatically be appended.

IMPLICATIONS OF THE FACILITY

The Envisaged Use of Macros

The primary purpose of the facility is to allow each installation to configure its
compiler to support any subset of CODASYL COBOL that is desired. The examples given
so far have shown how a compiler which falls short of CODASYL COBOL can be extended
by defining macros. Since a macro for a given construct has priority over the same
construct supported by the compiler itself it is also possible to override the implementation provided by the compiler. This is of use in avoiding compiler bugs and
non-standard implementations or in prohibiting or highlighting features which are
considered undesirable. For example, if the installation standards banned the use
of the ALTER verb a macro could be defined for ALTER which would produce a fatal diagnostic message but no Substitute Coding.

When transferring COBOL programs from one compiler to another, the Macros can be used to achieve compatibility. Any desired features of COBOL which are lacking on both compilers will normally be provided on the new compiler by the same macro as on the old compiler. But when the Substitute Coding includes incompatible features there will be a need for a new version of the macro which generates different Substitute Code. If the new compiler supports some features that were previously supplied by macros then the relevant macros can be discarded. If, on the other hand, the new compiler lacks some of the desirable features which were built into the old compiler then additional macros will be needed. In all cases the applications programs can remain unaltered.

In effect these steps will make the compilers upward compatible. But in some cases (such as a permanent move from an ANS 68 Compiler to an ANS 74 Compiler) it might be preferable to actually translate some of the constructs in the applications programs once and for all. Macros could also be used for this provided the compiler offers the option of outputting the applications programs with the macro calls replaced by substitute coding (in other words, acting like a pre-processor). Then, for example, all EXAMINE statements (ANS 68) could be replaced by the equivalent INSPECT statements (ANS 74).

The Macros required for the various purposes described above would be fairly standard and would probably be supplied by manufacturers, software houses or other bodies. There is, of course, nothing to stop the individual user from writing his own macros. If the user restricts himself to implementing a subset of CODASYL COBOL the process will no doubt be onerous in some cases but will be relatively straightforward. The user could go a step further and design his own extensions to COBOL. If done on any scale this would probably require a similar commitment and degree of central control as is necessary for setting up a data base.

Whoever writes the macros it is essential that they are thoroughly tested. Since the applications programmers regard the macros as part of the compiler the programmers must be able to place as much trust in the macro definitions as they would in a good compiler.

Impact on COBOL Development

The addition of the macro facility requires surprisingly few enhancements to COBOL. The proposed changes, which only effect the macro definitions, involve eight minor additions to the language and the relaxation of rules on four existing features. There is, however, a far bigger impact on the way that compilers are written.

With the help of Macros, users would be able to update their versions of COBOL at their own speed. If, for example, level 77's are dropped from the latest release of a compiler, the users could continue to use them via a macro for as long as was desired. This should remove from CODASYL the spectre of universal disapproval whenever they do anything to stop COBOL being upward compatible. Thus decisons to delete or amend features of the language could be based more on merit and less on the extent to which existing programs depend on the feature. In the other direction macros could also serve CODASYL members by allowing them to test some of the proposed additions prior to incorporation in the language.

The Macro facility could also simplify the present system of subsetting COBOL in the American National Standards. In a standard which contained the macro facility all the features could be divided into two classes - those which could be written as macros and the remaining set of "elementary" features. Hopefully this class of elementary features would be small enough to regard as the minimum subset. Unlike the present minimum subset this one would contain no redundant features and would be sufficient for any application for which full COBOL is sufficient. The rest of the language would merely provide a more convenient method of achieving the same results.

There would be two subsets in this Standard. One would be the minimum subset described above and the other would be the minimum subset plus the macro facility. With this macro facility the rest of the language could be provided by macros if necessary. Any of these remaining features which were supported directly by the compiler would not increase the power of language but could increase its efficiency.

Possible Drawbacks

There appear to be two potential drawbacks with macros. The most serious for COBOL is probably the danger that the language will diverge into a large number of incompatible dialects. At worst, there is the worry that every programmer might define a different set of ill-conceived macros for each program. The fact that all macros must be placed in a central library makes this worst situation infeasible. No doubt some of the larger installations will devise their own dialect of COBOL but these are the type of installation which is already doing just that by means of pre-processors.

The other drawback traditionally associated with Macros is in the area of efficiency. In particular Macros will tend to slow down compilations. This facility includes some features which should alleviate this problem: the macro definitions are executed not interpreted and unduly sophisticated features are avoided. Despite this it is likely that some users will find the overhead of macros too great just as some find certain current features of COBOL too extravagant. Such cases are likely to be in the minority and for other users the feasibility of macro processors with COBOL has been well demonstrated with the Cobra [6] and Meta-Cobol [7] pre-processors.

Planned Developments

A pilot implementation of the facility is in progress at UMIST. The objective is to demonstrate its functional feasibility. It is hoped that this will be followed by a full systems study and a complete implementation which would indicate its economic viability. The long term objective is its incorporation into CODASYL and then ANS

COBOL and its general availability in COBOL compilers.

Conclusions

There are problems associated with the use of Macros but they are ones which have been overcome to the satisfaction of current users of macro pre-processors. The incorporation of macros into COBOL overcomes shortcomings inherent in pre-processors and could offer all users the opportunity to select their own version of COBOL. This would allow COBOL, at last, to satisfy its design objectives of being an up-to-date portable language.

Acknowledgements

The macro facility described in this paper was designed by a working group consisting of K. H. Meyer of British Gas, A. Morrison of the Central Computer Agency, A. E. Sale of Alpha Systems Limited, J. E. Sawbridge of Plessey and Chaired by the author. In addition, J. Yow of UMIST is implementing the pilot system in consultation with R. M. Gallimore of UMIST and the author.

The opinions expressed in this paper are not necessarily shared by those named above.

References

1. CODASYL (1978). COBOL Journal of Development 1978, Canadian Government, Department of Supply and Services.

2. ANSI (1968), American National Standard Cobol X3.23-1968, American National Standards Institute.

3. ANSI (1974). American National Standard Programming Language Cobol X3.23-1974, American National Standards Institute.

4. BROWN, P. J. (1974). Macro Processors and Techniques for Portable Software, Wiley.

5. LEVENWORTH, B.M. (1966). Syntax Macros and Extended Translation. C.A.C.M., Vol. 9, pp.790-93, Nov. '66.

6. HAMILTON, J.G.A., FINLAYSON, E.D., HEYWOOD-JONES, A.H. (1973). Computer-aided Program Production, Datafair 1973.

7. ADR (1976). MetaCOBOL Concepts and Facilities, Applied Data Research.

THE REFERENCE STRING INDEXING METHOD

H.-J. Schek
IBM Wissenschaftliches Zentrum
Tiergartenstrasse 15
D-6900 Heidelberg

SUMMARY

The motivation for the reference string indexing method may be derived from the intention to retrieve any piece of information by specifying arbitrary p a r t s of it. Common restrictions such as the usage only of a certain set of descriptors or (complete) keywords in document retrieval systems or the specification of only certain (inverted) attribute values for queries in formatted files should be removed without loosing performance necessary for interactive usage.

The solution to be described is essentially based on the realistic assumption that the frequency distribution for the occurrence of character strings with a certain length, or words, or word sequences in textual files, and also for the occurrence of attribute values or value combinations in formatted files is n o t uniform but rather highly hyperbolic or "Zipfian". The same is valid also for the u s a g e of data, expressed as the "80-20"-law. Exploiting this assumption, a (small) set of "reference strings" is generated by a statistical analysis of collected queries or - if not available - by usage estimation with the original data. The inversion to these reference strings with respect to records or record clusters gives the reference string index.

Corresponding to the estimated usage frequency, a search argument may have been made available completely as a reference string or has to be decomposed into shorter reference strings. Therefore, the reference string access is adaptive with the consequence that a routine query may be answered faster than a non-routine one.

The reference string index may be applied as a new adapted index in information retrieval systems as well as in formatted files as single or multi-attribute index. In addition it can be applied for phonetic and general record similarity search.

1. Introduction

The reference string indexing method is related to problems often called partial match retrieval, associative search or multi-attribute retrieval. A directly accessible piece of stored information (record, block or cluster) which is represented as a string of characters shall be retrieved (and optionally subsequently updated) by specifying one or several a r b i t r a r y fragments (or substrings, fragmentary information) which must be contained in the desired record(s).

This provides a more general "accessing by contents" than by preassigned and often artificial keys. So, the main objective of the proposed access method is to support formal search functions summarized as functions for "partial match retrieval" Direct access becomes possible in cases where serial file scanning often has been the only solution until now. Examples for such searches are queries specifying values for not inverted attributes in formatted records, queries with partially specified attribute values, search with phonetic patterns, and nearest neighbour search. In text retrieval systems with automatic indexing, an important application of the reference string access method is the search with f r a g m e n t s of keywords which appear e.g. as c o m p o n e n t s of chemical substances or in c o m p o u n d w o r d s of a natural language. In formatted files the reference strings provide a p a r t i a l inversion either for a single attribute, or for several attributes as a new multi-attribute index, in both cases in adaptation to the data.

The idea of the reference string access method is easily explained by an example: The set of all records which contain somewhere the string 'CHROMAX' is contained in the set of all records which contain somewhere both strings 'MAX' & 'CHROMA' which, in turn is contained in the set with 'AX' & 'CHROM' & 'OMA'.

Obviously, many more decompositions would be possible. However, decompositions of this kind motivate the idea of introducing a set of standard strings which are used in every decomposition. These strings will be called reference strings or shorter "refstrings" and may or may not be meaningful strings of characters. The only purpose of a refstring is to give a reference into a record where it occurs. So it is assumed that each refstring has an inverted list which means that a list of accession numbers, unique keys, or logical pointers of those records is known which contain somewhere the corresponding refstring as a substring.

The automatic determination of a reasonable set of refstrings is the key problem and the main idea for its solution is to determine substrings which have a high r e f e r e n c e frequency, in other words, which have a high probability to be n e e d e d for a reference into the records. Estimates for these probabilities may be derived from an analysis of the queries used in the past and by extrapolation to future usage. For a first estimation of usage frequencies, letter and word resp. attribute co-occurrence statistics of the original data itself are evaluated to give an initial set of refstrings.

It is important to observe that two facts are exploited essentially in this approach. The distribution of the reference frequency of data as well as the distribution of the occurrence and co-occurrence frequency of letters or syllables in words or words in texts etc. is not uniform but rather hyperbolic. The first is known as "80-20-rule" /KNU73/ and the second as Zipf's distribution /ZI49/. This means that only a small quantity of data is used in most of the queries resp. it means that a small set of different strings covers the whole (textual) file. With respect to the refstring index it means that the size (number of different refstrings) can be kept small even if one prescribes a high mean selectivity.

The reference string indexing method has been developed starting in 1974 with the problem of similarity search in keyword lists including the misspelling case and phonetics /SCHE75/. In the more general context here and in its further extensions it has relations to the following literature: determination of "equifrequent character strings" and application for text searching and data compression /BA74, CLA72/, substring and pattern matching /AHO75, KNU74, MCR74, HA71/, partial match retrieval and related index or tree (trie) constructions /RI76, BU76, BE75, WO71, LU70/ , term association and statistical thesaurus constructions /STE74, SA68, LUS67/, general indexing and access methods /WE75, WA75, BAY73/. The following is a shortened and revised version of /SCHE77/.

2. The Reference String Index

In order to include the most general case it is assumed that the file P from which data have to be retrieved contains directly accessible records. Each record may be formatted in fields (attribute structure) or may be without format (free format case like in natural language texts) or it may be a mixture between both which means that certain attributes may contain a textstring as attribute value such as a title of a book. Without loss of generality it is further assumed that attribute values appear in their character representation. The reference string index is motivated by the intention to retrieve any record by specifying arbitrary parts of it. One or several strings which must occur in the matching records as substrings - disregarding attribute or word positions for the moment - shall be sufficient for retrieval.

2.1 Refstrings for Substring Matching

Obviously, a solution for the retrieval-by-parts problem consists in the solution of the substring or pattern match problem. However, in the context here methods are not allowed which scan the whole file, even though there are sophisticated methods. The reason is that one should avoid to transfer the whole file from external to internal storage. Therefore, one would like to have a fast method which allows to decide which records contain somewhere a certain substring x, or somewhat weaker, which may contain x with a high probability.

A first, unrealistic solution would be to provide an index for each possible substring which occurs in P. The size of such a complete substring index would be prohibitively high and is also not necessary as the following considerations show:

1. Only those substrings which have a high probability to be
   specified in queries should be included in the substring index
   called refstring index with the refstrings as entries.

2. If a search argument x is not contained in the refstring
   index but if x contains substrings y1,y2,...,yn which are
   refstrings, then every record which does not contain all
   yi,i=1,...,n must not contain x and can be excluded from
   processing.

3. In order to support an arbitrary substring search (meaning
   that at least one decomposition into refstrings is possible)
   either all single characters, or all character pairs, or all
   character triples are defined to be refstrings too and are
   included in the refstring index.

The second principle /HA71/ may introduce false drops: A record
containing all substrings yi of x will not necessarily contain
x. To express it more precisely: Let be $J(x)$ the set of all
record numbers of those records containing x anywhere. Then the
following corrolaries hold

C1: If y substring of x then $J(x) \subseteq J(y)$

C2: If x contains y and z as substrings then
    $J(x) \subseteq J(y) \cap J(z)$

C3: If x contains y and y contains z as a substring
    then, instead of (2) one has $J(x) \subseteq J(y)$
    (intersection with $J(z)$ is redundant).

C1 is easily be proved, C2 and C3 are consequences from C1.

The first and third consideration lead to the introduction of two
disjoint sets of refstrings, namely the set of basic refstrings
BRS and the set of additional refstrings ARS with the following
definitions.

Def. BRS: The set of basic reference strings $BRS=\{brs_1,...,brs_m\}$
          contains a l l strings of a certain length k which

occur in P (k=1, or k=2, or k=3).

Def. ARS: The set of additional reference strings
ARS={$ars_1,...,ars_n$} contains c e r t a i n
refstrings with length g r e a t e r k.

Because of this definition one has for the set of all refstrings
RS = BRS∪ ARS and BRS ∩ ARS = 0. The determination of ARS is
described in chapter 3.

The motivation for these definitions will be clear if one applies
RS for arbitrary substring searches: Let x be a search string
with length between k and m. Then the list of record numbers
J(x) which contain x as a substring is desired. For that purpose
relate to x the set of all substrings in x and denote it by X.
further, determine the refstrings in x

AXRS := X ∩ ARS, BXRS := X∩ BRS, XRS := AXRS ∪ BXRS

Because all substrings with lenght k of file P are in BRS one has

A1: If BXRS = Ø then x does not occur in P.

If this trivial case is excluded the set
XRS={$xrs_1,xrs_2,...,xrs_1$} has at least one element. In
application of (C2) using a l l refstrings one has

A2: $J(x) \subseteq J'(x)$: = $J(xrs_1) \cap J(xrs_2) \cap .....\cap J(xrs_1)$

In A2 redundant intersections may occur if one or several xrs_i's
are contained in one or several (longer) elements xrs_j. They may
be omitted applying (C3). Therefore, the set MXRS $\subseteq$ XRS is
introduced

MXRS = {mxrs | (mxrs∈ XRS) and mxrs is not substring of
any other element of XRS}

Obviously, MXRS contains all longest possible refstrings leading
to non-redundant logical operations. One may select other
non-redundant subsets SXRS⊂XRS but MXRS has special properties

A3: 1. J(MXRS) = J(XRS)
     2. J(MXRS) $\subseteq$ J(SXRS)

for all subsets SXRS $\subset$ XRS. In other words, the set MXRS leads
to the smallest possible number of records to be transferred
a n d   to the smallest number of false drops. Especially,

A4: If MXRS = {x} then J(MXRS) = J(x)

In this case the smallest possible number of block transfers is
also necessary. No false drops are encountered in this case.

2.2. Refstrings for Partial Match Retrieval

The application of the refstrings for partial match retrieval
(PMR) in the sense of /BE75/, also called a query of order s
/YA77/, is derived from the substring search described so far.

A query which specifies s attribute values $v_1, v_2, \ldots, v_s$ is
regarded as a substring match problem where all s substrings
$v_1, \ldots, v_s$ have to occur in one record. So, instead of a single
set related to one substring, one defines V_i to be the set of
all substrings in v_i, i=1,2...,s and uses as set X the union
$V_1 \cup V_2 \ldots \cup V_s$ All other definitions remain unchanged.

It is obvious that this method allows also to specify attribute
values itself only partially. This is important in cases where
attributes may consist of several keywords like a title of a book
or where compound attributes such as chemical formulas occur.
The refstring index supports a search with arbitrary attribute
components.

A further generalization of PMR beyond the definition in /BE75/
is again especially important for the specification of several
equivalent keywords in the free format case. For simplicity of
notion this problem is discussed for the case when one attribute
value v may have two equivalent values x and y. Therefore,
records have to be accessed containing somewhere x or y or both.

In order to avoid the union $J(x) \cup J(y)$ in the practically important case where x and y have common reference strings on defines

$$M.RS = MXRS \cap MYRS$$

Then the number of logical operations is reduced without changing the result by

$$J'(xvy) := J[M.RS] \cap (J[MXRS-M.RS] \cup J[MYRS-M.RS])$$

Note that the number of logical operations corresponds also to the number of secondary data transfers.

A final extension is the similarity search: Again the set MXRS is determined which contains all refstrings without the redundant ones belonging to $X := V_1 \cup V_2 \ldots \cup V_s$ where V_i again are the sets of all substrings in the given attribute values v_i. But instead of determining the records $J[MXRS]$ which have to contain e v e r y reference string from MXRS, a weaker requirement is set up: All records containing a sufficiently h i g h n u m b e r of reference strings - not necessarily all - are considered as candidate records for a refined similarity inspection. To be more specific, let $p(i)$ be a positive (weight) number, related to each reference string i, then a record j is a candidate if

$$p(i_1) + p(i_2) + \ldots + p(i_1) \geq \hat{p}.$$

The value $\hat{p}$ is a given threshold and the indices i_j, $j=1,2,\ldots,1$, belong to those refstrings from MXRS which occur also in the considered record.

3. Determination of Refstrings

The set of additional refstrings ARS is the better the smaller the number of inverted lists can be kept which have to be accessed and processed for the execution of all queries including update actions. Under limitation of storage and because of update processing it is reasonable to limit the set of refstrings to a set of such strings which are really needed and which have a high probability to occur in a query. A query which specifies rarely used keywords or attribute values will then be answered by the usage of the basic reference strings (perhaps somewhat slowlier) whereas a common query is answered fast and more directly with the aid of the additional refstrings.

Let be S a file which contains a sample of previous query arguments. If no query arguments are available the original file P or a sufficient large sample of it is used as S. Assume further that a delimiter character separates query arguments, attribute values, or keywords etc. A substring s from S is understood in the following as a substring of S which does not contain the delimiter character.

The algorithm for refstring determination to be described is motivated by the following considerations:

(1) The frequency of a certain substring from S is regarded as an indicator for the frequency of future usage.
(2) If the frequency of a certain substring is either too low then s is not considered worthwile being a reference string. If the frequency is too high, then s is a reference string only if the condition in the next rule (3) is satisfied.

In the low frequency case, establishing an inverted list is unnecessary because it is never or rarely used; in the high frequency case, the substring has a high probability to occur within a longer refstring.

(3) If a certain substring s is contained in a reference string rs as substring then s is considered to be a refstring too only if the frequency of s w i t h o u t the occurrences in rs is high enough.

Therefore, a decision whether a certain string is taken as refstring will not be based on the absolute reference frequency alone. One has to take into account the frequency of indirect references by longer refstrings. If e.g. the string ROM occurs always in connection with CHROM or BROM and these two have been included in the set of refstrings it is not worthwhile to have also ROM as refstring.

The algorithm needs two main steps: Refstring candidates are determined first for the refstring generation in the second step. The second step is described first.

3.1 Refstring Determination Using Candidates

Assume that sets Q_k of candidates q_k for refstrings with length k are available together with their absolute frequencies $f(q_k)$. Generally $f(s)$ denotes the frequency of a substring s in S, and RS_j denotes the set of refstrings with length j. Assume further that the maximum length of a candidate is denoted by m and that the set of refstrings RSj, j=m, m-1...,k+1 have been determined already.

For the determination of RS_k, take a $q_k \varepsilon Q_k$ and relate to it the two sets $RSL_{k+1}(q_k)$ and $RSR_{k+1}(q_k)$ being the sets of those refstrings $rs_{k+1} \varepsilon RS_{k+1}$ which have q_k as left resp. right substring. Similarly, relate to q_k the two sets $QL_{k+1}^-(q_k)$ and $QR_{k+1}^-(q_k)$ which contain those candidates $q_{k+1} \varepsilon Q_{k+1}$ which have q_k as left resp. right substring and which are n o t element from RS_{k+1}.

With these sets the i n d i r e c t reference frequency $irfl(q_k)$ resp. $irfr(q_k)$ of q_k is defined by

$$irfl(q_k) = \sum_{rs_{k+1} \varepsilon RSL_{k+1}} f(rs) + \sum_{q_{k+1} \varepsilon QL_{k+1}^-} \max [irfl(q_{k+1}), irfr(q_{k+1})] \qquad (1)$$

$$irfr(q_k) = \sum_{rs_{k+1} \varepsilon RSR_{k+1}} f(rs) + \sum_{q_{k+1} \varepsilon QR_{k+1}^-} \max [irfl(q_{k+1}), irfr(q_{k+1})] \qquad (2)$$

The weight $p(q_k)$ of each candidate q_k is defined by

$$p(q_k): = f(q_k) - \max[irfl(q_k), irfr(q_k)] \qquad (3)$$

and q_k will be selected to be a refstring if

$$p(q_k) \geq t. \qquad (4)$$

The value t is a given threshold. Following the definition, one proves that $p(\cdot)$ in (3) is nonnegative. p is an estimate for the reference probability of q being referenced alone and not within the context of a longer refstring.

In the definition of the indirect reference frequency (1,2) one recognizes two terms: the candidate q_k may be indirectly referenced by longer r e f s t r i n g s r_{k+1} or by candidates q_{k+1} which themselves are only indirectly referenced by any longer refstring rs_j, $j \gg k+1$. Therefore the indirect reference frequency of a q_k depends of the indirect reference frequencies on some q_{k+1}. So, the definitions (4,5) are complete by the additional one

$$irfl(q_m) = irfr(q_m) = 0$$

which states that the refstring candidates with maximal length m are not directly referenced.

A small example shall illustrate the formulas: let

Q = ABCDE(100,0,0),ABCDF(10,0,0),BABCD(120,0,0)
Q = ABCD(150,100,120   ),ABCE(100,0,0),BABC(200,120,0),ABCF(30,0,0)
Q = ABC(300, .)

The numbers in parantheses behind each string mean the absolute frequency and the indirect reference frequencies irfl and irfr. The underlined elements are refstrings. For a decision whether q =ABC should be a refstring, one determines

        RSL = { ABCE(100,0,0)}
        RSR = { BABC(200,120,0)}
        QL⁻ = { ABCD(150,100,120),ABCF(30,0,0)}
        QR⁻ = {   }

Corresponding to these sets the indirect reference frequencies for ABC are

        irfl = 100+max(100,120)+max(0,0) = 220
        irfr = 200+0   = 200,

and the weight is given by

$$p = 300 - \max (220,200) = 80$$

If the given threshold value for the weight is not greater 80, ABC is used as refstring.

3.2 Determination of Refstring Candidates

The method described so far assumes that refstring candidates Q_k, $k \leq m$ are available with their frequencies. A straight forward solution would be to use all possible substrings in S as candidates and to count their frequencies. A better solution is to restrict the counting to those substrings which have a chance of being a candidate. Because of (4) and due to $p(q_k) \leq f(q_k)$ one may exclude all substrings having an absolute frequency $f(q_k) < t$. Further one applies the inequality

$$f(x) \leq \min[f(y),f(z)]$$

for a string x which contains y, length k-1, as left and z, length k-1, as right substring. Based on these simple observations, a construction of Q_k, k=1,2,...,m, is given:

```
      set k:=0 and Q_1:=Alphabet;
loop: k:=k+1 until m
      count frequencies of q_k ε Q_k in S;
      redefine Q_k by deleting q_k ε Q_k with f(q_k)<t;
      define Q_{k+1} : q_{k+1} ε Q_{k+1} consists of a q_k' εQ_k and
                       a q_k' εQ_k as left resp. right substring;
      goto loop;
```

This algorithm guarantees that no substring s with length $k \leq m$ and with a frequency $f(s) \geq t$ is deleted. The proof is evident by assuming the contrary.

A little example explains the generation of candidates. Let
Q_3 = {CNO(100),H2O(500),NOH(150),CH2(50),2OC(60),CA5(200),HCL(300)}
and t=50. Then, only for the following combinations
Q_4 =CNOH(100), CH2O(50), H2OC(60)}
the frequencies in F have to be counted. The underlined numbers in Q_4 are only estimates (upper limits) for the final

frequencies. They are applied to improve the hash table generation necessary for counting. The combinations with high estimated numbers are used first when the probability for hash conflicts is low.

3.3 Determination of Refstring Combinations

Since a refstring may not be longer than the substrings in S between delimiters (refstrings do not go beyond attributes or single words), the frequency count of certain refstrings may be still too high. Therefore, the above algorithm is applied again but now one level higher: instead of analysing the co-occurrence of characters within strings in S now the co-occurrence statistics of refstrings within the next larger context are determined. Depending on S and the application this larger context is defined to be one query or one sentence of one document, generally one record in S.

The same technique as for the refstring generation is applied for refstring sequences: First, candidate sequences are determined up to a maximum length and then the refstring sequences are selected out from them using the formula (3). The sequence order and the adjacency condition can be removed depending on the application considered.

The described algorithm exploits the hyperbolic distribution already for the refstring generation. It has only a complexity of O(n) and seems to be suitable particularly for the production of longer refstrings and their combinations. The sort step in /BA75/needs O(n.logn) operations and the method described there seems to be more suitable for a small refstring set with short lengths.

3.4 Quantitative Results, Examples

As mentioned in the introduction the realistic assumption of having hyperbolic-like occurrence or usage statistics is essential for the number of refstrings. The fig. 1 shows the distribution of substrings with length 3 occurring in more than 35000 German words.

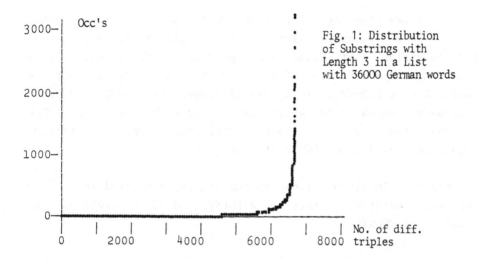

Fig. 1: Distribution
of Substrings with
Length 3 in a List
with 36000 German words

A refstring index to a list of 44000 chemical substances contains
only appr. 2000 refstrings with lengths between 3 and 5 and gives
a mean selectivity of 0.2 percent (Each refstring occurs in at
least 50 and, as a mean value, in 90 substances).

REFSTR	WEIGHT	REFSTR	WEIGHT
METH	50	ETHYL	888
MINE	62	HYDRO	814
AMIN	89	ATION	628
TION	139	77009	566
MINO	61	AMINO	360
LINE	101	YDROX	387
LINO	52	DROXY	378
IONE	108	-METH	336
HYDR	86	IDINE	340
DINO	55	DINE-	69
LING	86	LINE-	64
DING	83	PHENY	345
ROLI	85	HENYL	339
OLIN	101	AMINE	266
PHEN	73	OLINE	207
IDIN	111	THYL-	303
STER	141	ANALO	284
CHEN	72	HALOG	286
NYL-	70	-ACID	283
GENE	100	CHLOR	279
ATIC	82	RIDIN	271
RINE	168	-ANAL	277
RING	90	ALPHA	273
ETHI	75	STERO	172

REFSTR	WEIGHT	REFSTR	WEIGHT
TERM	572	=3750	162
MINI	450	=SYS7	115
ICPU	338	=SYS3	96
0000	19	=5320	71
SCPU	227	=3740	61
OFDE	197	=3270	48
SWIT	183	=2260	42
OTHR	179	=3420	43
DASD	170	=3600	37
ELAS	166	=S306	37
MEMO	146	=3330	25
0001	23	01000	15
LCPU	112	=3145	27
1000	15	00200	13
2000	11	00500	11
0100	14	00300	15
TAPE	92	02000	13
0002	19	OPABX	23
6000	11	=3135	22
0007	24	00030	13
0200	19	=2740	20
0020	12	=XX34	20
TCUS	33	=2365	13
0030	25	=3168	19

Tab.1 Sample of Refstrings with
Length 4 and 5 Generated
with a List of 44000
Chemical Words

Tab.2 Sample of Refstrings with
Length 4 and 5 Generated
with the 3000 DP Product
Records

Table 1 shows the beginning of the refstring list with length 4
resp. 5 from this application. Table 2 contains the beginning of
the refstring list from a DP products application. These
refstrings are generated from 3000 formatted records with 4
searchable attributes. In a current larger test, 3000 refstrings
have been generated to a formatted file with more than 30000
records and 9 searchable attributes each containing
organizational data on IBM customers.

The maximum length of these automatically generated refstrings
was m=5. Therefore strings like PHENY and HENYL could not be
combined to PHENYL.

4. Follow-on Problems of the Refstring Index

Critical points in the application of the refstring index are

  (1) the detection of refstrings in a partial match query
  (2) the number of refstrings and their inverted lists to be
      processed
  (3) the number of records in P to be accesssed, especially
      the ratio between the number of real matches and the
      number of candidates to be inspected.

Starting with the last point first, one observes that it contains
two separate problems: the overall number of accesses of P
regardless of matches or mismatches and the mismatch-ratio (false
drops). Both critical points are improved by the following.

4.1 Clustering by Refstrings

If one assumes that the refstrings are parts of records with a
suitably high reference frequency - and with this objective they
have been generated - and that a PMR-action qualifies more than
one record, then it is desirable that all qualifying record
candidates are collected in one or a few blocks to be transferred
from external to internal storage. Since access is organized by
refstrings, blocks have to be defined by common refstrings, in
other words, by common pieces of record contents. This is

usually called clustering. The following approach generalizes
the remarks in /SCHE75/:

Let be MXRS the set of the longest non-redundant refstring as
related to a string x and p(.) a positive weight function
defined for refstrings (e.g. proportional to reference
frequency). A positive valued function ri(x) called "reference
importance" is defined for each string x by

$$ri(x) = \sum_{rs\varepsilon\ MXRS} p\ (rs)$$

If one applies this function on P, records can be sorted in
descending order of this "importance". Roughly spoken, the first
record contains the longest and most often referenced reference
strings. A proximity measure between two strings $x\varepsilon P$ and $y\varepsilon P$ may
be defined by

$$prox(x,y) = \sum_{rs\varepsilon\ M.RS} p(rs)$$

MXRS resp. MYRS belongs to x and y resp. and M.RS contains
refstrings common to x and y. Obviously, prox is symmetric:

$$prox\ (x,y) = prox\ (y,x)$$
and satisfies the inequality
$$prox\ (x,y) \leq min\ (ri(x),\ ri(y)).$$

These functions will be used to define clusters:

```
          let 0<α≤1; k=0; next:=1
loop:     set r̂:=r_next; r̂i=ri(r̂); k:=k+1;
          (r_next is the record in P being not yet
                        element of a cluster)
          define next cluster C_k:={r_j εP|prox(r̂,r_j)≥αr̂i};
          redefine F by deleting records r_j εC_k from P;
          Goto loop;
```

One observes that this algorithm defines clusters with pairwise
disjoint elements. Each record is in exactly one cluster and
each cluster contains at least one record because
$prox(\hat{r},r) = \hat{r}i \geq \alpha\hat{r}i$ for each α. The factor α is responsible for the

size of the cluster. The number of clusters decreases for $\alpha \to 0$.

Such an algorithm has a complexity of $O(n^2)$ whereas sorting with respect to the ri-function needs only $O(n\log n)$. Experiments in /NU76/ show that conventional blocking after the ri-sort compares favorable with the original clustering regarding the number of data transfers.

More important for the access method using the refstring index is the fact that the inversion to be computed for the refstrings is related to the clusters not to the records. Therefore, the maximum number of inverted list entries is reduced to the number of clusters which can be an order of magnitude smaller than the number of records. The advantage of this method is that the secondary data overhead regarding storage size and processing time is reduced. Several matching records can be found in a single cluster. The disadvantage is the higher number of false drops In fact, a cluster candidate contains all specified refstrings but not necessarily within single records.

4.1.1 A Worst Case Simulation

To quantify the above remarks a worst case test was simulated in the following way: The file P was a list of 32000 different German words. Each word is a record. Clustering was introduced by using every NC words (NC=1,3,5,10) as one cluster after having sorted P alphabetically. Search arguments were simulated by selecting 500 fragments at random from the interior of the words. The refstring index contained only letter pairs as (basic) refstrings. No additional longer refstrings were allowed which again simulates the worst case behaviour of the access method with respect to the number of data accesses and false drops.

In the following table 3 the number of words in each cluster is NC=10. N-ACCESS is the number of primary data (cluster) accesses, and N-MATCH is the umber of found matching words in these clusters.

INPUT STRING (RANDOM)	N-ACCESS (CLUSTER)	N-MATCH (WORDS)
AUPTVERB	1	2
SOZIALET	4	4
EINTRAGU	5	2
INTERPRE	5	3
NERSCHAF	48	9
STRIENAT	4	2
STENSPAR	13	1
HEINKEHR	2	7
GROSSKRE	1	1
PERSONEL	12	5
NZUNGSAU	4	1
RWALTUNG	34	96
HANDELSP	6	4

Table 3: Extract of the Worst Case Test with Random Fragments, Length 8 in a Word Component Search Application

The table 4 shows the distribution of the access rate, defined as the number of cluster accesses divided by the total number of clusters. One recognizes that in 55 percent of the 500 queries with length-6-fragments and NC=3, the access rate is lower or equal 0.1 percent corresponding to 10 clusters or less.

ACCESS RATE INTERVAL (PERCENT)	NC=1 LENGTH					NC=3 LENGTH					NC=10 LENGTH				
	4	5	6	7	8	4	5	6	7	8	4	5	6	7	8
0-0.1	30	55	70	83	87	16	37	55	70	79	4	14	24	34	45
0.1-1	49	39	27	15	12	45	47	38	27	19	25	39	45	47	43
1-5	20	6	3	2	1	32	16	7	3	2	33	32	24	16	12
>5	1	0	0	0	0	7	0	0	0	0	38	15	7	3	0

Tab.4 Access Rate Distribution in the Worst Case Test

The match-access ratio is defined as the quotient between the number of matching words and the number of cluster accesses which are necessary to answer a fragment search. In table 5 the distribution of this quotient is shown for the 500 search fragments. The match access ratio can be greater 1 for NC greater 1.

MATCH	NC=1					NC=3					NC=5					NC=10				
ACCESS	LENGTH					LENGTH					LENGTH					LENGTH				
RATIO	4	5	6	7	8	4	5	6	7	8	4	5	6	7	8	4	5	6	7	8
> 1.	22	40	60	76	82	48	53	62	76	80	42	54	60	68	76	40	41	50	56	59
.75-1.	36	28	18	12	10	14	10	10	4	6	18	8	8	4	4	14	10	6	4	4
.5-.75	20	16	9	8	8	14	14	10	10	10	14	13	9	8	8	14	14	12	12	10
.25-.5	12	8	9	4	0	14	12	12	6	2	14	12	11	12	4	16	16	14	14	12
0-.25	10	8	4	0	0	10	11	6	4	2	12	13	12	8	8	16	19	18	14	15

Tab.5 Match Access Ratio in the Worst Case Test

4.2 Decomposition of a PMR-Query into its Refstrings

This problem does not exist in classical inverted file
approaches for unformatted and formatted DB-systems. Either a
keyword inversion or inversion for certain attributes has been
provided for or not. In the more general refstring inversion,
the attribute values or keywords specified have to be inspected
to determine which refstring inverted lists are applicable and
which are suited best. For the determination whether a certain
substring within an argument of a query is a refstring, a special
hashing technique has been applied based on hashing by division
and chaining /MA75/. The modification applied needs three bytes
for each refstring with length 3 and four bytes for each
refstring with length 4 or 5 including the link field and a chain
flag. So, hash tables for more than 1000 refstrings with length
4 or 5 may be stored in one page of 4k bytes. Details are
described in a separate note /SCHE77/.

Sequential application of the hash function on necessary string
positions within the specified query arguments may therefore lead
directly to the set MXRS. In order to reduce the number of
refstrings further, the string positions which have to be tested
may be selected in such a way that also highly correlated
refstrings, (not only completely redundant ones) are avoided. If
e.g. in the search argument ATOMIC the refstrings ATOM,TOMI,OMIC
are contained, one could omit TOMI and use only the intersection
of the lists between ATOM and OMIC because of the

three-characters overlapping TOMI.

In general, instead of testing each position, a number of s<k (e.g. s=k-1) characters is skipped after having found a refstring with length k.

The following examples show how a query in a component search application is transformed into refstrings. They belong to the list with 44000 chemical terms.

Query	Refstrings
AMINO ACID	AMINO ACID
BETA HYDROXYLASE	HYDRO BETA OXY YLA ASE
BETA-HYDROXY LASE	BETA- -HYDR LASE ROX XY
HYDROXY PROPYL METHYL	HYDRO PROPY METHY OXY YL
DICHLOR ACETAT	CHLOR ACETA DIC TAT
HALOGEN	HALO OGEN
CHLOROFORM	CHLOR FORM RO OF
SULFONAMID	AMID SUL FON LF NA
IONISATION	ATION IONI ISA
GEN-MUTATION	ATION GEN N-M UTA MU
FUZZY	ZZ UZ FU ZY

4.3 Processing of Inverted Lists

For certain basic refstrings the number of occurrences may be very high, (e.g. the strings ER, EN, NE occur very often in German language). In a question for all clusters, containing ER and EN and VE simultaneously, no longer, additional refstrings may be used. Therefore, intersection of these three inverted lists is necessary. Considerations due to Haerder /WE75/ with respect to storage and processing of inverted lists and own experiments resulted in an implementation of uncompressed bit lists a n d index lists. As a rule of thumb - gained experimentally by timings - an inverted list is implemented as index list if the frequency of the corresponding refstring is less than 0.2 to 0.5 % of the number of clusters. This is the same order of magnitude as commonly used for a decision whether access should be direct using the inverted list or sequential neglecting the access path. Haerder gives an upper limit of 1-5 % for direct access.

According to the splitting into bit and index lists, the

intersection between inverted lists is executed

1. as an operation between two bit lists or
2. as testing bits at given index positions in case of intersection between a bit and an index list
3. as an intersection between two index lists which is solved by locally changing one index list into a bit list and subsequent application of method 2.

The advantage of this approach is that the lists may be unordered - an important point because of update actions. Union is performed similarily but instead of testing bits at the index positions, one bit is set in the bit list which now contains the result of the union.

Corresponding to these possibilities for execution, complexity bounds may be found easily. The remarks are restricted to the intersection case. Assume that k refstrings rs_1, ... , rs_k have been found. Assume further that rs_i are sorted in ascending order of their frequencies denoted by m_i. So, rs_1 is the most selective refstring. If one denotes with ε the elementary operation to test or to set a bit at a certain position then the total number t_ε of elementary operations is between

$$m_1 \leq t_\varepsilon \leq (k-1)m_1 + \sum_{i=2}^{k} m_i$$

From timing results one finds that one elementary operation ε needs 22 microseconds CPU time on an IBM /370-145. This means that the intersection of 11 refstrings with occurrence frequencies $m_i = 100$ needs less than 44 milliseconds CPU time on this model confirming that CPU time for list processing is not a problem.

A critical point, however, is the number k of lists to be transferred from external to internal storage. One should note that the main advantage for additional longer refstrings is the reduction of the number of inverted lists to be processed. A further reduction is obtained by the elimination of refstrings from MXRS which are estimated to be highly correlated. Now, as a third possibility for the reduction of k during execution, let

453

ql resp. qc be the cost to read and process an inverted list resp. one cluster and denote by $rc(\hat{k})$ the number of cluster candidates after the processing of the first $\hat{k}$ list ($\hat{k} < k$). Assume further that an estimate β for the reduction of $rc(\hat{k}+1) = \beta \cdot rc(\hat{k})$ is known. Then inverted list processing may be terminated after $\hat{k}$ if

$$rc(\hat{k}) \leq \frac{ql}{qc(1-\beta)}$$

This formula is valid even for the optimistic estimate $\beta = 0$ (next intersection would lead to an empty list) or for the pessimistic $\beta = 1$ (next reduction does not reduce rc).

5. Applications

The access method by the refstring index is non hierarchical and not influenced by a special sequence of the clusters to be accessed. Several different applications are obtained by a special interpretation of the clusters and the related refstrings.

5.1 Refstring Indexing in Non-Formatted Data

Documentation systems with automatic indexing such as the STAIRS system generate a dictionary which contains each document string between delimiters apart from strings contained in a "negative list". In the application of such a system for a language with composita as for the German language but also for medical diagnosis texts or for descriptions of chemical substances etc., the problem arises that important components are contained in the interior of the dictionary strings which means that a query specifying only components of a dictionary entry may not be answered directly with the aid of the prepared inversion. A common solution to this wellknown problem is either to add c e r t a i n components to the dictionary or to establish a compound word relation within a thesaurus.

The disadvantage of such a method is that the components to be introduced have to be defined m a n u a l l y . They are

application-dependant and necessarily are not complete if new
documents containing new terms are introduced. Furthermore, one
knows from programs performing a compound word decomposition
/SCHO77, IZ77/ that a high number (7000-15000) components has to
be maintained for the German language. This number of "reference"
components even will be increased if documents are used
containing also chemical elements or technical compound
(artificial) words or if documents in several languages have to
be processed.

In this situation it seems to be a good solution to replace the
reference components by automatically generated reference
strings. They have the advantage to allow the specification of
a r b i t r a r y   search fragments independantly of a
language or of the occurrence of artificial terms.

A further problem in systems with automatic indexing is the high
number of dictionary entries (500000 is a usual size). It leads to
the question whether all these words are needed at all. In /HE74/
and /GE76/ it is pointed out that a saturation of dictionary
entries may not be observed. A high number of very short
inverted lists has to be administrated without being ever used.
Since each word may appear also in different (flexion) forms such
as 'atom', 'atoms', 'atomic', 'atomar', etc., a user being aware
of this fact would specify his query in the above example as
'atom**' in order to include all documents containing the keyword
atom with at most two additional characters at the end. In order
two answer this example question, 4 accesses to the 4 related
inverted lists and their union have to be performed.

On the other hand, the dictionary contains also few words which
are very frequent and which occur also frequently in a sequence
with another frequent word. A typical example may be the sequence
'pattern recognition'. If this sequence is specified in a query,
two inverted lists have to be transferred and their intersection
has to be computed. A solution to this problem by using
refstrings and refstring combinations as document index is
currently under investigation. Experiments known from the
literature (e.g. /SA68, LUS67/) show that statistical term
association give reasonable results, especially for the

application in a floating dictionary /STE 74/. The refstring index applied on documents is a generalization.

The following two examples from a terminal session are taken from experimental programs (PL/1 and ASSEMBLER) for a word component search running under VM-CMS on a /370-145. Virt. Time is the CPU time in milliseconds. The first example is taken from a word component search through a list with 32000 German words (NC=1).

```
SEARCH COMPONENT(S)     VERSICHERUNG RENTE GESETZ

VIRT. TIME      311

FOUND WORDS FOR     VERSICHERUNG RENTE GESETZ

KNAPPSCHAFTSRENTENVERSICHERUNGSGESETZES
RENTENVERSICHERUNGS-NEUREGELUNGSGESETZ
RENTENVERSICHERUNGS-NEUREGELUNGSGESETZE
RENTENVERSICHERUNGS-NEUREGELUNGSGESETZEN
RENTENVERSICHERUNGSAENDERUNGSGESETZ
RENTENVERSICHERUNGSGESETZ
RENTENVERSICHERUNGSNEUREGELUNGSGESETZE
```

In the second example the same programs search through a list with 120000 mixed language words. One cluster contains 5 words (NC=5)

```
SEARCH COMPONENT(S)     PHOSPH   SULF

VIRT. TIME    349

FOUND WORDS FOR     PHOSPH    SULF

BIS(TRIARYLPHOSPHORANYLIDENE)SULFAMIDES
BIS(TRICHLOROPHOSPHORANYLIDENE)SULFAMIDE
SULFIDPHOSPHORE
SULFIDPHOSPHOREN
TRICHLOROPHOSPHORANYLIDENE)SULFAMIC
ZINKSULFID-PHOSPHOREN
ZINKSULFIDPHOSPHORE
```

5.2 Refstring Index as Multi-Attribute Index in Formatted Data

Because attributes in formatted files may also consist of natural language words or personal names, addresses etc., the possibility for a search with components of attributes or arbitrary masking seem also relevant here: A refstring index can be provided for certain attributes to allow partially specified attribute values. In addition, the refstring index method may support queries which

specify more than one attribute value. The usual approach for
processing such queries is to prepare inverted lists for each of
those "important" attributes or to use combined indexes in the
sense of /LU 70/. A new proposal by the refstring inversion is
obtained in the following two ways:

1. If one attribute is more important than all others and often
   further attributes together with the dominant one are
   specified, than an improvement over the single attribute
   inversion is obtained if additional refstrings from other
   attributes are selected which occur often in combination with
   frequent refstrings of the dominant attribute. Compared with
   the combined index appraoch, this proposals combines only
   c e r t a i n attribute v a l u e s instead of whole
   complete attributes.

2. For several applications it seems to be a reasonable solution
   to regard the whole records or a subset of "searchable
   attributes" as "document" for a refstring generation.
   Therefore, one s i n g l e refstring index is valid for
   s e v e r a l record attributes. This means that for the
   computation of the direct record access the attribute
   information is neglected. Here, again only attribute value
   combiantions as refstring sequences will occur and not
   complete combined attributes.

This idea has been applied on a file with 3000 records containing
information on DP-products distributed over 8 attributes. Four
of them have been assigned to be searchable. Quantitative results
are summarized in table 6. They are obtained by a usage
simulation in the following two ways:

1. A record out from the 3000 is selected at random and two
   values of its four searchable attributes are selected again at
   random to give the arguments of a partial match query. This
   procedure is iterated to get 100 partial match queries.
   Obviously, these queries simulate a worst case usage because
   each attribute value has the same probability to be specified
   in such a query and therefore do not correspond to the
   prepared index.

2. The records are sorted corresponding to the function ri, (see 4.1). Then, the two most frequent attribute values are selected in every record, starting with the first record in the new sequence. These two values are considered as arguments of a partial match query. The procedure is iterated until 100 different partial match queries are found. In this case the usage simulation corresponds to the prepared index and is the favorable case .

The following table shows the results of these simulations with respect to secondary data accesses (IOS), primary data accesses (IOP), number of secondary data accesses if only pairs as refstrings are used (IOSB) and the related number of primary data accesses (IOPB). MATCH is the number of found matching records. If the number of candidate records is less than NCM during the processing of the query, the secondary data accesses are stopped and the primary data are accessed. Values of 1 and 10 for NCM are used in the tests.

	NCM	IOS	IOP	IOSB	IOPB	MATCH
worst case	1	2.8	10.2	5.7	10.2	10.1
favor.case	1	2.5	10.7	5.7	11.9	10.5
worst case	10	2.3	11.6	5.2	10.9	10.1
favor.case	10	2.0	11.4	5.0	12.4	10.5

Table 6: Refstring Index Applied as Multi-Attribute Index

The main influence is the reduction of secondary IOs due to longer refstrings, even in the worst case simulation. The optimal value for the number of secondary IOs is 2.0 which is obtained in the usual approach where all attributes are completely inverted. This value can be kept smaller only if combined indexes or combined refstrings are used.

Acknowledgement

I want to thank Barbara Ruhbach, Rainer Nussbaum and Hans Peter von Reth students at the Universities of Heidelberg and Mannheim for their great assistance in implementing the programs and evaluating the experiments.

References

AHO75   A. V. Aho, Margret J. Corasick, Efficient String Matching: An Aid to Bibliographic Search, Comm. ACM (1875), Vol. 18, No. 6, pp. 333-340

AHO74   A. V. Aho, The Design and Analysis of Computer Algorithms, Addison-Wesley Publishing Company, Reading, (Mass.) 1974

BA75   J. J. Barton, S. E. Creasy, M. R. Lynch, M. J. Snell, An Information-Theoretic Approach to Text Searching in Direct Access Systems, Comm. ACM (1974), Vol. 17, No. 6, pp. 345-350

BAY73   R. Bayer, E. McCreight, Organization and Maintenance of Large Ordered Indexes, Acta Informatica 1 (1972), pp. 173-189

BE75   J. L. Bentley, Multidimensional Binary Search Trees Used for Associative Searching, Comm. ACM (1975), Vol. 18, No. 9, pp. 509-517

BU76   W. A. Burkhard, Hashing and Trie Algorithms for Partial Match Retrieval, ACM Transactions on Data Base Systems, (1976), Vol. 1, No. 2, PP. 175-187

CLA72   A. C. Clare, E. M. Cook, M. F. Lynch, The Identification of Variable-Length, Equifrequent Character Strings in a Natural Language Data Base, Computer Journal Vol. 15, No. 3, pp. 259-262

GE76   F. Gebhardt, Wortstatistiken an groesseren Textsammlungen, Nachrichten f. Dokumentation, 2-1977, Hrsg. von der Deutschen Gesellschaft f. Dokumentation e.V., Seite 53-58

HA71   M. C. Harrison, Implementation of the Substring Test by Hashing, Comm. ACM (1971), Vol. 14, No. 12, pp. 777-779

HE74   R. Henzler, Quantitative Beziehungen zwischen Textlaengen und Wortschatz, Hrg. Zentralstelle fuer maschinelle Dokumentation, Frankfurt, Nr. ZMD-A-23, Beuth- Verlag, Frankfurt, 1974

IZ77   H. Izbicki, Composita Program, Documentation Draft, IBM Laboratory Vienna, March 1977

KNU73   D. E. Knuth, The Art of Computer Programming, Sorting and Searching, Addison-Wesley Publishing Company, Reading, (Mass.) 1973

KNU74   D. E. Knuth et al, Fast Pattern Matching in Strings, Technical Report No. STAN-CA-74-440, 1974

LUS67   G. Lustig, A New Class of Association factors, in Mechanized Information Storage, Retrieval and Dissemination, (ed. K. Samuelson), Proeedings of the FID-IFIP Conf., Rome, 1967, North-Holland Publ. Comp. Amsterdam 1968.

LU70   V. Y. Lum, Multi-attribute Retrieval with Combined Indexes, Comm. ACM, (1970), Vol. 13, No. 11, pp. 66-665

MAU75   W. D. Maurer, T.G. Lewis, Hash Table Methods, Computing Surveys, (1975), Vol. 7, No. 1, pp. 6-19

MCR74   E. M. McCreight, A Space-Economical Suffix Tree Construction Algorithm, JACM (1976), Vol. 23, No. 2, pp. 262-272

NU76   R. Nussbaum, Diskussion verschiedener Aehnlichkeitsanordnungen in grossen Wortlisten, Diplomarbeit Universitaet Mannheim, Institut f. Wirtschaftsinformatik, 1977.

SA68   G. Salton, Automatic Information Organization and

Retrieval, Mc Graw-Hill, New York, 1968

SCHE75   H.-J. Schek, Tolerating Fuzziness in Keywords by Similarity Searches, IBM Scientific Center., Heidelberg (1975), Technical Report TR 75.11.010 contained in Kybernetes 6 (1977) Special Issue on Fuzzy Systems

SCHE77   H.-J. Schek, The Reference String Access Method and Partial Match Retrieval, IBM Scientific Center Technical Report TR77.12.009.

SCHO77   G. Schott, Automatische Kompositazerlegung mit einem Minimalwoerterbuch, Vortrag bei der Fruehjahrstagung GMDS-GI, Giessen, April 1977

STE74   I. Steinacker, Indexing and Automatic Significance Analysis, Journal of the American Society for Information Science, (1974), Vol. 25, No. 4, pp. 237-241

WA73   R. E. Wagner, Indexing Design Considerations, IBM Systems Journal, (1973), No. 4, pp. 351-367

WE75   H. Wedekind, T. Haerder, Datenbanksysteme II, Reihe Informatik)18, Bibliographisches Institut Mannheim/Wien/Zürich, B.I.-Wissenschaftsverlag 1976

WO71   E. Wong, T. C. Chiang, Canonical Structure in Attribute Based File Organization, Comm. ACM, (1971), Vol. 14, No. 9, pp. 593-597

YA77   S. Yamamoto, S. Tazawa, K. Ushio, H. Ikeda: Design of a Balanced Multiple-Valued File Organization Scheme with the Least Redundancy, Proc. of the Very Large Data Base Conf., Tokio, Oct. 1977, p.230.

ZI49   G. Zipf, Human Behaviour and the Principle of Least Effort, Addison-Wesley, Cambridge, Mass. 1949.

WORKING SET SIZE REDUCTION BY RESTRUCTURING APL WORKSPACES

D. Kropp, H. Wrobel
IBM Scientific Center Heidelberg
Tiergartenstrasse 15
6900 Heidelberg, Germany

ABSTRACT

APL Workspaces usually have a random distribution of the defined APL functions. The physical location of a defined APL function within an APL workspace depends on its history of creation, modification, or copying. Furthermore in APL a good programming style results in a large number of small functions. The execution of an APL program in a virtual storage environment usually leads to comparatively large working set sizes because the APL functions are scattered over the workspace. However, the interrelation between the functions can be analysed and exploited to reorganize a workspace and thus reduce the working set sizes. Methods of restructuring an APL workspace are proposed. Results of an investigation on a large APL/CMS workspace using these methods are presented.

INTRODUCTION

Virtual storage computing systems have come into wide use in recent years. Most APL installations on system /370 machines today run under virtual storage operating systems, giving more freedom to the APL programmer because of the availability of much larger workspace sizes. Consequently large APL programs are being created. Running in a virtual storage environment they can have bad performance characteristics, as has been observed in other large programs. Under high machine load conditions a program will run most efficiently, if it has a high locality of reference corresponding with small working sets. Experimental techniques have been developed, and methods proposed to improve locality in a program by rearranging relocatable sections of code (1-4) and by following certain guidelines in program design (5).

Certainly one intention of virtual storage is to relieve the application programmer from having to deal with physical storage limitations. On the other hand, large programs in a virtual storage environment tend to have performance problems, unless attention is paid to locality of reference. For frequently used programs, the application programmer needs to be concerned with the efficient execution of his program.

APL systems more than any other programming language shield the programmer from the internals of his workspace. On the surface there seems to be no easy way to influence the structure of the internal representation of an APL program. But the fact that APL programs usually consist of many rather small Defined Functions suggests in principle the possibility of locality improvements of APL workspaces.

This paper presents a feasible procedure to get a controlled internal structure of an APL/CMS workspace. It describes how functional interrelations are used for restructuring a workspace. Some results from experiments with an APL workspace with 511 Defined Functions are presented.

REARRANGING AN APL WORKSPACE

Control of workspace structure

The internal structure of an APL workspace depends on the history of performing copy operations, erasing objects, and defining or modifying Functions. The internal structure can be controlled by using the)COPY,)PCOPY, and)GROUP commands. Investigating the effect of the)COPY and)PCOPY commands in APL/CMS shows that they can easily be used to control the internal relative location of Defined Functions and/or Global Variables. These commands, when executed for single objects or defined Groups of objects, append the copied objects to the content of the target workspace maintaining the sequence as specified in the Group. For economic reasons one will use Groups to rearrange a large number of objects. The normal APL capabilities to manipulate Group definitions are somewhat limited. Fortunately, the Stack Processor provides a means to define a Group from a large list of names. In APL/CMS, the)COPY command places the portion of the symbol table for the Group of objects copied adjacent to the Group, thus scattering the symbol table over the workspace. If one wants to avoid this one can copy a Group representing the whole wanted structure.

The restructuring method

The intention of restructuring is increasing the degree of locality of the program to be executed. Locality means "keeping a program's address-space references confined to as few pages as possible for as long as possible" (5). In terms of the working set good locality corresponds with small working set sizes. Locality and working set size of a program vary with time as a program proceeds through different execution phases.

Improvements in performance by restructuring can in particular be expected, when the relocatable sections of a program are smaller than the page size of the paging system. In an APL workspace the relocatable sections are the Defined Functions and the Global Variables. In the example discussed below the internal representation of an average of 8 Functions fills one page of 4 k bytes. This figure of about 8 Functions per page seems quite normal for common APL coding.

Two kinds of restructuring are usually distinguished: static and dynamic. Restructuring based on an analysis of the program prior to execution is called static. Quite elaborate static analysis of interrelations and cyclic structures can be performed and used for restructuring algorithms (e. g. 6).

If one wants to make use of the dynamic behaviour of the program one has to analyse run time data like instruction traces. By using dynamic restructuring substantial improvements of the paging behaviour are possible (1,3,4).

A dynamic approach like the one described in (1) applied to an APL workspace would be difficult and time consuming. So far we have confined ourselves to a static restructuring procedure working on the set of Defined Functions. The essential locality improvements may be achieved by an algorithm that separates code of disjunctive program phases into separate clusters and puts Functions as close as possible to the calling Functions. Obviously compromises have to be made, since many Functions are called from different parts in the program.

The algorithm to create the restructuring list was coded in APL. It works as follows: First the name of the main Function is put into the list. Next the names of all the Functions used by it are added to the list. For each Function in the list the names of the Functions it references are added to the list immediately after its name. If a name has already appeared above its possible position, it is not added. Duplicate names are removed starting from the bottom of the list. No attempt to detect loops is made. The example in Fig. 1 illustrates this process: Function 1 refers line by line to Functions 2,3,4 und 5. After name 1 the names 2,3,4,5 are appended to the list. Next Function 2 is analysed providing the names 6,10,7, and 4. 6,10, and 7 are inserted into the list between 2 and 3. Then Function 6 is analysed and so forth. The created list of this example is shown in Fig. 1.

Starting from a main Function, to which all other Functions are connected directly or indirectly, the process of creating the structured sequence of names can be done automatically. The program which does the restructuring also accepts a list of Function names the "subtrees" of which are to be excluded from the process. Thus existing knowledge of the overall structure of the program can be included in its restructuring.

Names of Functions which are called from several places and do not call other Functions are extracted afterwards into a separate list of names of "isolated" Functions. The objects of this list are comparable with objects one would have put into a root segment of an overlay structure.

From the lists of names Groups are defined and transferred to the workspace to be restructured. With the)PCOPY command then a restructured workspace can be built up in a Clear Workspace. Names contained in more than one Group make no problems. The)PCOPY command will copy a Function only the first time its name appears. From this follows that groupings considered to be most important (e.g. the "isolated" Functions) should be copied first.

EXPERIMENTAL RESULTS

The investigation was made with an APL workspace of 420 k bytes. This workspace consists of 511 Defined Functions covering about 245 k bytes and 17 permanent Global Variables needing 45 k bytes. The size of the available work area is 130 k bytes. The Gobal Variables were not considered to be put into a prefered place or sequence. The workspace contains an interactive program having a number of distinct phases.

Obviously the area accessible to improvements is that with 245 k bytes or about 60 pages of 4 k bytes. The experiment was run on a system /370 model 145 with the APL/CMS microcode assist feature. The paging technique used by VM/370 is demand paging with an LRU (Least Recently Used) algorithm. The real storage available to the program was controlled by use of the VM/370 LOCK command for a different user. Runs of 4 minutes virtual CPU time were measured for each defined real storage size. The VM/370 MONITOR facility was used to get the presented data.

The runs took place in a "locked out" virtual machine, saying it had to page against itself. A comparison is made of runs of the original workspace with runs of the restructured workspace. In Fig.2 are the "parachor" curves (5) shown derived from these runs, and in Fig.3 the performance improvements as paging ratios are given. Considering the numbers of page reads the greatest improvement is seen in the left part where paging starts to be excessive. This corresponds with the results in (1). More interesting with respect to "what is saved in a multi user situation" may be to compare pages of real storage for equal paging rates of the two workspaces. These savings are larger in the right part of the curves. In fact, the more desirable load situation for the programs is, when they can work right of the knee of the parachor curves. The real storage savings in numbers of pages lie between 8 and 11 pages which is a 14-17% improvement.   Taking into account that essential parts of the working set are formed by the APL interpreter, the APL variables, and some CP/CMS activity the improvement is more significant.

More detailed information can be seen from Fig. 4-5. They show the frequency distributions (in %) of "projected working-set sizes" as defined in CP. The comparison is made at 4 paging rates for interactive phases (Fig. 4a-d) and CPU-intense phases (Fig. 5a-d) which are distinguished by CP. The curves reflect a common structure of frequency of 2 prefered "projected working set sizes", and the curves of the restructured workspace are relatively displaced to the left. This can be interpreted as: smaller working set sizes are more frequent all through the different program phases.

These results show that static restructuring of an APL workspace has a reasonable effect. The algorithm used so far is a somewhat "ad hoc" solution, to get a first impression, what might be possible in restructuring APL workspaces. Clustering algorithms better exploiting the static overall interrelation between Defined Functions in some analogy to the "nearness matrix" of the dynamic approach of (1) seem possible. On the other hand, a program so intensively used like the APL interpreter itself might gain performance in a virtual storage environment by restructuring it either dynamically or with respect to frequency of usage of APL Primitive Functions. One also might think of some permanent feature in an APL system that does a static restructuring when a workspace is copied into a Clear Workspace.

In virtual storage environments so far the aspect of locality mostly is neglected by the user, probably, since restructuring methods usually are not easy to handle, and the effect, though actually present, becomes largely hidden in the multiprogramming situation. But in APL as well as in other programming languages with an automatic mechanism, available to the user as an option, large programs can be structured for better performance without special knowledge.

REFERENCES

(1) Hatfield, D.J. and Gerald,J.:Program Restructuring for Virtual Memory. IBM Systems Journal, 10, 3, 168-192, 1971.

(2) Hatfield, D. J.: Experiments on Page Size, Program Access Patterns, and Virtual Memory Performance. IBM Journal of Research and Development 16, 1, 58-66, 1972.

(3) Ferrari,D.: Improved Locality by Critical Working Sets. Communications of the ACM, 17, 11, 614-620, 1974.

(4) Baier,J.L. and Sager, G.R.: Dynamic Improvement of Locality in Virtual Memory Systems. IEEE Transactions on Software Engineering, SE-2, 1, 54-62, 1976.

(5) Morrison, J.E.: User Program Performance in Virtual Storage Systems. IBM Systems Journal, 12, 3, 216-237, 1973.

(6) Baer,J.L. and Caughey,R.: Segmentation and Optimization of Programs from Cyclic Structure Analysis. Proc. AFIPS 1972, Spring Joint Computer Conference, 40, 23-36.

FIGURES

Fig. 1:   Graph of "Function Call" interrelations and restructured Function names list.

Fig. 2:   Page Reads ("parachor" curves) and Page Writes plotted against available real storage. The thick lines are the curves of the restructured workspace.

Fig. 3:   Relative performance as ratio of (page reads of original workspace) to (page reads of restructured workspace).

Fig. 4a-d: Frequency distributions of "projected working set sizes" at interactive phases of paging rates A,B,C,D (Fig. 2).

Fig. 5a-d: Frequency distributions of "projected working set sizes" at CPU-intense phases of paging rates A,B,C,D (Fig.2).

467

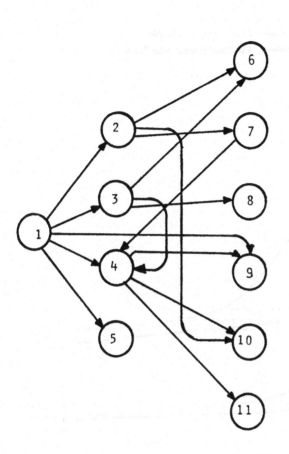

Fig. 1

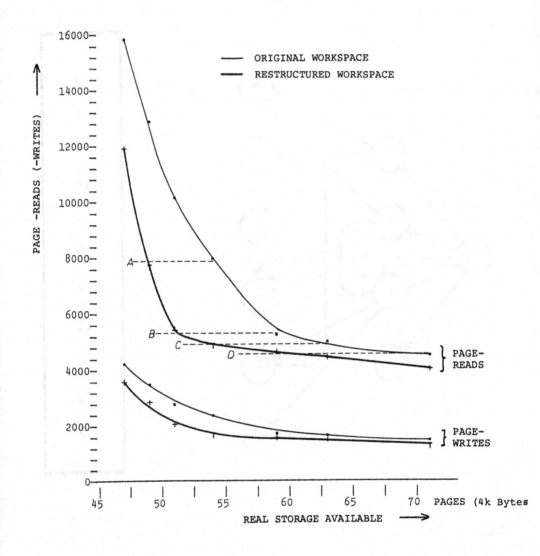

Fig. 2

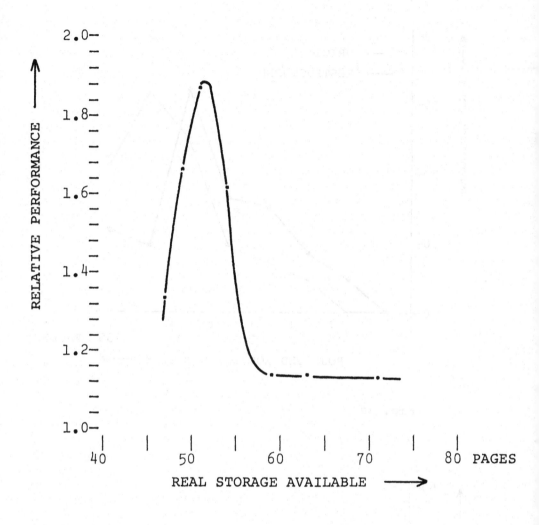

Fig. 3

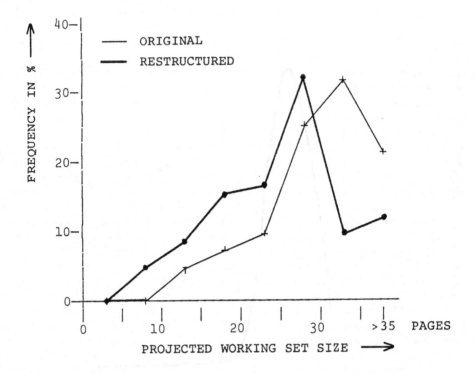

Fig. 4a

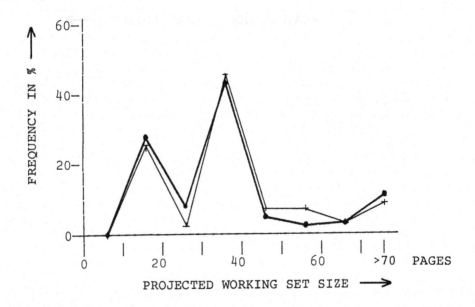

Fig. 5a

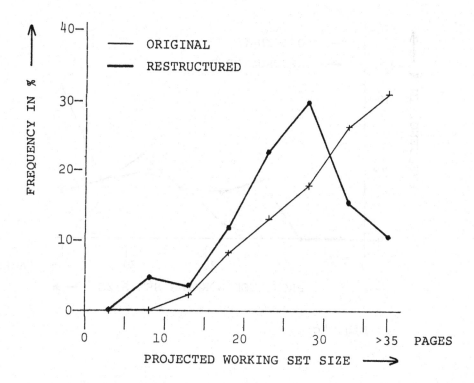

Fig. 4b

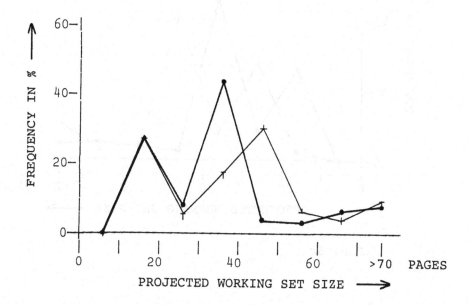

Fig. 5b

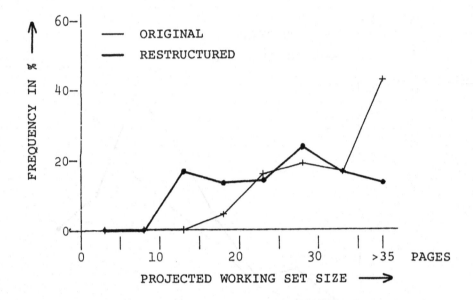

Fig. 4c

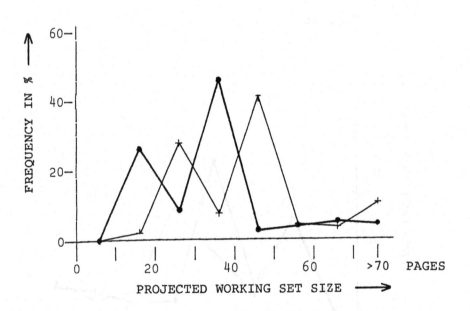

Fig. 5c

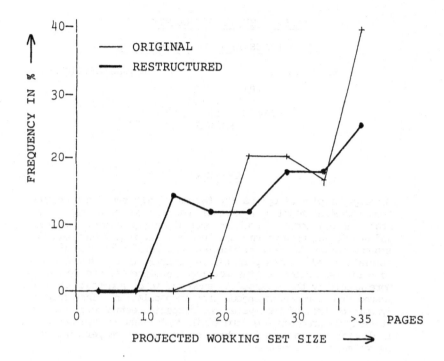

Fig. 4d

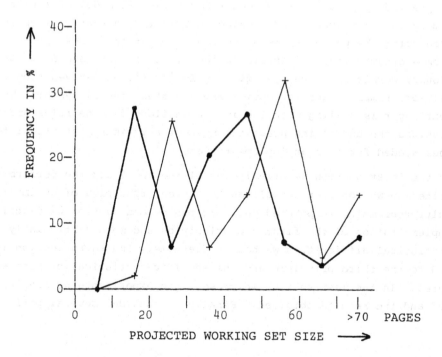

Fig. 5d

OPTIMAL MULTIPROGRAMMING :

PRINCIPLES AND IMPLEMENTATION

Marc BADEL   Jacques LEROUDIER

IRIA - LABORIA
B.P. 105
78150   Le Chesnay
FRANCE

ABSTRACT

Three principles of optimality for multiprogramming are derived
from a general model of a virtual memory computer system. They
state the existence of both an optimal multiprogramming degree
and an optimal program mixture in the multiprogramming set. The
implementation of these principles is carefully studied in a
simulation model which permits, in contrast to analytical models,
to relax assumptions on the workload submitted to the system.
More precisely the workload is supposed to be non-stationary and
issued from a non-homogeneous program population. Therefore a
dynamic control of the system needs special estimators. These
statistical problems are studied in detail and an implementation
is described. The results obtained on the system performance
improvement are presented.

INTRODUCTION

In previous papers [1, 2] we described a general model of a virtual
memory computer system and studied in both from a static viewpoint by
predicting the performance in the case of stationary assumptions and
from a dynamic one by controlling the system in the case of a non-stationary workload. More specifically in [3, 4], we derived from our model some simple rules for optimizing a virtual memory system. The present paper is dealing with the implementation of these rules and it emphasizes the underlying assumptions and the various statistical techniques needed for such an implementation.

In the first section we briefly recall the model and the performance
criterion we consider as well as the three principles of optimality for
multiprogramming we pointed out. In the second section we describe the
implementation of the first two principles and more particularly the
statistical difficulties we had to overcome. The third section is devoted to the third principle and the specific statistical problem encountered. In the last section we present the experiments we have conducted and the results obtained. Finally we consider several performance

criteria and we show they can be satisfied simultaneously.

It is worth noting that the statistical results we derive solve general problems occurring when estimating mean values of non-stationary stochastic processes. Therefore they can be widely used out of the context in which we apply them.

1 - THE MODEL AND THE PERFORMANCE CRITERION

1.1 - The model

The model represented in figure 1 has been described in detail in [1,2].

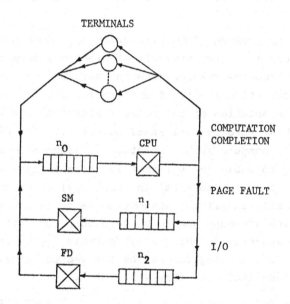

TERMINALS

COMPUTATION
COMPLETION

PAGE FAULT

I/O

n_0 CPU

SM n_1

FD n_2

Figure 1

It consists of a CPU, a secondary memory (SM) device (e.g., fixed head disk or drum), a file disk (FD) and a set of terminals. A queue of requests is associated with each device and the order in which these requests are satisfied is assumed to be FIFO. There is no queue in front of the terminals since each terminal is dedicated to a user (i.e. a user's program or a process).

If N is the number of terminals n_0, n_1, n_2 the numbers of processes, respectively, in the CPU queue, SM queue and FD queue, the degree of multiprogramming n is defined by

(1) $n = n_0 + n_1 + n_2$, $0 \leq n \leq N$

The degree of multiprogramming represents the number of processes (or programs) sharing the memory.

The processes behaviour is characterized by a computation time followed either by a page fault after which the process enters the SM queue or by an I/O request in which case the process enters the FD queue. Processes which terminate their service at the SM or at the FD return to the CPU queue.

Obviously, the degree of multiprogramming n does not remain constant, it varies according to the arrivals of processes from terminals and their departures from the multiprogramming set (because a program has completed its computation time). Therefore it is not possible to derive a steady state solution of the model (values of figures of merit such as queue lenghts, resource utilizations, etc...) for each value of n. However we will compute such a solution by assuming our model is near completely decomposable [5, 6], that is there is time enough between two variations of n (two terminal interactions) to ensure we reach stationarity. The main underlying assumption is concerned with terminal interactions which are supposed to be rare compared with internal interactions in the system. In this case, Courtois [5, 6] shows that the figures of merit we are looking for can be derived for a given n by computing them at steady state for this value of n.

In order to verify this assumption, we simulated our model and estimated our figures of merit according to two strategies :
- first by running n experiments, one for each value of n,
- secondly by running only one experiment, in which n can fluctuate according to the arrivals into or the departures from the system.

We found out identical results (confidence intervals overlapped) by these two approaches, thus we considered our model was actually near-completely decomposable.

Let T_{sm}, T_{fd} respectively be the mean service times of the SM and the FD, and T_{pf} and T_{io} the virtual mean times between page faults and I/O operations.

From the work-rates theorems due to Chang and Lavenberg [9], we can derive the following relations between T_{sm}, T_{fd}, T_{pf}, T_{io} and the CPU, SM and FD utilizations, respectively denoted by $A_{cpu}(n)$, $A_{sm}(n)$ and $A_{fd}(n)$:

(2) $\qquad A_{sm}(n) = \dfrac{T_{sm}}{T_{pf}} A_{cpu}(n)$

(3) $\qquad A_{fd}(n) = \dfrac{T_{fd}}{T_{io}} A_{cpu}(n)$

It should be noted that these equations can be written whithout any assumption on the distribution functions involved in the model. If we assume exponential distributions, the model is then a particular case of Jackson's network [7, 1] and the stationary solution of the system $p(n_0, n_1, n_2)$ is known and given by

$$p(n_0, n_1, n_2) = \frac{1}{G(n)} \left(\frac{T_{sm}}{T_{pf}}\right)^{n_1} \left(\frac{T_{fd}}{T_{io}}\right)^{n_2}$$

where
$$G(n) = \sum_{n_1, n_2} \left(\frac{T_{sm}}{T_{pf}}\right)^{n_1} \left(\frac{T_{fd}}{T_{io}}\right)^{n_2}$$

In [8] Buzen shows that the CPU utilization can be written as :

(4) $\qquad A_{cpu}(n) = \dfrac{G(n-1)}{G(n)}$

and an algorithm is proposed for computing $G(n)$, and hence $A_{cpu}(n)$.

The relations (2) and (3) assume there is no overhead, that is no CPU time is consumed when such an event as a page fault or an I/O occurs. If we denote respectively by θ_{pf}, θ_{io} and θ_{cpu} the mean overheads when a page fault, an I/O or a program completion happens, and by T_{cpu} the mean virtual time of program executions, it is shown in [10] that :

(5) $\qquad A'_{cpu}(n) = A_{cpu}(n) \ \dfrac{1}{1 + \dfrac{\theta_{cpu}}{T_{cpu}} + \dfrac{\theta_{pf}}{T_{pf}} + \dfrac{\theta_{io}}{T_{io}}}$

where $A_{cpu}(n)$ is the total CPU utilization and $A'_{cpu}(n)$ the only CPU utilization consumed by program users (without overhead).

Therefore the relations (2) and (3) become :

(6) $\qquad A_{sm}(n) = \dfrac{T_{sm}}{T_{pf}} A'_{cpu}(n)$

(7) $\qquad A_{fd}(n) = \dfrac{T_{fd}}{T_{io}} A'_{cpu}(n)$

It should be noted that since

$\qquad A_{cpu}(n) \leq 1$

we have :

(8) $\qquad A'_{cpu}(n) \leq \dfrac{1}{1 + \dfrac{\theta_{cpu}}{T_{cpu}} + \dfrac{\theta_{pf}}{T_{pf}} + \dfrac{\theta_{io}}{T_{io}}}$

It remains to characterize T_{pf} which depends on the program behaviour but also on the system and, in particular, on the memory allocation policy. Assuming that the memory is allocated to a program on a page-on-demand basis we shall use experimental evidence to relate the mean time T_{pf} between two consecutive page faults of a program to the amount of space allocated to this program. Belàdy and Kuehner [12] have proposed the following model to fit their measurements :

(9) $\qquad T_{pf} = am^k$

Furthermore, we shall assume that the available core memory is of size M and that it is equally shared among user programs which yields :

$\qquad m = \dfrac{M}{n}$

and

$\qquad T_{pf} = a(\dfrac{M}{n})^k$

We studied in [13, 14] (see also [1]) the different values of a and k for several programs. We pointed out that these two parameters are not independent but related to each other by :

(10) $\qquad k \simeq - \log a + 1 \text{ or } a \simeq 10^{-k+1}$

1.2 - The criterion under consideration

We are interested in two kinds of performance measures. We want, on the one hand, to have a good utilization of the different system resources and, on the other hand, to ensure a satisfactory system's response time.

These two imperatives can be conciliated by considering the ratio of the real execution time of n programs with no multiprogramming to the real execution time of n programs with a multiprogramming degree n. D(n), also called "dilatation" [2, 3, 10], measures the expansion of the real time which is obtained by multiprogramming n programs. We can evaluate D(n) by noting that D(n) is also the ratio of the program throughputs with and without multiprogramming (see [10]). Then, D(n) is expressed by :

$$(11) \qquad D(n) = A'_{cpu}(n) \ (1 + \frac{T_{fd}}{T_{io}} + \frac{T_{sm}}{T_{pf}(1)} + \frac{\theta_{pf}}{T_{pf}} + \frac{\theta_{pf}}{T_{pf}(1)} + \frac{\theta_{io}}{T_{io}})$$

where $T_{pf}(1) = a(\frac{M}{I})^k$

From (3), we can write :

$$(12) \qquad D(n) = A'_{cpu}(n) + A_{fd}(n) + A'_{cpu}(n) \ (\frac{T_{sm}}{T_{pf}(1)} + \frac{\theta_{pf}}{T_{pf}(1)} + \frac{\theta_{pf}}{T_{pf}} + \frac{\theta_{io}}{T_{io}})$$

So D(n) is not only a measure of the system response time, but also a measure of the resource utilizations, since it can be expressed as their sum. The term $\frac{T_{sm}}{T_{pf}(1)}$ comes from the existence of a virtual memory. It can usually be neglected, if the size of the memory is appropriate because $T_{pf}(1)$ is large compared to T_{sm}. The other terms $\frac{\theta_{pf}}{T_{pf}(1)}$, $\frac{\theta_{pf}}{T_{pf}}$ and $\frac{\theta_{io}}{T_{io}}$ come from the overheads : in general, they can be neglected too.

To summarize, if the system is well sized, the third term in (12) is negligible (if not, it is not worthwhile of optimizing anything), hence D(n) can be approximated by :

$$(13) \qquad D(n) \simeq A_{cpu}(n) + A_{fd}(n)$$

It should be noticed that equation (13) is similar to the one written in [2, eq. (7)] which was obtained with another approach. In [2] "dilatation" is defined as a measure of the parallelism between the resources of the system. Thus, our criterion takes into account the three following measures :

        - parallelism between system resources

- resource utilizations
- system's response time

which are shown to be related.

1.3 - Three principles of optimality

In [3] (see also [4]), we pointed out three principles for optimizing the "dilatation" D(n).

Principle I :

In a multiprogrammed virtual memory computer, optimum performance is achieved if the utilization of the secondary memory (SM) remains in the 50% region.

Principle II :

At the optimum working point the sensitivity of the degree of multi-programming to the utilization of secondary memory is minimum.

Principle III :

An optimal mixture of programs within the multiprogramming set is achieved when the mean virtual time between consecutive I/O requests on a specific device of program belonging to the multiprogramming set is equal to the mean service time of the I/O device.

We also indicated a possible implementation of the first two princi-ples. In the next section, we rapidly recall this implementation and the underlying assumptions, but we will insist upon the statistical problems we have encountered.

It was important to recall the properties of our model, because the proof of the relations (2), (3), (5), (6), (7) and (12), without any distributional assumptions, can only be found in [10], and these relations are essential for implementing the third principle (see §3).

2 - IMPLEMENTATION OF THE 50% RULE

The control of the degree of multiprogramming we have implemented is based on the first two principles. The structure of the whole model

481

is represented in figure 2.

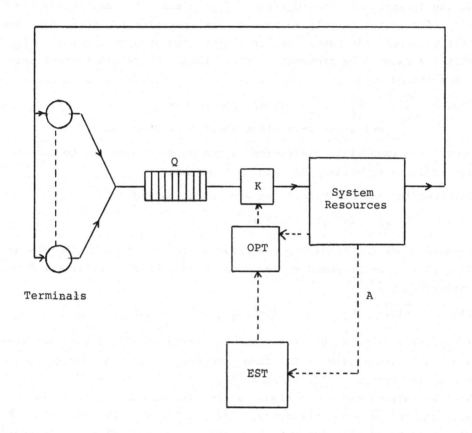

Figure 2

An estimator EST of system behaviour uses measurements on the secondary
memory device to estimate its utilization. An optimizer OPT computes
decisions by using the estimates provided by EST as well as the
knowledge of the system state. OPT furnishes commands to the switch
K controlling the access of users (waiting in the queue Q) into the
resource loop.

This control is described in detail in [3, 11]. It mainly consists of building a confidence interval for the SM utilization in order to decide the admission or the rejection of programs. Because of our assumption of non-stationarity we have to use special estimators which are able to forget the past. We first presented such estimators in [1, 2]; they are said to be exponentially smoothed. We recall that we estimate a mean value by :

$$(14) \qquad \hat{E}(X) = Y_i = \alpha X_i + (1-\alpha)Y_{i-1} \ , \ 0 < \alpha \leq 1$$

$(X_i$ being the i-th measurement of the variable X)

where the "smoothing" coefficient α can be choosen so as to determine the relative importance of "old" measurements.

In order to estimate $V(\hat{E}(X))$ we are led to estimate

$$(15) \qquad V(X_i) = E(X^2_i) - (E(X_i))^2$$

Because of the non-stationarity, we have to use smoothed estimators again. In [1, 2, 3] we proposed to consider a "smoothing" coefficient β for estimating $E(X^2)$:

$$(16) \qquad \hat{E}(X_i^2) = U_i = \beta X_i^2 + (1-\beta)U_{i-1} \ , \ 0 < \beta \leq 1$$

The problem is to ensure that the two estimates ($\hat{E}(X_i)$ smoothed with α and $\hat{E}(X_i^2)$ with β) age at the same "speed". If such a precaution is not taken, the expression (15) can be negative when the stochastic process X(t) is varying along the time. It is this point we discuss in this section, for we encountered some difficulties with it. We refer the reader to [2, 3, 11] for the other details of the 50% rule implementation.

A given α being selected for stationarity reasons, we are searching a β in order to get the expected aging effect.

It is clear that the relation between α and β depends on the stochastic process X(t) (here, the SM utilization) and more particularly on its first two moments. This explains that, while we found a satisfactory solution with $\beta = \alpha$ [1, 2] when X(t) was the dilatation (a sum of utilizations), this is not any more the case when X(t) is the SM utilization.

From (14) we have :

$$(17) \qquad V(Y_i) = \alpha^2 V(X_i) + (1-\alpha)^2 V(Y_{i-1})$$

hence

$$V(Y_i) = \alpha^2 V(X)(1 + (1-\alpha)^2 + \ldots + (1-\alpha)^{2(i-1)})$$

finally

$$(18) \qquad V(Y_i) = \alpha V(X) \frac{1 - (1-\alpha)^{2i}}{2 - \alpha}$$

Thus

$$(19) \qquad \lim_{i \to \infty} V(Y_i) = \frac{\alpha\, V(X)}{2 - \alpha}$$

Now we will study the behaviour of our estimators on a sudden variation of $E(X)$ and $V(X)$ in order to determine an order of magnitude of β with respect to α.

Assume this variation occurs at time t=0 and we know two unbiaised estimators E_0 and V_0 of $E(X)$ and $V(X)$ at this time. Assume too that the measurement indexing starts at t=0.

From (19), it comes

$$(20) \qquad V(E_0) = \frac{\alpha V_0}{2-\alpha}$$

and from (14), (16), (17) :

$$(21) \qquad E(Y_i) = \alpha E(X_i) + (1-\alpha)\, E(Y_{i-1})$$

$$(22) \qquad V(Y_i) = \alpha^2 V(X_i) + (1-\alpha)^2\, V(Y_{i-1})$$

$$(23) \qquad E(U_i) = \beta E((X_i)^2) + (1-\beta)\, E(U_{i-1})$$

with the relation :

$$(E(Y_i))^2 = E(Y_i^2) - V(Y_i)$$

If the system has been initiated a sufficiently long time ago, we may consider there is no bias due to its starting point, and equation (25) in [2] gives :

$$(24) \qquad E[\hat{V}(Y_i)] = \frac{\alpha}{1 - \alpha^2} (E(U_i) - E[(Y_i)^2]) + \frac{1 - \alpha}{1 + \alpha} E[\hat{V}(Y_{i-1})]$$

Let B_i be the bias (due to the variations of E_0 and V_0) on the variance of Y_i after i measurements. By definition, we have :

(25) $B_i = E[\hat{V}(Y_i)] - V(Y_i)$

Owing to the recurrence equations (21), (22), (23) and (24), we can compute B_i as a function of i and β only (with α given), if we know the initial conditions (at the time t=0) :

$$E(Y_0) = E(E_0)$$

$$E(U_0) = V(E_0) + E(E_0^2) = \frac{\alpha}{2 - \alpha} V_0 + E(E_0^2)$$

$$V(Y_0) = V_0$$

$$E(\hat{V}(Y_0)) = V_0$$

For each β, we can compute :

$$\max_i |B_i|$$

where i ranges from 1 to L, L being large enough so that B_i remains constant (B_i converges). We can take the smallest of these different values obtained for different values of β:

(26) $\min_\beta (\max_i |B_i|)$

Let β_0 be the value of β for which the minimum(26) is reached. Then β_0 is the value of β we are looking for, since it ensures that the bias introduced by the sudden variation at time t=0 is minimal.

By this way, we numerically determined convenient values for β. From our experiments, it seems that :

$$\beta = 0.8\alpha$$

is an acceptable value for α lying in the interval :

$$0.05 \leq \alpha \leq 0.20$$

and for a relatively large variation of the stochastic process X(t) (since X(t) is the SM utilization) :

mean variation : $0.33 \to 0.95$
variance variation : $4 \times 10^{-4} \to 13 \times 10^{-4}$

It is this value of β we have used to implement the 50% rule.

We have been bent on presenting these specific results because they appeared to be important even if we neglected them in a first approach

[1, 2].

We do not recall the optimization we obtain in applying the 50% rule to control the degree of multiprogramming (see [2, 3, 11]). In the sequel of the paper we assume this optimization has been achieved and we are trying to optimize the program mixture of the multiprogramming set by applying our third principle.

3 - IMPLEMENTATION OF THE THIRD PRINCIPLE

3.1 - Balancing the resource utilizations
Let us recall that the criterion we optimize reads (cf. (12) and (13)) :
$$D(n) \simeq A'_{cpu}(n) + A_{fd}(n)$$
or (cf. (7))

$$(27) \qquad D(n) \simeq A'_{cpu}(n)(1 + \frac{T_{fd}}{T_{io}})$$

With the 50% rule we are able to exhibit a n which maximizes (27), that is :

$$(28) \qquad D^* = D(n_0) = A'_{cpu}(n_0)(1 + \frac{T_{fd}}{T_{io}}) \geq D(n) \quad \forall \, n$$

Up to now we have always assumed the homogeneity of the program population, if this does not hold, there exist sub-populations with specific program behaviours. That means, in terms of our model, there exist programs with different T_{io} (mean virtual time between I/O requests).

Our third principle is stating that D^* (see (28)) is optimal if :

$$(29) \qquad T_{io} = T_{fd} \Longleftrightarrow \frac{T_{fd}}{T_{io}} = 1$$

that is, from (7), all the resource utilizations (except the SM utilization which is kept in the vicinity of 50%) are equal.

T_{fd} is an architecture parameter and it will be considered fixed. But T_{io} is a program behaviour parameter and, if there exist program sub-populations (it is mostly the case in the reality), by a convenient program mixture in the multiprogramming set we can create a resulting T_{io} such that it is equal to T_{fd}. It is the implementation of this program mixture control we describe in this section.

486

To validate this last optimization, we will extend the model studied by considering several I/O devices, formally m devices and three in our simulation model (see section 4).

The extended model is represented in figures 3 and 4.

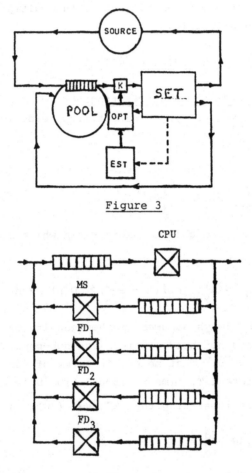

Figure 3

Figure 4

This model consists of :
1 - a program SOURCE modelling both interactive arrivals from a set of terminals and batch arrivals.
2 - a resource SET composed of one CPU, one paging unit and m I/O devices (see figure 4).
3 - a POOL of programs split into a FIFO queue for programs just ente-

ring the system and a BAG for the others.

4 - a module EST in charge of picking up measurements on the system state and the behaviour of each program submitted to the system (EST is also in charge of updating the different statistical esti- mates).

5 - a module OPT which defines an optimal degree of multiprogramming and a system "state".

6 - a module K which both decides the admission of a program into or its rejection from the SET and chooses this program.

To summarize , the global control we proposed is based upon
- the control of the degree of multiprogramming by applying the 50% rule.
- the control of the program mixture by applying our third principle.

Notice that if we consider m I/0 devices, the third principle reads :

$$(30) \qquad T_{io_j} = T_{fd_j} \qquad \forall\ j,\ 1 \leq j \leq m$$

3.2 - The implementation

From (30) we deduce :

$$(31) \qquad \frac{T_{fd_j}}{T_{io_j}} = 1 \qquad \forall\ j,\ 1 \leq j \leq m$$

where T_{fd_j} are constant architecture parameters.

At a first glance, comparing $\dfrac{T_{fd_j}}{T_{io_j}}$ to 1 seems to be equivalent to

comparing $\dfrac{T_{io_j}}{T_{fd_j}}$ to 1. For statistical reasons it is not true.

T_{fd} is a perfectly known constant, so it is quite similar to have it in the numerator or in the denominator. On the other hand T_{io_j} is an estimate which can vary from small values - very frequent I/O, thus T_{io_j} is often measured and well known - to large one -rare I/O, rarely and badly known. Therefore when T_{io_j} is ill determined (thus very disper- sed) its numerical value is very large and fluctuates heavily. In order to avoid this annoying effect we are led to consider the quantity

$\frac{1}{T_{io_j}}$ which is small when fluctuating, and therefore to compare $\frac{T_{fd_j}}{T_{io_j}}$ to

one.

Now the problem is to decide, at a given instant, what program we have to

- inject into the system (n_0 increases, $A_{sm}(n)$ is under 50%)
- reject from the system (n_0 decreases, $A_{sm}(n)$ is above 50%)
- exchange in the system (n_0 remains constant, $A_{sm}(n)$ is equal to 50%)

in order to respect the relation (31).

To perform the comparisons we use an absolute value norm which is relatively simple and cheap to implement.

The $\frac{T_{fd_j}}{T_{io_j}}$ estimates resulting from the system state (the mixture in the

multiprogramming set) are not directly evaluated. They are derived from the Chang-Lavenberg equations (see (7)) :

$$(32) \qquad \frac{T_{fd_j}}{T_{io_j}} = \frac{A_{fd_j}(n)}{A'_{cpu}(n)} \qquad \forall\, j,\ 1 \le j \le m$$

that is from resource utilization estimates which we previously talked about (see (14), (24) and [2, 3, 11]).

For the moment we will assume we are able to estimate (see section 3.3)

quantities such as $\frac{T_{fd_j}}{T_{io_{j,k}}}$ where $T_{io_{j,k}}$ stands for the mean virtual

time between I/O request of the program k for the I/O unit j.

If n is the current degree of multiprogramming and m the number of I/O devices in the system, we propose the following scheduling algorithm :

① - Candidate to enter the system (condition n < n_0)
It is the program in the POOL for which the following lower bound is reached :

$$(33) \quad \min_{k \in POOL} \sum_{j=1}^{m} \left| n \frac{A_{fd_j}(n)}{A'_{cpu}(n)} + \frac{T_{fd_j}}{T_{io_{j,k}}} - (n+1) \right|$$

② - <u>Candidate to leave the system</u> (condition $n > n_0$)
It is the program in the main memory (in the SET) for which the following
lower bound is reached :

$$(34) \quad \min_{k \in SET} \sum_{j=1}^{m} \left| n \frac{A_{fd_j}(n)}{A'_{cpu}(n)} - \frac{T_{fd_j}}{T_{io_{j,k}}} - (n-1) \right|$$

③ - <u>Candidate to be exchanged</u> (condition $n = n_0$)
First, a program is rejected into the POOL (more precisely into the BAG)
by applying ② and then a new program is injected into the SET by
applying ① (it can be the same as the one rejected).

It is noteworthy that quantities such as $\dfrac{T_{fd_j}}{T_{io_{j,k}}} = \dfrac{A_{fd_j}(n)}{A'_{cpu}(n)}$ may vary

from 0 to $+\infty$, but that they are bounded (lying in the interval $]0, 1]$)
when the resource j is underloaded ($A_{fd_j}(n) < A'_{cpu}(n)$) and otherwise

possibly infinite (see figure 5).

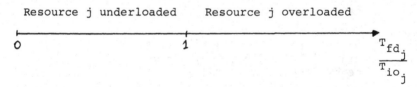

Figure 5

Therefore in (33) and (34) overloaded resources have heavier weights
than underloaded ones. This effect favours the underloading of a
resource. So to avoid that, we propose to weight a resource according
to its utilization.

Our weighting (see figure 6) is equal to 1 when the resource is overloaded and linearly increasing when the underloading appears. Note
that this weighting must also be increasing with the number m of resources in the system.

490

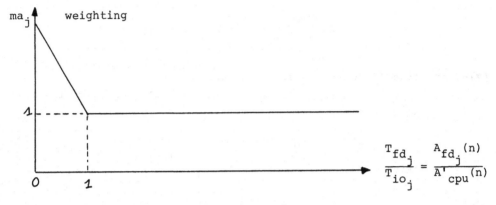

Figure 6

Thus we propose the weight q_j for each resource j :

$$
(35) \qquad q_j = \begin{cases} 1 \text{ if } \dfrac{T_{fd_j}}{T_{io_{j,k}}} \geq 1 \\[3ex] (1-ma_j)\ \dfrac{T_{fd_j}}{T_{io_{j,k}}} + ma_j \text{ otherwise} \end{cases}
$$

where a_j is a coefficient related to physical characteristics of the resource j. Then, expressions (33) and (34) become :

$$
(36) \qquad \min_{k \in POOL}\ \sum_{j=1}^{m}\ q_j\left|\ n\ \frac{A_{fd_j}(n)}{A'_{cpu}(n)}\ +\ \frac{T_{fd_j}}{T_{io_{j,k}}}\ -\ (n+1)\ \right|
$$

$$
(37) \qquad \min_{k \in SET}\ \sum_{j=1}^{m}\ q_j\left|\ n\ \frac{A_{fd_j}(n)}{A'_{cpu}(n)}\ -\ \frac{T_{fd_j}}{T_{io_{j,k}}}\ -\ (n-1)\ \right|
$$

The last problem we are faced before tackling the statistical one is dealing with the measurement initialization.

When a program arrives in the system from the SOURCE it is queued in the FIFO queue in the POOL of programs. It must be noted that we have no information on it. Two policies are possible. On the one hand we may guess we know some "a priori" information on programs and use it to initialize measurements, on the other hand we may front the hard reality and admit we do not know anything about program behaviour. It is this last hypothesis we assume.

In this context, we propose to initialize the $T_{io_{j,k}}$'s measurements of each program k with the T_{fd}'s value. That is we assume that an unknown program makes a "balanced" (in the sense of our third prinicple) use of the system resources. The aim of such an initialization is to favour the entrance of new programs only when the system is balanced and therefore when our criterion is optimized (cf. principle III).

3.3 - Estimation of the $T_{io_{j,k}}$'s

The control of the program mixture we propose needs to estimate utilizations such as $A_{fd_j}(n)$ and $A'_{cpu}(n)$, and quantities such as the $T_{io_{j,k}}$'s or the $\frac{1}{T_{io_{j,k}}}$'s (see § 3.2). Estimating utilizations has already been solved in previous papers [2, 3, 11] (see also § 2). We now are solving the estimation problem of the $\frac{1}{T_{io_{j,k}}}$'s.

On the contrary to the utilization case, the $\frac{1}{T_{io_{j,k}}}$'s cannot be evaluated at anytime since they need the occurrence of an event, here an I/O request. However we would like to dispose of $\frac{1}{T_{io_{j,k}}}$'s estimations at anytime, in particular, when a decision has to be taken.

In a first step, we will assume that the virtual time intervals between the I/O requests of a program k for a resource j are exponentially distributed. Let us consider the Bernoulli's variable X(t) defined by

$$X(t) = \begin{cases} X(t) = 1 \text{ if an event has occurred between the times t and } t + \Delta t \\ \\ X(t) = 0 \text{ otherwise} \end{cases}$$

where Δt is a time interval arbitrary small.

To be able "to forget the past", we propose to estimate the $\frac{1}{T_{io_{j,k}}}$'s by :

(38) $\quad [\widehat{\frac{1}{T_{io_{j,k}}}}] = L(t + \Delta t) = \alpha X(t) + (1 - \alpha\Delta t)L(t)$

where α is a smoothing coefficient.

If (38) is unbiaised, we must have :

$$E(L(t)) = \frac{1}{T_{io_{j,k}}}$$

so let us compute $E(L(t))$. It comes from (38)

$$E(L(t + \Delta t)) = \alpha E(X(t)) + (1 - \alpha \Delta t)E(L(t))$$

From the exponential assumption we get

$$E(X(t)) = \frac{\Delta t}{T_{io_{j,k}}}$$

Hence

$$E(L(t + \Delta t)) = \alpha \frac{\Delta t}{T_{io_{j,k}}} + (1 - \alpha \Delta t)E(L))$$

When $\Delta t \rightarrow 0$ we get the differential equation :

$$\frac{d(E(L(t)))}{dt} = \frac{\alpha}{T_{io_{j,k}}} - E(L(t))$$

Therefore

$$(39) \qquad E(L(t)) = \frac{1}{T_{io_{j,k}}} + Ce^{-\alpha t}$$

Thus if C is not equal to zero, our estimation is biaised but this bias is decreasing with t.

Notice that if $E(L(0)) = \frac{1}{T_{io_{j,k}}}$, then C is equal to zero, but in general

$\frac{1}{T_{io_{j,k}}}$ is unknown and the initialization which is performed is

$$E(L(0)) = 0$$

In this case $C = - \frac{1}{T_{io_{j,k}}}$.

Implementation of our estimator.

Let us consider a time interval $\left[t_\ell, t_m\right]$ in which no event has occurred, so (38) becomes :

$$L(t + \Delta t) = (1 - \alpha \Delta t)L(t)$$

by integrating we get :

(40) $L(t) = L(t_\ell)e^{-\alpha(t-t_\ell)}$ $\forall\ t\ \in [t_\ell,\ t_m]$

where $L(t_\ell)$ is known (in the limit $L(t_\ell)$ is $L(0)$, a given initial value).

The algorithm we propose is as follows. Assume t_ℓ is the last instant such as $L(t_\ell)$ is known, two possible cases :

① - An event occurs at time t (necessarily the first one after t_ℓ), then

$$L \leftarrow L \times e^{-\alpha(t-t_\ell)} + \alpha$$

$$t_\ell \leftarrow t$$

② - No event at time t, but we want to estimate $L(t)$, then

$$L \leftarrow L \times e^{-\alpha(t-t_\ell)}$$

$$t_\ell \leftarrow t$$

Thus an estimation of $L(t)$ is available in L at any time. Remark that the estimate L is decreasing with t for a given t_ℓ. That means $L(t)$ gets smaller and smaller if no event occurs. This effect is in complete agreement with what we want.

Validation of our approach

We will now consider the estimator L defined by our algorithm (see ① and ②) and study it when no assumption is made concerning the distribution of the virtual time intervals between events.

The occurrence of an event (in our context, an I/O request) is a discontinuity point for $L(t)$ and $E(L(t))$, since just after an event $E(L(t))$ is changed into $E(L(t)) + \alpha$(cf. ①. We will study $L(t)$ just after an event.

Let t_1, t_2, ..., t_n be the instants of event occurrences and t_0, the starting instant. Let us consider the time intervals between events τ_1, τ_2, ..., τ_n, where

$$\tau_i = t_i - t_{i-1}$$

From the point ① of our algorithm, we derive

(41) $L(t_i) = L(t_{i-1})e^{-\alpha\tau_i} + \alpha$

So

(42) $E(L(t_i)) = E(L(t_{i-1})e^{-\alpha\tau_i}) + \alpha$

If we assume that $L(t_{i-1})$ and $e^{-\alpha\tau_i}$ are independent, we get :

494

(43) $E(L(t_i)) = E(L(t_{i-1}))E(e^{-\alpha\tau_i}) + \alpha$

For the moment, let us assume that the τ_i's are i.i.d. (further we will relax this stationarity by tuning our forgetfulness parameter α) and f is their common density function. Then

$$\forall i \quad E(e^{-\alpha\tau_i}) = \int_0^\infty e^{-\alpha x} f(x)\,dx$$

If we note f^* the Laplace transform of f, we can write :

$$\forall i \quad E(e^{-\alpha\tau_i}) = f^*(\alpha)$$

Then the equation (43) becomes :

$$E(L(t_i)) = E(L(t_{i-1}))\, f^*(\alpha) + \alpha$$

Since $f^*(\alpha) < 1$, we can solve this recurrence equation :

(45) $\lim\limits_{i\to\infty} E(L(t_i)) = \dfrac{\alpha}{1 - f^*(\alpha)}$

Thus our estimator converges.

Notice that if the time intervals between events are exponentially distributed, we get :

(46) $\lim\limits_{i\to\infty} E(L(t_i)) = \dfrac{1}{T_{io_{j,k}}} + \alpha$

Therefore it is trivial to unbias the estimator.

Now we will express $f^*(\alpha)$ without any assumption concerning f.

The parameter α is a smoothing parameter, so we will assume it is close enough to zero to develop $f^*(\alpha)$:

$$f^*(\alpha) = 1 - \alpha E(f) + \ldots + (-1)^k \frac{\alpha^k}{k!} E(f^k) + O(\alpha^{k+1})$$

Hence it comes from (45):

$$\lim\limits_{i\to\infty} E(L(t_i)) = \frac{\alpha}{1 - (1 - \alpha E(f) + \frac{\alpha^2}{2} E(f^2) + O(\alpha^3))}$$

Thus

(47) $\lim\limits_{i\to\infty} E(L(t_i)) = \dfrac{1}{E(f)} + \dfrac{\alpha}{2} + \dfrac{\alpha}{2}\dfrac{V(f)}{(E(f))^2} + O'(\alpha^2)$

$E(f)$ is the mean virtual time between events, that is $T_{io_{j,k}}$. If we note K the variation coefficient of f ($K^2 = \dfrac{V(f)}{(E(f))^2}$), we obtain :

$$(48) \qquad \lim_{i \to \infty} E(L(t_i)) = \frac{1}{T_{io_{j,k}}} + \frac{\alpha}{2} + \frac{K^2 \alpha}{2} + 0'(\alpha^2)$$

Therefore we deduce from this equation :

1 - our estimation is biaised and in the Poisson case (K=1) the bias is α (see (46)).

2 - the bias is minimum ($\frac{\alpha}{2}$) when the time intervals are constant (K=0).

3 - in general, we can control the bias by choosing α. If B_{max} is the maximum bias we accept, from (48) we derive

$$\alpha \leq \frac{2B_{max}}{1 + K^2}$$

4 - we have to find out a trade-off between :
- the bias, a small α
- the distribution stationarity, a sufficiently large α to forget the past fast enough.

L is the estimator we have used to estimate the $\dfrac{1}{T_{io_{j,k}}}$'s. To avoid a too important overhead, we tabulated such expressions as $e^{\alpha(t_k - t_\ell)}$ when we implemented our control.

4. VALIDATING OUR CONTROL

4.1 Experiments on the mixture optimality

We built a simulation model as described in figures 3 and 4, and we implemented our control in it.

Having previously validated our control of the degree of multiprogramming (see [2, 3, 11]), in these experiments, we are mostly concerned with the control of the program mixture.

Of course the improvement obtained by a suitable program mixture greatly depends on the heterogeneity of the program population submitted to the system. Therefore to measure this improvement we considered

extreme conditions. Three kinds of workloads are generated by the SOUR-
CE. All of them are similar concerning their memory (T_{pf}) and CPU
(T_{cpu}) behaviours, but they differ on their I/O behaviour. The LOAD I
uses the I/O device 1 ($T_{io_{1,k}}$ = 30 msec.) without using the other

devices ($T_{io_{2,k}}$ = $T_{io_{3,k}}$ = ∞). The LOAD II and III are defined in the

same manner, that is :

   LOAD II $T_{io_{1,k}}$ = $T_{io_{3,k}}$ = ∞ and $T_{io_{2,k}}$ = 30 msec.

   LOAD III $T_{io_{1,k}}$ = $T_{io_{2,k}}$ = ∞ and $T_{io_{3,k}}$ = 30 msec.

In other respects the I/O devices characteristics are identical
(T_{fd_1} = T_{fd_2} = T_{fd_3} = 100 msec.).

In such a context, it is clear that none of the three LOADS is a "good"
worload for the system and clear too that we need to combine them.

We have defined an experiment in which the three loads are not naturally
mixed together. The SOURCE sequentially and cyclicly generates programs according to the characteristics of the loads I, II and III. The
mean time between each load is 4 secondes.

We performed this experiment twice. In the two cases, we controlled
the degree of the multiprogramming with the 50% rule, but on the one
hand the program mixture is not controlled and on the other hand it is.
The result obtained are presented in figures 7 and 8.

We plot the instantaneous "dilatation", measured in the system, as a
function of the time.

In the first experiment (without control) the loads are not naturally
mixed together and since they are equivalent from the dilatation viewpoint ($T_{io_{j,k}}$'s symetry in the D(n) expression), this criterion is uni-
formly low, fluctuating around 1.3 (hardly better than in monoprogram-
ming).

In the second experiment, the optimization due to the program mixing is
effective from the arrival of a program which is of a different "type".
So, according to the arrivals of loads I, II and III, the dilatation
is gradually increasing from 1.5 to 2.7 and to 3. Therefore we double

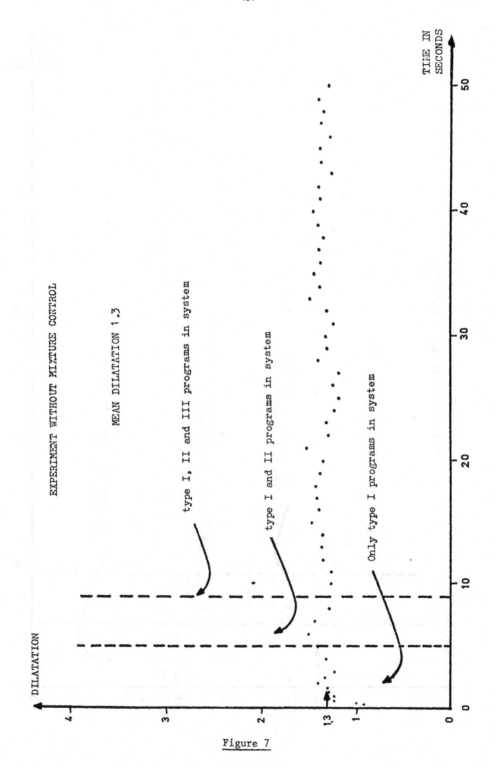

Figure 7

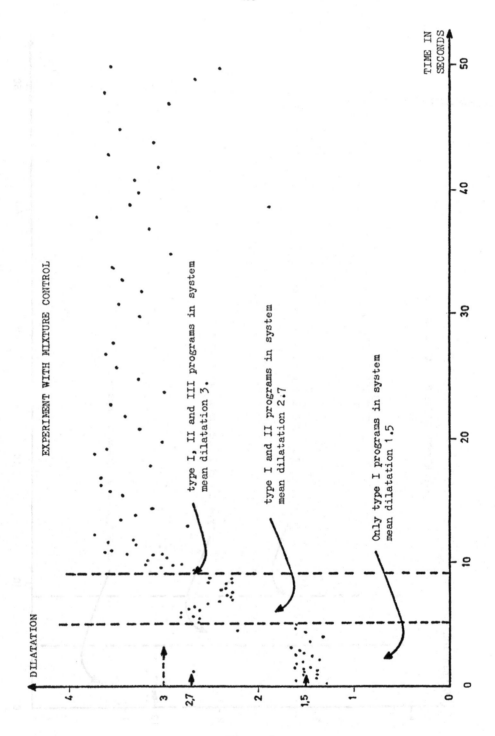

Figure 8

the system performance.

Two remarks are noteworthy. On the one hand the mixture control is near instantaneous, so we may conclude that our estimator L of the $\frac{1}{T_{io_{j,k}}}$'s is quite relevant and the tuning of α is right. On the other hand the criterion fluctuations increase with the mixture efficiency (injection and rejection of program coming from an heterogeneous population). Therefore, in such a context, it is utopian to control the system by looking for a criterion maximum (see [2]). We have to use an indirect optimization such as the one provided by the 50% rule (see also [4]).

4.2 - The "losers"

Up to now the proposed scheduling algorithm aims to optimize a unique performance criterion : the dilatation.

The dilatation enables to measure the usage of the different system resources. So its optimization will ensure a good use of all the system components. With an assumption of homogeneous program population, we showed in [2, 11] that the mean program response time is also optimized.

Since this assumption does not hold any longer, this means that programs can be "forgotten" if their behaviours do not fit the "state" of the system. To avoid this phenomenon detrimental not to the system performance but to the user's satisfaction, we must introduce a new criterion in the optimization in order to take into account this satisfaction.

In fact, we would like to consider an arbitrary number of criteria. As a matter of fact a computer system consists of several layers with several scheduling policies. The layer we consider is very close to the physical resources, but the scheduling algorithm we propose has to be able to take into account the constraints (economical costs, priorities, etc.) coming from other layers. It is this problem we attempt to solve in this section. In a first step we will restrict ourselves to the losers' problem. In this view we have to
- recognize a "forgotten" program
- to favour such a program by an appropriate mechanism.

In order to recognize a forgotten program we propose to compare its "efficiency" (see [2, 11]) with the system "efficiency". We recall that program efficiency between the two instants t_{i-1} and t_i is the sum of the service times provided to the program during this time interval divided by $t_{i-1} - t_i$. We show in [2, 11] that the "dilatation" is the sum of the program efficiencies. So we derive the system efficiency as:

$$(49) \qquad e_s = \frac{D(n)}{n}$$

To take into account the services provided to the programs in POOL, we generalize (49) by considering :

$$(50) \qquad e_s = \frac{D(n)}{\ell}$$

where $\ell = n + $ number of programs in POOL.
If e_k is the efficiency of program k, let us consider

$$(51) \qquad s_k = \frac{e_k}{e_s}$$

Notice that the greater the coefficient s_k is, the better the program k has been served. So we propose to define a penalty for each program k such that :

$$(52) \qquad p_k = C\frac{s_k}{\gamma_k}$$

where C and γ_k $(0 < \gamma_k \leq 1)$ are fixed positive constants.
Thus the relations (36) and (37) become respectively :

$$(53) \qquad \min_{k \in POOL} \; (\sum_{j=1}^{m} q_k | n \frac{A_{fd_j}(n)}{A'_{cpu}(n)} + \frac{T_{fd_j}}{T_{io_{j,k}}} - (n+1) | + p_k)$$

$$(54) \qquad \min_{k \in SET} \; (\sum_{j=1}^{m} q_k | n \frac{A_{fd_j}(n)}{A'_{cpu}(n)} - \frac{T_{fd_j}}{T_{io_{j,k}}} - (n-1) | + p_k)$$

Therefore the proposed penalty penalizes programs better served than the system state permits and it favours the others.

The coefficient C enables to tune the penalties weights with regard to the dilatation optimization. If C is close to zero, then the sole dilatation is taken into account, but if C is of the order of m (the number of I/O units) , only the other criteria are considered. In our mind, C is a coefficient fixed at the system generation for a given site and a given application.

For the moment let all the γ_k be equal to one, for example (all the programs are equivalent).

To tune the coefficient C in order to avoid the appearance of forgotten programs in the system, we performed several experiments. The coefficient C must not be too small so that the forgotten programs are effectively favoured, but not too large so that the dilatation is any more considered.

The best value we have found for C is 1.5 : the system performance is not changed but the users are better served. We give in figure 9 the results we obtained in the same conditions as previously described (see § 4.1).

			Without penalty (C = 0)	With penalty (C = 1.5)
Number of programs processed during the experiment			52	66
Probability that the system efficiency is < 0.01	Type I	0.47		0.42
	Type II	0.50		0.25
	Type III	0.31		0.20
	Together	0.44		0.31
Mean efficiency of entirely processed programs			0.082	0.123
Mean dilatation during the experiment			3.1	3.1
Number of programs in POOL at the end of the experiment	Type I	19		6
	Type II	20		9
	Type III	11		2
	Total	50		17

Figure 9

4.3 - The other scheduling levels

Other levels or other layers in the system can have their own scheduling policies. Because we consider a level which is very close to the physical resources, the other levels must be able to interact with it. It is the aim of the coefficients γ_k.

Each γ_k ($0 < \gamma_k \leq 1$) is associated with a program k. This coefficient indicates the degradation of services we admit for a program. As a matter of fact, the efficiency of a program k is not directly compared to the system efficiency e_s, but to $\gamma_k \, e_s$.

We have limited the γ_k's variations to the interval $]0,1]$ in order to avoid that other scheduling levels can completely ruin physical system performance.

5 - CONCLUSION

This paper deals with the implementation of three principles of optimality for multiprogramming which state the existence of both an optimal multiprogramming degree and an optimal program mixture in the multiprogramming set. Their implementation has been realized in a simulation model which permits, in contrast with analytical models, to rèlax assumptions on the workload submitted to the system. In the present case, the workload is supposed to be non-stationary and issued from a non-homogeneous program population. Therefore a dynamic control of the system needs special estimators. These statistical problems are studied in detail and an implementation has been realized. The results obtained on the system performance improvement show that the performance can be doubled in the case of a very heterogeneous workload and the "thrashing" avoided in all cases.

Moreover it is noteworthy that the statistical results we derive solve general problems occurring when estimating mean values of non-stationary stochastic processes. Therefore they can be widely used out of the context in which we applied them.

Finally by considering several performance criteria our study is close to the real world. So we are now interested in the implementation of our solutions in a virtual memory system which is being built on a CII-IRIS 80 computer.

REFERENCES

[1] BADEL M., GELENBE E., LENFANT J., LEROUDIER J., POTIER D. - "Adaptive optimization of the performance of a virtual memory computer" - Proceed. of the Workshop on Computer Architecture and Networks - IRIA - North-Holland Publishing Co. - August 1974.

[2] BADEL M., GELENBE E., LEROUDIER J., POTIER D. - "Adaptive optimization of a time-sharing system's performance" - IEEE Proceed. on Interactive Computer Systems - June 1975.

[3] LEROUDIER J., POTIER D. - "Principles of optimality for multiprogramming" - ACM-SIGMETRICS and IFIP W.G. 7.3. Conference Harvard University - March 1976.

[4] DENNING P.J., KAHN K.C., LEROUDIER J., POTIER D., SURI R. - "Optimal multiprogramming" - Acta Informatica - Vol.7 - Fasc.2 - Dec. 1976.

[5] COURTOIS P. - "On the near-complete decomposability of networks of queues and stochastic models of multiprogramming computing systems" MBLE Report - Brussels - Belgium - November 1971.

[6] COURTOIS P.J. - "Decomposability, instabilities and saturation in multiprogramming systems" - Communications of ACM - Vol.18 - n°7 - July 1975.

[7] JACKSON J.R. - "Jobshop-like queueing systems" - Management Science, 10, 1, October 1963.

[8] BUZEN J.P. - "Computational algorithms for closed queueing networks with exponential servers" - Communications of ACM - 16, 9, Sept. 1973.

[9] CHANG A., LAVENBERG S.S. - "Work-rates in closed queueing networks models" - Operation Research, 22 , 1974.

[10] POTIER D., LEROUDIER J., BADEL M. - "Un modèle d'analyse des performances d'ordinateurs multiprogrammés à mémoire virtuelle" - Rapport IRIA-LABORIA n° 152 - January 1976.

[11] LEROUDIER J. - "Systèmes adaptatifs à mémoire virtuelle" - Thèse d' Etat - Grenoble University - May 1977.

[12] BELADY L.A., KUEHNER C.J. - "Dynamic space-sharing in computer systems" - Communications of ACM - Vol. 12 - n°5 - May 1969.

[13] BURGEVIN P., LEROUDIER J. - "Characteristics and models of programs behavior" - National Conference ACM 76 - Houston - Texas - Oct. 1976

[14] BURGEVIN P., INGELS Ph., LEROUDIER J. - "Analysis of program behaviour" - Rapport IRIA-LABORIA n° 237 - June 1977.

A SIMULATION MODEL FOR A LOOP ARCHITECTURE DISTRIBUTED COMPUTER SYSTEM

S. Baragli and S. Valvo

Ing. C. Olivetti e C.

Progetto Centrale Sistemi

IVREA (TO)

ABSTRACT

Moving from a practical problem, the preliminary definition of a loop network, a simulation model has been built and will be described in this paper. For its characteristics of flexibility, modularity and programmability, this model has been used also as a tool for documenting the evolving project and as a design aid.

At the end, some of the results obtained are presented in graphical form: the loop utilization and performance at different transmission speeds, and the system's response time versus load variations.

INTRODUCTION

This paper is concerned with a simulation model that has been built as an aid for the project of a loop-connected distributed computer system for office automation applications.

Aloop network can be built between N terminals by systematically connecting the output of one to the input of the next and the output of the last to the input of the first (FRA74). In the case studied, it has been decided to let messages flow in only one direction (e. g. clockwise).

Many processors (the hosts) use this facility to communicate; they are attached to the loop through special hardware interface modules, called nodes. Any node, if not in transmitting mode, must always receive bytes from one side and retransmit them to the other, thus permitting messages circulation.

As two or more hosts could try to put messages on the line at the same time, some form of loop control must be provided to prevent mutual interference. This control can be either centralized (transmission by any node must be authorized by a special unit, the loop controller) or distributed. In this project, distributed control has been chosen for reliability and cost considerations.

At least two control strategies are possible:

a) when a node wants to transmit, it inserts the message between two other mes- sages, delaying the second for as much time as necessary (REA76);

b) a particular "go-ahead" message circulates around the loop, allowing only one node at a time to trnsmit (FAR72, NEW69).

In the first case, a hardware mechanism must be provided to switch the incoming message in a delay buffer when so needed; the second strategy was deemed more simple and was adopted for this project. So, when a node has to trnsmit a message, it must follow the sequence:

i) wait for the "go-ahead" item to arrive;

ii) destroy this item and transmit the message instead of the item;

iii) transmit another "go-ahead" token, thus releasing the line.

In loop networks, the amount of time by which any message is delayed is called node delay. In the case studied, the node delay must be fixed in time; for the dif- ferent trials simulated with the model, this delay has been made equivalent to the transmission time of one or more bytes.

At the software level, communication is process-oriented; there is only one set of communication primitives between processes, whether the processes are allocated

506

on the same processor or not. To give an example, let us suppose that the process
A, allocated on host n. 1, wants to send a message to the process B, that belongs to
the host n. 2. The following actions will take place:

1) the interprocess communication facilities in the o. s. nucleus of host n. 1 determine, in a transparent way, whether the process B is local to the host n. 1;

2) as process B is not local, a command is issued to the node n. 1, in order to send
the message through the loop;

3) if the process B is being addressed for the first time, the node has no way of
determining that the process B is on host n. 2; so the message is sent on the line
with the address 00 (broadcast message);

4) every node in the loop receives the message; obviously, only the node n. 2 passes it to its host; afterwards, there will be a reply message, containing the physical
address of its sender (n. 2);

5) as the reply message reaches the node n. 1, this node becomes informed that the
process B is allocated on the host n. 2; this information will be used, if necessary,
by the node n. 1, but not by process A. This makes easier the reconfiguration of
the system.

At a first approach to the definition of the real system, the need arose for an estimat
of the minimum line transmission speed and consequently of the technology required
for the interface transmitters. It was decided to carry out this estimate using a
simulation model, written in the language SIMULA67 (BIR73). This model has not
been used for this estimate only, but it grew with the project.

A project's development usually consists of:

i) the choice of the architecture to be utilized;

ii) the logical partitioning of the system into different modules, with the definition
of the respective interfaces;

iii) the implementation of these modules.

The simulation model can be present in this process if it represents, at an high
abstraction level, all of the system's parts. Then it can:

a) compare the performances of different system's architectures;

b) verify the consistency of the interface definitions between the various system
parts;

c) provide a context to test an already implemented module; then cycle back to point
b) and re-evaluate the throughput if this module behaves somehow differently
from the specifications;

d) suggest, as a consequence of points b) and c), structure modifications in order to maintain or enhance performance;

e) document the project evolution.

The system studied has been designed for office automation applications: typical hosts are interactive intelligent terminals, printers, "smart copiers" etc.

Microprocessors will be used both for the hosts and the nodes, as they are low cost components and there is no need for high calculating speed (the results obtained show that the required line speed is of the order of 19200 baud).

THE SIMULATION MODEL.

The modelling facilities.

The simulation model has been implemented with a program written in the language SIMULA67, running on an IBM 370/168. We have taken advantage of the various modelling facilities included in the system-defined classes SIMSET and SIMULATION.

The class SIMSET allows the definition of circular two-way lists and of the associated elements, that can be manipulated by special primitives, such as INTO, OUT (to insert or remove the last member of the queue), CARDINAL (to obtain the number of elements in the queue) etc.

The necessary concepts for discrete event modelling are provided in the system class SIMULATION which is itself prefixed by SIMSET so that all the latter concepts are available. A simulation program is composed of a collection of processes which undergo scheduled and unscheduled phases. When a component is scheduled, it has a time associated with it. This is the time at which its next active phase is scheduled to occur. When the active phase of one component ends, it may be rescheduled (marked with the time when its next active phase will be executed), or else descheduled. In either case, the scheduled component in the system marked with the smallest time for its next active phase is resumed. Thus, the currently executed component (which may be referenced by a call on the procedure CURRENT) always has the least time associated with it; this is taken as the simulation time itself, and the simulation time jumps discretely forwards from one value to another when a new component becomes active.

In the model of the loop, three kinds of processes have been defined and have been included, together with the procedures acting on them, in different modules: the hosts, the nodes and the loop itself.

The structure of the model.

A development tool of this kind should, if possible, be independent of structure variations; these are expected to come about as the project proceeds, through technology improvements or feedbacks like those described in the introduction, point d).

Consequently, the main goals have been modularity, flexibility and programmability of the model.

Modularity has been achieved by partitioning the implementation of the simulation model into small, simple modules, which have evident meaning and functions (see fig. 1).

The model itself is logically partitioned into three "big" modules, that communicate through well-defined interfaces. The first module simulates the functions of the loop hardware, the second those of the nodes, the last those of the host set (fig. 2).

Hence it is possible, for instance, to precise or to modify in some way or another the node structure, with the sole care of not changing the interfaces with the loop and the hosts.

Interfaces definitions are built over the following interprocess communication primitives and objects:

a) SEND (Q, E, P) :

enqueues element E into the queue Q and activates process P, if this was suspended on that queue;

b) WAIT (Q) :

tests queue Q : if empty, the process executing the WAIT operation is suspended;

c) DELAYED-ACTIVATION (P, T, PN, F) :

causes process P to be activated at current time + T and passes to it a buffer pointed by pointer PN; if the buffer contains useful information, the flag F is set to TRUE;

d) LIST1 objects :

they are queues with an associated binary semaphore.

As to flexibility, the simulation model is placed on an high abstraction level, if compared with the object to be simulated: the structure of the SIMULA program modules is as simple as possible; they can "faithfully" reproduce the system only because they heavily rely on the use of parameters.

To give an example, let us consider the BASE-MODULE, about which more details

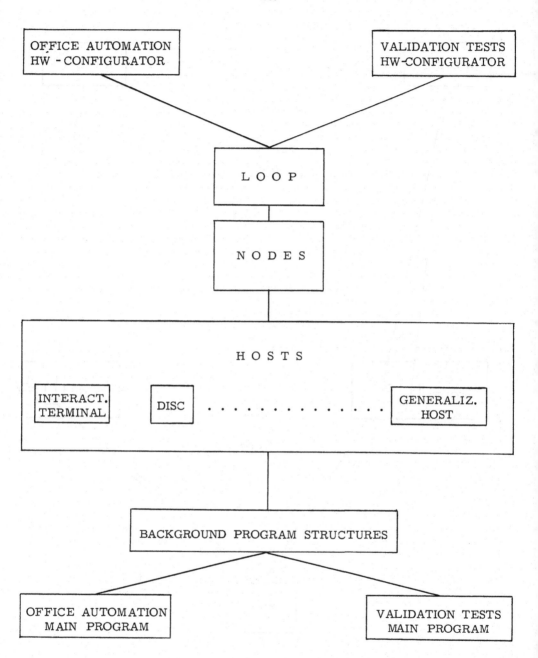

Fig. 1 - Program partitioning.

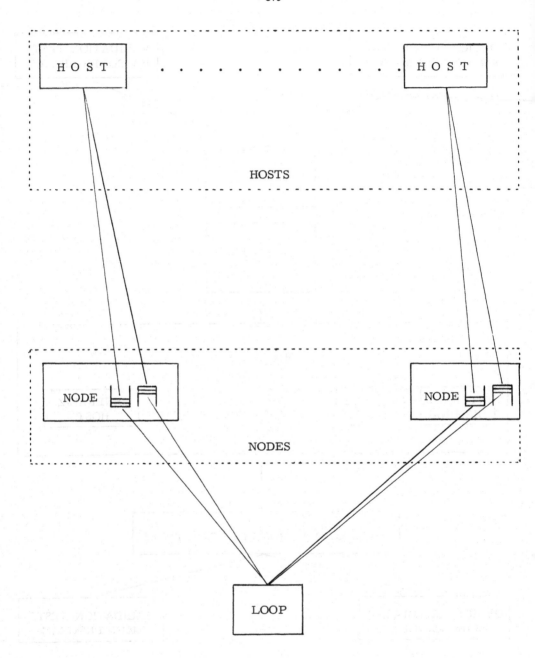

Fig. 2 - Model partitioning.

will be given in the following.

Its main part is a process called BASE-BLOCK, whose sole function is that of receiving and sending messages; in so doing it makes use of the information stored beforehand in a table. Given a chain of BASE-MODULEs, the simulation of arbitrarily complex communication protocols between them can be carried out simply by modifying, through associated procedures, the data in the tables.

All the variables meaningful for the system description have been written in the model as parameters, so as to be easily modifiable from the outside. In particular, among these there are the loop speed, the number of nodes, the "go-ahead" message length, the node delay; so it is also possible to investigate relations between these variables.

The concept of the simulation model's programmability must be understood in a particular sense. Many persons, other than those who have developed this tool, must have access to it, whenever it is to be used as a design aid for R&D groups like ours. This means that it is mandatory to hide the specific mechanisms of SIMULA, through the building of ad hoc procedures; by calling them, the SIMULA-unskilled user can tailor the model over his needs and execute the tests he wants.

But this process needs not be pushed too forward, because, anyway, the users will master programming techniques.

The procedures we have defined may be divided into two groups: those intended for the assignment of parameters to the system and for its configuration (examples 1, 2) and those intended to describe the communication protocol (example 3).

Examples:

1) NODE-NUMBER (10) : establishes that there will be 10 nodes on the loop network;

2) ASSIGN ("DISC", 3, 150, 30): the third terminal will be a disc, with mean access time of 150 msec $\pm$ 30 msec;

3) PROTOCOL ("BSC II", 4, 7, 10): the system-defined BSC II protocol will be used between the host n. 4, acting as master and terminals n. 7 - 10, working as slaves.

At present, it is not planned to implement any procedure intended to simplify for the user the definition of new protocols; yet this is quite possible, in principle. There is, instead, another possibility: a SIMULA programmer can, simply by writing a new class prefixed by BASE-BLOCK, redefine the message analysis (e.g. , ignoring the table).

The implementation of the host module.

Among the three modules, a brief description will be given of the host module, that is to say of the one intended to simulate the host set. Four classes (the class concept is a feature of the SIMULA67, see BIR73) are defined in it:
INTERACTIVE-TERMINAL, DISC, PRINTER, GENERALIZED-HOST.

Of course, different objects of these classes can be built by assigning different values to their parameters; for instance, in an INTERACTIVE-TERMINAL can be defined: time required for the transaction-initiated operations, mean time between transactions, length of the inquiry message to be sent to the discs, number and addresses (node numbers) of the discs etc.

In a DISC object can be assigned: the mean access time, the minimum and maximum length of the messages to be sent to the operators etc.

In a PRINTER object we can select the speed.

At present, the actions of these three classes are those one would expect from an operator, a disc and a printer once one has exactly defined the transactions and the background printing activity.

The tests on the simulation model have been carried out with protocols derived from the analysis of a typical office automation application. On the contrary, as the name suggests, the structure of the GENERALIZED-HOST class is neither application- nor peripheral-oriented; actually, it has been written in the hope of making easy the definition of a group of GENERALIZED-HOSTS, connected through the loop network, with different functions; and to allow the implementation of any communication protocol among them.

In this class two process types are defined (see fig. 3):

a) SW-BLOCK;

b) BASE-BLOCK.

A SW-BLOCK continuously sends commands to its associated BASE-BLOCK; it can do this either by synchronizing itself on the replies from the BASE-BLOCK, or a-synchronously, with exponentially-distributed intermessage times.

The BASE-BLOCK analyzes the received commands one at a time; for any of them it decides whether, in what quantity and for whom it has to send messages through the loop. On receiving the respective replies, it repeats the same analysis, that may cause it to send other messages on the line, or to close the cycle by replying to the SW-BLOCK, and waiting for another command.

If a GENERALIZED-HOST object is to be activated only on requests coming from the loop, it must obviously be deprived of the SW-BLOCK process; the BASE-BLOCK will receive commands directly from the line.

The two processes cooperate via SEND and WAIT primitives; they share an area that contains, among other objects, two input and two output queues (for messages coming from and addressed to the loop and the SW-BLOCK); the BASE-BLOCK uses a particular table (see preceding paragraph) for obtaining all the information necessary to the message analysis.

The table, when so needed, may also specify different reply strategies for the same message, with the respective probabilities.

The configurator concept.

In the program there are some classes intended to play the role of system's configurators. It may be useful to explain with some details the function of these classes.

At present, there are an hardware configurator (the only one enrighted to assign values to system's parameters) and a software one. First of all, the configurator for a part of the system must create and link all the objects belonging to that part.

As the execution of the simulator program begins with the activation of the configurators, an automatic separation between the phases of system's generation and actual simulation is obtained.

Let us say here that the simulation model's items can be divided into two groups: "static" (intended to represent hardware pieces, processes etc.) and "dynamic" (corresponding to messages). The generation phase provides the creation of the static parts, whilst, during the simulation phase, dynamic parts (messages) are produced and acted upon by the static parts. An approach of this kind is called "machine-based".

By now, it should be clear that the definition of the configurator classes simplifies the modification not only of the values of the model's parameters, concentrated in the HW-CONFIGURATOR, but also of the system's structure: this may be accomplished, in a localized manner, by re-defining only the actions of the HW-CONFIGURATOR. To clarify this point, here a list follows of the operations involved in a GENERALIZED-HOST (a SW-CONFIGURATOR) creation and activation (refer to fig. 3):

a) the HW-CONFIGURATOR generates a GENERALIZED-HOST object, then it links this object with a node and passes to it all the parameters; among these parameters some will be needed to connect the SW-BLOCK and BASE-BLOCK processes;

b) the GENERALIZED-HOST, now acting as a SW-CONFIGURATOR, creates and links the aforesaid processes and their common data area (e.g. table for message analysis); this is local to the GENERALIZED-HOST, and becomes available to the processes because it is passed them as a common parameter by the SW-CONFIGU-RATOR;

c) at this point, the HW-CONFIGURATOR compiles, inspecting the host data area, the table entries, thus defining the communication protocol for that host.

This operation is accomplished through the execution of specific procedures local to the HW-CONFIGURATOR.

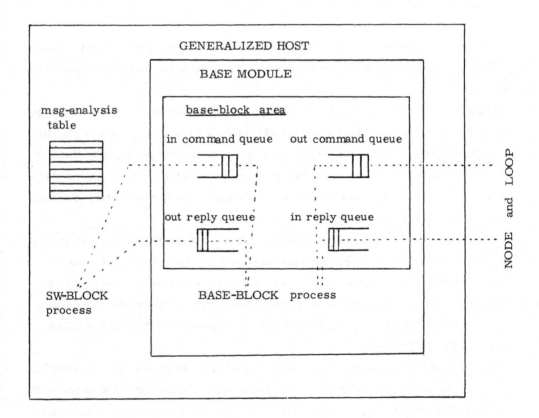

Fig. 3 - Generalized host's structure.

RESULTS.

First of all, the model has been used to obtain the numbers that have been required in order to achieve the primary goal: the preliminary definition of the system for a typical office automation utilization. In the following pages some of the more typical curves (fig. 4, 5, 6) are given.

A simple mathematical model has been written for the loop network; it relates some of the more meaningful variables : transmission speed, node delay and number, utilization of the loop; this analysis and the simulation results both point to the utilization of the loop as a faithful measure of the system's level of crowding.

The utilization of the loop is defined as:

$$O = \frac{\text{MESSAGE TRANSMISSION TIME}}{\text{TOTAL TIME}}$$

The curves in fig. 6 give a visual confirmation of the preceding statement: they refer to a simple, though unrealistic, situation in which the load on the network (number of transactions per minute, of lines to be printed etc.) varies with time, reaching only a minimum or a maximum value.

The broken line shows the load variations: in the second curve, every "X" represents a sample of the loop's utilization and the resulting curve follows the preceding one with a short delay: from this one can have an estimate of the response time of the system.

In the condition of the maximum load, the system is at its stability limit: correspondingly, we can see that the third curve (sampling the number of elements in the mostly used transmission queue) oscillates strongly. The oscillations are caused by a sort of "noise" in the input to the system (e. g. the arrival times for the transactions are randomly varied with a Poisson distribution; only the mean interarrival time can be assigned); so there is a good probability that, at a certain moment, the fluctuations in the message production rates of the various nodes sum up to cause an overcrowding, with a resulting increase in the number of messages waiting to be transmitted. This can be seen as a "micro-breakdown" of the system .

Shortly afterwards, usually, the equilibrium returns, but it takes a certain time to reduce the number of elements in the queues, unless an opposite fluctuation occurs.

At present, it has not yet been possible to compare these curves with the actual behaviour of a real system, since the implementation of the loop is still in progress.

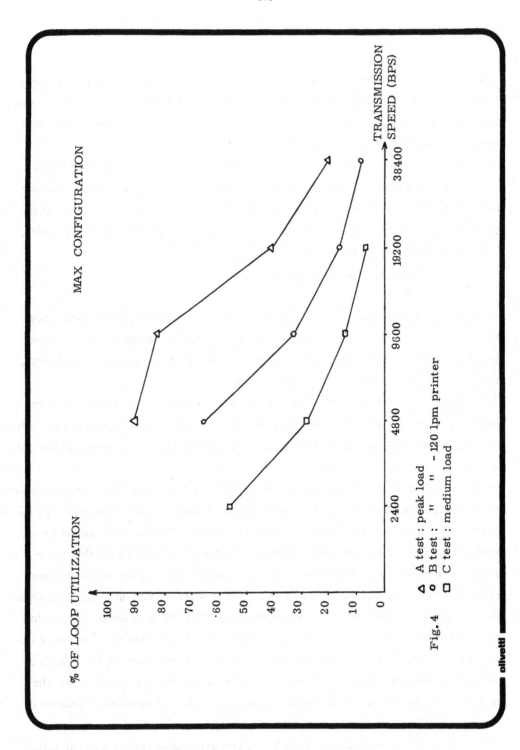

Fig. 4

△ A test : peak load
o B test : " " – 120 lpm printer
□ C test : medium load

517

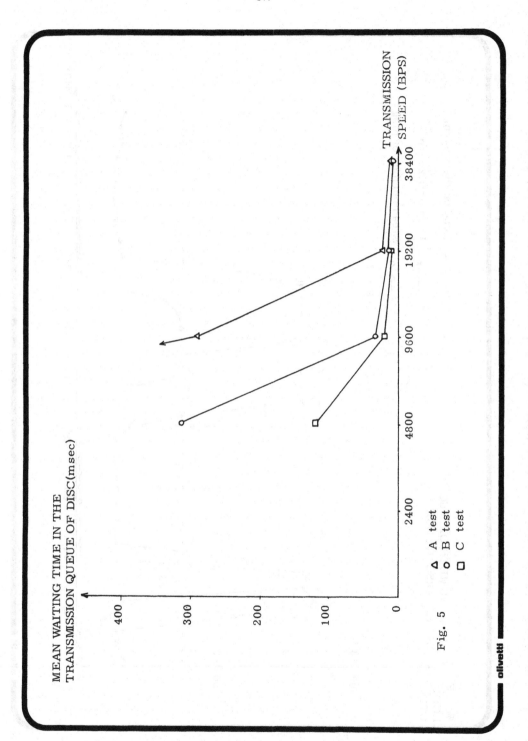

MEAN WAITING TIME IN THE
TRANSMISSION QUEUE OF DISC(msec)

TRANSMISSION SPEED (BPS)

△ A test
○ B test
□ C test

Fig. 5

olivetti

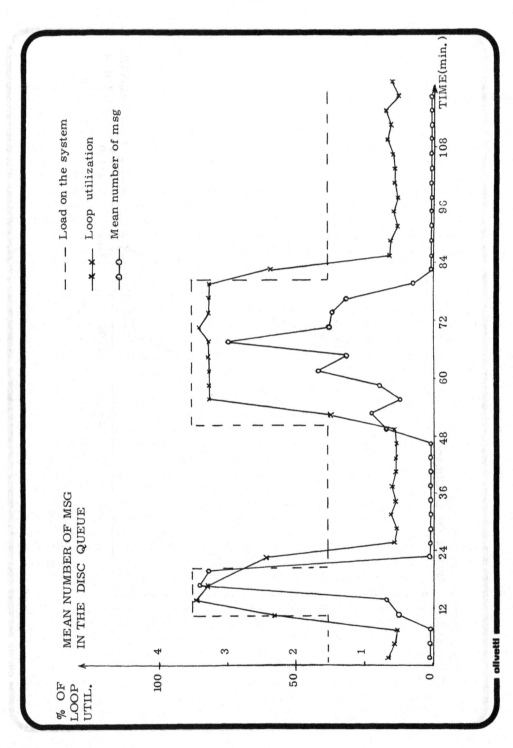

Fig. 6

CONCLUSIONS.

The proposed simulation model has already achieved some of its goals: its flexibility has allowed us to obtain quickly the required curves and to show the system's behaviour with quite different configurations, transmission speeds, node delay etc.

At present, the work on the model proceeds in the direction of software simulation and of the study of other possible applications of the architecture.

From the first results, some ideas have been suggested; as it has been shown in the preceding paragraph, the utilization of the loop is a good measure of the system's level of crowding. In a real system, it is desirable to avoid the micro-breakdowns; then one can imagine a prevention technique based on the estimate of the loop's utilization, easily made by every node. When this measure signals that the system has reached a nearly critical state, the nodes with less urgent messages or with a small number of elements in their transmission queues can decide to temporarily slow down their transmission rates.

The approximate length of the various modules of the simulation program are given: loop and nodes modules, 80 code lines each; host module, 370 lines, of which 100 have been written for the GENERALIZED-HOST. The total length of our program is 800 lines, not including the procedures written to simplify the model's programming. These data show that the proposed simulation approach is simple to implement, and that the model's modularity and flexibility are obtained with a modest program length.

The simulation program has been run on an IBM 370/168, as we have already remarked; the required CPU time of this computer is, of course, a function of simulated time but also of the memory used, the kind of tests etc. The mean ratio between simulated time and 370 CPU time is approximately 40.

BIBLIOGRAPHY.

BIR73 : Birtwistle, Dahl, Myhrhaug, Nygaard - SIMULA BEGIN - Petrocelli Ch. 1973.

CAS76 : G. Casaglia, S. Copelli, N. Lijtmaer - DISTRIBUTED CONTROL FOR LOCAL NETWORK OF MINIS : A DESIGN APPROACH USING MICROPROCES-SORS - Euromicro Symp. (p. 119), oct. 1976.

FAR72 : D. J. Farber, K. Larson - THE STRUCTURE OF A DISTRIBUTED COM-PUTER SYSTEM - THE COMMUNICATION SYSTEM - Proc. Symp. on Comp. Comm. Network and Teletraffic, Pol. Inst. of Brooklin Press, apr. 1972.

FRA74 : A. G. Fraser - LOOPS FOR DATA COMMUNICATIONS - Bell Lab., Comp. Science Tecn. Report n. 24, dec. 1974.

GOR69 : G. Gordon - SYSTEM SIMULATION - Prentice-Hall, 1969.

KLE75 : L. Kleinrock - QUEUING SYSTEMS - Wiley and sons, 1975.

NEW69 : E. E. Newhall, W. D. Farmer - AN EXPERIMENTAL DISTRIBUTED SWITCH-ING SYSTEM TO HANDLE BURSTY COMPUTER TRAFFIC - Proc. ACM Symp., Pine Mountain, Georgia, oct. 1969.

PIE72 : J. R. Pierce - NETWORK FOR BLOCK SWITCHING OF DATA - Bell Syst. Techn. Journal, jul.-aug. 1972.

REA76 : Reames, M. T. Liu - DESIGN AND SIMULATION OF THE DISTRIBUTED LOOP COMPUTER NETWORK (DLCN) - Proc. 3rd Ann. Symp. on Comp. Architecture, jan. 1976.

ZIE75 : B. P. Ziegler - THEORY OF MODELLING AND SIMULATION - Wiley and sons, 1975.

ACKNOWLEDGEMENTS.

We wish to thank N. Lijtmaer, of the Istituto per l'Elaborazione dell'Informazione, Pisa, and ing. F. Tisato, Politecnico di Milano, for their suggestions on the best approach to the problem and ing S. Brandi, Olivetti, for having helped us in the de-finition of the program structure.

SCHEDULING DEPENDENT TASKS FROM AN INFINITE STREAM

IN SYSTEMS WITH NONPREEMPTIBLE RESOURCES

Wojciech Cellary

Institute of Control Engineering
Technical Univerity of Poznan

60-965 Poznan , POLAND

ABSTRACT. An infinite stream of independent jobs composed of sets of dependent tasks, which are being fed into a uniprocessor computer system with nonpreemptible resources is considered. A joint approach is presented, to optimization of a given system performance measure, namely mean flow time of jobs, and the solution of the system performance failure problems, namely the determinacy of the set of tasks, deadlock and permanent blocking. For deadlock avoidance the approach is applied which radically reduces overhead involved without losing the benefit of improved resource utilization.

1. INTRODUCTION

In this paper we will consider an infinite stream of jobs which are being fed into a system. The jobs are independent of each other, i.e. there are no precedence constraints put on the order of their execution. However, we will allow every job to be a set of dependent tasks. Thus, from the operating system point of view, we deal with the infinite stream of dependent tasks.

As is well known, this is an important assumption commonly met in practice. First, usually programs associated with a complex job are prepared in the form of subroutines which can be executed in parallel with some precedence constraints. Second, it is possible to detect in a continuous program, some program blocks which can be executed in parallel. This problem was considered in [1,2] . Of course, in general, the division of a job into tasks improve system efficiency.

We will consider a uniprocessor computer system with many units of nonpreemptible resources of one kind. As is known, in such systems two general problems must be taken into account. The first one is the optimization of a given system performance measure like for example mean flow time. The second one is the solution of the system performance failure problems, namely the determinacy of the set of tasks, deadlock, and permanent blocking.

Commonly, these two general problems are solved separately. On the one hand, there are some approaches to the solution of the particular system performance failure problems [1,2,6,7,8,9,10] which do not take into account any system performance measure. On the other hand, there exist well known algorithms for scheduling tasks on processors tending to optimize a given system performance measure, which either do not take into account any additional resources, ar allow the usage of additional resources under such restrictions that the problem of system performance failures does not arise. These analyses only admit a hierarchical operating system structure. At the lower level, nonpreemptible resources are allocated with regard to system performance failures. Then, at the higher level, tasks with granted additional resource requests are scheduled to optimize a given system performance measure. Let us note, that such approach concerns only the optimal allocation of processor time but not the usage of additional resources.

However, in the last year some papers have appeared dealing with joint approach to the solution of the system performance failure problem and the optimization of a given system performance measure [3,4,5] , that is approaches to all system performance failures which explicitly take into account a given system performance measure. In particular, in [5] algorithms for task scheduling in systems with many nonpreemptible resources of one kind were considered, when an infinite stream of independent tasks is being fed into the system; these algorithms tend to minimize mean flow time. For the solution of the deadlock problem, they use a new approach to deadlock avoidance presented first in [8] and modified in [5] . This paper develops the results obtained in [5], since it concerns stream of dependent tasks, and the minimization of the mean flow time of jobs instead of tasks. As we have mentioned, such model is more realistic in practice and moreover, allows system efficiency to be improved.

In Section 2 of this paper we select approaches to the solution of the peformance failure problems.

In the third Section, some definitions are introduced, and tests used in the deadlock avoidance method are presented. These tests remain generally the same as those presented in [5] , but their presentation is necessary for other parts of this paper.

In the next Section we distinguish all situations which can occur in the system from the resource allocation point of view, and present algorithms for their servicing. The last Section contains some conclusions.

2. SELECTED APPROACHES TO THE SOLUTION OF SYSTEM PERFORMANCE FAILURE PROBLEMS

In this section we will select approaches to the solution of system performance failure problems. Let us start with the problem of the determinacy of the set of tasks.

As is known, the set of tasks is called non-determinate if the results produced by independent tasks depend on the speed and order in which these tasks are executed [7] . The determinacy problem concerns the cases, when independent tasks read from and write to common storage locations. In our system this problem concerns tasks composing one job, since we asume /which is usually done in practice/ that the storage locations reserved by any two jobs are mutually exclusive of each other. The solution of this problem consists in the introduction of the additional, proper precedence constraints among tasks. Algorithms determining which precedence constraints should be added in the set of tasks are presented in [7] . As we have mentioned, in our approach, we assume arbitrary precedence constraints among tasks composing a job. Thus, we allow for the solution of the determinacy problem.

Let us pass now to the deadlock problem. As is well known, deadlock is the system state in which the progress of some tasks is blocked indefitely because each task holds nonpreemptible resources that must be acquired by others in order to proceed. From three general approaches to deadlocks, detection and recovery, prevention, and avoidance, we have selected the avoidance approach since it is characterized by the highest system throughput. Moreover, a cost /i.e., overhead involved/ paid in this approach, which is relatively high can be significantly reduced [8] .

The main idea of deadlock avoidance consists in the application of a so called safety test to examine, on the basis of prior knowledge of the task resource claim, i.e. the strict upper bound on the resource

requests of a task, whether the granting of a given request involves
deadlock danger or not. For this examination a so called safe sequence
of tasks is created. The number of resources free at any moment /i.e.,
not allocated to any task/ must be enough to enable the remaining
requests of the first task in the safe sequence to be granted. The sum
of the resources free and allocated to the first task must be enough
for granting the remaining requests of the second task in the safe sequer
ce, and so on. It is proved [9] that if a safe sequence of tasks can
be created then no tasks are deadlocked. The safety test consists in
simulating request granting and an attempt to create a safe sequence
containing a task that requests resources, in order to answer the question whether this request can be really granted or not.

The large overhead involved in avoidance methods results from the
necessity to apply the safety test in every case of resource request
and almost every case of resource release, and from the number of tasks
making up every safe sequence. A way of significantly reducing overhead
without losing the benefit of improved resource utilization is the application of the necessary condition for deadlock as an admission test.
Such an approach was presented in [8] . The admission test precludes
many unsafe resource allocation states and as a result, significantly
reduces the number of applications of the safety test. A valuable property of the admission test is that it depends entirely on claims which
are constant, and not on numbers of resources allocated to tasks which
vary. Thus,application of this test is reduced to the moment when a new
task enters the system or a task is completed. The safety and admission
tests presented in [8] concern testing resource allocation states. They
were modified in [5] to test state transition, and thus testing is reduced to comparison of only two numbers. These tests will be presented
in the next section.

The last but not least system peformance failure problem is permanent blocking. As is well known, this problem concerns tasks whose
resource requests will never be granted because of a steady stream of
requests from other tasks which are always granted before those of a
blocked task. In our strategy for solving this problem we will consider
the blocking of whole jobs. We asume knowledge of the "blocked-free"
time for every job and periodically examine whether the time elapsed
since the arrival of a job exceed its blocked-free time. If so, the
minimum number of resources is reserved, for the completion of every
task composing the blocked job. This approach to the permanent blocking
problem may be considered also as an optimization of the secondary system performance measure, namely lateness after blocked-free time.

3. BASIC DEFINITIONS

Every task T in the system will be characterized by the following.

1°. Claim - C(T) - which represents the strict upper bound on resources used simultaneously;

2°. Rank - R(T) - which represents the difference between the claim of task T and the number of resources currently allocated to T;

3°. Priority - $\pi(T)$ - calculated as follows.
Let J(T) be a set of tasks composing a job which contains task T, let S(T) be the set of successors of task T, and let $\tau(T)$ denotes the remaining performance time of task T, i.e., the run-time needed by task T to completion.
Then the priority of task T

$$\pi(T) = \sum_{T_i \in S(T)} \tau(T_i) + \tau(T) - 10 \times \sum_{T_i \in J(T)} \tau(T_i)$$

Let us briefly discuss this formula.

The component $- 10 \times \sum_{T_i \in J(T)} \tau(T_i)$ concerns the priorities of jobs, i.e. it decides about the precedence of the choice in the case of tasks that belong to different jobs. In general, the job with the shortest remaining performance time will be taken first for servicing, so that the number of jobs residing in the system will be minimized as well as the mean flow time of jobs.

On the contrary, the component $\sum_{T_i \in S(T)} \tau(T_i) + \tau(T)$ of priority $\pi(T)$ decides about the precedence of the choice in the case of tasks composing the same job, since the values of remaining component are equal for all tasks composing a job. In this case the higher priority is associated with the task whose completion allows, in accordance with precedence constraints, for the starting of the subset of tasks with the greatest, global performance time. An example of task priorities in a job is presented in Fig. 1.

Summarizing, every task in the system will be characterized by its claim, rank and priority. Moreover, we assume that every job in the system is characterized by a blocked-free time which is a time elapsed from job arrival, after which the job is treated as permanently blocked.

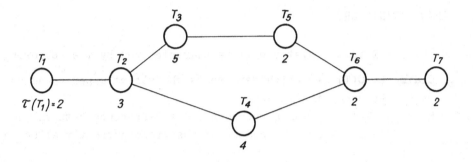

point in time	T_i	$-10 \times \sum\limits_{T_j \in J(T_i)} \tau(T_j)$	$\sum\limits_{T_j \in S(T_i)} \tau(T_j) + \tau(T_i)$	$\pi(T_i)$
job arrival	T_1	-200	$+20$	-180
completion of T_1	T_2	-180	$+18$	-162
completion of T_2	T_3	-150	$+11$	-139
	T_4	-150	$+8$	-142
completion of T_3	T_4	-100	$+8$	-92
	T_5	-100	$+6$	-94
completion of T_4	T_3	-110	$+11$	-99
completion of T_5	T_6	-40	$+4$	-36
completion of T_6	T_7	-20	$+2$	-18

Fig. 1. An example of task priorities.

Now let us divide tasks residing in the system into certain classes.

A task which has at least one resource allocated to it will be denoted by a competitor. A task all of whose predecessors are completed, and which are passed successfully through the admission test, but holds no resources will be denoted by a candidate. Candidates and those competitors whose last resource request did not pass successfully through the safety test will be denoted by waiters. Finally, tasks all of whose predecessors are completed, but which did not pass successfully through the admission test will be denoted by potential candidates.

Let us pass now to the safety and admission tests [5] . First, let us define the promotion of a task T,which consists in allocation a resource to task T and decrementing its rank by one. The allocation state of a task can be described by the number of promotions starting from the fictitious initial state in which no resources are allocated to it /i.e., $R(T) = C(T)$ / up to its current rank. Let the number of resources be equal to t. The allocation state of the system can be stored in a vector denoted by $\bar{x}$, where x_k is the total number of promotions of all competitors from rank = k to rank = k-1 , /k=1,...,t/.

As an example let us consider the following system:

$$t = 6 \qquad CLAIM = (3,4,5) \qquad RANK = (2,1,4)$$

then

$$\bar{x} = (0,1,2,1,1,0) \ ,$$

since task T_1 was promoted from rank = 3 to rank = 2; task T_2 from rank = 4 to rank = 3, from rank = 3 to rank = 2, and from rank = 2 to rank = 1; and task T_3 was promoted from rank = 5 to rank = 4.

For the formulation of the safety and admission tests let us define the following vectors:

$$\bar{t} = (t,t-1,t-2,...,1) ;$$
$$\bar{q} = (q_1,q_2,...,q_t) ,$$

where

$$q_k = t - k + 1 - \sum_{j=k}^{t} x_j \ ;$$
$$\bar{s} = (s_1,s_2,...,s_t) ,$$

528

where

$$s_k = t - k + 1 - \sum_{j=k}^{t} y_j \quad,$$

and y_j is the number of tasks whose claim $= j$;

and pointers :

$$I_q = \begin{cases} \min_{q_k=0} \{k\} & \text{, if there exists such a k that } q_k = 0 \text{ ;} \\ t+1 & \text{, otherwise ;} \end{cases}$$

$$I_s = \begin{cases} \min_{s_k=0} \{k\} & \text{, if there exists such a k that } s_k = 0 \text{ ;} \\ t+1 & \text{, otherwise.} \end{cases}$$

In our example :

$$\bar{t} = (6,5,4,3,2,1) ,$$
$$\bar{q} = (1,0,0,1,1,1) ,$$
$$\bar{s} = (3,2,1,1,1,1) ,$$
$$I_q = 2 ,$$
$$I_s = 7 .$$

The safety and admission tests can be formulated as follows [5] .

Theorem 1.   Let the initial state be safe. A state transition is safe if and only if

$$R(T) < I_q$$

where R T is the rank of a task which request a resource.

Theorem 2.   Let the initial state be safe. If the state remains safe after the addition of a task T to the set of candidates, then

$$C(T) < I_s$$

where $C(T)$ is the claim of task T.

4. TASK SCHEDULING AND RESOURCE ALLOCATION ALGORITHMS

In this section we present algorithms for all the situations of resource allocation in the system, that require specific servicing. These algorithms, in reference to task scheduling, are generalizations of those in [5].

Situation 1 : The occurence of a request from a competitor or the arrival of a new candidate.

In this situation we must examine whether the request can be granted safely or not. As result from Theorem 1 this examination consists in the comparison of only two numbers $R(T)$ and I_q. If it is successfull, the resource is allocated, and the following algorithm is performed to calculate the new values of $\bar{q}$, I_q and $R(T)$.

Algorithm 1

```
I_q, i := 1;
while  i ≤ R(T)  do
  begin
     q_i := q_i - 1;
     i := i + 1
  end;
while  q_{I_q} > 0  and  I_q ≤ t  do  I_q := I_q + 1;
R(T) := R(T) - 1;
```

As an example let us consider the system state:

$$t=6 \qquad CLAIM = (4,2,6) \qquad RANK = (3,1,4)$$

then

$$\bar{t} = (6,5,4,3,2,1) \,,$$
$$\bar{q} = (2,1,1,0,0,0) \,,$$
$$I_q = 4 \,.$$

In accordance with Theorem 1, we can grant only the request of either task T_1 or T_2. The allocation of a resource to task T_3 could cause deadlock, since even after the completion of T_2, there may be not enough resource to complete either T_1 or T_3 . After granting the request of T_1 we obtain:

$$RANK = (2,1,4),$$
$$\bar{q} = (1,0,0,0,0,0),$$
$$I_q = 2.$$

If a request cannot be granted safely, the competitor is included in the set of waiters and must wait for the release of resources.

The arrival of a new candidate is equivalent to its first request, and as a result to a request from a competitor.

Situation 2 : The release of resources.

In this situation, first, the new resource allocation state must be calculated. We use the following algorithm.

Algorithm 2

Let A be the number of released resources.

```
i := 1;
while  i ≤ R T   do
  begin
    q_i := q_i + A;
    i := i + 1
  end;
I_q, R T := R T + A;
while  A > 0  do
  begin
    q_i := q_i + A;
    i := i + 1;
    A := A - 1
  end;
while q_{I_q} > 0  and  I_q ≤ t  do  I_q := I_q + 1;
```

For example let us consider the release of one resource /A=1/ by task T_3 in the following system:

$$t=6 \qquad CLAIM = (4,2,6) \qquad RANK = (3,1,4)$$

$$\bar{q} = (2,1,1,0,0,0),$$
$$I_q = 4.$$

After the execution of Algorithm 2 we obtain :

$$RANK = (3,1,5),$$
$$\bar{q} = (3,2,2,1,1,0),$$
$$I_q = 6.$$

After updating the resource allocation state we should try to grant the requsts of waiters, which up to now could not be granted because of the state transition safety test.

Algorithm 3

1°. Search for waiters with ranks less than I_q. If no such task can be found, stop the algorithm.

2°. Choose the task with maximum $\pi(T)$ and grant its request /use Algorithm 1/.

3°. Repeat from step 1°.

In the first step of Algorithm 3, we search for tasks whose requests can be granted safely. The searching procedure can be improved if we keep the set of waiters ordered in ascending rank order. Let us note that this algorithm is applied only for waiters and not all the tasks whose precedence constraints are fulfilled. Thus, the number of examined tasks is greatly reduced. Moreover, it is easy to prove that this algorithm will have to repeated at most as many times as the number of released resources.

Situation 3 : The arrival of a new potential candidate.

In this situation we must check whether the new potential candidate can be moved to the set of candidates or not, using the admission test. Moreover, if the new potential candidate did not pass successfully through the admission test, we will try to replace a candidate, with an appropriate claim and lower priority, in the set of candidates by a new potential candidate. Such replacement is beneficial for mean flow time, and is always possible since candidates do not have resources allocated to them.

For servicing of this situation, we will formulate two admission
tests: the first calculated for the set of competitors $/\bar{s}$, $I_s/$ and
the second for both competitors and candidates $/\bar{s}^*$, $I_s^*/$. The use of
these tests will be presented in Algorithm 5. However, first let us
formulate an algorithm used for updating values of $\bar{s}$ and I_s after
addition/deletion of a task to/from the set of competitors as well as
$\bar{s}^*$ and I_s^* after the addition/deletion of task to/from the set either
of competitors or candidates.

Let T be the new potential candidate.

Algorithm 4

Boolean variable AD equal to <u>true</u> denotes addition, <u>false</u> -
- deletion;

I_s, i := 1;
<u>while</u> i $\leqslant$ C T   <u>do</u>
  <u>begin</u>
    <u>if</u> (AD) <u>then</u> s_i := s_i - 1 <u>else</u> s_i := s_i + 1
  <u>end</u>;
<u>while</u> $s_{I_s} > 0$ <u>and</u> $I_s \leqslant$ t <u>do</u> I_s := I_s + 1;

The algorithm for the admission of a new potential candidate is as
follows.

Algorithm 5

1°. If $C(T) < I_{s*}$, move T to the set of candidates and use
Algorithm 4 for the calculation of the new values of $\bar{s}^*$
and I_{s*} ; then stop the algorithm.
2°. If $C(T) \geqslant I_s$ stop the algorithm.
3°. From among candidates with claims greater than or equal to
C T find task T_r with priority $\pi(T_r)$ at the minimum. If
no such task can be found, or $\pi(T_r) \geqslant \pi(T)$ then stop the
algorithm; otherwise replace T_r by T in the set of candidates
and update the values of $\bar{s}^*$ and I_{s*} using Algorithm 6. Then
stop the algorithm.

Algorithm 6

```
i := C(T) + 1;
while i ⩽ C(T) do
  begin
    s*_i := s*_i + 1;
    i := i + 1
  end;
I_s* := i;
while s_{I_s*} > 0 do I_s* := I_s* + 1;
```

Situation 4 : The completion of a competitor.

In this situation we should:

1°. Update values of $\bar{q}$, I_q, $\bar{s}$, I_s, $\bar{s}*$, I_{s*} using Algorithms 2 and 4.

2°. Try to move a potential candidate to the set of candidates /it can be proved that there will be at most one such potential candidate/ - using Algorithm 7.

3°. Add the tasks whose precedence constraints are now fulfilled to the set of potential candidates - using Algorithms for servicing Situation 3.

4°. Try to allocate resources released by the completed competitor using Algorithms for servicing Situation 2.

Algorithm 7

1°. Search for a potential candidate T with $C(T) < I_{s*}$ and with $\pi(T)$ at the maximum. If no such task can be found, stop the algorithm.

2°. Move T to the set of candidates and use Algorithm 4 for calculating new values of $\bar{s}*$ and I_{s*}. Then, stop the algorithm.

Situation 5 : The detection of a permanently blocked job.

In this situation we must activate a special strategy of resource allocation to grant the requests of tasks composing the blocked job. This strategy consists in the division of system resources into two parts: the one part ensures the completion of a blocked job, and the remaining resources are used for servicing requests from other jobs. The servicing of the blocked job and remaining jobs is independent, and performed in accordance with Situations 1 to 4 but for a reduced number of resources.

The first problem in Situation 5 which can be distinguish is the determination of the number of resources necessary for the blocked job. It can be shown that this number is equal to the maximum rank of tasks which are not yet being completed, composing the permanently blocked job. Let us note that the rank of a task which is not a competitor is equal to its claim. This number should be updated after the completion of every task composing the blocked job, since it should be at the minimum.

The next step of the permanent blocking prevention strategy consists in waiting for the completion of all competitors. Then, as was mentioned above, the resources are divided and servicing of the blocked job is performed independently.

It could seem that the completion of all competitors should not be necessary. However, we want to preclude a situation in which the remaining after division resources are allocated to tasks which cannot complete because of the lack of resources, and thus block these resources during permanently blocked job servicing.

Let us note that the strategy presented here is less conservative than Holt's well known strategy [10] , since it allows the parallel servicing of the blocked and non-blocked jobs.

5. CONCLUSIONS

The two general problems of task scheduling in systems with non-preemptible resources: first, the optimization of a given system performance measure, and second the solution of the system performance failures problems /determinacy of the set of tasks, deadlock and permanent blocking/, are commonly solved separately. The algorithms presented in this paper represent a joint approach to both of these problems, which may lead to better results, being obtained. They deal with the case of infinite stream of dependent tasks and thus concern a wide class of practical situations. The algorithms are presented in the form which allows for direct implementation in operating systems.

REFERENCES

1. Baer, J.L., A survey of some theoretical aspects of multiprocessing, Computing Surveys vol. 5, No 1, 1973.
2. Bernstein, A.J., Analysis of programs for parallel processing, IEEE Trans. Comp. vol EC-15, 1966, No 5.
3. Cellary, W., On resource allocation policies in uniprocessor systems with nonpreemptible resources, MTA SZTAKI Tanulmanyok 69, 1977.
4. Cellary, W., Resource allocation strategies in computer systems with nonpreemptible resources, Foundations of Control Engineering, vol. 2, No 3, 1977.
5. Cellary, W., Task scheduling in systems with nonpreemptible resources, in: H. Beilner and E. Gelenbe, Medelling and Performance Evaluation of Computer Systems, /Proc. of the III International Symposium/, North Holland Publishing Co., 1977.
6. Coffman, E.G., Jr., M.J. Elphick, A. Shoshani, System deadlocks, Computing Surveys vol. 2, No 3, 1971.
7. Coffman, E.G., Jr., P.J. Denning, Operating Systems Theory, Prentice Hall, Englewood Cliffs, N.J., 1973.
8. Habermann, A.N., A new approach to avoidance of system deadlocks, Revue Francaise d'Automatique, Informatique et Recherche Opérationelle 9. sept B-3, 1975
9. Habermann, A.N., Prevention of system deadlocks, Comm. ACM vol 12, No 7, 1969.
10. Holt, R.C., Comments on prevention of system deadlocks, Comm. ACM, vol. 14, No 1, 1971.

SCHEDULING PREEMPTABLE TASKS ON UNRELATED PROCESSORS
WITH ADDITIONAL RESOURCES TO MINIMIZE SCHEDULE LENGTH

Roman Słowiński

Institute of Control Engineering, Technical University of Poznań,
60-965 Poznań, Poland

ABSTRACT The problem considered is that of scheduling n preemptable tasks on m parallel processors, when each task requires for its processing a processor and one resource unit from the set of additional resources. The processing times of a task on different processors are unrelated. We present the method for solving this problem which is composed of two stages. In the first stage, a linear programming problem is solved giving the minimum schedule length and optimal task processing times on particular processors. On the basis of this solution, in the second stage the optimal schedule is constructed taking into account the resource constraints. Theorems are proved concerning the feasibility of the second stage algorithm, and the upper bound on the number of preemptions in the optimal schedule. The cases of independent and dependent tasks are considered.

1. INTRODUCTION

In recent years we have been able to observe increased interest in scheduling problems associated with a certain model of a multiprocessor computing system /see [7] for a survey/. Much effort has been applied to problems concerned with cases where each task only requires one processor for its processing. In this paper, we consider an augmented multiprocessing model which allows for the possibility that certain tasks may require the use of various limited resources during their processing. Some special cases of this model were studied in [2,4,5,6,8] for various performance measures. However, all the previous studies assumed the processors to be identical. Even under this assumption, almost all problems are NP-complete [7] and hence they are computationally intractable. We shall be concerned with a problem which seems to be also NP-complete, where processors are unrelated, i.e. the processing times of tasks on different processors are arbitrary. This problem, without additional resource constraints, has been considered in [3,9].

In Section 2, we describe the model of the computing system and give some basic definitions. In Section 3 and 4, a two-stage method for solving the problem is presented for the cases of independent and dependent tasks, respectively. Also in Section 3, theorems concerning the feasibility of the second-stage algorithm, and the upper bound on the number of preemptions in the optimal schedule are proven. Section 5 contains some final remarks.

2. THE MODEL OF THE COMPUTING SYSTEM

Let us describe the model of the computing system considered in this paper. Three finite sets are given, which are the main components of the model:

- the set of tasks $\mathcal{J} = \{T_1, T_2, \ldots, T_n\}$,
- the set of unrelated processors $\mathcal{P} = \{P_1, P_2, \ldots, P_m\}$,
- the set of additional resources $\mathcal{R} = \{R_1, R_2, \ldots, R_p\}$.

Each task T_j requires for its processing a processor and one unit of a specified resource. Let S_j be the set of processors which may execute T_j. Associated with each task T_j is the vector $\bar{t}_j = [t_{ij}]$, where t_{ij} is the time required for the execution of T_j by processor $P_i \in S_j$, provided that the specified additional resource unit is allotted to T_j. The processing of each task may be arbitrarily interrupted and restarted later without any time penalty, possibly on another processor. Moreover, permissible task orderings are determined by a set of precedence constraints given in the form of a "task-on-edge" directed acyclic graph with only one origin and only one terminal. The graph nodes /events/ are numbered from 1 to N in such a way that node j occurs not later than node k, if j<k. Such an ordering is always possible but may not be unique in a given precedence graph. We shall assume that only one ordering is imposed for a given problem.

Each processor P_i is able to process at most one task at a time. The set of tasks which may be processed on P_i will be denoted by C_i.

For each resource R_k there is a bound B_k which gives the total number of the resource units available at any given time. The set of tasks which require resource R_k will be denoted by D_k.

The objective is to minimize the finishing time /schedule length/ T of the set of tasks subject to the imposed constraints.

3. SCHEDULING INDEPENDENT TASKS

Let us consider the case of independent tasks, i.e. those which can be processed simultaneously.

We shall present a two-stage scheduling method which is a generalization of the method given in [3,9] for the model without additional resources. In the first stage, we find a generating schedule that is one which minimizes the schedule length T and gives optimal processing times of tasks on particular processors, but which does not necessarily fulfil the feasibility condition that some parts of a task are not executed simultaneously on more than one processor. In the second stage, on the basis of generating schedule, the optimal schedule is constructed which ensures that the feasibility condition and resource constraints are satisfied.

THE FIRST STAGE

Let $x_{ij} \in \langle 0, t_{ij} \rangle$ be the total processing time of task T_j on processor $P_i \in S_j$, $j=1,2,\ldots,n$. In order to find the generating schedule, we have to solve the following linear programming /LP/ problem.

Minimize T
subject to:

$$T - \sum_{T_j \in C_i} x_{ij} \geq 0 \qquad\qquad i=1,2,\ldots,m \qquad /1/$$

$$T - \sum_{P_i \in S_j} x_{ij} \geq 0 \qquad\qquad j=1,2,\ldots,n \qquad /2/$$

$$B_k T - \sum_{T_j \in D_k} \sum_{P_i \in S_j} x_{ij} \geq 0 \qquad k=1,2,\ldots,p \qquad /3/$$

$$\sum_{P_i \in S_j} x_{ij}/t_{ij} = 1 \qquad\qquad j=1,2,\ldots,n \qquad /4/$$

$$x_{ij} \geq 0 \qquad \text{for all } P_i \in S_j, \quad j=1,2,\ldots,n \qquad /5/$$

Condition /1/ ensures that the active time of any one processor will not exceed T, condition /2/ - that each task is completed by time T, and condition /3/ - that the time of using any resource type will not exceed T. Condition /4/ ensures that each task is completed.

Solving the above LP problem we obtain the optimal values of x_{ij}, $P_i \in S_j$, $j=1,2,\ldots,n$, which minimize T. However, we do not know the task part start times which make the optimal schedule. Below we present an algorithm which constructs an optimal schedule in polynomial time.

THE SECOND STAGE

In the second stage, knowing the generating schedule, we shall construct the optimal schedule.

Let X denote the $m \times n$ matrix of nonnegative elements which are the optimal values of x_{ij}, $P_i \in S_j$, $j=1,2,\ldots,n$, obtained in the first stage. Column j /task T_j/ of matrix X will be called critical if $\sum_{i=1}^{m} x_{ij} = T$. Similarly, resource R_k will be called critical if $B_k T = \sum_{T_j \in D_k} \sum_{i=1}^{m} x_{ij}$. Let us also define the $m \times m$ diagonal matrix Y of nonnegative processor idle times: $y_{ii} = T - \sum_{j=1}^{n} x_{ij}$, $i=1,2,\ldots,m$.
The columns of Y will represent dummy tasks which do not require additional resources. We shall denote by Z the $m \times (n+m)$ matrix composed of matrices X and Y as indicated below:

$$Z = \left[\begin{array}{|c|c|} \hline X & Y \\ \hline \end{array} \right]$$

Let us introduce the set NC, called the generating set, containing m positive elements of matrix Z which are:
- exactly one element in each critical column,
- exactly one element in each of B_k columns representing tasks requiring the critical resource R_k,
- no more than one element in the remaining rows and columns.
The resource requirements of tasks represented in NC cannot exceed the resource constraints, i.e.

$$\left| \left\{ T_j : \underset{i}{\exists} z_{ij} \in NC \wedge T_j \in D_k \right\} \right| \leq B_k, \quad k=1,2,\ldots,p. \qquad /6/$$

For the set NC we have to calculate the parallel processing time DELTA of the task parts represented in NC.

The construction of the optimal schedule proceeds in the following way:
1° Find the generating set NC.
2° Calculate DELTA from the following formula

$$DELTA = \begin{cases} z_{min} & \text{if } T - z_{min} \geq z_{max}, \\ T - z_{max} & \text{otherwise,} \end{cases}$$

where $z_{min} = \underset{z_{ij} \in NC}{min} [z_{ij}]$, $\quad z_{max} = \underset{z_{ij} \notin NC}{max} [z_{ij}, \underset{k}{\underset{T_j \in D_k}{\sum}} \sum_{i=1}^{m} z_{ij}/B_k]$.

3° Decrease T and all $z_{ij} \in NC$ by DELTA. If $T = 0$ then end the procedure, otherwise go to step 1°.

It can be seen that set NC is constructed in such a way that at the
end of each iteration the elements of matrix X, as well as T, fulfil
conditions /1/ - /3/. Let us also note that for each set NC, DELTA is
chosen such that either one of the positive elements in the matrix Z is
reduced to zero, or one more column or resource type becomes critical.
Each of these events may occur a finite number of times which ensures
that the optimal schedule will be obtained in a finite number of iterations. However, in order to prove this, we have to demonstrate the existence of NC for each schedule fulfilling conditions /1/ - /3/ /in particular, for the generating schedule/.

<u>THEOREM 1</u> For each schedule fulfilling conditions /1/ - /3/, $n \geq m$ and
$T > 0$, there exists a generating set NC.

<u>PROOF</u> Let us construct an $(m+n) \times (m+n)$ matrix V as follows

$$V = \begin{bmatrix} X & Y \\ W & X^T \end{bmatrix}$$

where W is an $n \times n$ diagonal matrix of nonnegative elements:

$$w_{jj} = T - \sum_{i=1}^{m} x_{ij} \qquad j=1,2,\ldots,n.$$

As can be seen, each row sum and column sum of V is equal to T. Thus,
in matrix $\frac{1}{T} V$, each row sum and column sum is equal to one. Since all
elements of the square matrix $\frac{1}{T} V$ are nonnegative, this is a doubly
stochastic matrix which is a convex combination of permutation matrices,
as follows from the Birkhoff - von Neumann theorem [1]. It is evident
that any one of the permutation matrices in such a convex combination
can be identified with a generating set NC if it satisfies the resource
constraints /6/ and contains B_k elements representing tasks $T_j \in D_k$, for
each critical resource R_k. Condition /3/ ensures that at any time within
the schedule length T, it is possible to find no more than B_k tasks
which use resource R_k; thus, at least one of the permutation matrices
in the above convex combination is identified with a generating set NC. ◇

Let us now pass to the problem of the bound on the number of preemptions in the optimal schedule. From the linear programming formulation posed in the first stage, follows that for $m > 2$, in the optimal
basic feasible solution, there will be no more than $2n+m+p$ positive
variables. In fact, there will be no more than $n+v_1+v_2+v_3$ positive variables, where v_1, v_2, v_3 are the numbers of inequalities /1/, /2/ and /3/
correspondingly, in which variables transforming them into equalities

are equal to zero. In other words, v_1 is the number of processors with zero idle time, v_2 is the number of critical tasks, and v_3 is the number of critical resources. But $1 \leqslant v_1 \leqslant m$, $0 \leqslant v_2 \leqslant m-1$ and $0 \leqslant v_3 \leqslant g$, where $g \leqslant m$ is the maximum number of resource types for which $\sum_k B_k \leqslant m$. Hence, in the optimal solution, there will be no more than $n+2m+g-1$ positive variables, one of which is T.

Thus, if $n > m$, there exists a generating schedule with no more than $n+2m+g-2$ positive x_{ij} values. If we could construct an optimal schedule without introducing additional preemptions, then the upper bound on the number of preemptions in the optimal schedule would be equal to $2m+g-2$. However, the second-stage algorithm generally introduces additional preemptions. We shall now establish an upper bound on this number.

First, let us make a certain modification to the matrix Z with the objective of reducing the number of preemptions in the optimal schedule. This modification is also beneficial for the running time of the algorithm for finding the generating set NC, which will be discussed later. The idea of this modification is to replace all the tasks /including the dummies/ using the same resource type, which are assigned to only one and the same processor in the generating schedule, by a new task. At the end of the second stage, we have to create a schedule for the original set of tasks by reassigning the time intervals DELTA obtained for the new tasks, to the tasks which they replaced.

THEOREM 2 The upper bound on the number of preemptions in the optimal schedule is equal to $2m^2-4m+m(g-1)+m(v_N-1)+2$, where $g \leqslant m$ and $v_N \leqslant \min\left[n-2m-g+2,\ mp\right]$.

PROOF The modified matrix Z' will contain v_o+v_N columns /tasks/ and no more than $v_o+v_N+v_1+v_2+v_3-1$ positive elements, where v_o is the smallest number of original tasks and v_N - the maximum number of new tasks. Hence, $v_o \leqslant v_1+v_2+v_3-1$, and v_N is bounded by $\min\left[n-v_o,\ mp\right]$. Since each iteration of the schedule construction procedure determines the parallel processing time of m task parts, the procedure will terminate with $\left[m\,(\text{number of iterations}) - (v_o+v_N)\right]$ preemptions. Hence, we may obtain a bound on the number of preemptions by bounding the number of iterations.

We already know that after each iteration, either one of the positive elements in the matrix Z' is reduced to zero, or one more column becomes critical, or finally, one more of the resource types becomes critical. Exactly m elements become zero in the last iteration. Thus, there will be no more than $v_o+v_1+v_2+v_3+v_N-m$ iterations of the first kind, $m-v_2$ iterations of the second kind, and $g-v_3$ iterations of

542

the third kind, hence at most $v_0+v_1+g+v_N$ iterations in all.

It is resonable to assume that no one of the new tasks is critical /if such a task exists, the problem can be reduced to m-1 processors/. Under this assumption, we can reduce the bound on the number of itera- tions by m. In the last iteration, m columns are critical, and some of them were critical at the beginning /having then at least two non-zero elements/. If a column becomes critical, and there exists exactly one positive element in that column, then at least one element of matrix Z' was reduced to zero in that iteration. If, however, the critical column has at least two positive elements, then in a certain iteration, when the number of positive elements in that column is reduced to one, at least two elements of matrix Z' are reduced to zero simultaneously. Thus, the total number of iterations is overestimated by at least m. It follows that the bound on the number of preemptions is equal to $m(v_0+v_1+g+v_N-m) - (v_0+v_N)$. Since $v_0 \leqslant 2m-2+g$, $v_1 \leqslant m$, and $n \geqslant m$, we obtain the thesis. ◇

Let us now pass to the description of the algorithm for finding the generating set NC. The algorithm makes use of the modified matrix Z'. Let us define a zero-one matrix A of the same size as matrix Z':

$$a_{ij} = \begin{cases} 1 & \text{if } z'_{ij} > 0, \\ 0 & \text{otherwise.} \end{cases}$$

For simplicity of computation, it is better to find the set NC in ma- trix A. The elements of A selected for the set NC will be marked with the symbol △, to represent an assignment of the processor in that row to the task in that column. The block diagram of the algorithm is shown in Fig.1 and Fig.2.

This algorithm was programmed in Fortran for an ICL 1900 computer [10]. It may easily be shown that the presented algorithm finds the set NC in $O(v_N m^2)$ time.

EXAMPLE The following example steps through the algorithm for finding a generating set NC. The following fictitious data are given:

R_k: R_1 R_2 R_1 R_2 R_1 R_2

$$A = \begin{matrix} 1 & 0 & 0 & 0 & 0 & 1 \\ 0 & 1 & 0 & 1 & 1 & 1 \\ 1 & 1 & 1 & 0 & 0 & 0 \\ 0 & 1 & 1 & 0 & 0 & 0 \\ 0 & 0 & 0 & 1 & 1 & 1 \end{matrix}$$

* = critical column /task/

$B_1 = 3$
$B_2 = 2$

The construction of a generating set NC proceeds in the following way:

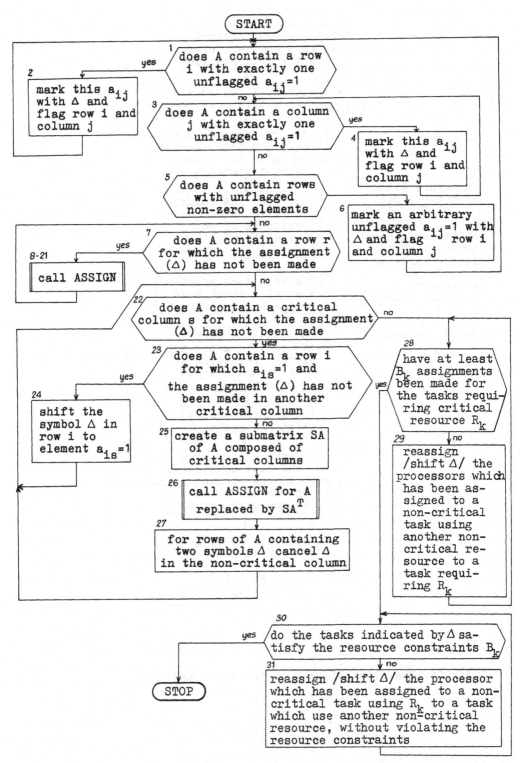

Fig. 1. The block diagram of the algorithm for finding a generating
set NC.

544

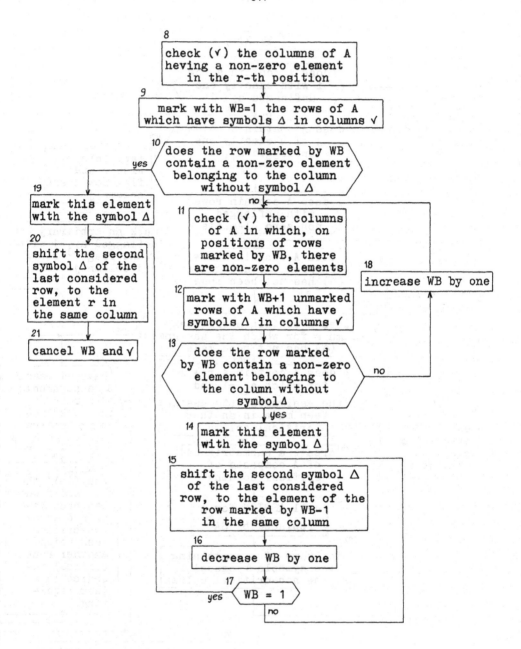

Fig. 2. The block diagram of procedure ASSIGN.

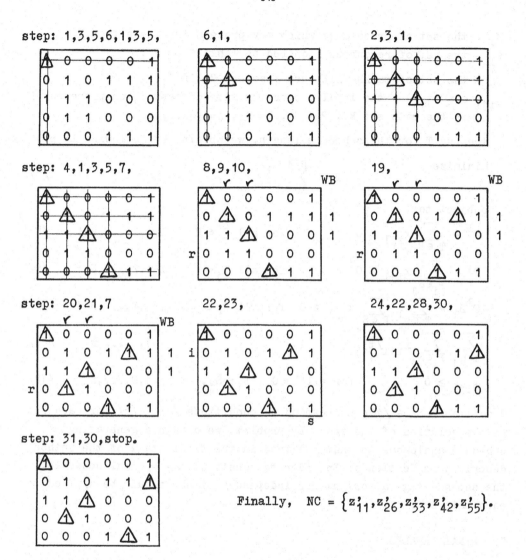

step: 1,3,5,6,1,3,5,   6,1,   2,3,1,

step: 4,1,3,5,7,   8,9,10,   19,

step: 20,21,7   22,23,   24,22,28,30,

step: 31,30,stop.

Finally, $NC = \{z'_{11}, z'_{26}, z'_{33}, z'_{42}, z'_{55}\}$.

4. SCHEDULING DEPENDENT TASKS

Let us now consider the case where, in the task set $\mathcal{T}$ precedence constraints are given in the form described in Section 2.

Let F_1 denote the set of all tasks which may be processed between the occurence of node l and $l+1$ in the precedence graph. Sets F_1, $l=$ $=1,2,\ldots,N-1$, will be called the main sets. Since the parts of tasks performed in particular main sets are independent, we may apply the approach proposed for the case of independent tasks to them. Let us introduce some additional definitions:

T^l - the schedule length for the set F_l,

G_j - the set of the numbers of main sets containing task T_j,

H_1 - the set of processors which may process the tasks from F_1,

K_1 - the set of resources required by tasks from F_1,

D_k^l - the set of tasks which use resource R_k in F_1,

$x_{ijl} \in \langle 0, t_{ij} \rangle$ - the total processing time of task T_j on processor P_i in the main set F_1; $P_i \in S_j$, $l \in G_j$, $j=1,2,\ldots,n$.

Keeping T for the schedule length, we obtain the following LP problem.

Minimize
$$T = \sum_{l=1}^{N-1} T^l \qquad /7/$$

subject to:

$$T^l - \sum_{T_j \in F_1} x_{ijl} \geq 0 \qquad \text{for all } P_i \in H_1, \quad l=1,2,\ldots,N-1 \quad /8/$$

$$T^l - \sum_{P_i \in S_j} x_{ijl} \geq 0 \qquad \text{for all } T_j \in F_1, \quad l=1,2,\ldots,N-1 \quad /9/$$

$$T^l B_k - \sum_{T_j \in D_k^l} \sum_{P_i \in S_j} x_{ijl} \geq 0 \quad \text{for all } R_k \in K_1, \quad l=1,2,\ldots,N-1 \quad /10/$$

$$\sum_{l \in G_j} \sum_{P_i \in S_j} x_{ijl}/t_{ij} = 1 \qquad j=1,2,\ldots,n \quad /11/$$

$$x_{ijl} \geq 0 \qquad \text{for all } l \in G_j, P_i \in S_j, \qquad j=1,2,\ldots,n \quad /12/$$

Conditions /8/ - /10/ correspond to conditions /1/ - /3/ for each F_1. As the solution of the above LP problem, we obtain a generating schedule analogous to that obtained in the first stage of the method described in Section 3. In order to construct the optimal schedule, the second-stage algorithm for independent tasks should be applied to each main set.

5. FINAL REMARKS

Scheduling problems, in which additional resource constraints are taken into account, are almost all NP-complete, and thus computationally intractable. Models of computer systems involving additional resources, however, are better adapted to practical situations and for this reason they are worth considering. In this paper one such problem has been examined, when tasks need a processor and one resource unit from the set of additional resources for their processing. This may be, for example, a situation in which a task requires a processor and an input/output device.

The method described finds an optimal schedule in two stages: in the first, an LP problem is solved, and in the second, which is polynomial in time, an optimal schedule is found. This method can also be applied in the case of dependent tasks, when the set of nodes /events/ in the precedence " task-on-edge" graph is ordered.

REFERENCES

1. C. Berge: Théorie des graphes et ses applications, Chapt. X, Dunod, Paris 1958.
2. J. Błażewicz: Mean flow time scheduling under resource constraints, Preliminary Report PR-19/77, Technical University of Poznań /Poland/ March 1977.
3. J. Błażewicz, W. Cellary, R. Słowiński, J. Węglarz: Deterministic problems of scheduling tasks on parallel processors. Part I. Sets of independent tasks, /in Polish/ Podstawy Sterowania 6, June 1976, 155-178.
4. M.R. Garey, R.L. Graham: Bounds for multiprocessor scheduling with resource constraints, SIAM J. on Computing 4, 1975, 187-200.
5. M.R. Garey, R.L. Graham, D.S. Johnson, A.C.-C. Yao: Resource constrained scheduling as generalized bin packing, J. Combinatorial Theory Ser. A, 21, 1976, 257-298.
6. M.R. Garey, D.S. Johnson: Complexity results for multiprocessor scheduling under resource constraints, SIAM J. on Computing 4, 1975, 397-411.
7. R.L. Graham, E.L. Lawler, J.K. Lenstra, A.H.G. Rinnoy Kan: Optimization and approximation in deterministic sequencing and scheduling: a survey, Report BW 82/77, Mathematisch Centrum, Amsterdam, October 1977.
8. K.L. Krause, V.Y. Shen, H.D. Schwetman: Analysis of several task-scheduling algorithms for a model of multiprogramming computer system, J.ACM 22, 1975, 522-550, 24, 1977, 527.
9. E.L. Lawler, J. Labetoulle: Scheduling of parallel machines with preemptions, Proc. of the IXth Internat. Symp. on Mathematical Programming, Budapest, August 1976 /to appear/.
10. R. Słowiński: Algorithm FEASCHE - computer program in Fortran, Preliminary Report PR 10/77, Technical University of Poznań,/Poland/ May 1977.

DEADLOCK AVOIDANCE IN GRAPH-STRUCTURED TASK SYSTEMS

by

M.Di MANZO, A.L.Frisiani

Istituto di Elettrotecnica
Università di Genova,Italy

and

G.Olimpo

Laboratorio per le Tecnologie Didattiche
Consiglio Nazionale Ricerche, Genova,Italy

SUMMARY

Existing models for deadlock detection and avoidance give practical solution only in the case of chains of independent tasks.
In this paper we propose a non enumerative approach to deadlock avoidance when the workload consists of a graph-structured task system. The avoidance algorithm is based on an extention of the Coffman and Denning deadlock model.

1. Introduction

The problem of deadlock avoidance has been widely examined by several authors [2,4,5,7]. However, most researches have been concerned mainly with systems of independent processes, and only a few results are known for the case of interacting processes [5] . Moreover, the deadlock detection and avoidance methods which have been defined for graph-structured task systems are essentially enumerative, and therefore quite time consuming; for practical implementation purposes sufficient and rapidly verifiable conditions are needed. On the other hand, the extensive development of parallel processing techniques and multiprocessor architecture is increasing the interest in process coope-ration as a very basic tool for programming methodologies and computer systems opera-tions; so, we feel that deadlock avoidance techniques for large systems of interacting processes should be improved.

The aim of this paper is to present a practical method for deadlock avoidance when the workload is a system of interacting processes that can be structured as a graph of tasks. With respect to the basic assumptions our model is quite close to that developed by Coffman and Denning [2]; we made this choice because the model in [2] is perhaps the most widely known and the most suitable for the extension to the case of our interest. Therefore in the following we will often refer to [2], even if a number of fundamental definitions are reported here in order to allow a self-contained reading of the paper. The paper is organized as follows. In section 2 a set of basic definitions is present-ed, and a theorem is proved which allows the detection of deadlock in a general graph structured task system; in section 3 we discuss the problem of deadlock avoidance in the simplified case of a tree structured task system; in section 4 an approach to dead-lock avoidance with graph structured task systems is suggested.

2. Basic definitions and theorem

We define a <u>task system</u> to be a pair $G = (I, \lessdot)$, where $I = T_1, T_2, \ldots, T_n$ is a set of tasks, and $\lessdot$ is a partial ordering (precedence relation) on I. Given two tasks T and T', $T \lessdot T'$ means that task T is to be completed before task T' begins. A task system can be represented by a <u>precedence graph</u>, where each vertex is a task and the vertices corresponding to two tasks T and T' are connected by a directed edge iff $T \lessdot T'$. A <u>path</u> of length k through the precedence graph G is a sequence of edges $(T_{r_1} T_{r_2})$ $(T_{r_2} T_{r_3}) \ldots (T_{r_{k-1}} T_{r_k})$ passing thorough vertices $T_{r_1}, T_{r_2}, \ldots, T_{r_k}$.
For i and j such that $1 \leqslant i < j \leqslant k$, T_{r_i} is a <u>predecessor</u> of T_{r_j} and T_{r_j} is a <u>successor</u> of T_{r_i}; if $j = i+1$ then T_{r_i} is an <u>immediate predecessor</u> of T_{r_j} and T_{r_j} is an <u>immediate successor</u> of T_{r_i}. Let S_i be the set of immediate successors of T_i and P_i be the set of immediate predecessors of T_i. A task with no predecessors is an <u>initial task</u> and a task with no successors is a <u>terminal task</u>. The level of task T_i (written $\ell(T_i)$) is k if the length of the longest path from T_i to any terminal task is k. An <u>execution sequence</u> of a system G of n tasks is a sequence of task initiations and terminations $\alpha = a_1 a_2 \ldots a_{2n}$ satisfying the following conditions:

1. For every task $T \in I$ the symbols $\overline{T}$ (task initiation) and $\underline{T}$ (task termination) appear exactly once in α .
2. If $a_i = \overline{T}$ and $a_j = \underline{T}$, then $i < j$.
3. If $a_i = \underline{T}$ and $a_j = \overline{T}$, and $T \lessdot T'$, then $i < j$.

More details can be found in [2].
Since a task represents a computation unit during which total resource requirements don't change, the only significant events are task initiations and terminations. If the physical system consists of m resource types[*], it is possible to define for each task T_i a <u>request vector</u>[**] $\hat{q}_i = (q_{i1}, q_{i2}, \ldots, q_{im})$, representing the number of units of each resource type which must be allocated to the task T_i before its initiation, a <u>release vector</u> $\hat{r}_i = (r_{i1}, r_{i2}, \ldots, r_{im})$, representing the number of units of each resource type which are released to the system by task T_i on its termination, and a <u>transfer vector</u> $\hat{t}_{ij} = (t_{ij1}, t_{ij2}, \ldots, t_{ijm})$ for each $T_j \in S_i$, representing the number of units of each resource type which are transferred to task T_j by task T_i on its termination. We suppose that every terminated task holds no unit of resource, and hence the following relation must be satisfied, for $1 \leqslant j \leqslant n$:

$$\hat{q}_j + \sum_{\forall i : T_i \in P_j} \hat{t}_{ij} = \hat{r}_j + \sum_{\forall k : T_k \in S_j} \hat{t}_{jk} \qquad (2.1)$$

---

[*] Obviously we will consider only those resource types which satisfy all necessary conditions for deadlock [2].

[**] From here on, vectors will be distinguished by the symbol $\wedge$.

Given an execution sequence $\alpha = a_1 a_2 \ldots a_n$, a corresponding state sequence
$\sigma = s_0 s_1 s_2 \ldots s_{2n}$ can be defined, each state s_k specifying the amount of available
resources and the number of resources units allocated to and requested by each task
after the event a_k. Hence, each state s_k is defined by $2n+1$ vectors $\hat{v}(k)$, $\hat{P}_i(k)$
and $\hat{Q}_i(k)$, $1 \leq k \leq n$, where vector $\hat{v}(k)$ specifies the number of available units of
each resource type, and vectors $\hat{P}_i(k)$ and $\hat{Q}_i(k)$ specify respectively the number of
units of each resource type hold and requested by task T_i after the event a_k. Vectors
$\hat{P}_i(k)$ and $\hat{Q}_i(k)$ are defined as follows[*]:

1. $\hat{P}_i(k) = 0$ if $\exists j : (a_j = \underline{T_i}) \wedge (j \leq k)$

2. $\hat{P}_i(k) = \sum_{\forall \ell : T_\ell \in P_i^*} \hat{t}_{\ell i}$ if $\not\exists j : (a_j = \overline{T_i}) \wedge (j \leq k)$

       where P_i^* is the subset of P_i such that $T_\ell \in P_i^*$ iff

       $T_\ell \in P_i$ and $\exists h : (a_h = \underline{T_\ell}) \wedge (h \leq k)$.

3. $\hat{P}_i(k) = \sum_{\forall \ell : T_\ell \in P_i} \hat{t}_{\ell i} + \hat{q}_i$ if $\exists j : (a_j = \overline{T_i}) \wedge (j \leq k)$ and

       $\not\exists h : (a_h = \underline{T_i}) \wedge (h \leq k)$.

4. $\hat{Q}_i(k) = \hat{q}_i$ if $\exists h : (a_n = \underline{T_j}) \wedge (h \leq k)$ for each $T_j \in P_i$ and

       $\not\exists \ell : (a_\ell = \overline{T_j}) \wedge (\ell \leq k)$.

5. $\hat{Q}_i(k) = 0$ otherwise.

In the initial state s_o:

1. $\hat{v}(0) = \hat{R}$, the vector specifying the total amount of system resources.
2. $\hat{P}_i(0) = 0$ for $1 \leq i \leq n$.
3. $\hat{Q}_i(0) = q_i$ if T_i is an initial task, $\hat{Q}_i(0) = 0$ otherwise.

Given an execution sequence $\alpha = a_1 a_2 \ldots a_k \ldots a_{2n}$, event $a_k = \overline{T_i}$ is <u>allowable</u> if
$\hat{Q}_i(k-1) \leq \hat{v}(k-1)$. Obviously task terminations are always allowable events. An execution
sequence is <u>valid</u> if all initiation events are allowable.
We give now a very intuitive definition of deadlock:

<u>definition 1</u>: Let G be a system of n tasks, $\alpha = a_1 a_2 \ldots a_k$ a partial valid execution
sequence, I a set of tasks such that $T_i \in I$ iff $\hat{Q}_i(k) > 0$. State s_k contains a
<u>deadlock</u> if there is at least one task $T_i \in I$ such that it is impossible to find a
valid partial execution sequence $\alpha' = a_1 a_2 \ldots a_k a_{k+1} \ldots a_p$ defined as follows:

---

[*] From here on we define a vector $\hat{v} = (v_1, v_2, \ldots, v_j)$ to be a <u>null vector</u> $(\hat{v} = 0)$ if
$v_i = 0$ for $1 \leq i \leq j$; a <u>positive vector</u> $(\hat{v} > 0)$ is a not null vector such that $v_i > 0$
for $1 \leq i \leq j$; a <u>negative vector</u> $(\hat{v} < 0)$ is a not null vector such that $v_i \leq 0$ for $1 \leq i \leq j$.

$$1. \quad \alpha \text{ is prefix of } \alpha'$$
$$2. \quad a_p = \overline{T}_i$$

Task T_i is then said to be <u>deadlocked</u>.

Definition 1 says that the system contains a deadlock if we are unable to initiate a task which is ready to be executed, all its predecessors being terminated. Clearly, if a task is deadlocked, all its successors are also blocked, and a whole subgraph cannot be executed.

A deadlock detection procedure based directly on definition 1 is necessarily enumerative. Therefore, we must look for necessary and sufficient conditions which can be more easily verified. Such conditions are stated in the following theorem:

<u>Theorem 1</u>: Let G be a system of n tasks and $\alpha = a_1 a_2 \ldots a_k$ a partial valid execution sequence. Suppose that there is a non empty set D of indices such that for each i in D:

$$\hat{Q}_i(k) \not\leq \hat{v}(k) + \sum_{j \notin D \vee D^*} \hat{P}_j(k) \tag{2.2}$$

where D^* is the set of indices such that $\ell \in D^*$ iff T_ℓ is a successor of T_i, for any $i \in D$. Then s_k contains a deadlock and every task T_i, $i \in D$, is deadlocked.

Theorem 1 is an obvious generalization of Denning's definition, but its proof is quite cumbersome and so it is not discussed here (see Appendix). This theorem is interesting mainly because it proves that the definition of deadlock given in [2] can be slightly modified to cover the case of graph structured task systems and gives a simple method to detect deadlock situations. However, our specific goal is avoidance rather than detection, and to avoid deadlock each state must be checked for safeness, according to the following definitions:

<u>Definition 2</u>: Let $\alpha = a_1 a_2 \ldots a_k$ be a partial valid execution sequence: state s_k is safe if there exists at least one valid complete execution sequence having α as a prefix.

In the following sections a further step will be made, looking for sufficient conditions for safeness. We are restricting ourselves to sufficient conditions because we are unable to find necessary and sufficient conditions that are not essentially enumerative.

3. Tree structured task systems

We will consider at first a special case, that of a tree structured task system. This model can represent a situation in which a process, at a particular stage of its activity, starts the execution on one or more processes, which run independently of each other, starting in turn new sets of processes and so on.

A sufficient condition for safeness can be based on condition (2.2). In fact, suppose that if task T_i can be initiated, then the whole subtree G_i, having T_i as root, can be terminated; this means that there is at least one partial execution sequence, involving the whole set of tasks of subtree G_i, that is valid if the resources granted to T_i become available to its successors. In such a case, if no task asking for resources is deadlocked, then the system is in a safe state. These remarks may be formalized as follows:

<u>Theorem 2</u>: Let G be a tree structured task system, $\alpha = a_1 a_2 \ldots a_k$ a partial valid execution sequence and I a set of indices such that $i \in I$ iff the immediate predecessor of T_i is terminated but $\underline{T_i} \notin \alpha$. Each task T_i is the root of a subtree G_i. Let $\hat{M}_i^*$ be a vector representing the minimal amount of resources that must be allocated to subtree G_i to guarantee that all tasks belonging to G_i can be terminated. Let $\hat{Q}_i^*(k) = \hat{M}_i^* - \hat{P}_i(k)$, for each $i \in I$. If no set of indices $D \subset I$ can be found, such that, for each $i \in D$:

$$\hat{Q}_i^* \nleqslant \hat{v}(k) + \sum_{j \notin D} \hat{P}_j(k) \qquad (3.1)$$

then state s_k is safe.

<u>Proof</u>. If no set D exists satisfying (3.1), then at least one index $i_1 \in I$ exists, such that $\hat{Q}_{i_1}^* \leqslant \hat{v}(k)$ (otherwise a set $D = I$ would satisfy (3.1)). This means that the available resources suffice to complete subtree G_{i_1}. On completion, G_{i_1} release all allocates resources, which are $\hat{P}_{i_1}(k)$ plus all subsequently obtained units up to the maximum $\hat{M}_{i_1}^*$. Therefore, the amount of available resources is now $\hat{v}(k) + \hat{P}_{i_1}(k)$, but, by hypothesis, at least one index $i_2 \in I$ exists such that $\hat{Q}_{i_2}^* \leqslant \hat{v}(k) + \hat{P}_{i_1}(k)$; so subtree G_{i_2} can be completed and so on.

Condition (3.1) is only sufficient, as can be proved by a simple counter-example. Consider the following system:

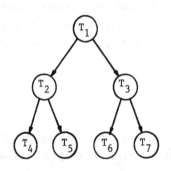

$\hat{R} = (3,3)$

$\hat{q}_1 = (2,2), \hat{r}_1 = (0,0), \hat{t}_{12} = \hat{t}_{13} = (1,1)$

$\hat{q}_2 = (1,1), \hat{r}_2 = (0,0), \hat{t}_{24} = \hat{t}_{25} = (1,1)$

$\hat{q}_3 = (1,1), \hat{r}_3 = (2,1), \hat{t}_{36} = (0,1), \hat{t}_{37} = (0,0)$

$\hat{q}_4 = (1,0), \hat{r}_4 = (2,1)$

$\hat{q}_5 = (2,1), \hat{r}_5 = (3,2)$

$\hat{q}_6 = (3,1), \hat{r}_6 = (3,2)$

$\hat{q}_7 = (2,2), \hat{r}_7 = (2,2)$

Let $\alpha = \overline{T}_1 \underline{T}_1$ be the partial execution sequence. Then state s_2 is defined by:

1. $\hat{v}(2) = (1,1)$
2. $\hat{P}_2(2) = \hat{P}_3(2) = (1,1)$, $\hat{P}_i(2) = (0,0)$ for $i \notin \{2,3\}$
3. $\hat{Q}_2(2) = \hat{Q}_3(2) = (1,1)$, $\hat{Q}_i(2) = (0,0)$ for $i \notin \{2,3\}$
4. $\hat{M}_2 = \hat{M}_3 = (3,2)$
5. $\hat{Q}_2(2) = \hat{Q}_3(2) = (2,1)$

It can be easily verified that the set of indices $D = \{2,3\}$ satisfies (3,1) but the system is not deadlocked, because the sequence

$$\alpha' = \overline{T}_1 \underline{T}_1 \overline{T}_3 \underline{T}_3 \overline{T}_2 \underline{T}_2 \overline{T}_4 \underline{T}_4 \overline{T}_5 \underline{T}_5 \overline{T}_6 \underline{T}_6 \overline{T}_7 \underline{T}_7$$

is a valid execution sequence.

The central point in theorem 2 is the evaluation of the vectors $\hat{M}^*$'s.
A truly minimal vector is quite difficult to define for two main reasons. First, it is not guaranteed that an unique minimal vector exists; in the above example we could define a different vector $\hat{M}_3^{*'} = (2,3)$, which is as minimal as $\hat{M}_3^*$, because if two units of the first type and three of the second type were allocate to subtree G_3 all tasks of G_3 could be completed, and $\hat{M}_3^{*} \not> \hat{M}_3^*$ (as well as $\hat{M}_3^{*} \not> \hat{M}_3^{*'}$). Secondly, the problem of finding a vector $\hat{M}_i^*$ for subtree G_i is essentially the same as sequencing all the tasks of a tree to verify if, for a given set of system resources, the initial state is safe. To be sure that a set of resources is minimal, we must prove that substracting even one unit of resource we make the initial state unsafe; this test could be performed in a non enumerative way if necessary conditions for safeness were known, but theorem 2 gives us only sufficient conditions. Therefore, to avoid enumerative methods, we will look for a minimal vector $\hat{M}_i \geq \hat{M}_i^*$, trying to define $\hat{M}_i$ as close as possible to $\hat{M}_i^*$. A recursive method to compute $\hat{M}_i$'s can be based on the following theorem:

Theorem 3: Let G a tree structured task system, G_i a subtree having task T_i as root, $\{G_{i_1}, G_{i_2}, \ldots, G_{i_n}\}$ the set of subtrees having as roots the immediate successors of T_i, and T_s the immediate predecessor of T_i.
Let I be the set of indices such that $j \in I$ iff $T_j \in S_i$. Let $\hat{v}_i$ be a vector of resources such that no set of indices $D \subset I$ can be found having the following property:

$$\hat{M}_j - \hat{t}_{ij} \not< \hat{v}_i + \sum_{k \notin D} \hat{t}_{ik} \quad \text{for each } j \in D \tag{3.2}$$

Choose $\hat{v}_i$ such that no vector $\hat{v}_i' < \hat{v}_i$ can be found which satisfies the above requirement. Then

$$\hat{M}_i = \hat{v}_i + \hat{t}_{si} + \hat{q}_i - \hat{r}_i \tag{3.3}$$

The proof is trivial and will be omitted.
The computation of $\hat{M}_i$ can be performed only once for every subtree G_i of G, because $\hat{M}_i$ is independent of the specific execution sequence, as stated by (3.3) (in fact, no

reference is made to any state of the sequence σ). To compute $\hat{M}$'s the best approach is to find the immediate predecessors of terminal tasks, each identifying a subtree, to evaluate the resource requirements of this first set of subtrees, and to go on in this manner reducing step by step the tree to its root.

4. Graph structured task systems

In the case of graph structured task systems theorem 2 cannot be applied because of the complexity of precedence relations; therefore a different approach to safeness verification is needed.

Informally, a criterion for safeness can be stated as follows. Given a task system G, suppose that a partial valid execution sequence α_k has been found, containing the terminations of all tasks at a level greater than j; suppose also that, if all tasks at a level j are terminated, then a valid completion sequence can be found. Then the state s_k is safe if no task at level j is deadlocked. By recursion a general rule to verify safeness can be stated, which implies the possibility of executing the task system level by level.

Consider, for instance, the following task system

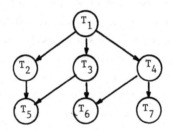

and suppose it can be executed level by level, that is the following execution sequence is valid:

$$\alpha^*: \quad \overline{T}_1 \underline{T}_1 \overline{T}_2 \underline{T}_2 \overline{T}_3 \underline{T}_3 \overline{T}_4 \underline{T}_4 \overline{T}_5 \underline{T}_5 \overline{T}_6 \underline{T}_6 \overline{T}_7 \underline{T}_7$$

Consider now the partial execution sequence $\alpha: \overline{T}_1 \underline{T}_1 \overline{T}_2 \underline{T}_2 \overline{T}_5$; the activation of task $\overline{T}_5$ does not lead to an unsafe state only if the amount of available resources after its termination allows us to execute tasks T_3 and T_4.

We will now formalize the above criterion, proving at first the following lemma.

lemma 1: Let G be a system of tasks, $\alpha = a_1 a_2 \dots a_k$ a partial valid execution sequence I a set of indices such that $i \in I$ if $\hat{Q}_i(k) > 0$. Suppose $\hat{q}_i - \hat{r}_i \gtrless 0$ for each $i \in I$. Let $I_p \subset I$ be the subset of I such that $i \in I_p$ iff $\hat{q}_i - \hat{r}_i < 0$ and $I_n \subset I$ the subset of I such that $j \in I_n$ iff $\hat{q}_j - \hat{r}_j \geqslant 0$. If no set of indices $D_p \subset I_p$ exists such that, for each $i \in D_p$:

$$\hat{q}_i \not< \hat{v}(k) + \sum_{\substack{j \in I_p \\ j \notin D_p}} (\hat{r}_j - \hat{q}_j) \qquad (4.1)$$

and no set of indices $D_n \subset I_n$ exists such that, for each $i \in D_n$:

$$\hat{r}_i \not< \hat{v}(k) + \sum_{j \in I_p} (\hat{r}_j - \hat{q}_j) + \sum_{j \in D_n} (\hat{r}_j - \hat{q}_j) \qquad (4.2)$$

then a partial valid completion sequence α_1 exists, such that $\underline{T}_i \in \alpha_1$ for every $i \in I$.

Proof. If no set D_p exists such that (4.1) is satisfied, there is at least one task T_{i_1} , $i_1 \in I_p$, such that $\hat{q}_{i_1} \leqslant \hat{v}(k)$; otherwise a set $D_p \equiv I$ would exist. This means that task T_{i_1} can be initiated with the currently available resources. Let I_{p_1} be the subset of I_p made of all indices $i \in I_p$ except i_1 . No set $D_p = I_{p_1}$ exists such that (4.1) is satisfied; therefore there is at least one task T_{i_2} , $i_2 \in I_{p_1}$, such that $q_{i_2} \leqslant v(k)+$ $-q_{i_1} + r_{i_1}$. But the quantity $\hat{v}(k) - \hat{q}_{i_1} + \hat{r}_{i_1}$ represents the amount of available resources after the termination of T_{i_1}, and hence task T_{i_2} can be initiated after the termination of T_{i_1}. Proceeding in this manner it is possible to prove that a valid partial execution sequence $\alpha_2 = \overline{T}_{i_1} \underline{T}_{i_1} \overline{T}_{i_2} \underline{T}_{i_2} \ldots$ exists, such that $\underline{T}_i \in \alpha_2$ for each $i \in I_p$. After the execution of all tasks T_i, $i \in I_p$, the amount of available resources is

$$\hat{v}' = \hat{v}(k) + \sum_{j \in I_p} (\hat{r}_j - \hat{q}_j) > \hat{v}(k)$$

The execution of each task T_i, $i \in I_n$, does not increase the available resources, being $\hat{r}_i - \hat{q}_i \leqslant 0$. If no set D_n exists such that (4.2) is satisfied, there is at least one task T_{ℓ_1} such that

$$\hat{q}_{\ell_1} \leqslant \hat{v}' - \sum_{\substack{j \in I_n \\ j \neq \ell_1}} (\hat{q}_j - \hat{r}_j)$$

otherwise a set $D_n = I_n$ would exist. Therefore task T_{ℓ_1} can be initiated with the resources which are available after the execution of all tasks T_ℓ, $\ell \in I_{n_1}$, being $I_{n_1} =$ $= I_n - \{\ell_1\}$. But no set $D_{n_1} \equiv I_{n_1}$ exists, so that at the end a partial valid execution sequence $\alpha_3 = \ldots .\overline{T}_{\ell_2} \underline{T}_{\ell_2} \overline{T}_{\ell_1} \underline{T}_{\ell_1}$ can be found, such that $T_\ell \in \alpha_3$ for each $\ell \in I_n$. Therefore, $\alpha_1 = \alpha_2 \| \alpha_3$.

Conditions (4.1) and (4.2) are also necessary if the set of possible sequences α_1 is limited to sequen ces containing no events $\overline{T}_i$ or $\underline{T}_i$ with $i \notin I$. The proof is trivial and will be omitted. Also the following corollary can be trivially proved:

corollary 1. Let G be a system of tasks, $\alpha = a_1 a_2 \ldots a_k$ a partial valid execution sequence, I a set of indices such that $i \in I$ iff $\hat{Q}_i(k) > 0$. Suppose that a subset $I' \subset I$ exists such that $\hat{q}_i - \hat{r}_i \not\gtrless 0$ for each $i \in I'$. For each $i \in I'$ substitute $\hat{r}_i$ with a new vector $\hat{r}'_i = (r'_i = (r'_{i1}, r'_{i2}, \ldots, r'_{im})$ such that $r'_{ij} = r_{ij}$ if $q_{ij} \geqslant r_{ij}$, and $r'_{ij} = q_{ij}$ if $\hat{q}_{ij} < \hat{r}_{ij}$. Then lemma 1 can be applied but conditions (4.1) and (4.2) are never necessary.

Suppose now that a task system G can be executed level by level, that is a valid execution sequence $\alpha^* = a_1^* a_2^* \ldots a_{2n}^*$ exists such that if $a_i^* = \underline{T}$, $a_j^* = \overline{T}'$ and $\ell(T) > \ell(T')$, then $i < j$. Let $\alpha = a_1 a_2 \ldots a_k$ be any valid partial execution sequence, S' a set of integers such that $s \in S'$ iff:

1. $\exists i : \overline{T}_i \in \alpha \qquad \ell(T_i) = s$, and

2. $\exists j : \overline{T}_j \notin \alpha$ and $\ell(T_j) = s$.

and S a set of integers such that $s \in S$ iff $\min\{S'\} \leqslant s \leqslant \max\{S'\}$. For each $s \in S$ let I_s be a set of indices such that $i \in I_s$ iff $\overline{T}_i \notin \alpha$ and $\ell(T_i) = s$. Modify every vector $\hat{r}_i$, $\forall i \in I_s$, $\forall s \in S$, according to the assumptions of corollary 1, if $\hat{q}_i - \hat{r}_i \not\gtrless 0$. Let I_{sp} be the subset of I_s such that $i \in I_{sp}$ iff $\hat{q}_i - \hat{r}_i < 0$, and I_{sn} be the subset of I_s such that $i \in I_{sn}$ iff $\hat{q}_i - \hat{r}_i > 0$. Under these assumptions, the following basic theorem can be proved:

<u>Theorem 4</u>: For every $s \in S$, if no set of indices $D_{sp} \subset I_{sp}$ exists, such that, for each $i \in D_{sp}$:

$$\hat{q}_i \not\leqslant \hat{v}(k) - \sum_{\substack{s' \in S \\ s' > s}} \sum_{t \in I_{s'}} (\hat{q}_t - \hat{r}_t) + \sum_{\substack{j \in I_{sp} \\ j \notin D_{sp}}} (\hat{r}_j - \hat{q}_j) \qquad (4.3)$$

and no set of indices $D_{sn} \subset I_{sn}$ exists, such that, for each $i \in D_{sn}$:

$$\hat{r}_i \not\leqslant \hat{v}(k) - \sum_{\substack{s' \in S \\ s' > s}} \sum_{t \in I_{s'}} (\hat{q}_t - \hat{r}_t) + \sum_{j \in I_{sp}} (\hat{r}_j - \hat{q}_j) + \sum_{j \in D_{sn}} (\hat{r}_j - \hat{q}_j) \qquad (4.4)$$

then the state s_k is safe.

<u>Proof</u>. Let s_{max}, s_{min} be respectively the maximum and minimum of set S. By assumption, for every $i \in I_{s_{max}}$, if $T_j \lessdot T_i$, then $\overline{T}_j \in \alpha$; therefore task T_i is ready to be initiated or will be ready after a finite time interval. In any case, the initiation of T_i is constrained only by the availability of resources. But, by lemma 1, if conditions (4.3) and (4.4) are satisfied, a partial valid execution sequence $\alpha_{s_{max}}$ can be found, such that $\underline{T}_i \in \alpha_{s_{max}}$ for every $i \in I_{s_{max}}$. Hence all tasks at level $s_{max} - 1$ will be

ready for initiation after a finite time form event a_k. But, by lemma 1, if conditions
(4.3) and (4.4) are satisfied, a partial valid execution sequence $\alpha_{s_{max}-1}$ can be
found, such that $\underline{T}_i \in \alpha_{s_{max}-1}$ for every $i \in I_{s_{max}-1}$, and so on. Therefore, if conditions (4.3) and (4.4) are satisfied, all task at a level $s \geqslant s_{min}$ can be completed with
the currently available resources; let $\alpha_1 = \alpha_{s_{max}} \| \alpha_{s_{max}-1} \| \dots \| \alpha_{s_{min}}$ be the va-
lid partial completion sequence containing the terminations of all tasks T_i, for
$i \in \bigcup_{s \in S} I_s$. Consider now the partial execution sequence α_2 containing all and only
all the terminations of all task T such that $\bar{T} \in \alpha$ but $\underline{T} \notin \alpha$; a termination event
does not decrease the amount of available resources, and therefore a new valid partial
execution sequence α' can be built merging α_1 and α_2. If $\alpha^*_{s_{min}}$ is the prefix
of the valid execution sequence α^* such that $\underline{T} \in \alpha^*_{s_{min}}$ for every T such that $\ell(T) \geqslant$
$\geqslant s_{min}$ and $\underline{T'} \notin \alpha^*_{s_{min}}$ for every T' such that $\ell(T') < s_{min}$, then the two sequences
$\alpha \| \alpha'$ and $\alpha^*_{s_{min}}$ contain exactly the same events; therefore, if α'^* in the comple-
tion sequence such that $\alpha^* = \alpha^*_{s_{min}} \| \alpha'^*$, then α'^* is also a valid completion sequence
for the partial execution sequence $\alpha \| \alpha'$, that is $\alpha \| \alpha' \| \alpha'^*$ is a valid execution se-
quence, and consequently the state s_k is safe.

The last problem is to find sets D_p and D_n in lemma 1, D in theorems 1 and 2, and D_{sp}
and D_{sn} in theorem 4. This problem is easily solved because:

1. All sets can be searched independently of each other.
2. The search always involves a set of tasks which, upon termination, do not increase
   (decrease) the amount of available resources; therefore the order of allowable initiation events is not important [2].

Then the simplest algorithm for the detection of sets D_p and D_n in lemma 1 is the
following:

Algorithm 1:

Step 1: detection of set D_p
   1. Initialize $D_p = I_p$ and $\hat{v} \leftarrow \hat{v}(k)$.
   2. Search for any index $i \in I_p$ such that $\hat{q}_i \leqslant \hat{v}$; if none is found goto 4.
   3. $D_p \leftarrow D_p - \{i\}$; $\hat{v} \leftarrow \hat{v} + (\hat{r}_i - \hat{q}_i)$; goto 2.

Step 2: detection of set D_n
   4. Initialize $D_n = I_n$ and $\hat{v} \leftarrow \hat{v} + \sum_{j \in I_n} (\hat{r}_j - \hat{q}_j)$.
   5. Search for any index $i \in I_n$ such that $\hat{r}_i$ $\hat{v}$; if none is found terminate
      the algorithm.
   6. $D_n \leftarrow D_n - \{i\}$; $\hat{v} \leftarrow \hat{v} + (\hat{q}_i - \hat{r}_i)$; goto 5.

If the number of tasks in $I_p \cup I_n$ is z, the running time of the algorithm is $O(m \cdot z^2)$.
Using more sophisticated techniques, it is possible to speed up the algorithm, reaching

a running time of $O(m \cdot z)$ [2]. Sets D_{sp} and D_{sn} in theorem 4, and set D in theorems 1 and 2 can be found in a very similar fashion.

5. Conclusions

The formulation of the deadlock detection and avoidance problem made by Coffman and Denning has been extended here to cover the more general case of a graph structured task system. The proposed approach preserves the basic property of the simpler models developed for independent chains of tasks, i.e. it allows the definition of non-enumerative avoidance algorithms. The conditions on which such algorithms are based are only sufficient, and therefore it can happen that a resource request is not accepted even if it could be granted without entering an unsafe state; however, without this approximation, we are unable to design algorithms which are not essentially enumerative.

APPENDIX

Proof of theorem 1. Suppose that at least one index $i \in D$ exists such that it is possible to find a partial valid execution sequence $\alpha' = a_1 a_2 \ldots a_k a_{k+1} \ldots a_p$, α being a prefix of α' and $a_p = \overline{T}_i$. Let the sequence $\alpha'' = a_{k+1} \ldots a_p$ be a subset of α'. Suppose also that α'' does not contain the initiation of any task T_j, $j \in D$. If this is not the case, and a set of events $\{a_{k_1}, a_{k_2}, \ldots, a_{k_p}\}$ exists such that $a_{k_s} \in \alpha''$ and $a_{k_s} = \overline{T}_{i_s}$ for $i_s \in D$, $1 \leqslant s \leqslant p$, it is always possible to impose new values to p and i such that $a_k = a_{k_1}$, $i = i_1$ and $\alpha'' = a_{k+1} \ldots a_{k_1-1}$. Task T_i is initiated, hence $\hat{Q}_i(p-1) \leqslant \hat{v}(p-1)$. Let A be the set of indices such that $j \in A$ iff α'' contains the event $\overline{T}_j$, and B the set of indices such that $j \in B$ iff α'' contains the event $\underline{T}_j$; then:

$$\hat{v}(p-1) = \hat{v}(k) + \sum_{j \in B} \hat{r}_j - \sum_{\ell \in A} \hat{q}_\ell =$$

$$= \hat{v}(k) + \sum_{j \in B'} \hat{r}_j + \sum_{\ell \in B \wedge A} \hat{r} - \sum_{u \in A'} \hat{q}_u - \sum_{s \in B \wedge A} \hat{q}_s =$$

$$= \hat{v}(k) + \sum_{j \in B'} \hat{r}_j - \sum_{u \in A'} \hat{q}_u + \sum_{\ell \in B \wedge A} (\hat{r}_\ell - \hat{q}_\ell)$$

being $B' = B \wedge (B \wedge A)$ and $A' = A \wedge (\overline{B \wedge A})$. If $\hat{P}_{\ell, fin}$ is a vector specifying the total amount of resources held by task T_ℓ after its initiation (and before its termination), then we can write:

$$\hat{v}(p-1) \leqslant \hat{v}(k) + \sum_{j \in B'} \hat{r}_j + \sum_{\ell \in B \wedge A} (\hat{r}_\ell + \hat{P}_\ell(k) + \sum_{s \in B} \hat{t}_{s\ell} - \hat{P}_{\ell, fin}) =$$

$$= \hat{v}(k) + \sum_{j \in B'} (r_j + \sum_{s \in B \wedge A} \hat{t}_{js}) + \sum_{\ell \in B \wedge A} (\hat{r} + \sum_{s \in B \wedge A} \hat{t}_{s\ell} +$$

$$+ \hat{P}_\ell(k) - \hat{P}_{\ell, fin}) \leqslant$$

$$\leqslant \hat{v}(k) + \sum_{j \in B'} \hat{P}_j(k) + \sum_{\ell \in B \wedge A} \hat{P}_\ell(k) = \hat{v}(k) + \sum_{j \in B} \hat{P}_j(k)$$

Therefore it can be stated that:

$$\hat{Q}_i(k) = \hat{Q}_i(p-1) \leqslant \hat{v}(k) + \sum_{j \in B} \hat{P}_j(k)$$

By definition $B \wedge D = \emptyset$, and consequently $B \wedge D^* = \emptyset$; hence $\hat{Q}_i(k)$ $\hat{v}'k) + \sum_{j \in D \wedge D^*} \hat{P}_j(k)$, which contradicts the theorem.

So far we have proved that (2.2) is a sufficient condition for deadlock detection. We will now prove that it is also necessary. Let D be a set of indices such that $i \in D$ iff no valid partial execution sequence α_i can be found leading to initiation of T_i and having, as a prefix, α (every task T_i, $i \in D$, is deadlocked). Then, in the best case, it is possible to find a partial valid execution sequence having α as a prefix and containing the termination of all tasks T_j, $j \notin D \vee D^*$. Let $\alpha' = a_1 a_2 \ldots a_k a_{k+1} \ldots a_p$ be such a sequence. Being T_i, $\forall i \in D$, a deadlocked task, $\hat{Q}_i(p) \leqslant \hat{v}(p)$, and no task T_j exists such that $\hat{Q}_j(p) \leqslant \hat{v}(p)$, because only tasks T_i, $i \in D$, are asking for resources in state s_p. Therefore set D satisfies condition (2.2). This means that, if state s_k contains a deadlock, a state s_p, $p \geqslant k$, will be necessarily found, in which condition (2.2) is satisfied.

6. References

Coffman, E.G.jr., Elphick, M.J., Shoshani, A. "System Deadlocks", Computing Surveys 2, (1971), 67-78.
Coffman E.G.jr., Denning, P.J. "Operating Systems Theory", Prentice Hall, 1973.
Habermann, A.N. "Prevention of System Deadlock", Comm. ACM 12, (1969), 373-377.
Havender, J.W. "Avoiding Deadlock in Multitasking Systems", IBM Syst. J. 7, (1968), 74-84.
Hebalkar, P.G. "A Graph Model for Analysis of Deadlock Prevention in Systems with Parallel Computation" , Proceed. IFIP Congress (1971), 168-172.
Howard, J.H.jr. "Mixed Solutions for Deadlock Problem", Comm.ACM 16, (1973), 427-430.
Llewellyn, J.A. "The Deadly Embrace - a Finite State Model Approach", Computer Journal 16, (1973), 223-225.

SYNCHRONIC ASPECTS of DATA TYPES :

Construction of a non-algorithmic solution of the Banker's problem.

Piero R. Torrigiani

Gesellschaft für Mathematik und Datenverarbeitung mbH Bonn
Institut für Informationssystemforschung
Schloss Birlinghoven - Postfach 1240
D-5205 St. Augustin 1

Abstract

By means of two simple data structures, namely the "cell" and the "interrupt", increasingly complex structures are defined. The specification of these structures is given in a novel extension of the path notation which allows to study the synchronic characteristics of the types we define. These structures are shown to be sufficient to specify a non-algorithmic solution to dynamic - Banker's like allocation problems.

0. Introduction

The specification of systems involving concurrency is a hard task. Among the available tools, we shall use the object-oriented notation for path expressions presented in [1] : it uses a SIMULA like "class" concept and the syntactic extensions of the path notation introduced by P.E. Lauer and the author in the 1-st part of [2] . Interesting characteristics of this notation appear when they are used for the description of synchronic aspects of data structures. In this paper, by means of composition and modification of two simple structures, we construct a non-algorithmic solution to the problem of dynamically allocating units of a resource. The solution is non-algorithmic in the sense that it is described by a system rather than by an algorithm: no computation is performed in order to decide the order of allocation for concurrent requests, nor any global state of the system drives any decision making. An algorithmic representation is implicitly motivated

by considerations that have to do with a particular implementation.
This, invariably obscures the concurrency which is inherent to the problem and which might be present in the solution. Hence we feel that it is
important to obtain as a first step, a representation which is in the
form of an abstract, concurrent system. For example, in the case of the
Banker's problem a solution in the form of an algorithm, imposes a global synchronization or the requests for resources, which originate from
essentially concurrent processes. This restriction arises from the nature of the scheme used for describing the solution. In our representation, this restriction will be absent.

In the next sections the notation is shortly introduced and explained
through those examples which will be then used in the last section to
construct the solution to the Banker's problem.

1. The notation

Programs in our notation consist of class definitions, class instantia-
tions, paths and processes. As we are interested in the synchronic aspects of the structure defined by the classes, their definition will be
only in terms of path expressions [3]; the paths defining a structure
will state the possible orders of operations that an instantiated class
provides to its environment i.e. to the processes and to the other system components. Some operations defined into a class may be possibly
used only for internal synchronisation, hence they should not be known
to the environment; this is obtained by means of a special declaration,
called operation part, in the class where all exportable operations of
the class (separated by commas) are listed and specified.

Informally we can say that paths (and eventually processes) are grouped
together into a class to which a name is given; a use of this name in
the body of a program generates an instance of that class which is also
thereby given a name; the generation of an instance of a class consists
in replacing the instantiation or call of that class by a copy of the
paths (and eventually processes) which constitute it, after properly
prefixing the instance-name to the occurrences of operation names in
these paths (and processes). The generation of instances is assumed to
take place at compilation time; after compilation programs will appear
in the usual path notation. The semantic of path programs is given in
terms of transition nets in [4] and we will show the nets that our programs produce. (In the figures showing the nets, the double lines belong to the processes, the dotted lines include instances of classes
and exportable operations. When unnecessary the internal paths of instances are not shown). The grammar defining our notation can be found

in the appendix A2 in [1] .

2. Examples

Some examples should serve as introduction to the notation. In fig. 1
we see that the class specifies a very simple data structure consisting
of a single "cell" where two operations are defined as exportable namely "deposit" and "remove"; the path which associates these operations
(fig. 2) states that every "remove" on any instance of that cell must
follow a previous "deposit" on the same instance of that cell, and that
the first operation, again on any instance must be a "deposit". Mutual
exclusion between processes accessing (both for reading and writing) is
thus achieved together with the fact that no information will be lost.
Various extensions of this structure can be found in [2] where path
programms are given which define various buffers and buffer usage.
Other applications of our notation are studied in [5,6,7] .
In the previous example nothing has been said about a fair use of the
cell; fairness is by itself an interesting problem and the notation we
use for our class definition has been proved useful in this sense [1,2] .
We have to mention that our approach is more similar to ideas concerning the use of arbiters than to those where fairness is obtained by the
use of no arbiters [8,9] i.e. we are going to have no restriction on
the relative speeds of a limited non-fixed number of "parallel" processes and we are going to have an arbitration hypothesis to be used in
conflict resolution; (cf. [1] 3-rd part):
"If two or more operations in a path are mutually excluded those which
are shared by processes have priority (of activation) over the others".
This hypothesis doesn't solve the starvation of the previous example as
it doesn't distinguish between processes in a conflict; in the next
section this problem is shortly described.

3. Sequencing through data definition.

Data structures can be defined in such a way that their operations provide the means for sequencing processes on other data. Figure 3 shows
the structure needed for the "fair" sequentialization of an operation
on a cell.
For those who are already familiar with the macro path notation, the
that   example should be quite clear. In any case, we wish to recall
how the arbitration hypothesis is needed for the correct behaviour of
this scheme; i.e. in order to guarantee that, for $1 \leq j \leq k$, the conflict
between "INT(j).skip" and "INT(j).accepted"will be always solved in
favour of the last after "INT(j).request" took place. (see fig. 3,4).
It is important to notice that now we have the means to distinguish

```
class cell;
  operation deposit, remove endop;
  path deposit; remove end;
endclass;
...
cell X;
...
process...;X.deposit;... end;
process...;X.deposit;... end;
process...;X.remove;... end;
```

Fig. 1

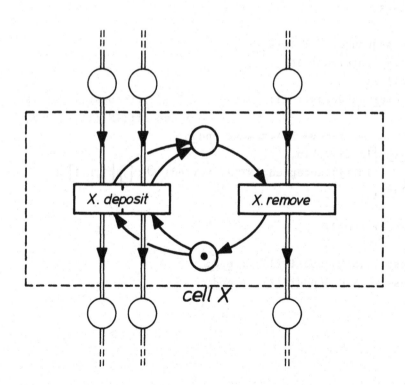

Fig. 2

with the help of indices the processes themselves.
Other discipline of sequentialization of operations, e.g. priority
sequentialization, have been considered extensively in [1] ; here we
need only to consider the following path, which, when substituted in
the sequencer would yield a simple priority scheme:

path (INT(1).skip
      ;(INT(i).skip i), (INT(i).accepted;INT(i).served) 2,k,1),
                (INT(1).accepted;INT(1).served) end;

class interrupt;
    operation skip, request, accepted, served endop;
    path skip, (request;accepted;served) end;
endclass;
...
class sequencer (k: integer);
    array interrupt INT (k);
    cell X;
    operation deposit (i: integer) = INT(i).request;INT(i).accepted;
                                    X.deposit;INT(i).served,
              remove = X.remove endop;
    path [INT(j).skip,
          INT(j).accepted;INT(J).served) ∂ ; [j] |1,k,1] end;
endclass;
...
sequencer SEQ(2);
...
process...;SEQ.deposit(1);...end;
process...;SEQ.deposit(2);...end;

Fig. 3

565

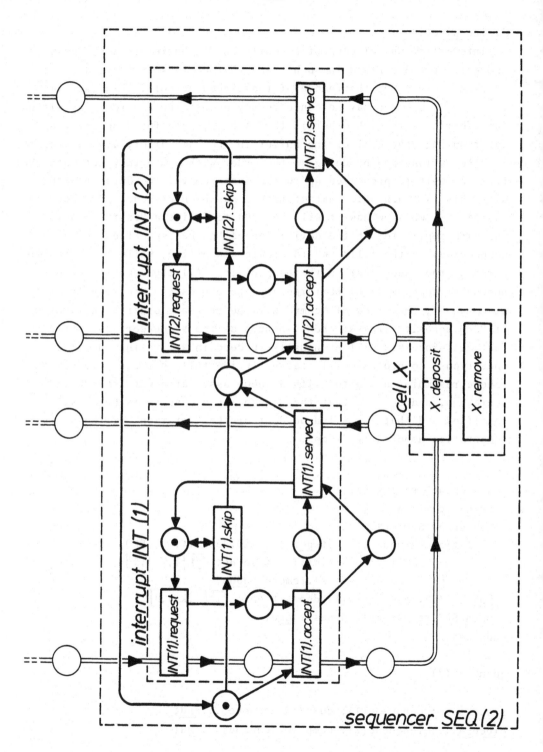

Fig. 4

4. The pipeline

An interesting way of connecting cells into a buffer is a scheme where
the cells are organised in such a way that every successive deposit
must be preceded by the remove of the preceding cell; this fact implies
that more processes may be concurrently depositing and removing without
interferring with each other if they access, both for depositing and
for removing, the cell of the buffer in the same order, and each of them
in alternation, and by doing so the order with which they activate the
first deposit is preserved up to the last remove. The data structure
with this a synchronic characteristic has been called a pipeline,
in fig. 5-6 we show how it will be used to define a "fair" queue.
The path replicator in the definition of "queue" specifies the interconnections of the cells "W" to create a pipeline, while the path syn-
chronize the "sequencer" with the pipeline. All the computation denoted by "comp" will be performed by the processes in a mutually exclusive manner; the order of "comp" will be the same order of activation
of "SEQ.deposit" (any of them), hence, as the sequencer guarantees
that all processes will eventually succeed in activating one of such
operations, no process will starve. Incidentally one can note how this
scheme resembles a monitor with a queue associated with the "wait"
operation, and observe that the sequencing mechanism is here explicitly
specified and that it is independent of the "duration" of the protected
computation denoted by "comp".

```
class queue (k: integer);
   array (cell) W (k);
   sequencer SEQ (k);
   operation enqueue (i: integer ) = SEQ.deposit(i); SEQ.remove;
           [W(i).deposit;W(i).remove ∂ ; [i] |1,k-1,1] ; W(k).deposit
              dequeue = W(k).remove endop;
   [path W(i).remove;W(i+1).deposit end; [i] |1,k-1,1]
   path SEQ.remove;W(1).deposit end;
endclass;
...
queue Q (2);
...
process...;Q.enqueue(1);"comp";Q.dequeue;...end;
process...;Q.enqueue(2);"comp";Q.dequeue;...end;
```

Fig. 5

567

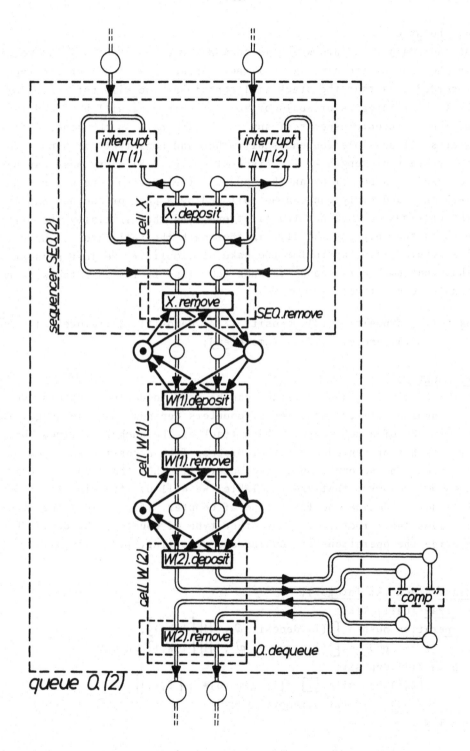

Fig. 6

5. The stack

Another data structure which is commonly used is the stack; as we did
in the case of the queue we will not concentrate at all on the type of
recorded data that the stack will contain and we will not hence define
a stack of integers or reals or so; our main goal will be the definition
of the synchronic aspects of any stack. The path in the stack definition
states all possible sequences of pushes and pops that can happen to an
empty stack of length k; i.e. (see also fig. 7,8) every push can be
followed by a pop or by another push if the previous one was not the
k-th one, and every pop can be followed by a push or another pop if
the previous one was not the first one, and that initially the first
push is the only possibility. In that path the replicator is used in
its general form; again for the sake of simplicity we just present
what that path corresponds to when the class is instantiated with k
equals, e.g., three:

path Z(1).deposit; (Z(2).deposit; (Z(3).deposit;Z(3).remove) ;
        Z(2).remove) ; Z(1).remove end;

6. Counters

Both the pipeline and the stack can be used to define counters. We shall
distinguish between two types of counters depending on the usage. The
first type of a counter will keep track of the number of times some
process has prevented some other process from accessing some common
resource. The second type of a counter will show the number of units of
a common resource that are available. We will use the pipeline scheme
in order to define the first type of counter. (fig. 9,10). The stack
provides the structure for the other type of counter. The cells "R"
provide the operations of reading and resetting. (fig. 11,12).

class stack (k: integer);
    array (cell) Z(k);
    operation push = [Z(i).deposit ⍺ , [i] [1,k,1],
            pop = [Z(i).remove ⍺ , [i] | 1,k,1] endop;
    path Z(1).deposit;
        [Z(i).deposit; [i] Z(i).remove)* ; | 2,k,1]
                Z(1).remove end;
endclass;

Fig. 7

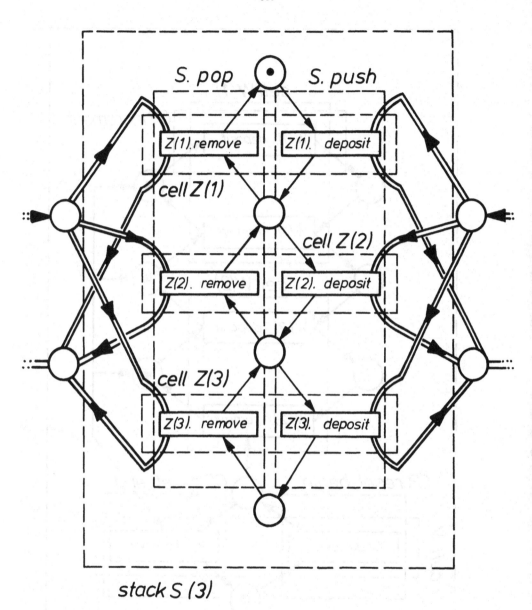

stack S (3)

Fig. 8

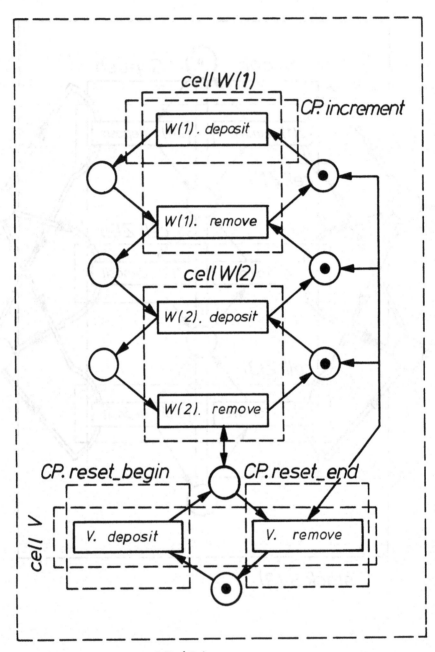

Fig. 9

```
class counter_up_to (k: integer);
  array (cell) W (k);
  cell V;
  operation reset_begin = V.deposit,
            reset_end   = V.remove
            increment   = W(1).deposit endop;
 [path W(i).deposit;reset_end* ;W(i).remove end 𝒶;  ⅈ | 1,k,1];
 [path W(i).remove;reset;reset_end* ;W(i+1).deposit end 𝒶; ⅈ|1,k-1,1];
path V.deposit;W(k).remove * ;V.remove end;
endclass;
```

Fig. 10


```
class readable_counter (k: integer):
  array (cell)  W(k);
  array (cell)  R(k-1);
  operation increment = [ W(i).deposit 𝒶 , ⅈ | 1,k,1],
            decrement = [ W(i).remove  𝒶 , ⅈ | 1,k,1],
            read(1: integer) = R(1+1).deposit endop;
  path (R(k+1).deposit;R(k+1).remove),
  [(W(i).remove;((R(i).deposit;R(i).remove)* ⅈ ) * ;W(i).deposit)
     𝒶, | k,1,-1] end;
endclass;
```

Fig. 11

7. A more complex example

With the apparatus we have we are ready to construct a system for the
dynamic allocation of a resource to concurrent processes. It is worth
remembering that the current programming languages (e.g. Concurrent
Pascal) do not give the programmer tools for constructing solutions to
this kind of problems, in spite of the extensions that have been recently proposed.[11,12] . A nice example of this kind of a problem is
well-known Banker's problem which is due to Dijkstra. The solutions that
have been proposed so far [13,14,15] are basically sequential solutions
to a problem which inherently involves concurrency. Briefly stated,
these solutions analyze the "present" states of a fixed number of concurrent processes to determine if there is a "sequence in which" these

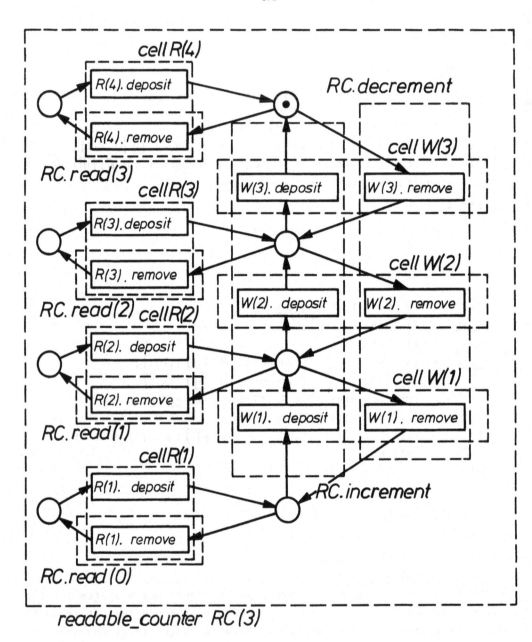

Fig. 12

"processes can be completed, one at a time if necessary".([15] p.124).
Consequently, a state in the Banker's problem is defined to be "safe if
it is possible for the Banker to enable all his present customers to
complete their transactions within a finite time." ([15] p.43). This
means that in these solutions what is guaranteed is that the present
customers will complete their transactions. We shall now develop a
solution to the Banker's problem which will permit an arbitrary and
varying number of customers to concurrently enter and leave the "problem
space". Our solution will also guarantee freedom from starvation. We
shall build stepwise a system which will be by construction always in
a safe state. (We will not allow any process to ask for loans bigger
than the capital of the bank). In order to do so let us make first some
simple observations: a) we have to avoid deadlock, b) we have to avoid
starvation. It is true that b) implies a), but the converse, in general,
is not true. Deadlock is consequence of "wrong" choices, and it can be
solved by forcing the banker to make the "right" ones; but because of
the unknown rate of requests of loans the right choice could indefinite-
ly favour some processes over others violating b). Let us first observe
the program shown in figure 13. It shows a bank and its customers; the
state of the vault of the bank is represented by the readable counter C.
Access to the vault (both for getting and giving back units, i. e.
C.decrement and C.increment respectively) is protected by the interrupts
RQ (ReQuest) and RL (ReLease), which may be viewed as the "cash-counters"
of the bank. The Banker's behaviour is shown in figure 14.a. He applies
a deadlock-free sequentialization of the operations on the vault. In
simple words, one could imagine our banker sitting on the token in
figure 14.a and moving from window to window in the bank serving the
customers; from this figure we can see easily that he tries first his
best to collect money into the vault, and only when he cannot do it any
longer, looks at how much he has got and accordingly moves to the
"correct" counter and accomplishes the requests of his customers. The
Banker gives a single unit (enables C.decrement through the activation
of RQ(j).accepted for some j) to exactly one of those who asks for an
amount less than or equal to his current cash creating a new situation
where for sure a customer (possibly the same one) will get units
guaranteeing that at least one customer will get his last unit and will
give the loan back. In fact in that program the long path is a modification of the path of the sequencer; here first of all there are two arrays
of interrupts to be visited: the "RL" ones used for incrementing the
counter, and the "RQ" ones used for decrement it (cf. the operation
part); secondly we observe that the main cycle of the path is the one
on the "RL" interrupts, which are visited in a priority manner

```
class bank (k: integer);
  array (interrupt) RQ,RL (k);
  readable_counter A (k);
    operation get_loan(i: integer) =
        RQ(i).request; RQ(i).accepted;C.decrement;RQ(i).served,
            give_back (i:integer) =
        RL(i).request;RL(i).accepted; [C.increment ∂] ; [z] |1,i,1] ;
        RL(i).served
endop;
```

```
path  [(RL(I).skip [; [ (C.read (1-1)
      [ ;(RQ(j).skip, (RQ(j).accepted;RQ(j).served))
            [j| 1-1,1,-1] ) ∂, [1] | k+1,1,-1] [z] k+1-i,k,-1] [i] ),
        (RL(i). accepted;RL(i).served)) | k,1,-1]   end;
endclass;
...
bank B (3);
...
process...;B.get_loan(3);B.get_loan(2);B.get_loan(1);...
        ...;B.give_back(3);...end;
process...;B.get_loan(2);B.get_loan(1);...
        ...;B.give_back(2);...end;
process...;B.get_loan(1);...;B.give_back(1);...end;
```

Fig. 13

(cf. section 3); when all "RL.skip"'s are activated in a row all "read" will be enabled and depending on the contents of the counter the "QR" interrupts are visited one after the other starting from the i-th one if the counter's content was i. See also figure 14.a,14.b. Due to the fact that the processes are concurrent with one another it may happen (in the "worst" case) that only the process asking for a loan of one unit goes on making the others waiting forever; in fact the other two processes conflict with each other on the request of the secon unit, and the three of them conflict on the third. Again in simple words we could say that at the counter it is true that our Banker sees only one customer at a time and it is true that he (the Banker) is not serving a "wrong" customer, but only from his point of view: i.e. he is sure he will always have enough money to proceed in his task, but the confusion at the cash-counters may be such that the Banker is actually serving a "wrong" customer from a customers' point of view; e.g. at the

readable_counter C(3)

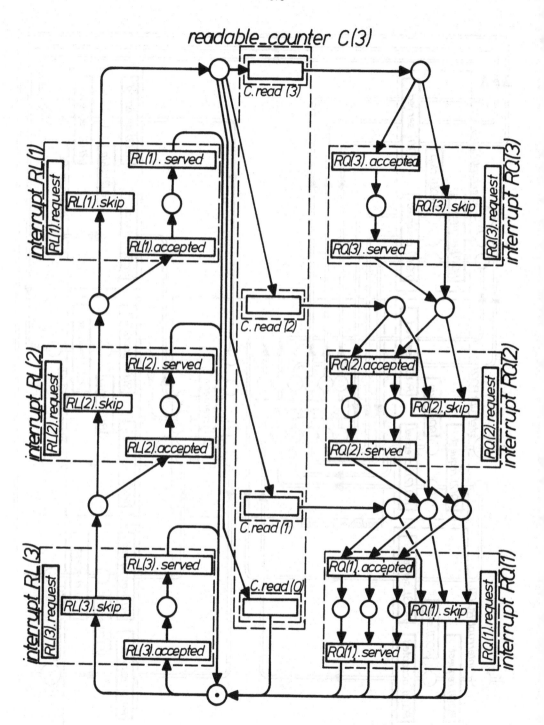

Fig. 14.a

576

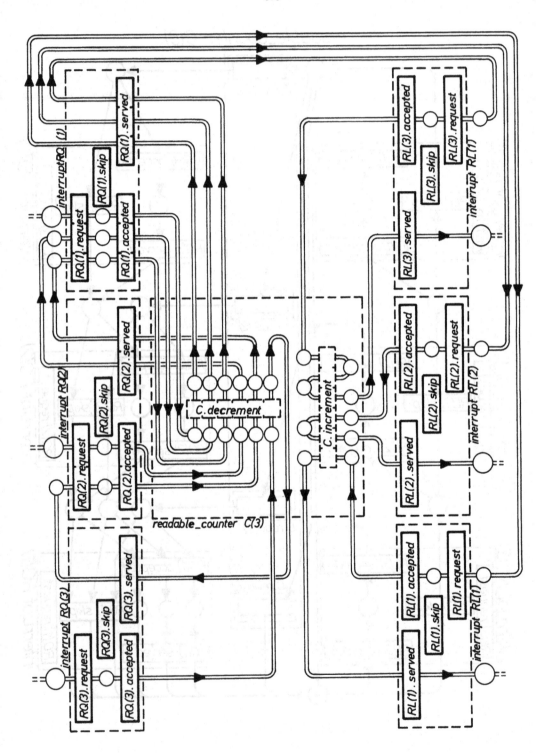

Fig. 14.b

577

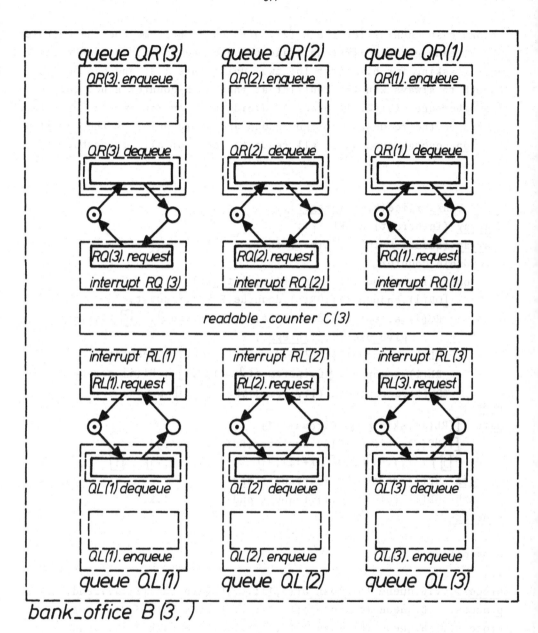

Fig. 15

578

second cash-counter (at the RQ(2).interrupt) there may be the customer
who already got his first unit and the one who has not got anything yet
and is asking for his first (of two) unit; this last one can be fast
enough to always win the conflict at that cash-counter hence letting
the other one starve. In order to avoid this conflict we can use "fair"
queues on the requests for the second and third unit. By the way if we
put long enough queues for all requests (and for all giving back pos-

```
class bank_office(k,m: integer);
  array (interrupt) RQ,RL (k);
  array (queue) QR(m-1),QL(m-1) (k);
  readable_counter C (k);
    operation queue_for_loan_of(i,j:integer)
        [QR(1).enqueue(j);QR(1).dequeue;RQ(1).request;
        RQ(1).accepted;C.decrement;RQ(1).served ];   [1] | i,1,-1],
            give_back(i,j:integer) =
        QL(i).enqueue(j);QL(i).dequeue(j); RL(i).request;
        RL(i).accepted; [C.increment]; z 1,i,1] ; R1(i).served
        (RL(i),accepted;RL(i).served))|k ,1,-1]  end;
endop;
path  [(RL(i).skip [;[ (C.read(1-1)
      [;(RQ(j).skip, (RQ(j).accepted;RQ(j).served))
        [j] | 1-1,1,-1]] ], [1] | k+1,1,-1] z  |k+1-i,k,-1]   [i]  ),
  path  QR(i).remove;RQ(i).request end;
  path  QL(i).remove;LQ(i).request end   ;  i  1,k,1  ;
endclass;
...
bank_office B (3,6);
...
process...;B.queue_for_loan_of(3,1);...;B.give_back(3,1)...end;
process...;B.queue_for_loan_of(3,2);...;B.give_back(3,2)...end;
process...;B.queue_for_loan_of(2,3);...;B.give_back(2,3)...end;
process...;B.queue_for_loan_of(2,4);...;B.give_back(2,4)...end;
process...;B.queue_for_loan_of(1,5);...;B.give_back(1,5)...end;
process...;B.queue_for_loan_of(1,6);...;B.give_back(1,6)...end;
```

Fig. 16

sibilities) we can allow that more than one process can concurrently
request the same ammount of units without this starvation. The program
is shown in fig. 16. The differences with the previous program consists
only in the introduction of the queues QR and QL of the obligation to
the processes to use their operations before interrupting the bank,
and of the paths which connect the queues to the interrupts. See also
figure 15. In that program whenever a process has been accepted the
first interrupt it is not yet guaranteed that all its other interrupts
will be accepted and served, not because of "unfair" conflict resolution
(this is in fact taken care of by the queues) but because the speeds
of the processes may be such that (after the initial situation) the
counter will never reach again high values; i.e. the Banker will never
have again the whole capital and will not be able to guarantee in a
finite time high requests. In order to avoid this it is necessary that
from time to time the bank office refuse to serve new customers and
terminates serving those who have already got part of the loan or which
are by the way "inside" the bank office. This must be done with some
care. If we simply "lock out" of the bank some new customers, when we
open it again no guarantee can be given to all that they can in fact
enter it before the new closing time! Actually the closing period is
going to be period in which new loans request should be waiting while
those which had been started are completed, hence all we need is some
more space to let the new coming customers wait and a way to distinguish
between new and old customers. This is achieved by substituting the
queues "QR" with a more complex mechanism composed by two fair queues,
one for the old and one for the new customers, and a special sequencer
composed out of two normal sequencers and two interrupts, the first of
which is connected to a counter, in such a way that all customers which
go through that interrupt will be counted and eventually blocked
up to when the reset on that counter is done, the programs in
figure 17,18 show this mechanism: the substitution of the queues with
the new ones in the bank office produce the programm in figure 19 which
completely solves the problem. In fact in that programm one can notice
that only the "new" customers increment the counter, (cf. the operation
part of figure 19,18 and the paths in figure 17) that is every process
will be counted only once i.e. when he is going to obtain the first
unit of his loan; as the reset of all counters is done every time the
readable counter is found to held is maximum value, (after the activation
of C.read(k+1) and before visiting all QR queues: cf. the long path in
figure 21) the Banker let new customers wait only when he judges that
he served too many people without having had his whole capital back.
(This judgement is simply a function of the length of the counters).

```
class multiple_sequencer_and_counter(j,k,m: interger);
  cell M; cell N;
  array sequencer SEQ(k)  (j);
  array counter_up_to B(m) (j-1);
  array interrupt NT(j);
  operation reset_begin = N.deposit,
            reset_end = N.remove,
            deposit(i,j) = SEQ(j).deposit(i); SEQ(j).remove;
                           NT(j).request; NT(j).accepted;
                           M.deposit; NT(j).served,
            remove = M.remove endop;
  path [ NT(i).skip, (NT(i).accepted;NT(i).served ∂ ;
       [i] | 1,j,1] end;
  [ path SEQ(i).remove;NT(i).request end ∂ ;  [i] |1,j,1] ;
  [ path NT(i).request;B(i).increment end ∂ ;  [i] |1,j-1,1];
  [ path N.deposit;P(i).reset_begin;B(i).reset_end;N.remove end ∂;
       [i] | 1,j-1,1]
endclass
```

Fig. 17

```
class double_queue (k,m: integer);
  array (cell) F (k-1); queue QOLD. (k); queue QNEW  (k);
  multiple_sequencer_and_counter MSEQ(2,k,m);
  operation reset_begin = MSEQ.reset_begin;
            reset_end = MSEQ.reset_end,
            enqueue_and_count(i: integer) = QNEW.enqueue; QNEW.dequeues
            MSEQ.deposit(1,i); MSEQ.remove; [F(1).deposit;F(1).remove∂;
            [1] |1,K-2,1];  F(k-1).deposit,
            enqueue (i: integer) = QOLD.enqueue; QOLD.dequeue;
                                   MSEQ.deposit(2,i);MSEQ.remove;
                              [F(1).deposit;F(1).remove ∂ ;
                     [1] | 1,k-2,1] ; F(k-1).deposit,
            dequeue = F(k-1).remove endop;
  path SEQ.remove;F(1).deposit end;
  [ path F(i).remove;F(i+1).deposit end ∂ ; [i]|1,k-2,1] ;
endclass;
```

Fig. 18

```
class new_bank_office(k,l,m: integer);
  array (interrupt) RO,RL (k);
  array (double_queue) QR(1-1,m) (k);
  array (queue) QL(1-1) (k);
  readable_counter C (k);
  cell R;
      operation queue_for_loan!of(i,j:integer) =
          QR(i).enqueue!and_count{j);QR(i).dequeue;RO(i).request;
          RQ(i).accepted;C.decrement;RQ(i).served;
          [QR(1).enqueue(j);QR(1).remove;RQ(1).request;
          RQ(1).accepted;C.decrement;RQ(1).served𝔞;  [1|i-1,1,-1],
              give_back(i,j:integer) =
          OL(i).enqueue(j);OL(i).dequeue(j); RL(i).request;
          RL(i).accepted;[C.increment𝔞; [z]|1,i,1] ; RL(i).served
endop;
path  [(RL(i).skip
    [;[ (C.read(1-1)[ [ ;R.deposit;R.remove[z]|2k+1-1,k+1,1]
      ;(RQ(j).skip, (RQ(j).accepted;RQ(j).served))
      [j]|1-1,1,-1])𝔞, [1]|k+1,1,-1][z]|k+1-i,k,-1] [i] ),
      (RL(i).accepted;RL(i).served))| k-1,1,-1] end;
  [path QR(i).remove;RQ(i).request end;
  path QL(i).remove;LQ(i).request end;
  path R.deposit;OR(i).reset!begin;OR(i).reset_end;R.remove end
      𝔞; [i]|1,k,1];
endclass;
...
bank_office B (3,6);
...
process...;B.queue_for_loan_of(3,1);...;B.give_back(3,1)...end;
process...;B.queue_for_loan_of(3,2);...;B.give_back(3,2)...end;
process...;B.queue_for_loan_of(2,3);...;B.give_back(2,3)...end;
process...;B.queue_for_loan_of(2,4);...;B.give_back(2,4)...end;
process...;B.queue_for_loan_of(1,5);...;B.give_back(1,5)...end;
process...;B.queue_for_loan_of(1,6);...;B.give_back(1,6)...end;
```

Fig. 19

8. Conclusions

Through the simple synchronic characteristics of two structures, through their collection into classes and their connection by paths, new structures have been defined whose synchronic properties define the use of commonly used data types (queues, stacks, buffers) as well as of new types (sequencers, etc.). With this technique a well known allocation problem has been solved, taking into account the specific difficulties it involves: namely the avoidance of deadlock, and the avoidance of the two types of starvation. The first type of starvation overcome by "fair" sequentialization of equivalent conflicting operations, the second by the proper use of counters which could also be adjusted (playing with their length) to optimize the system behaviour accordingly to the expected request-distributions.

References

[1] P.R. Torrigiani, P.E. Láuer: An object oriented notation for path expressions, in AICA 77, Vol.3, pp.349,371, Pisa, 1977.

[2] P.E. Lauer, P.R. Torrigiani: Towards a system specification language based on paths and processes, Computing Laboratory, University of Newcastle upon Tyne, Technical Report Series, N., 1976.

[3] R. Campbell: Path Expressions: a techniques for specifying process synchronization, Ph.D. Thesis, University of Newcastle upon Tyne, August, 1976.

[4] P.E. Lauer, R. Campbell: Formal semantics for a class of high-level primitives for coordinating concurrent processes, acta informatica 5, 1975, pp. 247,332.

[5] R. Devillers: Non starving solutions for the Dining Philosophers problem, ASM/30, Computing Laboratory, University of Newcastle upon Tyne, 1977.

[6] R. Devillers, P.E. Lauer: Some solutions for the Reader/Writer problem, ASM/31, Computing Laboratory, University of Newcastle upon Tyne, 1977.

[7] P.E. Lauer, M.W. Shields: Abstract specification of resource accessing disciplines: adequacy, starvation, priority and interrupts,

Workshop on <u>Global description methods for synchronization in real-time applications</u>, AFCET Paris, 1977.

[8] K. Lautenbach: Ein kombinatorischer Ansatz zur Beschreibung und Erreichung von Fairness in Scheduling-Problemen, in <u>Applied Computer Science</u>, Hanser-Verlag, München, 1977.

[9] C.A. Petri: Modelling as a communication discipline, in <u>3rd international Symposium on Modelling and Performance Evaluation of Computer Systems</u>, Bonn, Oct. 1877.

[10] R. Devillers, P.E. Lauer: A general mechanism for the local control of starvation: application to the dining philosophers and to the reader/writer problem, ASM/32, Computing Laboratory, University of Newcastle upon Tyne, 1977.

[11] A. Silberschatz, R.B. Kieburtz, A. Bernstein: Extending Concurrent Pascal to allow dynamic resouree management, in <u>Proceedings of the 2nd international conference on Software Engineering</u>, San Francisco, 1976.

[12] P. Ancillotti, M. Boari, N. Lijtmaer: Dynamic management in a language for real time programming, in <u>AICA 77</u>, Vol.1, pp.335-348, Pisa, Oct. 1977.

[13] E.W. Dijkstra: De Bankiers Algorithme, EWD116, Math. Dep. Technological u., Eindhoven, The Netherlands, 1965.

[14] A.N. Habermann: Prevention of system deadlocks, in <u>CACM</u> 12, N.7, 1969.

[15] P. Brinch Hansen: <u>Operating System Principles</u>, Prentice Hall Series in Automatic Computation, Englewood Cliffs, 1973.

The Impact of Technology on Information System

W.K. Liebmann

IBM Laboratories, Böblingen, Germany

1. Introduction

An information system receives information in the form of data and commands, it analyzes this information against a preprogrammed set of algorithms, and it issues the results of this analysis - again in the form of data and commands - for further use. This general definition of the tasks of an information system leads automatically to a description of its major hardware components. The system needs input equipment to receive information, it needs a data processing unit to execute the analytical portion of the task, and it needs output equipment to inform the user of the result of the analysis.

To perform the analytical tasks, the data processing unit usually requires fast access to a large collection of background information related to a particular problem. In modern information systems, these background data are stored and organized in data banks. Thus, in addition to input, output and data processing equipment, external storage facilities for data banks become the fourth major hardware component of an information system.

The major hardware components are linked together by equipment designed to facilitate the communication between them. Fig. 1 shows schematically such a basic hardware configuration of an information system.

The base technology for all information system components is quite different. The input/output devices rely heavily on mechanical or display technologies, the external storage devices rely almost exclusively on magnetic media technologies, while the central processing units (CPU) are essentially an embodiment of the silicon large scale integrated circuitry (LSI).

In analyzing the impact of technology on the various components it is, however, quickly noted that also in input/output and in external storage devices, LSI is the major technological driving force, and that in CPU, input/output and external storage alike, information systems progress is most dramatically stimulated through the very rapid evaluation of the silicon integrated circuit technology.

In assembling a report on the impact of technology development
on information systems it appears thus most prudent to first evaluate the state of the art and the most likely evolutionary directions of the silicon integrated circuit technology and then - with
this background - proceed to an impact assessment of the major information system hardware components.

2. The development of the silicon integrated circuit technology

Silicon - in the form of a great variety of oxides abundent in
the earth crust - has a number of physical and chemical properties,
which make it uniquely suited for the design and the fabrication of
integrated circuits. Silicon is a semiconductor whose band gap of
1.1 eV is such that through addition of appropriate doping elements
electronic devices structures, like transistors, diodes resistors,
capacitors, can be obtained which operate efficiently at ambient
temperatures, for instance in the range between -20 and 100°C.
Silicon can easily be (grown into) the form of very pure, defect
free single crystals, is machinable with standard tools to obtain
very perfect surfaces of a particular crystal orientation and is
mechanically stable enough to withstand handling in a modern technology environment. Most important, however, is the fact that, when
placed into an oxidizing atmosphere at elevated temperature, a very
dense silicon oxide layer grows on the surface of the semiconductor,
which adheres well to the base material and which forms an effective diffusion barrier to most of the doping elements which are of
technical interest with respect to the generation of the desired
electronic properties in the bulk semiconductor. Such elements can
only penetrate into the semiconductor surface where the oxide layer
has been selectively removed and where the bare semiconductor surface has been exposed to the doping atmosphere. Hydrofluoric acids
are efficient etchants to remove the oxide layers with very little
attack to the base material. A variety of photoresist materials is
available which are resistant to the attack of the hydrofluoric
acids. By microphotographic means a pattern can be transposed unto
the surface where certain sections of the oxide are shielded from
acid attack through the photoresist, while in other sections the
oxide has been exposed in the photoresist developing process and
can then be removed in the subsequent etch.

The combination of all these properties has rendered to the silicon
an exceptional position among all materials which find application

in modern technology, and if one looks at its pervasiveness in our high technology world and assesses its impact on our total lives, then it might very well be appropriate to call our time the age of silicon.

Silicon semiconductor components made their entry into commercial application in the fifties; the first integrated circuits, where several components were functionally interconnected on the same silicon chip a few square millimeters in area appeared in the sixties. Today, several thousand circuit elements are integrated into the same area.

The driving force which brought about such rapid technological development was of course entirely economic in nature. The basic manufacturing unit which is processed through the semiconductor lines of today's electronic component manufacturers, is a "wafer" a slice of silicon 75 to 125 mm in diameter (Fig. 2). In a given environment, the cost of processing such a wafer through the various diffusion, oxidation, evaporation and photoresits steps is essentially a constant, independent of the amount of electronic function contained on the wafer. Thus, the more circuit functions can be integrated into a certain silicon area, the cheaper the individual function is going to be.

In driving towards higher and higher integration densities in order to optimize production cost, several very positive side effects appear. Device dimensions and distances between devices become smaller, thus reducing capacitances and carrier travel times, all leading to better circuit performance or power/performance. The number of silicon chip to package connections per functional unit decreases, which in turn leads to better unit reliability, and the number of silicon chips, which must be packaged to obtain a certain function, and consequently their packaging cost decreases.

There are two forces which counteract the rapid progress of LSI: one stems from the fact that in designing so many functions into one individual unit like a silicon chip, the design complexity increases so rapidly that design errors are bound to occur, which will only be detected in a functionality test of the completed hardware, and which lead to expensive and time-consuming design iterations.

The second force stems from the increasing test complexity since a great number of LSI circuits, whose functionality must be individually verified in test, are only accessible through a very small number of input/output contacts. Both sectors, the design and the test, however, are amenable to data processing aids, and with continually improved price/performance of DP equipment, both obstacles to design progress can be overcome efficiently. LSI-designs can be software-simulated and error checked before they are modelled in hardware, which in effect eliminated the need for hardware engineering change recycle. With appropriate combinatorial analysis and iterative test patterns, also the most complex array of internal functions can be tested to a satisfactory level through just a few input/output pads. As an example: At the end of our semiconductor manufacturing line approximately 500 000 test patterns are applied through 23 pads to IBM's 2000 bit random access monolithic memory chip [1]. The test robot requires a few milliseconds to perform this test, which identifies the good and the bad chips on a wafer. Approximately 30 000 test combinations are required to establish functionality a chip with a few hundred circuits and 100 pads; such test patterns can be generated automatically from a particular logic design and can be applied to the chip in a small fraction of a second.

Invention an evolution in three areas stimulated the fast progress of the silicon integrated circuit technology:

1. The improvement of photolithographic dimensions. With today's photolithographic equipment and environmental control, geometrical shapes with dimensions approaching the wave lengths of visible light can be transposed into a semiconductor wafer. Electronbeam Lithography and y-ray lithography [2] [3] will extend the dimensional capabilities downward by orders of magnitude. With both technologies the desired device pattern will be directly enscribed into the photoresist-covered silicon wafer (instead of first generating a photomask from which the pattern is transferred to the wafer), and both technologies incorporate the potential to register subsequent exposures to the previous-one automatically which leads to a minimization of the dimensional tolerances that must be provided to account for misregistration of one photomasklayer against the previous one.

2. The integrated advancement of semiconductor processes and semi-

conductor device design. Ten years ago a circuit designer, a
semiconductor device designer and a semiconductor process designer were three independent agents who communicated with each
other through a set of rules: Today, they work as one team, often
this team is integrated by one person with detailed knownledge and
experience in all three areas, and synergism takes place to advance
the total state of the art faster than the contribution of the
individual components would otherwise have permitted. Typical
examples of this synergism is the one device FET (field effect
transistor) dynamic monolithic memory cell [4] as shown in Fig. 3.
This invention, which is the basis for practically all of today's
cost-performance FET monolithic memory chips was stimulated by
increasing the bit-density on the silicon to increase the bit
productivity and decrease the monolithic memory cost. It leads to
a bit cell area which is only 8 unit square, where the dimensional
unit is the minimum photolythographic line width of the particular
semiconductor process. 4 unit square is already required to achieve
a pattern of two lines crossing each other. Its practical implementation depends on a semiconductor technology which produces sufficiently low surface and junction leakage currents to retain the
capacitive charge at the storage node.

In modern semiconductor processes this control of leakage currents
is so good that the time intervals after which the stored charge
must be replenished are very long compared to the time required to
actually recharge the capacitor, which means that leakage current
and recharge considerations have essentially become negligible
factors in the operation of one device monolithic memory arrays.
Similarly, the desire for smaller and smaller cell areas and
consequently less and less stored charge aroused the inventiveness
of many circuit engineers to design very efficient FET sense circuits with which the small signals originating from just a 100 000
or so electrons could be reliably sensed [7].

Another example for this synergism is the superintegration of MTL
(Merged Transistor Logic) [6]. Fig. 4 shows the semiconductor layout and the circuit schematic of a bipolar MTL - random access
static memory cell. This cell requires only 30 unit squares of
silicon area, is considerably faster than the FET - one device cell
and is static, which means: it maintains its charge as long as the
power of the system is turned on and does not require refresh

during the operation. The dense layout was made possible with a bipolar semiconductor process which provided inverse n-p-n transistors
and lateral p-n-p transistor of sufficiently good characteristics
to obtain cell functionality and stability, a circuit design which
replaced the large area current limiting resistors with lateral
p-n-p transistors and a semiconductor layout highly adaptive
to the particular structure of the bipolar semiconductor technology design. Integrated circuits have come a long way from their
beginnings where the individual components of a semiconductor circuit were all merely buried side by side into the silicon and then
interconnected at the silicon surface.

3. The design of reliable metal/insulator systems for dense, low
   impedance interconnections on the silicon chip itself and between
   the chip and its carrier. Particular examples are the use of double
   layer polysilicon for interconnection and functionality of FET circuits [8], the use aluminum/quartz systems for multilayer chip interconnections [9] and the chrome-copper-gold-lead-tin system for chip
   to carrier interconnection [10].

The LSI progress has been most dramatic in the area of monolithic
memories since the functional and structural regularity of a memory
array is most compatible with the capabilities of the semiconductor
technology. Monolithic memories entered the data processing market
on a large scale with the introduction of the monolithic main
memories in the IBM systems 370-135 and 145 [9], which were announced
to the market in 1971 and 1970 respectively. The storage unit then
was a silicon chip containing 128 bits on about 20 mm². Today,
16K bits integrated on essentially the same area are available in
large productions quantities [8], the introduction of 64K bit chips
is announced [12], and serial readout monolithic memory chips in
CCD (Charge-Coupled-Devices) implementation carrying 128K bits are
available. At the high performance end, 1K bit chips with chip
access time of 10 Nsec are available.

Performance and density can be traded against each other over a
wide performance range as shown in Fig. 5 where the cell area of
certain monolithic memories is shown as a function of chip access time.

Projecting into the future the serial access monolithic memories (e.g. CCD`s) will continue their density lead by a factor 2 to 3 over random access memories because they are less leakage current sensitive and allow the design of simpler on-chip signal sensing circuits. In both memory types, random access and serial memories, the limits of the capabilities of the silicon technology are far from exhausted, and the rapid increase in memory chip density will continue, especially since after the introduction of electronbeamlithography no further physical boundaries to further advancement are apparent. The ultimate limits of RAM density is, of course, a subject of professional speculation. Liebmann [a] estimates that this limit in silicon will be approximately 10^7 bits/cm^2, while Mitterer [13] suggests that we can reach 10^9 bits/cm^2. In either case their is plenty of opportunity for further density improvement, which will result in further dramatic memory cost reduction, if the monolithic memory market has the power to absorb all the memory bits. But, right now, there is no saturation in sight. Even with the most dramatic cost reductions, however, it remains questionable whether the monolithic memories will ever be a real cost competitor for the disk storage devices which will probably continue to show better bit costs by an order of magnitude.

The most interesting competing technology to monolithic memories are the magnetic bubble memories [14]. In certain materials with a strong magnetic anisotropy (Orthoferrites, Hexaferrites, Garnets), small cylindrical domains can be formed through the application of an external magnetic field, where the magnetization direction is opposite from the magnetization direction of the bulk material. Fig. 6 shows schematically a magnetic bubble in a slice of anisotropic material. The bubble can be moved by applying externally a magnetic field gradient: it will move into the direction of the lowest external field. The presence of a magnetic bubble at a certain place at a certain time can then be the definition for a digital "one", the absence of the bubble can be digital "zero".

To initiate motion of the bubble along an exactly predetermined path, the gradient fields are structured along the surface by certain patterns shaped from soft magnetic materials, like for instance permalloy. Fig. 7 shows the motion of magnetic bubbles under the influence of an external rotating field in a "T"-bar environment.

The bubble memories are serial access memories by the nature of their storage mechanism. Their performance is gated by the mobility of the bubble through the bulk material. It will always be considerably slower than the performance of semiconductor memories. There is no inherent density - and thus probably cost advantage - of the conventional magnetic bubbles vis-à-vis semiconductor memories; bit-densities of $10^8/cm^2$ are probably achievable. This limit may be extended to considerably higher densities with the "Bubble Lattic File" [16] where the bubble size can be reduced to the order of magnitude of the crystal lattice of the base material itself. This concept, however, has currently only an advanced technology character.

The main advantage of bubble memories over semiconductor memories is the fact that they are non-volatile read-write memories, which retain their information content also if the power supply current is switched off. They will find initially their major application in those areas where small quantities of non-volatile storage are required, too small to afford the high entry cost of a rotating disk storage device. This could be in point of sale terminals, in small distributed processors which only periodically are linked to a larger central data processing installation, in portable equipment etc. It is very unlikely that the bubbles will displace the monolithic memories from computer main store or control store applications. Even though the bubbles are considerably slower than monolithic memories, they are much faster than rotating disk storage and as such good candidates to take over part of the "fixed head file" market, especially in a hierarchical memory organization (see chapter 4).

Silicon chips for logic applications today carry several hundred to several thousand circuits. In these logic chips depending on the design methodology an interesting trade-off between design effort and manufacturing cost can be made. This trade-off capability is characterized through the three basic design principles by which a certain logic function can be implemented with silicon integrated circuits.

1. A logic design, which represents the desired system function in an optimum manner, is translated circuit by circuit into a semiconductor layout. The result is a semiconductor design of very high circuit density and very good performance. The chip, however, does only represent the particular single function in a "custom design", and is not adaptable to other functions. Any change in the function or any error in circuit or semiconductor design will

result in a full E/C (engineering change) cycle, with associated
time delay and cost. The custom design approach relies heavily on
density and performance optimization through "hand-honing" and it
is thus not very amenable to computer automatization. The design
of such a single part number consequently is expensive and only
worth while if the total quantity of pieces which are to be manufactured in this custom design is very large.

2. On the opposite side of the semiconductor design spectrum is the
"master slice" approach. The attempt is made here to generate an
universal semiconductor layout which can implement every desired
logic function. This is achieved by generating a standard array
of logic circuits of the silicon chip and by providing means to
interconnect these circuits differently for each application. The
advantage here is - of course - that the semiconductor design only
must be done once for a multitude of part-numbers and that in the
event of an E/C only the interconnection pattern on the surface
is affected. The disadvantage that the universality of the circuit
design and the flexibility required to wire widely varying functions results in a loss of circuit density on the chip, and thus
in higher manufacturing cost per circuit. The master slice approach is most efficient in the case of many different part-numbers
where the quantity of the individual part number is low.

3. The third approach is the implementation of logic functions with
the help of a "microprocessor" where the attempt is made to combine the advantages of good manufacturing cost from custom design
and high volume for a single part number with the flexibility to
implement different logic functions. A microprocessor is a small
computer, consisting of a central arithmetic unit, associated
control and required input output circuit, scaled down so that
everything fits on one silicon chip. In that sense the microprocessor is not a fixed function, but will develop into more powerful computing units as the semiconductor technology progresses,
as could be seen from the appearance of originally one bit microprocessors to now 16 bit microprocessors [17]. The function of the
microprocessor is adapted to the particular logic task through an
appropriate micro-control program which is stored in an accompanying read-write or read-only store. In manufacturing cost, the
microprocessor ranges between the custom design which optimizes
circuit utilization, and the masterslice which optimizes flexibi-

lity. The major advantage of the microprocessor is, however, that
it clearly separates the function of the design of a silicon integrated circuit chip from the implementation of the function,
which the microprocessor later is to perform, and thus permits to
take advantage of the functional and cost benefits of the silicon
integrated circuit technology without the need to be a semiconductor expert or without disclosing the details of the application
to the semiconductor designer or manufacturer. The semiconductor
producer, on the other hand, can sell one standard part to a wide
variety of customers, with the cost advantages brought through
economy of scale and without the need to worry about the application details of his customers. Resulting from this almost ideal
situation for all parties concerned is a rapid expansion of microprocessor usage into all technical sectors of our lives.

In silicon, it appears quite feasible to integrate 2×10^7 logic circuits
per cm^2. The circuit count of very large data processing units, like
for instance the large IBM /370 CPU's (Central Processing Units) is
in the order of magnitude of 10^5 circuits. Comparison of semiconductor
capability and circuits required to implement certain functions show
that soon we will have very powerful processing units on one silicon
chip. Since there are no homogeneous islands of several thousand circuits in logic designs, these highly dense logic chips will carry a
functional mixture of logic circuits interdispersed with read-only and
read-write arrays or with circuits which facilitate input/output communication or testing. The master-slice, custom design and microprocessor design approach will thus merge, and on future logic chips we
will find master-slice like sections in areas with a high frequency of
engineering changes (for instance: control circuits) and there will be
custom design for other areas like data flow logic or imbedded arrays.
In any event, the "microprocessor" of the late nineteen-eighties might
very well represent a data processing power of several MIPS (million
instructions per second) on a single chip.

In summary, the technological progress of LSI will continue. There is
no saturation of the market or the technological capabilities in sight.
Also, there is no real competitive technology to displace the silicon
from its predominant role. For logic and memory of very large central
electronic complexes the "JOSEPHSON-TUNNEL DIODES" [18] might eventually
qualify. With their limited operating range at superconducting temperatures, it is difficult to imagine that Josephson-devices will ever
find the pervasiveness which silicon devices enjoy today.

3.   Input/Output

The interface between the user of an information system and the
system itself is the input/output equipment. The user could of
course be a machine or a robot, but for the purpose of this discussion it will be assumed that the user is a human being. This
human user interacts with the information system in a request/
response mode: he addresses a problem statement to the user and
expects after a reasonable length of time an answer. In an interactive man-computer environment (and I will limit my comments to
this environment), this answer will normally be used to formulate
the next problem statement, which receives again an answer, and so
on until in moving from general to specific the desired information is obtained. For such interactive operation an interactive
input/output terminal is required, with whose help the person can
communicate its problem statement to the information system, and
which can present the system's answer in a manner a human being
can understand. For practical purposes, the only mechanism at our
disposal to communicate to the system are our tactile facilities,
the most efficient way to receive the answers from the system is
by means of our visual input channel.

There are certainly other means for human-computer interaction.
Voice input and audio output are feasible but at present not very
efficient. Voice input to the system becomes difficult because of
the inherent complexity in the computer analysis of the received
message since the human language has a wide spectral distribution
and is rich in dialects and synonyms. Voice input nevertheless is
potentially useful as long as the application can get along with
a relatively limited and simple vocabulary. Portable audio input
terminals for stockroom control, where the vocabulary essentially
exists only of partnumbers - e.g. the digits zero to nine - and
simple commands, might be feasible. Audio output is considerably
simpler than voice input, since the analysis of the output message can now be delegated to a very efficient data processing
system, namely the human brain, who can easily cope with the problems of a synthetic computer language. But also in audio output
the application will be limited to the transmission of relatively
simple messages since our interpretive capability of audio signals
is far inferior to our visual interpretative capabilities. Special
applications of voice output are - however - quite successful,

like for instance the use of a synthetic language to communicate between a computer and a blind person [19]. Long range proposals for man-computer interaction include the use of devices for direct electronic coupling to the human brain and transfer directly from the brain waves, with a direct computer interpretation of the ELECTRO-ENCEPHALOGRAM [20]. This input/output channel, though theoretically very broad and fast, can however probably be discounted for practical use in the near future.

The most versatile I/O (input/output) terminal for todays interactive operation is the CRT (Cathode Ray Tube) terminal (Fig. 9). It consists basically of the following components (Fig. 10)

- a keyboard for message input using the human tactile facilities
- processing equipment which interpretes this message and converts it to a bit stream suitable for data processing
- a communications facility to communicate the bit pattern according to a certain protocol to the central processor of the information system, and to receive messages back
- a facility to convert the message into a signal pattern suitable for display on the cathode ray tube,
- the display head itself.

These basic elements are supplemented with devices for control and checking of the terminal and the communications operation.

The CRT-terminals have some unique advantages which make them ideally suited for their task:

- The keyboard essentially taken from a standard typewriter keyboard, is ideally adapted to the tactile capabilities of the human hands.
- Its operation is quiet. Several terminal operators working in close proximity, do not disturb each other excessively.
- The CRT is versatile and can display numerals, alphanumeric characters and even graphics. Size of characters and fonts can easily be changed under program control.
- There are no moving parts, resulting in very good reliability.
- Production costs are low through technology commonality

with the display devices used in modern television.

Technological deficiencies are:

- The image which is generated on the CRT screen is volatile
  and needs continuous refreshing. The CRT thus requires an
  image buffer from which the information to be displayed
  can be retrieved approximately 50 times per second, to
  guarantee flicker free operation.
- The CRT requires very high operational voltages, and the
  power consumption is high.
- The weight of the CRT is high.
- The dimensions are bulky due to the distance required between the cathode ray generation and deflection devices and
  the screen. Bulky dimensions and great weight render the
  CRT not very suitable for portable devices.
- The maximum display capability is approximately 2000 characters on the screen surface.

Technological progress in the keyboard and the display section
of the CRT terminal is largely exhausted. There is room for very
solid product engineering and subsequent cost reduction, but it
is unlikely that dramatic new technoligical developments will occur. The real technological progress in the CRT-terminals will be
stimulated through the progress of the silicon integrated circuit
technology. And here it will not so much be a reduction in the
price of the CRT unit, but it will be an extension of the terminal
functionality which can be obtained for a certain price. The various steps which characterize this expansion of CRT functionality
as driven by the silicon integrated circuit technology are clearly
discernable:

- First the hard wired logic functions were replaced by monolithic read only memories, the read only memories were personalized by the supplier who now had the flexibility to
  quickly react to new market requirements just through re-
  personalization of the ROS (read-only store), but without
  changing the base design. The character set could be updated
  easily, the CRT could be quickly adapted to different languages.

- Additional functional flexibility was obtained by executing more and more of the logic and control functions of the system with the help of microprocessors. At very low cost, a standard terminal could be adapted to new emerging functional requirements. The technical absolescence of a certain terminal type was delayed. The user himself could alter the functional characteristics of a terminal by exchanging certain program modules, stored on monolithic read-only chips.

- Next was the advent of the "intelligent" terminal, where enough processing power was added to the CRT so that either different options in existing programs could be exercised, or that the user could develop new programs for specific functions. An example would here be the formatting of the screen under user program control.

- Further advancement of price/performance of semiconductors led then to the completely programmable terminal controller, integrated into the CRT-unit, with its own read-write memory of sufficient size to offload certain processing functions from the central processing unit to the terminal and thus reducing the communications traffic between the host and the terminal.

- Along with expanded functional capabilities went the improvement of circuits for error detection or correction, for data compression or expansion for processor/terminal communication, and for control of the communications traffic and protocol.

All these advancements have made todays CRT-terminals very flexible data input/output station with enough processing power to format the message to the central processing unit, to check it for completeness, to verify its compatibility with the communication protocol, and to detect input errors and in many cases correct, or at least flag them. The format in which the answer from the processing unit is to be displayed can be adjusted to the specific requirements; often repeated program routines can be stored in the terminal under a certain function key, and new routines can be programmed directly at the terminal. The semiconductor integrated circuitry has made the CRT-terminal into a data processing station in its own right.

The most promising competitive technologies are the gas plasma panels [21] light emitting diode arrays [22], and liquid crystal devices [21]. The advantage of these technologies over the CRT-technology is that they do not require refreshing, operate flikker-free, they have a flat screen, are compact built, use low voltage and permit selective erasing of part of the screen. They are, however, not cost competitive with the CRT's because they cannot share in large consumer market of todays television industry.

In summary then, the CRT-terminal is a uniquely suited interactive input/output device to a communication system. Its major technological progress is characterized by expanded functionality and flexibility with the help of modern integrated semiconductor circuits. No major breakthroughs are to be expected in the CRT base technologies.

External Storage

External storage for communication systems is characterized by the increasing need for larger and larger quantities of on line data. These are data which are always at the disposal of the system and do not require a preparation or set-up time (for instance: fetching a data tape from the library and installing it into a certain tape unit). Such on-line data are stored on large disk storage devices (DASD) or on on-line mass storage devices, like for instance the IBM 3850 or the CDC 38500 [24]. I will direct my analysis towards the disk devices, because many of the technological conclusions reached for them do also apply to modern on-line tape devices.

Rotating disk devices are used to store the operating system of the central processing unit, user data and data banks. The largest disk storage devices today have a storage capacity of several hundred megabytes [25]. Parallel to the requirement for more on-line data is the trend towards the "non-removable" disk, where the data storage device cannot be removed from the drive mechanism and transported to another drive spindle. This trend to non-removable disks very well also accomodates the engineering requirements for high precision parts and cleanliness of the atmosphere surrounding the disk, to assure an error free operation.

The key elements of a rotating disk storage device are shown in Fig.
11. Several concentric drives are mounted on a drive spindle and rota-
te at constant speed. The disks are covered with a thin layer of magnetic material into which or from which the read-write head writes or
reads the stored data. To gain maximum efficiency in the read/write
process the gap between the read/write head and the magnetic medium
must be as small as possible. This is achieved by flying the aerodyna-
mically shaped head on a thin air cushion which is less than one micro-
meter in thickness. The data are recorded on concentric tracks; there
may be a thousand tracks on one disk surface. Under program control the
head moves from one track to another to deposit or fetch data. The critical performance parameters of the device is the access time which is
the time elapsed between a program command to fetch certain data and
their availability at the system channel.

This access is determined first of all by the time it takes to mechanically move the head from one track to the desired next track, and
the rotational delay from the time in which the head gets unto the track
until the starting address of the desired data set has rotated under
the head.
The portion of the access time which is due to the arm moving from
track to track can be eliminated by assigning one head to each track.
Because of the geometrical restrictions the consequence is a much wider spacing between tracks and consequently a reduced storage capacity
of the disk and a higher cost per bit. Such "fixed head files" are only
affordable for small sections of storage where a very high access speed
is essential.

Analyzing the impact of technology on DASD devices is best done by
analyzing the impact on the major functional characteristics of the
storage unit. The characteristic parameters are:

- The total storage capacity
- The cost per bit
- The access time
- The rate with which data can be transferred
  from the disk once the access to the start
  of data set has been made.

The storage capacity is determined by the number of tracks which can
be accomodated per unit diameter of disk, the total diameter of the

disk, and the "bit cell length", that is the length along a track
which is required to store one single bit. The parameters determining
the bit cell length are displayed in Fig. 12. They are the head to
disk spacing, the head gap size and the disk coating thickness. The
development of these parameters as a function of time, together with
the development of the bit cell length is also shown in Fig. 12. The
bit cell length changed from $5x10^{-3}$ cm/bit for IBM's 1301, which was
announced in 1961, to $5x10^{-4}$ cm/bit for IBM 3350 announced in 1975.
These dimensions have all reached the size of the magnetic particles
and the limits of machineability and manufacturability. There will
certainly be continuous product engineering improvement, but further
dramatic progress of these parameters cannot be expected. Here again
the LSI circuitry can help to overcome technological deficiencies,
especially through better error correction circuitry or better sense
amplier circuits integrated into the head, to sense the small magnetic
signal.
The number of tracks per unit diameter is determined through the positioning accuracy of the head on the track. The mechanical head positioning devices are optimized to a level where again dramatic improvement is difficult to project. The major improvement here also can come
with the help of LSI, to optimize the sense circuitry which signal the
correct positioning of the head on the track.

The size of the disk and thus the total number of tracks is limited
by the fractural strength of the material (centrifugal forces) and
problems in the machineability of the very large disks. This dimension is also unlikely to improve very much.
In summary then, only gradual advancement from todays storage densities of several hundred thousand bits per cm^2 is likely to occur.

The cost per bit is determined by how many bits can be accomodated
within the fixed costs of one drive spindle, one power supply and control unit. It is clear from the storage density arguments that low
bit costs are most easily achieved with very large storage capacity
sizes, while it will remain difficult to produce small capacity devices for small computing devices with still low bit cost.
The bit cost will gradually continue to improve with improving bit
density and through additional leverage of cheaper, more reliable LSI
circuitry.

The access time can also be improved through better control circuitry, rather than through the improvement of the mechanical elements. In modern disks, for instance, the motion of the head is controlled by microprocessors, who determine the acceleration of the arm, its maximum speed, and they decelerate the arms motion optimally so that the head gets to stop exactly on the desired track, without any overshoot.

Access time improvement is of course possible through hierarchical disk storage structures, through combination of small sections of fixed head file storage, backed-up by large capacity mobile head storage devices. Serial solid state memories (CCD or magnetic bubble devices) may here replace the fixed head files.

The data rate will also show gradual improvements in the same manner as the "bit cell length" decreases since the rotational speed of the disks is close to its technical maximum.

All together, the LSI circuitry will be the major contributor to DASD improvements, through better control and checking, better error correction, through integration of several disk drives into one large storage subsystem, improved electronics for the control of the channel traffic to the central processing unit. These advances of the electronic portion of the rotating disk storage units are one of the prerequisites which make the efficient operation of large data banks possible.

5. The Central Processing Unit

The central processing unit which consists of the central arithmetic unit, the main memory and control store, service and maintenance facilities, channels to external storage and communication facilities to the various input/output devices, has shown the most dramatic price/performance improvement. Since 1960, the price to execute one microinstruction has dropped by more than a factor of 100 [27]. There is no saturation in sight, and driven by the advancement of the silicon integrated circuitry, the price/performance improvement in the CPU will be much more dramatic then either in Input/Output devices or in external storage.

A major impact was the replacement of core-memories in the CPU's mainstore with monolithic memories. The largest monolithic mainmemories on todays CPU's have a capacity up to 16 megabytes of

read-write storage. The driving forces which provoked this change -
despite the enormous technological success of the magnetic core
memory technology - were the monolithic memories better cost, reliability, lower power consumption and volume, and their technological commonality with the logic circuitry surrounding the actual storage array.

In addition to better absolute costs, the monolithic memories also
had the advantage that their costs are practically independent of
the memory size (core memories always had to carry the fixed overhead of the expensive silicon integrated ciruitry for driving,
sensing and decoding, which could best be amortized over very large
bit quantities) so that also small, independent memory units could
be designed which could be distributed through-out the CPU, embedded
directly into the logic functions.

This fact stimulated very much the design of microprogram controlled
CPU's, where the microprogram is stored in monolithic read-write
control stores. Many engineering changes or new functional requirements of the control functions can now be handled on the microprogram level, that is through software, without the need to redesign the CPU hardware. Monolithic memories can easily be designed
with facilities for automatic error correction and detection which
effectively hides all technological deficiencies from the user.

The cost/performance trade-off capabilities in monolithic memories,
as shown in Fig. 5, has lead to the design of memory hierarchies,
where a main memory of several megabytes in size is buffered by a
smaller cache memory of much higher performance. The bit capacity of
the caches are usually only a few percent of the main memory (example: IBM 370-168: Cache: 32 Kilo-Bytes Main-Memory: 8 Megabytes)
which can accordingly be more expensive. The operation of the hierarchical memories then is such that the CPU addresses first the
cache, when a certain information is required. When the information
is in the cache, it can be presented to the CPU after a very short
time (IBM 370-168: Cache cycle time: 80 Nsec Cycle time: the time
delay between two subsequent memory addressing operation). Only if
the information is not in the cache, then the main memory is addressed and the required block of information - a page - is loaded
into the cache. In the 370-168, the corresponding main-memory cycle
is 320 Nsec. Since many data processing operations are sequential

in nature, the program structure can be organized in such a manner that the cache will contain the required information in the majority of cases, so that the performance of the memory hierarchy is determined by the cache performance, the cost is determined by the main-memory cost. In order to obtain a balanced CPU/memory system the cache cycle should be equal to the CPU cycle, which with the help of monolithic memories is easily achievable.

The advent of inexpensive LSI logic circuits has increased the processing power of the CPU's. Example: IBM 370-168: announcement: 1972 logic technology: 3-4 circuits per chip, 2.3 MIPS, IBM 3033, announcement: 1977 logic technology: 40 circuits per chip, 4.8 MIPS. In addition to just increasing the processor power, the silicon integrated circuitry enabled a considerable CPU task differentiation. Separate processing units for Input/Output processing, the maintenance and service subsystem, or the storage subsystem became affordable, introducing a large amount of parallelism into the CPU operation and thus increasing its throughput. These subsystems carry their own compliment of read-write memory to store the specific segments of the control program, which is necessary to perform their assigned function. Only when their complement of control program is exhausted will they have to go back to the CPU main memory for new control information, and very little interference with the general bus traffic of a CPU will occur.

Some processing tasks may be completely delegated from the CPU to the Input/Output area. Such distributed processing reduces the communication traffic between the CPU and the Input/Output units. The central host will then only be addresses when access to the central data bank is required.

All these facts are elements of the same development: a sharply decreasing cost of data processing. This makes it possible to divert some of the CPU's processing power away from problem solving tasks to assisting the user in his interaction with the data processing system. These "Ease of use" features can be designed to facilitate hardware or software error diagnostics, to aid in application programming, to assist or instruct the user in the operation of the system, or to add security features, which protect the data and programs stored in a computer against the misuse by unauthorized persons.

The trend will be to add so much "ease of use" of the appropriate cost that the power of an efficient information system is available to every authorized person, at all times, at all places. To avail himself of these services, the user does not need to be a computer expert, nor does he have to go through an expensive, time consuming training period. The information system itself will interactively teach the user what he needs to know [26].

6. Summary

- The technological progress of information systems is largely determined by the advancement of the silicon integrated circuitry. The cost/performance progress of this technology shows no signs of saturation in the foreseeable future.

- The silicon integrated circuit technology will continue to dramatically advance the price/performance of the central processing complexes of an information system.

- The advances in Input/Output Units will be more gradual, since much of the enhancement potential of the mechanical technology is exhausted. Here also the LSI technology will be a major contributor to technological progress.

- These different rates of technological progress will cause a shift in cost emphasis away from the CPU to the Input/Output systems and to external storage subsystems.

- The increased CPU processing power can accomodate more powerful operating systems and application programs which will stimulate the use of information systems by a much wider range of non DP professional users.

References

1. R. Remshardt,
   U.G. Baitinger: IEEE J. Solid State Circuits, Vol. Sc 11 (1976), No. 3, Page 352-259

2. F.L. Thompson: Sol. St. Technol. 17, (1974), No. 7, Page 21-30

3. D.L. Spears,
   H.I. Smith: Sol. St. Technol. 15, (1972), No. 7 Page 21-26

4. L.M. Terman: Proc. IEEE 59, (1971), Page 1044-58

5. K.U. Stein,
   H. Friedrich: IEEE J. Sol. St. Circuits, Vol. Sc 8, No. 5 (1973), Page 319-23

6. S.K. Wiedmann: European Solid State Device Conference, Munich 1976

606

7. K. Horninger: Digital Memory and Storage , W.E. Proebster, ED., Vieweg 1978, Page 121

8. C.N. Ahlquist et al: IEEE J. Solid State Circuits, Vol. Sc-11, No. 5, (1976), Page 570-74

9. P.B. Ghaie, W.R. Gardner, D.L. Crosthwait: IEEE Trans. Reliability, Vol. R2, No.4, (1973), Page 186

10. P. Totia, R. Sopher: IBM J. Res. and Dev., (1969), Page 220

11. W.K. Liebmann: Digital Memory and Storage, W. Proebster, ED., Vieweg 1978, Page 135

12. H. Yoshimura et al: Digest IEEE ISSCC 1978, Page 148-9

13. R. Mitterer: Digital Memory and Storage, W. Proebster, ED., Vieweg 1978, Page 97

14. F.H. De Leeuw: IBID, Page 203

15. A.J. Perneski: IEEE Trans. Magn. Mag-5, 554 (1969)

16. O. Voegeli et al: AIP Conf. Proc. 24, 617 (1975)

17. M. Suzuki et al: Digest IEEE ISSCC 1978, Page 206-207

18. P. Wolf: Digital Memory and Storage, W. Proebster, ED., Vieweg 1978, Page 247

19. J.A. Kutsch, Jr.: Nat. Comp. Conf. Proc., Vol. 46 (1977) Page 357-62

20. C. Fields: IEEE Trans. Prof. Com. VPC-20, No. 1 (1977) Page 2-6

21. G. Chodil: Proc. S.I.D. Vol. 17/1 (1976) Page 14-22

22. B. Kazan: IBID., Page 23-29

23. L.A. Goodman: IBID., Page 30-38

24. E. Lennemann: Digital Memory and Storage, W. Proebster, ED., Vieweg 1978, Page 65

25. P. Wentzel: IBID., Page 33

26. W.K. Liebmann: Proc. 9. Workshop, Institut für Produktionstechnik und Automatisierung, Fraunhofer Gesellschaft, Boeblingen, Nov. 1977.

27. L.M. Terman: Scientific American, Vol. 237, No. 3 (1977), Page 163-177

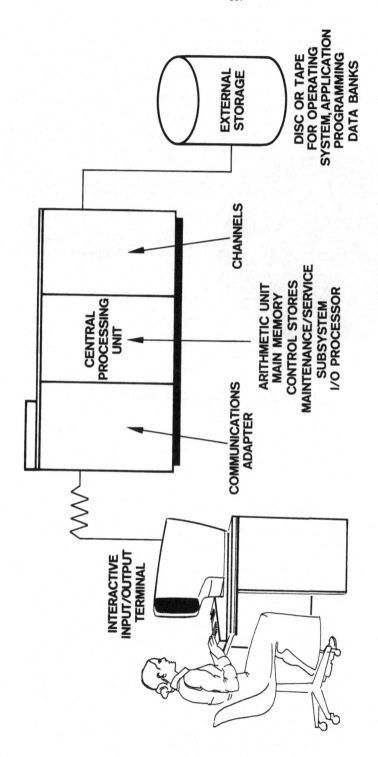

607

INTERACTIVE
INPUT/OUTPUT
TERMINAL

COMMUNICATIONS
ADAPTER

CENTRAL
PROCESSING
UNIT

CHANNELS

EXTERNAL
STORAGE

DISC OR TAPE
FOR OPERATING
SYSTEM, APPLICATION
PROGRAMMING
DATA BANKS

ARITHMETIC UNIT
MAIN MEMORY
CONTROL STORES
MAINTENANCE/SERVICE
SUBSYSTEM
I/O PROCESSOR

FIG. 1:   HARDWARE COMPONENTS OF
           AN INFORMATION SYSTEM

608

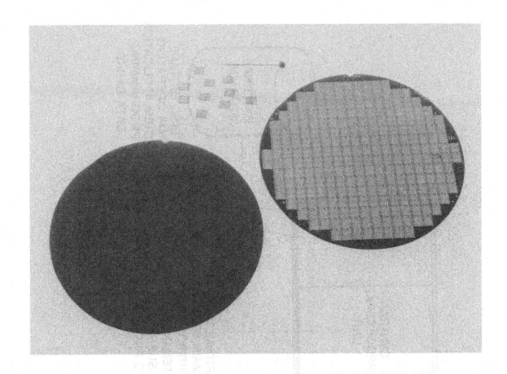

FIG. 2:   SILICON WAFER

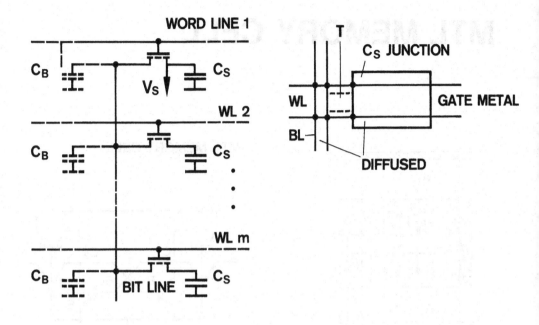

FIG. 3:   ONE DEVICE MONOLITHIC MEMORY CELL,
CIRCUIT AND LAYOUT EXAMPLE
(STEIN A. FRIEDRICH [5.])

MTL MEMORY CELL

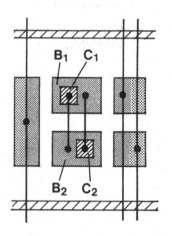

MTL MEMORY CELL

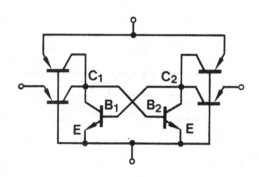

READ OPERATION

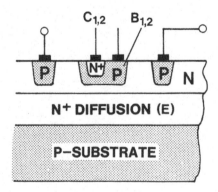

FIG. 4: MTL BIPOLAR MEMORY CELL (WIEDMANN [6.])

611

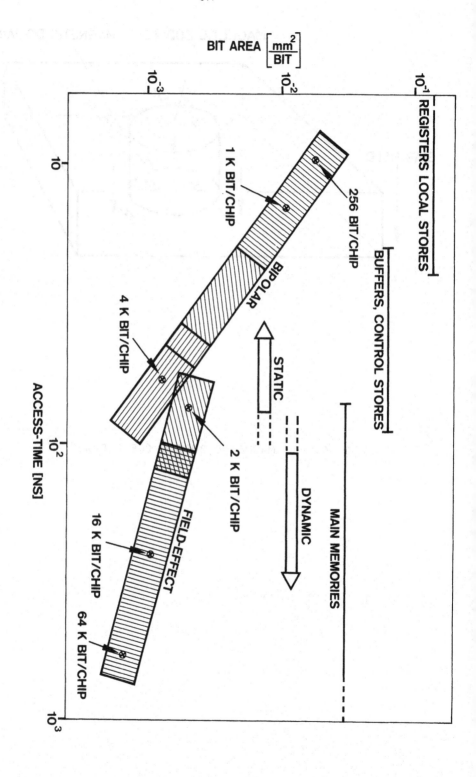

FIG. 5: BIT AREA REQUIREMENT OF MONOLITHIC
MEMORIES AS A FUNCTION OF ACCESS TIME

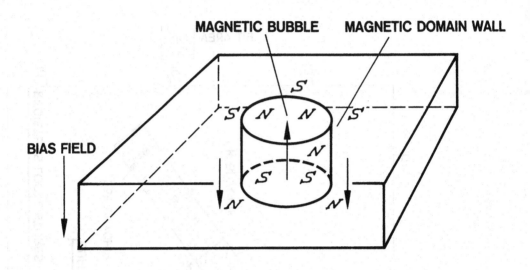

FIG. 6:   MAGNETIC BUBBLE (DE LEEUW [14.])

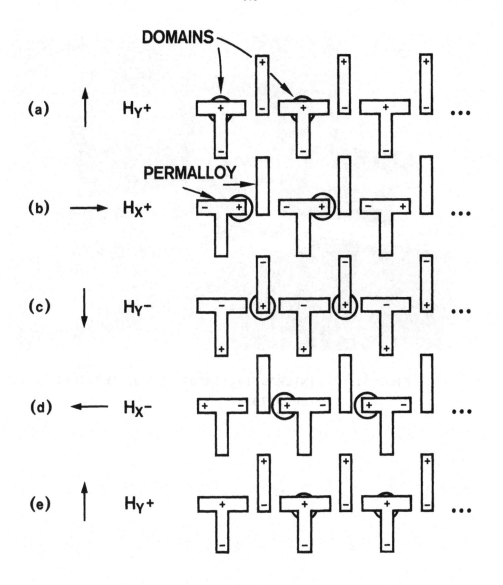

FIG. 7:   ROTATING FIELD DRIVE OF MAGNETIC BUBBLES
IN T-BARS: ROTATING FIELD: H_{Y+}, H_{X+},
H_{Y-}, H_{X-}. THE EXTERNAL FIELD IS LOWERED
AT THE POSITIONS OF THE POSITIVE POLE
(PERNESKI [15.])

614

FIG. 9:   INTERACTIVE INPUT/OUTPUT TERMINAL:
IBM 3270

615

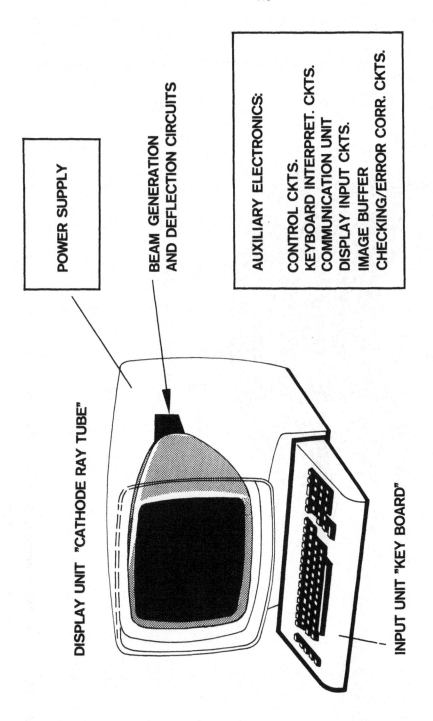

POWER SUPPLY

BEAM GENERATION
AND DEFLECTION CIRCUITS

AUXILIARY ELECTRONICS:

CONTROL CKTS.
KEYBOARD INTERPRET. CKTS.
COMMUNICATION UNIT
DISPLAY INPUT CKTS.
IMAGE BUFFER
CHECKING/ERROR CORR. CKTS.

DISPLAY UNIT "CATHODE RAY TUBE"

INPUT UNIT "KEY BOARD"

FIG. 10:   MAJOR HARDWARE COMPONENTS OF AN
           INTERACTIVE DISPLAY INPUT/OUTPUT UNIT

616

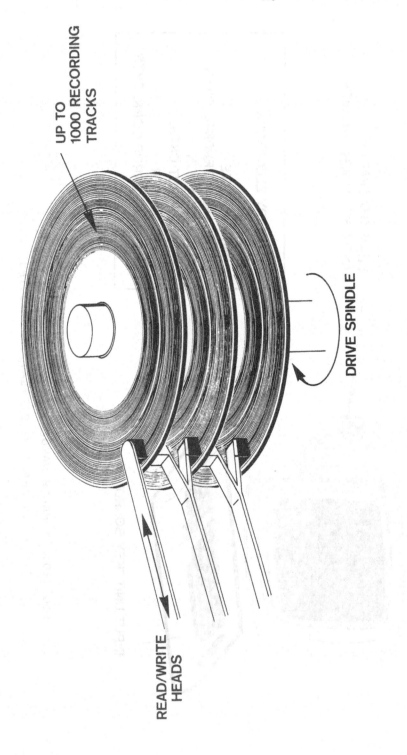

UP TO
1000 RECORDING
TRACKS

DRIVE SPINDLE

READ/WRITE
HEADS

FIG. 11: MAJOR COMPONENTS OF A ROTATING
DISC STORAGE DEVICE

INTERACTIVE SYSTEMS AS IF USERS REALLY MATTERED

Warren A. Potas
Brown University Visiting Research Associate
Computer and Automation Institute of the Hungarian
Academy of Sciences

Budapest, Hungary
(Correspondence: 308 North Stockton Ave., Wenonah, NJ 08090/USA)

ABSTRACT

Emphasis in the design of interactive systems is still too often
concentrated on definition of data objects and operators, while too
little attention is paid to the needs, perceptions and reactions of
eventual users. Full consideration of the psychological fit between
users and interactive systems, principally reflected in programmed
interaction possibilities and communication facilities, is desirable
so as to increase the efficiency of the man-machine system, and, more
importantly, to maximize the sovereignty, creativeness and satisfaction of the final user. In this paper, fundamentals of man-machine
symbiosis are presented, and requirements for system personality,
malleability, and feedback/response are analyzed. Through adoption
of these criteria early in the design process, it is argued that
users will more readily accept interactive computer systems as desirable tools of their increasingly information-dominated trades.

OBJECTIVES

Efforts in recent times to improve productivity have frequently involved introduction of interactive computer systems. In the past,
application of raw computing power was often sufficient to insure increased productivity. Although dramatic increases in hardware processing power continue to be made, underlying data bases and the range
of required tasks continue to grow even faster. Attention must be
turned to the more intangible elements associated with the link between people and computer systems in order to continue to improve
productivity. Systems must be created which augment people's creative capacity, which ease rather than hinder manipulation of data
objects, and which are enjoyably satisfying to users.

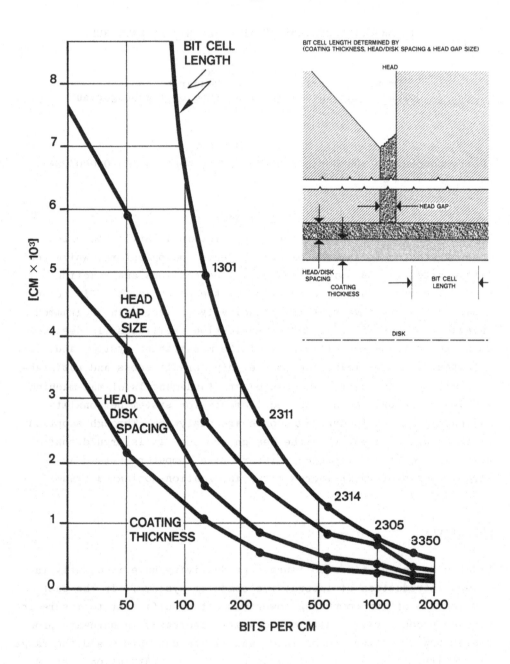

FIG. 12: "BIT CELL LENGTH" FOR VARIOUS
DISC PRODUCTS

This paper is not the first to discuss themes relating to user and computer system interaction. However, while many interactive computer systems are increasingly well designed in such aspects, the majority of systems in use and even under development do not rate very highly in man-machine symbiosis characteristics. It is in fact the author's critical evaluation of a well-intentioned large and important interactive graphics system for part programming numerically controlled machines [1] which helped convince the author of the necessity for a paper bringing together in one source a general analysis of use-oriented interactive system requirements.

BACKGROUND

This paper draws on the author's previous experience in the design and implementation of large-scale interactive systems, both graphic and non-graphic (notably [2] and [3]), and hands-on opportunity to examine numerous other interactive systems at a number of universities and research institutes. It is the immediate result of the author's participation in a research exchange program in Hungary during which human factors of Hungarian systems under development were evaluated and extensive study of computer literature relevant to the topics discussed herein was undertaken (ref. [4] and [1]).

An indication of the lack of attention paid final users in the design of interactive systems (whether graphic or non-graphic) and the low esteem with which such considerations are often regarded is the omission in most published papers of discussion regarding a system's user communication facilities and interaction characteristics. Emphasis invariably is rather given to the nature of data objects and the abstract nature of data manipulations provided. (Good examples of exceptions are [5] and [6].) This obviously makes dissemination of experiences in handling the user interface very difficult, especially since most papers are written before final users ever interact with the system reported. A notable exception is [7], which describes a FORTRAN package for interactive graphics and then presents a large section detailing user-identified suggestions for improvement (incorporation of which was planned in a subsequent revision).

On the other hand, several outstanding papers concerned with humane design of interactive systems do exist. Hansen [8] presents a set of 14 User Engineering Principles grouped into four major categories: 'Know the User', 'Minimize Memorization', 'Optimize Operations', and 'Engineer for Errors'; Robert Miller [9] examines the nature of 17 basic response time requirements in man-computer interactive systems and makes some sage general observations concerning such requirements; and David [10] discusses human physiological factors affecting information transmission, including human identification and discrimination capabilities, reaction time, pattern perception, and information rates.

Finally, a classic paper by George Miller [11] provides specific bases for analyzing human communication channels in terms of information capacity limits, data which can be used to better integrate man-system communication. While Miller's paper should be studied in its entirety, its central points are reviewed below.

One of Miller's main points is that there seems to be a human sensory channel capacity of approximately 7 items - people have the capability to differentiate among only 7 categories of unidimensional information (span of absolute judgment). Each added *dimension* of information (independently variable attributes) can increase the total number of information categories that can quickly be perceived, but at a decreasing rate. A second major point is that human short-term memory seems to be able to hold about 7 separate items (chunks) (span of immediate memory). However, more *information* can be held if it is encoded - that is, if each item is a large, meaty chunk. Miller summarizes, "The number of bits of information [3 bits ⇒ ∿7 categories] is constant for absolute judgment and the number of chunks of information [∿7] is constant for immediate memory. The span of immediate memory seems to be almost independent of the number of bits per chunk, at least over the range that has been examined to date."

A model for describing creativity might thus be the process of combining bits into chunks, transferring chunks from short-term memory into long-term memory, and, through closure, freeing the mind for processing a similar number of now more complex and information-rich items. Closure is defined by psychologists as the subjective sense

derived from completing a thought process or activity. In any involved
activity there are a large number of minor and major closures occurring,
each accompanying achievement of a sub-goal or a larger task.

FUNDAMENTALS OF MAN-MACHINE SYMBIOSIS

Two principles noted by van Dam [12] serve to highlight the issue that
user-oriented system design must emphasize the user from the very inception of project planning and design. First, people can get used
to anything and even come to espouse it. Second, user mastery of any
system inhibits desire for change in that system, no matter how cumbersome and ill-conceived it may seem to an independent critic or even
intellectually to the user himself.

If systems are to be designed to embody a high degree of user optimality, then such requirements must be factored into the design process from the very outset and cannot be left for possible *ex post
facto* consideration some time after a system is otherwise completed.
(Invariably the growth of a vested user community precludes or hin-
ders modification of a system.) User considerations must be dealt
with as an organic part of the entire design cycle.

Drawing an analogy between an interactive system and a physical tool
provides a useful way of viewing issues involved in system design.
A mechanical implement is accepted and used because it makes a task
easier. Given a choice of similar but unfamiliar tools, one will
usually be chosen (given similar direct cost to the user) because it
'works best' or is 'pleasantest to use' for the purpose. Relevant
criteria for the construction of a successful tool are a parsimonious
design, naturalness of use and aesthetic appeal. A tool is not very
successful if it diverts the attention of the user *from* the task he
intends to do and *towards* the tool itself. The better designed a
tool is, the fewer demands it should place on the user's thought and
attention processes (after a possible familiarization period).

The matter becomes more complicated when the tool is designed to provide a high level of support for the manipulation of information and
ideas; the manner in which information is presented and information

processing capabilities are provided must take into account the complicated nature of human thought processes. Thus the characteristics
of such data tools must relate to the *psychology* of the user as well
as subsume the characteristics stated above which are applicable to
mechanical tools.

Some of the most significant attributes of systems which are optimally accommodative for the user are environmental richness, a logical
and rational personality, appropriate user feedback, and the ability
to be fine tuned by the user to match his developed skill in using a
given system. These issues are discussed in further detail below.

SYSTEM PERSONALITY

First and foremost, systems should be regular, demonstrating rational
and consistent exteriors, regardless of whatever machinations may be
necessary under the surface. Every action or response by a system
should follow logically from a user's actions and the prior state of
the system. There should be no possible user action capable of de-
railing a system from its planned state graph. Whenever possible,
identical or similar groupings of user choices should always appear
in the same manner: a particular list of options should be listed
in the same order whenever it appears; a particular set of lightpen-
nable settings should always appear in the same spot on a graphics
terminal. Given a choice of user manipulable input tools (e.g., keyboard, lightpen, joystick, function keys), the most expressive and
appropriately manipulated one for conveying the user's specific intent should in each instance be enabled, although frequency of required switching among tools should be minimized. System messages
should be crisp - as succinct as possible while presenting complete,
unambiguous information. Language and terminology should be used
precisely.

Second, systems should be subservient without being obsequious. Many
major systems are designed so as to appear to be ever champing at the
bit, goading the user with questions and commands to which he is forced
meekly to succumb. Users may justifiably feel harassed when confronted
by such lock-step systems. Preferable are systems which make the user

sovereign. The Hungarian part programming system studied by the author and mentioned earlier [4] is one example of a system under development whose design is felt to allow an inadequate degree of user sovereignty. Rather than terminate a response with an imperative or a direct interrogative, for example, a system should indicate the state of operation it has reached and the options available for moving the system into another state. The user should always feel in control over the tool that is serving him.

Third, a system should be intelligent. It should make full use of the computational power available. The user should never be called on to enter any information which could be derived from data he has already supplied or to enter data in a bizarre (to the user) format just because it is actually an internal data format. To state a succinct rule, whatever a system *can* do to help the user and reduce his workload, it *should* do. As part of this philosophy, a system should scrupulousy check all data supplied by a user for every possible detectible error.

In his paper concerned with computer aided design systems, Allan [13] offers several thoughtful pointers generally relevant as desiderata for interactive system tools. "When considering the problems of the design of a design system, the concern arises that one may not only devise a system focused outside man on the end product, but one may also so rigidly structure the enquiry, the analysis, and the decisions, that all hope of freedom of decision is gone for the ultimate user or designer who it is hoped will be served. What is really sought is a system with two characteristics: (1) the system itself will not structure the user's view of the environment of his problem, the formulation of the problem, or the solution of the problem; (2) the design and decision processes will be designer-oriented, and an integral part of each problem itself."

SYSTEM MALLEABILITY

One of the problems which confronts both the designer and the user of an interactive system is the effect of a user's learning curve. A designer's dilemma is whether to create a system with an environment for the naive user which is radically simple and has copious instructions and detailed error messages, but which should be capable of easy mastery; or rather an environment for the experienced user which is terse and which requires that the user learn more and have to remember more before sitting down to utilize the system. Under the first alternative, the user need have little knowledge of the system before using it at least reasonably successfully, while in the latter case there exists a high learning barrier which must be overcome before any success can be achieved with the system. Of course, a system which is initially easy to learn and use may well seem dreadfully slow and boring once learned.

As summarized earlier, the research of George Miller [11] and others demonstrates that encoding of data is enormously significant in the ability of persons to mentally process increasingly complex data. Conversely, as people deal with ever larger chunks of information, tasks or data which originally seemed forbidding and complex become clearer and easier to grasp. An interactive system environment is no exception, so that as a user becomes familiar with a system his perception of it will become re-organized from small fragments to more encompassing concepts. In line with this theoretical analysis, there is specific evidence [14] that trial and error self-teaching is a more effective way for the naive user to learn the use of an interactive system (while also reinforcing his confidence) than a regimen requiring exhaustive study of manuals before any attempt can be made to use the system.

The above arguments provide powerful reasons why systems should be designed in such a way that a user can alter a system's interface, adapting it for each successive level of his mastery. Such adaptable systems would also provide for the myriad idiosyncracies of different individuals and allow creation of personally tailored work environments. If not minimized at the outset, the degree of system prompting could thus be subject to user adjustment. By allowing the terseness of system generated communications to be regulated to any desired minimum, efficiency, throughput and user satisfaction can be increased.

Numerous techniques for personalizing and extending interactive systems are available. At the simplest level, users can be allowed to redefine keywords and names, using any abbreviations or substitutes they like. The use of system 'shells' to hide increasingly sophisticated levels of command repertoires from beginning users have often been successfully employed. The ability of users to control the interactive environment via mode setting is also an important technique, encompassing, for example, defaults for output or command parameters and controls on both the timing and terseness of system feedback. Parameterized command macros which allow value and textual substitution and provide internal control structures, also parameterized, are another means of effectively allowing users to create unique, powerful commands as required. Specific personal environment-setting macros could automatically be invoked by a system upon identification of the user at the beginning of each session. Provision of user libraries for such macros and user-controlled preprocessors for input and post-processors for output are other possibilities.

Alternatively, drawing on extensible language concepts, systems may be structured to allow the development of command tools based on key underlying data objects and operators having relatively simple initial structure. The user could then build his command repertoire from these elements, or, if a system provided some initial set of commands built upon a basic internal command language, could redefine or enhance such commands. Although not specifically termed an extensible language, the success of APL may properly be partly attributed to its malleability, its richness of basic elements and structures which can be used to create small chunks incorporating immense complexity.

SYSTEM FEEDBACK/RESPONSE

Human communications environments typically include two logically distinct communications paths - a data path and a feedback path. In a conversation, the voice of the person speaking constitutes the data path (although often accompanied by 'body English' which adds bandwidth) while non-verbal expressions of the listener provide feedback -

reflecting interest, boredom, comprehension, disagreement, etcetera, back to the speaker.

In order to function well, people require feedback from their environment. Feedback serves two important purposes. First, it operates as a carrier signal to let both parties know that they are in fact still communicating. Secondly, it provides an otherwise quiet channel on which error conditions (non-comprehension, for example) can be indicated. If a listener signals non-comprehension (with a puzzled expression, for example), a speaker can modify his conversation in an attempt to restore comprehension on the data channel; conversely, if the problem is large enough, the data channel may need to be brought more heavily into the process of restoring positive feedback (as by question and answer). Adequate data and feedback channels are of immense importance for the success of interactive systems.

In discussing system response and feedback design, the crucial factor of psychological closure in human thought processes must be acknowledged. As already discussed, there are a limited number of concepts (approximately 7) which a person can deal with at any one time using his short-term memory. The optimal information processing efficiency of the user can be reached when his mental activity is directed at the attempt to purge or resolve the limited number of concepts fecund in his short-term memory. Whenever a subjective sense of completion takes place, closure is said to have occurred. External interruptions, unless expected and related to ongoing mental activity, have a tendency to derail such activity. However, interruptions or delays which occur after a closure are much less disruptive, and the greater the closure, the more significant can the interruption or delay be without interfering with mental activity or generating annoyance.

Several significant ramifications for the design of interactive systems result. As regards reporting errors to the user, immediate error feedback is generally desirable in order to exploit the associations inherent in temporal proximity. However, error reports should avoid triggering premature closure as part of a policy that the user not be unduly penalized for mistakes. Notice of the computer not listening (e.g., line dead) is an extreme example of an error condition requiring immediate attention of the user, while keystroke errors are conditions whose reporting should sometimes be delayed. Moreover, users frequently become aware of their errors even before they are reported by the system. Indeed, the time taken to elaborately report an error which the user already knows about can be far more disruptive of mental activity than the simple fact of the error:

Some well-received systems, as a result, have treated separately these two types of errors. For example, in the sophisticated and user oriented FRESS text editing system [15], common errors made by a user while editing are noted merely by the printing of a question mark. If the user does not know what the problem is (an infrequent occurrence), he can type a question mark back and receive the full text of the error message. However, in those cases where he already knows the error, the level of interruption of mental processing will be very low. For those unusual error conditions which a user could not reasonably be expected to anticipate, full messages are given at first.

On the other hand, many existing systems, as well as systems now being developed, delay the reporting of errors far beyond when they are detectible and when knowledge would be most useful to the user. The reviewed Hungarian part programming system [4] solicits information in large batches, delaying any checking until all questions in a batch have been answered by the user. As a result, valid data must be retyped along with the bad item if there is an error, and furthermore, data is sometimes solicited which is inappropriate given either an error condition or the nature of the data being entered.

As regards the quantitative aspect of system response time, much has been written - reference the notable paper by Robert Miller [9]. User's satisfaction is related to the degree of correlation between expected response and actual response. Irregular response conditions tend to restrict creativity instead of releasing it. It is desirable to avoid response delay in cases other than for major closure. Happily, computation requirements for the period preceding closure are frequently low but of high priority, while those following major closure are much higher, but with lower priority. It has been suggested [16] that a deliberate delay of 2 to 4 seconds may be desirable following an individual's closure on a major task before a system should respond with such messages as error reports. This seems reasonable, as the process of refreshing the mind after much mental activity often involves a review of the preceding activity, and the user may become aware of his own mistakes at this time.

AFTERWORD

While much of the specific content of this paper has been concerned with improving the efficiency and effectiveness in interactive system user communications, the attempt has also been made to stress the importance, as the root idea behind these efforts at tool building, of the desire to develop systems which are successful in augmenting and amplifying the creative potential of the user. Obviously, application of the guidelines outlined in this paper to the design of a man-machine system which reduces the role of the man to that of an efficient robot (in other words a role reversal in which the man is now the tool rather than the machine), expanded creativity is not obtained and the purpose behind these proposals is not realized.

Negroponte stresses this theme in his excellent paper dealing exclusively with these issues [17]. He "laments the absence of any effort to amplify creativity through computer aided design. Current systems attest to this deficiency by offering no precedent of a person using a computer to be creative, let alone be more creative than he or she would be without it."

While the generality of this charge may be debated, it is argued here that the enlightened application of suggestions made herein may move interactive systems a step in the right direction.

REFERENCES

[1]. Warren A. Potas, "Computer Aided Design: A Proposal for a Synergistic Graphic Drafting System Based on an Evaluation of the Hungarian 'DIALOGUS' System," Tanulmányok 52/1976, Computer and Automation Institute (Hungarian Academy of Sciences), Budapest, 1976.

[2].   W. David Elliott, Warren A. Potas, and Andries van Dam, "Computer Assisted Tracing of Text Evolution," Fall Joint Computer Conference 1971 Proceedings, AFIPS Press, Montvale, NJ, pp. 533-540.

[3].   W. David Elliott, Warren A. Potas, and Peter A. Rigsbee, "Implementation of a Simulation Environment for the AN/UYK-17 Signal Processing Element," NRL Report, Naval Research Laboratory, Washington, DC (20375), 1975.

[4].   Gyula Pikler and Vera Simon, "A General Dialogue System for Interactive Graphic Programming of NC Machines and CAD Systems," presented at IFIP/IFAC PROLAMAT 1976 Conference, Stirling, Scotland.

[5].   L. Mezei and A. Zivian, "ARTA, an Interactive Animation System," IFIP Congress 1971 Proceedings, North Holland Publishing, Amsterdam, 1971, volume TA-3, pp. 104-108.

[6].   R. J. Hubbold, "TDD - An Interactive Three Dimensional Drawing Program for Graphical Display and Lightpen," Advanced Computer Graphics (R. D. Parslow and R. Elliott Green), Plenum Publishing, London, 1971, pp. 1035-1045.

[7].   G. A. Butlin, "A FORTRAN Package for Interactive Graphics," Advanced Computer Graphics (R. D. Parslow and R. Elliott Green), Plenum Publishing, London, 1971, pp. 947-965.

[8].   Wilfred J. Hansen, "User Engineering Principles for Interactive Systems," Fall Joint Computer Conference 1971 Proceedings, AFIPS Press, Montvale, NJ, pp. 523-532.

[9].   Robert B. Miller, "Response Time in Man-Computer Conversational Transactions," Fall Joint Computer Conference 1968 Proceedings, Thompson Book Company, Washington, DC, pp. 267-277.

[10].   Edward E. David, Jr., "Physiological and Psychological Considerations," On-Line Computing (Walter J. Karplus), McGraw-Hill, NY, 1967, pp. 107-128.

[11].   George A. Miller, "The Magical Number Seven Plus or Minus Two: Some Limits on Our Capacity for Information Processing," Psychological Review, vol. 63, no. 2, March 1956, pp. 81-97.

[12].   Andries Van Dam, private communication, Brown University, Providence, RI, 1976.

[13].   John J. Allan III, "Foundations of the Many Manifestations of Computer Augmented Design," Computer Aided Design (J. Vlietstra and R. F. Wielinga), North Holland Publishing, Amsterdam, 1973, pp. 27-54.

[14].   T. C. S. Kennedy, "Some Behavioral Factors Affecting the Training of Naive Users of an Interactive Computer System," International Journal of Man-Machine Studies (1975), number 7, pp. 815-834.

[15].   "FRESS - File Retrieval and Editing System," Reference Manual, Brown University Center for Computer and Information Sciences, Providence, RI, 1970.

[16]. James Martin, Design of Man-Computer Dialogues, Prentice-Hall, Englewood Cliffs, NJ, 1973.

[17]. Nicholas Negroponte, "On Being Creative with Computer Aided Design," IFIP Congress 1977 Proceedings, North Holland Publishing, Amsterdam, 1977, pp. 695-704.

ACKNOWLEDGMENTS

It was my privilege to represent Brown University in Budapest as a participant in an international research exchange program arranged by Andries van Dam of Brown University and Jozsef Hatvany of the Computer and Automation Institute, Budapest.

I wish to express my appreciation to Jozsef Hatvany and Andries van Dam for their encouragement and efforts in making this program possible and to the host institutions as well as to the Hungarian Academy of Sciences, the Hungarian Cultural Relations Institute and the U. S. National Science Foundation for their funding of the program.

I am grateful to all my colleagues at the Computer and Automation Institute for their generous assistance and advice of both a technical and personal nature during the course of my program in Budapest.

I particularly wish to thank Andries van Dam for his generous assistance and advice in developing this paper.

MAN-COMPUTER DIALOGUES FOR MANY LEVELS OF COMPETENCE

P.A.V. Hall

Information Systems Division

SCICON Ltd.

Sanderson House

49 Berners Street

London, W.1.

ABSTRACT

Man-computer dialogues are viewed as languages and the relationship between programming languages and menu plus form dialogues is shown. Syntax notation is used for dialogue description and design. It is shown how to construct a man-machine system where the user can switch freely between programming language style of interface and a dialogue of menus and forms.

1 INTRODUCTION

Interactive systems are now very common (eg. 9,10,12,13,15). The systems vary in degree of sophistication they demand of the user, ranging from programming languages like BASIC or APL to systems based on menus, forms, etc. to systems based on natural language. The various alternative forms of man-computer dialogue have been surveyed in the very useful book by Martin (10).

In many systems there is a need to accommodate both experienced users, who require a very succinct language to enable them to work fast and without frustration, and naive users who need a lot of assistance. It is desirable for the naive user to progress to advanced status without having to learn a new advanced dialogue separate from their beginner's dialogue. HELP commands, abbreviations for English-like keywords, and MACROS (eg. 10,13) are widely used: these ideas are useful, and they do accommodate a range of competences, but can we not go further ?

I found myself involved in discussions around the design of an interactive system, facing advocates of a programming language approach and advocates of a menu plus form approach. It then occurred to me that these approaches are very closely related, and it should be possible to enable the user to freely switch from menus or forms to programming language within a single system. Thus I set out to formulate my proposals in detail: this paper is the result.

For the purposes of exposition, I have invented a library retrieval problem: this example is typical of many interactive systems concerned with on-line retrieval of information (eg. 10). Section 2 introduces the example, and later sections draw upon this example.

The first consequence of looking at dialogues as languages is that all the computing industry's experience of languages and compilers can be applied. In particular, the language can be described by a syntax, and this is what has been done in Section 2.

A BNF notation has been used in Section 2, but diagrams similar to state-transition networks (14) are used later, and are important.   The interaction can be designed initially as a programming language, and from this the dialogues derived: this is done in Section 2 for the example, and general rules are described in Section 3. The syntax description includes both man and computer generated symbols - the design of the dialogue is essentially a matter of deciding how much the man contributes and how much the computer contributes.   Only the combination of both computer and human contributions has meaning, spelling out the intentions of the man and the required actions of the computer.

A naive user is more at home with a dialogue of menus and forms, but as he gains competence he would like to short-cut the verbosity and slowness associated with the dialogue and key in directly his instructions.   This shift of behaviour is readily accommodated by viewing the dialogue as a language.   The basic ideas are given in Section 4, and various software issues are discussed in Section 5 and 6.

2   EXAMPLE   -   Library Administration

The books held in the library are recorded in the table BOOKS, with each book descr- ibed by AUTHOR, TITLE, PUBLISHER, YEAR of publication and library ACCESSION - NUMBER. People who are entitled to use the library are recorded in table SUBSCRIBERS with their NAME and ADDRESS, while actual loans are recorded in the LOANS table.   Thus there are three tables with columns as below:-

BOOKS (AUTHOR, TITLE, PUBLISHER, YEAR, ACCESSION-NUMBER)
SUBSCRIBERS (NAME, ADDRESS)
LOANS (NAME, ACCESSION-NUMBER)

Note that this can be thought of as relational system in as much as its information is tabular (4,11) but as will be seen below only limited operations will be allowed.

In use, the three tables can be independently updated or interrogated to answer such questions as "does the library have any books by SMITH?" or even "who has on loan that book by SMITH?".   To this end the simple query/update language of Figure 1 is desig- ned, as a conventional programming language.

Clearly we could expect the user to type commands into the system according to the syntax of Figure 1.   Instead we could design a menu/form driven dialogue offering the same facilities.   The menus and forms are shown in Figures 2 and 3.   These are derived directly from the syntax of Figure 1.   Each menu or form is displayed with a heading showing the complete dialogue so far, so that, for example, having made the following choices, menu 1: UPDATE, menu 2: DELETE, menu 3: BOOKS, when form 1 is dis- played, it will have a heading UPDATE DELETE BOOKS.

3   DESIGN OF DIALOGUES FROM A PROGRAMMING LANGUAGE

In the preceding example, we saw how an interactive system was specified by initially

633

FIGURE 1 Syntax and Semantics of the library administration update/query language

⟨command⟩ ::= UPDATE ⟨update⟩ /QUERY ⟨query⟩
⟨update⟩ ::= DELETE ⟨deletion⟩ /INSERT ⟨insertion⟩
⟨deletion⟩ ::= ⟨table - descriptor⟩
⟨insertion⟩ ::= ⟨table - descriptor⟩
⟨table - descriptor⟩ ::= BOOKS ⟨books - descriptor⟩
                    /SUBSCRIBERS ⟨subscribers-descriptor⟩
                    /LOANS ⟨loans-descriptor⟩
⟨books-descriptor⟩ ::= AUTHOR (⟨string⟩ /?) TITLE
     (⟨string⟩ /?) PUBLISHER (⟨string⟩ /?) YEAR
        (⟨number⟩ /?) ACCESSION-NUMBER (⟨number⟩ /?)
⟨subscribers-descriptor⟩ ::= NAME (⟨string⟩ /?) ADDRESS
                              (⟨string⟩ /?)
⟨loans-descriptor⟩ ::= NAME (⟨string⟩ /?)
            ACCESSION-NUMBER (⟨number⟩ /?)
⟨string⟩ ::= ⟨character⟩ *
⟨character⟩ ::= ⟨digit⟩ /A/B/.../Z/./-/,/space
⟨digit⟩ ::= 0/1/2/3/4/5/6/7/8/9
⟨number⟩ ::= ⟨digit⟩ ⟨digit⟩ *
⟨query⟩ ::= ⟨table-descriptor⟩

In the syntax notation,   * means arbitrarily many repetitions of the construct.

SEMANTICS OF UPDATE DELETE

The table selected is searched for the entries described; all entries matching on the field values supplied, with a '?' matching anything, are found; all these entries are deleted from the table.

Semantics of UPDATE INSERT

The table selected has added to it the new entry specified; where the field value is not known, a '?' can be entered - this will have the property that on later searches, a match will always be obtained on this field.

Semantics of QUERY

The table selected is searched for the entries described, as for UPDATE DELETE: the entries found are displayed for the user in tabular form, with a suitable mechanism for handling tables too large for a single screen.

---

specifying the user interface using conventional syntax methods with semantic annotation, with the interactive dialogues being derived from the syntax. In this section, we abstract general rules for the design of man-computer dialogues from a programming language. The details of the dialogue will necessarily depend upon the device through which the interaction takes place, and as in all design situations, these constraints will influence the early stages of the design process and the design will be iterative.

In the design of conventional programming languages, one usually distinguishes two phases. In the first phase, basic functional capabilities are decided; what information structures are to be manipulated and what manipulations are to be permitted. In this first phase, all the essential information that the user must supply (eg. numerical values) will be identified for each command, but no consideration of how the user will supply this information is undertaken. In the second phase, the details of how the facilities are presented to the user are considered: are values identified explicitly by a label which describes them or implicitly by position, and so on. User convenience and the capability of the computer to analyse the language are both considered at this second stage.

Figure 2 Menus for Library Administration update/query dialogue

MENU	OPTIONS	NEXT MENU/FORM
1	UPDATE	Menu 2
	QUERY	Menu 3
2	DELETE	Menu 3
	INSERT	Menu 3
3	BOOKS	Form 1
	SUBSCRIBERS	Form 2
	LOANS	Form 3

Figure 3 Forms for Library Administration update/query dialogue

FORM	FIELD TITLE	INITIAL DISPLAY	ENTRY
1	AUTHOR	?	string
	TITLE	?	string
	PUBLISHER	?	string
	YEAR	?	string
	ACCESSION-NUMBER	?	number
2	NAME	?	string
	ADDRESS	?	string
3	NAME	?	string
	ACCESSION-NUMBER	?	number

In systems design, the first phase corresponds to functional specification, while the second phase corresponds to user interface description: In formal approaches to language design (eg. 1), these two phases are known as "abstract syntax" and "concrete syntax" respectively.

In principle, in designing a dialogue system, we should progress our language design as far as functional specification only, and use that as the basis for the design of the dialogues. However, it is useful to design the language with a concrete syntax as if it were a conventional programming language since, as we will see in the next section, we will want experienced users to take short cuts by enabling them to revert to the related programming language.

Thus we take as our starting point a language which has been fully specified in its concrete syntax using normal language specification conventions (eg. 6).

Four basic rules suffice in guiding our design of the dialogue.

RULE 1 - Syntax of form ⟨class 0⟩ ::= ⟨class 1⟩ / ⟨class 2⟩ /.../ ⟨class n⟩
Make a menu one option per class on the right hand side. The syntax class names may
themselves not be adequate to guide the user, and extra guidance may be necessary.
This guidance could take the form of short descriptions of the classes, possibly only
displayed on demand, but the guidance could employ language terminal symbols like
"DELETE" etc. which are meaningful to the user. The case
      ⟨class 0⟩ ::= SYMBOL·1 ⟨class 1⟩ / SYMBOL-2 ⟨class 2⟩.../SYMBOL·n ⟨class n⟩
is especially useful since the terminal symbols SYMBOLi can be used to denote the
options, as was done in the menus of our Example of section 2.

In other cases, terminal symbols from further down the production sequence could be
used, as in
      ⟨arith - exp⟩ ::= ⟨add - exp⟩ / ⟨sub - exp⟩ / ⟨arith - exp⟩ /
                         / ⟨div - exp⟩ / ⟨number⟩ / (⟨arith - exp⟩)
      ⟨add - exp⟩ ::= ⟨arith - exp⟩ + ⟨arith - exp⟩
      ⟨sub - exp⟩ ::= ⟨arith - exp⟩ - ⟨arith - exp⟩
      ⟨mult - exp⟩ ::= ⟨arith - exp⟩ X ⟨arith - exp⟩
      ⟨div - exp⟩ ::= ⟨arith - exp⟩ ÷ ⟨arith - exp⟩
when the options list for the ⟨arith - exp⟩ menu could be:
+,-,X,÷,NUMBER, ( )
NOTE: in doing this, we are really digging behind the concrete syntax and are looking
at the abstract syntax and from this we are creating an alternative concrete syntax
equivalent to the first.

RULE 2 - Syntax of form:= ⟨class 0⟩ ::= ⟨class 1⟩ ⟨class 2⟩ ... ⟨class n⟩
No explicit menu for ⟨class 0⟩ is necessary, but the interactions for ⟨class 1⟩ ,
  ⟨class 2⟩ ... have to be successively worked through to gather the user requirements for ⟨class 0⟩ .

In our example of section 2, there is no example of this rule. However, it could
have been applied to all those productions which led to a form. For example, the
syntax production for books-descriptor could have been rewritten as
      ⟨books-descriptor⟩ ::= ⟨author⟩ ⟨title⟩ ⟨publisher⟩
                             ⟨year⟩ ⟨accession-number⟩
      ⟨author⟩ ::= AUTHOR (⟨string⟩ /?) ... etc ...
and the production for ⟨books-descriptor⟩ demands Rule 2 - to obtain the information
for a ⟨books-descriptor⟩ , each of ⟨author⟩ , ⟨title⟩ and so on, have to be acquired in turn.

These are the basic rules for a dialogue, and a dialogue could be constructed entirely
from these. However, in many cases,the user can input a collection at one go - this
leads to prompts and forms as in Rule 3. It is also important to keep the user aware
of his progress - hence Rule 4.

RULE 3 - It often happens that the possibilities for completing a syntax class, such
as ⟨number⟩ , are well known and that this can be input in one go. Thus for
      ⟨class⟩ ::= ⟨number⟩
a suitable prompt should enable the user to input a number without further menus.
For more complex primitives like dates, some rules of formation and a few examples
may be necessary to guide the user. Even syntax classes like arithmetic expressions
could be handled in this way.

Where the syntax class demands a set of such responses, we find the requirement for a
form: each separate response needs to be separately prompted in a way that is meaningful to the user and as in Rule 1, terminal symbols taken from the syntax would be especially useful. The example of section 2 has three such forms.

RULE 4 - at all intermediate positions keep the user notified of the story so far.
The easiest way to do this is to maintain on the display the command that he has built
up so far. In a complex interaction sequence, it is very easy to get lost.

In this section, we have progressed from a programming language to a dialogue of menus and forms and similar. This process can be carried through in the reverse direction, and given a dialogue of menus, forms, etc., a programming language can be very easily derived. As we shall see in the next section, we will require both language and dialogue, and the closer these are to each other the better. Ideally, they should be designed together, rather than one first and the other afterwards. Note that Black (2) comes very close to doing what we have done in this section, but then drifts away from the essential linguistic nature of a dialogue.

4 ENABLING USER SHORT CUTS

A completely interactive system can be very tedious to use. For a new user of a system, the menus and forms and other aids are very useful, and little prior learning is necessary in order to be able to effectively use the system. However, once the user obtains knowledge and experience, it can be very irksome to have to wait while a screen full of information is displayed, to have to exercise an option by positioning a cursor, and so on. A command language becomes attractive - the user can pace the system, providing that he does not have to type excessive language keywords, and he will be able to work as fast as his typing skills permit.

In section 3, we partially recognised this requirement by Rule 3, allowing complex information groups which would be readily comprehended by the user (eg. numbers) to be keyed directly into the computer. This capability can be fully generalised to allow the user complete freedom to switch to the command language and back to dialogues as required. To see how this can be done, let us return to our example.

Example revisited

The language and dialogue of the example of section 2 can be represented by a state diagram. One form of this is shown in Figure 4. In drawing this diagram, we have had to slightly modify the syntax to break-up the productions for the descriptions. Starting at state one, the first symbol of the dialogue selects a state-transition to move to the next state, and successive symbols in the dialogue sequence cause successive state transitions. The dialogue sequence continues until State 28, the terminal state has been reached. Figure 5 relates the states to menus and forms.

In the diagram of Figure 4, four sub-diagrams have been used - these form "sub-routines", but are not essential, since the sub-diagram could have been repeated at each place it is used, to give a single state diagram.

In general, from any state in the state diagram, a sequence of symbols will "drive" the system to a new state. Thus from state 2, the sequence INSERT LOANS NAME 'JONES' drives the system to state 25.

Consequently, all we have to do to allow the user to make shortcuts and bypass the dialogues is to allow him to key in sequences of symbols at any point and allow a suitable menu or form or other prompt at the end.

637

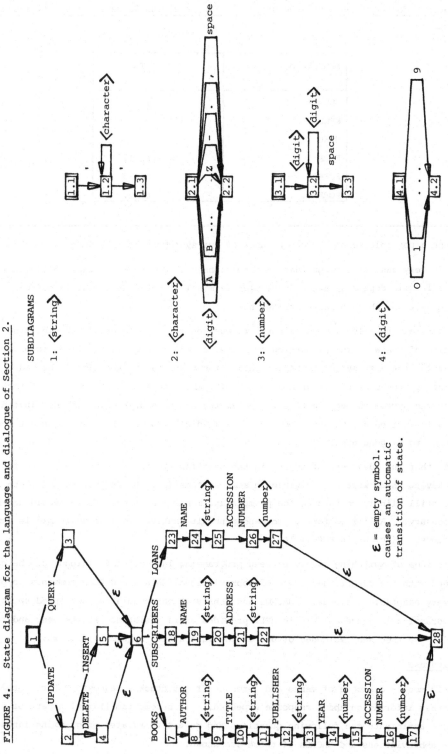

FIGURE 4.   State diagram for the language and dialogue of Section 2.

SUBDIAGRAMS

1: ⟨string⟩

2: ⟨character⟩

3: ⟨number⟩

4: ⟨digit⟩

ε = empty symbol.
causes an automatic
transition of state.

Figure 5 Correspondence between states of Figure 2, and Menus and forms of Tables 1 & 2

MENU/FORM	STATES
MENU 1	1
MENU 2	2
MENU 3	6
FORM 1	7 to 17
FORM 2	8 to 22
FORM 3	23 to 27

Hence a possible scenario for the user of our system is as in Figure 6.

It is very easy to arrange that continuations such as those of Figure 6 can be made. Details will depend upon the particular terminal. There is a cost in software, for we have to be able to parse strings now.

In general, to allow user shortcuts, we must build a generator/parser embodying the notion of state. In any particular state, a menu or form is displayed to prompt the user for the next symbol to be generated, the next state transition to be made. Optionally, the user can input a sequence of symbols to skip a few menus or forms, driving the system through several states as the symbol sequence is parsed; in the state finally arrived at, the next menu or form is displayed for the user to select the next symbol to be generated.

Note that the idea here of typing in the next few symbols is very similar to the idea of keying ahead which is possible in some systems (eg. IBM VM/CMS) though these systems still blindly go through the prompts and menus even though these should not be necessary. A limited form of command/menu alternatives is also available in some systems. (eg. DEC's PDP-11 IAS).

This idea of switching freely between programming language and dialogue can be combined with the established ideas of abbreviations, MACROs and HELP commands to provide a very adaptable dialogue. MACROs and abbreviations enable further speed-up at the programming language end, while HELP provides extra assistance at the menu and form end. We then see two modes of use of the interaction system as follows:

Naive user

Basic mode - dialogue of menus and forms. If get stuck - request HELP to obtain further information about response expected. If become familiar with part of dialogue - anticipate the following questions and revert to related programming language. Use abbreviations and MACROs if these are known.

Figure 6 Possible Scenario for Use of the system with user short cuts

STEP	USER ACTION	SYSTEM RESPONSE
1	signs on	Menu 1, in state 1
2	Selects UPDATE	Menu 2, in state 2
3	Selects INSERT and continues with LOANS NAME 'JONES'	Form 3 partially completed with NAME 'JONES', in state 25
4	Completes form with ACCESSION-NUMBER 103X	Detects error, in state 26 and recovers to prompt user with ? or number
5	Selects number	Menu 0,1,2,3,4,5, 6,7,8,9, in state 26/3.1/4.1
6	Selects 1	Menu 0,1,2,3,4,5, 6,7,8,9,Sp in state 26/3.2/4.1
7	Selects 0, etc.	cycles in state 26/3.2/4.1 until space input
8	Selects space to complete form and dialogue	In State 28 system makes INSERTion requested, and returns to state 1 for a new dialogue

Experienced user

Basic mode - programming language. If get stuck, with partially completed command or statement - system automatically throws up a menu or form to assist continuation. As gain experience - use abbreviations and other short forms, and where appropriate, define MACROs to further reduce keying effort, and thus speed-up communication.

Note that these "modes" are modes of user behaviour and not modes of the computer system. MACROs could be invoked directly from the dialogue, and requests for HELP could be made as part of a programming language style command.

Finally, a formal remark about the scope of this technique. Figure 4 is a simple finite state diagram for a finite state machine. We could draw this diagram because the language concerned was regular (6,8). For context-free languages, such diagrams become more complex, either becoming AND/OR graphs (7), or more usefully becoming state transition networks (14) which are similar to our Figure 4 but allow recursive sub-diagrams. Thus with context-free languages the notion of state becomes more

complex.   This does not set a limit on the computer's capabilities, but could set a
limit on the user.   During a dialogue, a user has to keep track of where his dialogue
is taking him.   With complex languages, especially of a context-free nature, we may
be leading him into trouble.

5   ERROR RECOVERY

At any time, the system may discover an error, or the user on inspecting his dialogue
so far may discover he has made a mistake.   In both cases, there is a need to be able
to backtrack to an earlier point in the dialogue to recover from the error.

The very simplest form of recovery available to the user is the editing of a response
prior to its"transmission" to the computer.   This editing could be local in the hardware of the terminal, or it could involve software;   transmission could be simply the
transmission of information from terminal to computer, or it could be by software
directing the message to the segment of software that decodes commands.   The act of
transmission is a commitment by the user that the message transmitted is intended,
and recovery after this is more complex.   The user requires an ESCAPE or CANCEL key,
to stop the processing he has initiated.   The CANCEL could CANCEL the complete comm-
and, returning the user to the base state of waiting-for-a-new-command, or it could
cancel just the last fragment of the command, with the user having to CANCEL many
times if he wishes to revert to base state.   A nice feature here could be the ability
to edit the command or command-fragment just cancelled, and re-transmit this: this
requires that the computer recovers the command and sends it to the user who then
edits it using the same mechanisms as if it were an original command.

For the computer, the backtracking requirement is much simpler than the error-recovery
requirement for batch compilation (6 ch 15) for it is not necessary to be able to
recover and continue syntax checking as in batch compilation.

Error recovery can be expensive in implementation (see Section 6), so one has to
seriously consider not supplying this - if the user makes mistakes, then he must live
with them, paying a penalty in delay or the need to undo the damage caused (eg. incorrect update).

6   IMPLEMENTATION CONSIDERATIONS

The software to control a dialogue of menus and forms is closely related to the software for compilation/interpretation.   Both kinds of software must have the language
syntax incorporated either in the code or in a table.

In a compiler, the syntax is used for parsing - the user supplies a complete sequence
of symbols, a "statement", which the software analyses using its syntax in order to
understand it: internally a parse tree or equivalent will be formed.   In a dialogue
system the software uses the syntax generatively;   to build up a statement, each time
in the syntax where a choice is possible, this choice is given to the user (Section 3,
Rule 1 and Rule 3).   As the sequence of menus and forms are worked through, the syst-

em builds up a complete statement of what the users requirements are, performing actions on the users behalf as sufficient information is obtained.   In the dialogue system, parsing as such is not performed, but the progress of the dialogue must again be recorded as a parse tree or equivalent.

It is then relatively easy to combine software for dialogue control and compilation/interpretation.   A single syntax table would be used, and this table would include all syntax for the programming language together with all abbreviations and short forms, and would include all menus and forms as appropriate together with any HELP text.   This syntax description would not necessarily be much larger than that required separately for dialogue control and compilation, since the menus and forms could be directly derived from the language syntax, and HELP text could be shared so as to satisfy several needs.

Error recovery requires a lot of software support.   The easy part is the backtracking through the states to an earlier position in the dialogue: this requires only a suitable representation of the syntax, and the actual command being recovered from.   The difficult part comes in stopping any actions being undertaken by the computer (eg. lengthy calculations or listings), and in undoing previous actions.   Cancellation of actions requires that the system is left in a tidy state, as if the action had never happened.   The most difficult part here is the undoing of "updates": a solution here resides in the use of "spheres of recovery" (3,5) or equivalent.   However, one has to accept that recovery arbitrarily far back in time will not be possible because of storage limitations and because some updates will already have been picked up and used by other users.

From the preceding discussion, we see that the implementation of an interactive system allowing free movement between programming language and a dialogue of menus and forms is more complex than the adoption of one single mode of interaction, but it is not as expensive as implementing the two modes as separate and independent modes of interaction.

It is worth commenting on Black's system (2).   While this does not give the user the flexibility we have been striving for, the software does contain interesting features. During the dialogue, a "statement" of what the user has done is built up and can be edited;   a command is code-generated for some other processor to execute.

7 CONCLUSIONS

This paper has shown that programming languages and man-computer dialogues are closely related.   Simple rules allow a programming language to be converted into a dialogue of menus and forms, and vice versa.   Section 2 showed a simple example of an interactive system with both styles of interface, and Section 3 showed rules of construction from one to the other.   Given such rules of conversion, conventional syntactic descriptions can be used for dialogue.

Having shown the relationship between programming language and dialogue, the next step
in sections 4, 5 and 6 was to show how, within a single system, the user could shift
freely from programming language use of the system to a dialogue of menus and forms,
and back again.   Traditional aids such as abbreviations, HELP, MACROs could be included to give a very flexible system capable of accommodating many levels of competence
in their users.   This facility has a cost in software which is comparatively small -
the question is, are the benefits worth the cost involved ? For specialised systems
like airline reservations, the answer is surely no, but for systems like management
information systems where the user population may be very varied, the answer must surely be yes.

8 REFERENCES

1. Becik H., Bjorner D., Henhapel W., Jones C., and Lucas P. "A Formal Definition of
    a PL/I subset" IBM Vienna Labs, report TR 25.139.
2. Black J.L. "A general purpose dialogue processor" National Computer Conference,
    1977. pp 397-408.
3. Bjork L.A. "Recovery Scenario for a DB/DC System" ACM 1973 Proceedings Vol. 28,
    pp 142-147.
4. Date C.J. "An Introduction to Database Systems" Addison - Wesley 1975.
5. Davies C.T. "Recovery Semantics for a DB/DC System" ACM 1973 Proceedings Vol 28,
    pp 136-141.
6. Gries P. "Compiler Construction for Digital Computers" Wiley, N.Y. 1971.
7. Hall P.A.V. "Equivalence between AND/OR Graphs and Context—Free Grammars" Comm.
    A.C.M. July 1973, pp 444-445.
8. Hopcroft J. and Ullman J. "Formal Languages and Their Relation to Automata" Addi-
    son-Wesley, 1969.
9. Infotech "Interactive Computing" State of the Art Report 10, Infotech 1972.
10. Martin J. "Design of Man-Computer Dialogues" Prentice Hall 1973.
11. Martin J. "Computer Database Organisation" Prentice Hall, 1975.
12. Pritchard J.A.T. "Selection and Use of Terminals in On-Line Systems" National
    Computing Centre 1974.
13. Watson R.W. "User Interface Design Issues for a Large Interactive System" Procee-
    dings of National Computer Conference 1976, pp 357-364.
14. Wood W.A. "Transition Network Grammars for Natural Language" Communications of
    A.C.M. Vol. 13, No. 10, 1970, pp.591-606.
15. Zloof M.N. "Query be Example" Proceedings of National Computer Conference 1975,
    pp 431-438.

DESIGN OF INFORMATION SYSTEMS FOR ARABIC

Dr. Patrick A.V. Hall
Ismael A. Hussein

SCICON International
P.Ó. Box 3883
RIYADH
KINGDOM OF SAUDI ARABIA

ABSTRACT

The design of information systems for the handling of Arabic text is considered.
The final objective is assumed to be the storage of information in Arabic, with the
requirement later to retrieve the information using Arabic.

Initially, those aspects of Arabic which are of concern in the design of computer
based information systems are surveyed. Arabic script is examined for its division
into words and its letters and their various forms. Particular attention is paid to
the use of spaces and decorative bars. Names of particular importance in many information systems, and Arabic names are examined in depth.

Then systems design in all its aspects is considered. Peripheral design and the design of internal codes are first considered, weighing the alternative choices against
the impact these would have on the retrieval algorithms. Then the various algorithms
for searching using indexes, comparing text for identity, and comparing for sorting
are examined. Also algorithms of secondary concern, searches for appropriate matches,
are discussed.

Finally, it is argued that codes and character forms should be standardised. Recommendations can be made concerning the handling of Arabic text and Arabic names, but it
is not proposed that these should be the basis for standardisation.

1. INTRODUCTION

This paper began with an anecdote. We were told a story of a group of European computing experts who built a demonstration system for Arabic users - it would store, in
Arabic, a person's name, and some associated information, and enable the person to
retrieve the associated information later using the person's name as a key. The first
Arab they tried, was called Ali (علي): he typed his name and other information, then
details of his friends. He then switched across to retrieval to recover the information about himself - and failed ! What had gone wrong ? Had the computer broken
down, or the software developed on bug ? Nothing of the sort: an Arab was using a
computer system which had been designed on non-Arabic principles. Ali had typed his
name with a flourish the first time, as علی, but left out the flourish the second
time, typing علي. In Arabic this variation is natural, in European languages, it
is unthinkable.

We took the anecdote to heart.   Maybe we as European technologists were in danger of building into our computer system too many assumptions about European languages, and not recognising the differences that exist in Arabic.

So we started to look at Arabic without making assumptions to discover what Arabic was really like, and to design afresh a computer system for Arabic.   One of the authors is a specialist in computing, but being European necessarily has all sorts of in-bred assumptions about the nature of language and computer information systems which may be invalid for Arabic.   The other author is an Arab and a specialist in liguistics.   Together we hope we have been able to sow the seeds of a fully Arabic Computer System.

In section 2, we survey the aspects of Arabic that seem particularly relevant to computer systems.   In section 3, we then examine the design of computer systems for Arabic, discussing the options available from the internal codes, to the peripherals, to the data capture and retrieval mechanisms.   Finally in section 4 we examine systems design from the point of view of standardisation.

2.   SURVEY OF ARABIC

2.1   Arabic characters and Text

European languages have developed a script in which text can either be cursive as in normal handwriting, or it can be non-cursive as in most printing and typing.   In cursive handwriting all letters within a word join together with breaks between words. In non-cursive printing or typing, letters within a word are packed close together, while between words pronounced spaces occur.

By contrast, Arabic is always cursive in both handwriting and printing.   Some letters join together, other do not.   Breaks in the cursive flow can occur within words, and must occur between words.

Our primary concern in investigating Arabic is to discern how Arabic would "naturally" be typed into the computer by untrained personnel, and how it should be printed for easy reading by untrained personnel.   Then, in designing our computer system we must arrive at a balance between making the computer handle Arabic naturally, and training the users to work effectively within the limitations imposed upon Arabic by the computer system.

2.1.1   Character Form

Arabic letters vary in shape in accordance with their position within a word.   While in general letter forms vary a lot (eg. 6) it is useful to adopt the approach of recognising four basic potential forms of letter - initial, medial, final or isolated, depending upon the position of the letter in a cursive sequence of letters.   Cursive sequences correspond to words or fragments of words.

Arabic letters are of two types.   Some of them can join the following letters and are called "inseparable letters" (Mitchell's terminology). Others never join the following letter and are called "separate letters". Each of the inseparable letters may appear in one of four forms. Separate letters can appear only in two forms. Figure 1 shows the different forms of Arabic letters together with their pronunciation in English.

There is the tendency to use only two forms of the inseparable letters on typewriters and printers, the initial and the isolated forms, and only one form of the separate characters - the isolated.

Let us consider one particular typewriter in detail. Figure 2 shows the different forms of the Arabic letters, as they appear on this typewriter. In addition, this typewriter has the diacritic symbols, shadda, madda and hamza.

Recently there have been several attempts to simplify the Arabic characters, using only one form, the initial form, for each letter (see the paper by Almajid, 1978). These developments are clearly of great interest to the computing profession and should be followed closely.

Written Arabic basically records the consonants of the language, together with the long vowels. The short vowels are added by the reader, but can be written using diacritic marks. There are several other diacritics in use in Arabic. In writing and printing, these diacritics are usually omitted unless they are necessary to avoid ambiguity. On typewriters, and some printing presses, it is only possible to insert some of the diacritics, and the practice is to insert a parenthetical remark to indicate a missing diacritic if this is necessary.

2.1.2 The Extension Bar

In Arabic handwriting (see Figure 3 (a)) it has become accepted practice to be able to extend the line joining adjacent letters at the discretion of the caligrapher. This idea has been carried through into typed and printed Arabic, a special extension bar character being used for this purpose. Figure 3(b) illustrates.

This extension bar is used to enhance readability, to justify the text being written on both the left and the right, or for artistic purposes. This extension bar has no linguistic significance at all.

2.1.3 Widths of Characters

The Arabic characters do not have equal widths. Some characters could take a space five times as big as others. On most typewriters two character widths are adopted.

For example, on one particular typewriter about forty percent of the forms of the letters are given single space widths (1/16 inch) while the other sixty percent of the forms of the letters are given double space widths (1/8 inch).

However, some characters run together, so that on occasions the characters must be separated by the appropriate use of the extension bar or the space character.

FIGURE 1. The different forms of the Arabic Letters, as used in printing (from Mitchell and Barber, 1972)

	isolated	final	medial	initial	English symbol	name
1.					aa/oo (')	'alif (hamza) (see below)
2.					b	baa'
3.					t	taa'
4.					th	thaa'
5.					j	jiim
6.					H	Haa'
7.					kh	khoo'
8.					d	daal
9.					dh	dhaal
10.					r	roo'
11.					z	zaay
12.					s	siin
13.					sh	shiin
14.					S	Soad
15.					D	Doad
16.					T	Too'
17.					DH	DHoo'
18.					9	9ayn
19.					gh	ghoyn
20.					f	faa'
21.					q	qoof
22.					k	kaaf
23.					l	laam
24.					m	miim
25.					n	nuun
26.					h	haa'
27.					w; uu	waaw
28.					y; ii	yaa'

FIGURE 2. The forms of the Arabic Letters appearing on an OLIVETTI LINEA 98 Arabic Typewriter.

	isolated	final	medial	initial	English Name
1.					alif
2.					baa'
3.					taa'
4.					thaa'
5.					jiim
6.					Haa'
7.					khaa'
8.					daal
9.					dhaal
10.					raa'
11.					zaay
12.					siin
13.					shiin
14.					Saad
15.					Daad
16.					Taa'
17.					DHaa'
18.					9ayn
19.					ghayn
20.					faa'
21.					qaaf
22.					kaaf
23.					laam
24.					miim
25.					nuun
26.					haa'
27.					waaw
28.					yaa'

2.1.4   The Space Character

Keyboard devices, such as typewriters, are equipped with a space character.   Under what circumstances is this used in Arabic ?   We have already seen in the last section that the space character may be used to enhance legibility, to prevent letters running into each other.   But we would also expect the space character to be used for the separation of words, a grammatical use of the space.   For the moment we take the concept of word as understood - the concept is explored further in section 2.2.1.

In handwriting, Arabic is written without leaving significant gaps between words with only breaks in the flow being breaks in the joining of successive letters, either between words or within words.   Figure 3(a) illustrates.   In printing, however, it is normal to leave significant gaps between words, as seen in Figure 3(b).   In typing the practice seems to be somewhere in between;   the space character is, or should be, used between words, but the gap created may not appear significant visually.

From the linguistic point of view, however, gaps between words are not necessary, these being purely graphically devices inserted to improve the appearance and readability of the text.

A problem related to that of the insertion of gaps into text, is the continuation of text onto the following line.   Where can the text be broken in order to continue onto the following line ?   In English the rule is either between words, or within a word between syllables with the use of a hyphen.   In Arabic, the rule is only to break a line of text at the division between words.

2.1.5   Arabic Numbers

Arabic numbers are used the same way as the English numbers, with units on the right. Normally we write Arabic numbers in the same order that we read them, as we do in English, that is we start from the most significant digit.   However, when transcribing a number, one would proceed from right to left.

In Arabic, the decimal point is written with a comma (as in German), using the letter 'raa' for this purpose and not the Arabic comma.   However, when numbers exceed a thousand, the numerals can be grouped with the groups separated again by a 'comma'. Thus numbers can be ambiguous, and the ambiguity must be resolved by the context in which the number appears.

2.1.6   Dates

Two systems of dates are used in the Arab world.   The Gregorian system used in Europe and America is in general use, but additionally the traditional Islamic Hajjerian calendar is used, especially in Saudi Arabia.

The Hajjerian system of dating is based upon the phases of the moon, there being 12 lunar months per Hajjerian year.   This means that the Hajjerian year is slightly shorter than the Gregorian year which is based upon the phases of the sun.   The Hajjerian new year falls approximately ten days earlier each Gregorian year.

FIGURE 3. Samples of Arabic Text: (a) handwritten (b) printed . In both cases note the allignment left and right by extending the line joining adjacent letters, and note the ansence of clear gaps between adjacent words. Both passages have been repeated in (c) and (d) to mark the word breaks with a solidus.

(a) handwritten, from a newspaper advertisement

علماً بأنه لن يتم تمديد هذه الفترة
مرة اخرى، وسوف يجري بيع أسهم
المتخلفين بعد خمسة عشر يوماً من

(b) printed,

بهذه المناسبة ينبغى ان نذكر بشعور الامتنان العميق وعظيم التقدير لما يتقدمه لنا صاحب
السمو الملكى الرئيس العام لرعاية الشباب وسمونائبه الكريم من توجيهات وارشادات واهتمام بالغ
بتشجيع الشباب السعودى من اجل احياء وتطوير الفنون التشكيلية وايجاد المناخ الفنى الملائم
لمسايرة النهضة الكبرى التى تشمل مملكتنا الحبيبة .

(c) handwriting showing the word breaks

علماً/بأنه/لن/يتم/تمديد/هذه/الفترة/
مرة/اخرى/ وسوف/يجري/بيع/أسهم/
المتخلفين/بعد/خمسة/عشر/يوماً/من/

(d) printing showing the word breaks.

بهذه/المناسبة/ينبغى/ان/نذكر/بشعور/الامتنان/العميق/وعظيم/التقدير/لما/يتقدمه/لنا/صاحب/
السمو/الملكى/الرئيس/العام/لرعاية/الشباب/وسمو/نائبه/الكريم/من/توجيهات/وارشادات/واهتمام بالغ/
بتشجيع/الشباب/السعودى/من/اجل/احياء/وتطوير/الفنون/التشكيلية/وايجاد/المناخ/الفنى/الملائم/
لمسايرة/النهضة/الكبرى/التى/تشمل/مملكتنا/الحبيبة/.

FIGURE 4. Examples of Arabic Words. The word is a compound object, and nouns have attached to them the article, prepositions, possessive pronouns, and the 'word' "and".

people	ناس	book	كتاب
the people	الناس	his book	كتابه
with the people	بالناس	like his book	ككتابه
like the people	كالناس	and his book	وكتابه

Since both systems are defined in terms of astronomical phenomena, we would expect a very exact correspondence between the dates of one system and the dates of the other system.   But this is not so.   Each calendar system is concerned for the correspondence between two astronomical phenomena - days (i.e. rotations of the earth on its axis) and lunar or solar periods.   Both systems must regulate themselves so that a whole number of days fit into the larger astronomical period on which they are based. The Gregorian system regulates itself by the prediction of astronomical events, with the number of days per year varying on a way which is intended to maintain the calendar in step with the phases of the sun.   By contrast the Hajjerian system regulates itself by astronomical observation, with the onset of successive months being determined by the observation of the new moon.   Occasionally the expected number of days in a month has to be adjusted: the adjustment may only be announced after a few days, then becoming retroactive to the beginning of that month.

The nett effect of these different methods of regulating the calendar is that an exact correspondence between predicted Hajjerian dates and predicted Gregorian dates is not possible, though correspondence to within a day is predictable.   For historical dates a precise correspondence between the Gregorian calendar and the Hajjerian calendar is possible, but this correspondence will not be possible by calculations based solely upon astronomical laws.   Rather a system of tabulated dates with interpolation between using astronomical laws is necessary.   However, for many purposes an exact correspondence is not necessary, and in this case a correspondence based solely upon astronomical laws can be used, giving an error of one or two days at most.   Note, however, that for the period of any retroactive adjustment there will be some uncertainty concerning the precise date recorded on documents - was the date recorded, adjusted, or not ?

2.2 Some Aspects of Grammar

2.2.1 What is a word ?

The notion of a word in Arabic is relatively complex, since an addition to an Arabic word can become a part of that word.   Look at the examples of Figure 4: the most surprising example is probably the inclusion of "and" into the word that follows.

2.2.2 Consonantal Roots

Most Arabic words are made by adding different prefixes or suffixes to what is called the root of a word.   This root is often made of three consonants.   (eg. Tritton).

Figure 5 shows some examples of words derived from a single root.   Note that in some variations, consonants actually occur between the consonants of the root.

2.2.3 Alphabetical Order for Arabic Words

The letters of the Arabic alphabet are tradionally ordered as in Figure 1, starting with Alif and ending in yaa.   To this must be added the rule that hamza precedes alif, and that the compound character laam-alif ﻻ ﻷ is treated as a laam follo-

FIGURE 6. Examples of Arabic names, to illustrate the variety of problems that might be encountered when analysing names.

(a) Ismael Abdelfatah Hussein Salih   اسماعيل عبد الفتاح حسين صالح

(b) Ahmad Hassan Asseeri   احمد حسن عسيري

(c) Abu Ja'far Mohammed ibn Musa Al Khowarizmi   ابو جعفر محمد بن موسى الخوارزمي

(d) Suraj Aldin Abd Rub Alnaby Saud Alkabeer   سراج الدين عبد رب النبي سعود الكبير

(e) Mohammed Haffez Mansour   محمد حافظ منصور

FIGURE 5. Examples of words derived from the consonantal root د ق ع

To assemble	انعقد
To make a contract	تعاقد
To believe	اعتقد
Knot	عقدة
Braid weaver	عقّاد
Lieutenant Colonel	عقيد
Contractor	مقاول
Faith, belief	عقيدة
Bunch	عقود
Belief	اعتقاد
Dogma, principle	معتقد
Complicated, complex	معقّد
Pacreas	معقد
Tied, bound	معقود

wed by an alif.

Traditionally all Arabic words in an Arabic dictionary are ordered in lists in accordance with the alphabetical order of their consonantal root.   So when looking up the word

MAKTUUB   مـكتو ب

we cannot find it under meem because we have to extract its consonantal root which is

K T B   ب ت ك

so we will find it under K - T - B.

For telephone directories and similar works it is already established practice to arrange these alphabetically in the European manner, and there have been some attempts to do this for Arabic dictionaries.

2.3 Arabic Names

It is worth considering how names are used in an information system.   Names are given to people as a label with which to refer to the person; but names are more than just a label, they tell you something about the person.   They tell you the nationality or ethnic and linguistic group to which the person belongs, and by analysing the name, the names will tell you the sub-group within the ethnic group from where the person comes - his family title or clan, the area in which he was born, his father's and grandfather's names, perhaps even his mother's name.   Thus a person's name carries with it a lot of information, an important aspect of the manipulation of names in an information system is the analysis of the name and the division of it into its information components.   You want to be able to answer the question "who is this person ?"

It is worth distinguishing between an individual's legal name and his informal names. With modern administration society people acquire a legal name at birth, and this name can only be changed by properly constituted legal processes.   The most common of these is marriage, but there are processes in most administrations whereby a person can voluntarily change his or her name, and whereby titles can become incorporated in the person's legal name.

2.3.1 Examples of Arabic Names

Consider the examples shown in Figure 6, where Arabic original and English transliterations are shown together.

The traditional structure of Arabic names is a declaration of lineage; each child is given a personal name, and the person's full name consists of his personal name followed by that of his father then his grandfather, and so on.   The first example of Figure 6 is the full name of one of the authors recording that Ismael's father was Abdelfatah, grandfather was Hussein, and so on.   Names such as these can go back for as many generations as desired.   The connective "bin" (often also transliterated as "ibn" or "ben") meaning'son of' is frequently used, as in the second example.

It is becoming increasingly common today for Arabs to use a family-name, presumably as a result of contact with the west.   This family name can either be a statement of the place or origin;   example (b) is Ahmad son of Hassan from the Asseer province of Saudi Arabia, while example (c) is Mohammed son of Musa from the village of Khowarizm. The family name is usually written last, and often prefixed by an "Al" (or "El") as in example (c).   Most commonly the family name appears to have been acquired by choosing one of the names in the sequence of lineage (typically the last out in the sequence written down - thus the author Ismael of this paper has on occasions been known as Mr. Hussein, and Mr. Salih).   In formal administration systems one is typically asked for four names, the last of which is taken as the family name.   The names in the Riyadh telephone directory are sequenced by family name: about half these have the prefix "Al".

There are some bemusing anomalies to the use of "Al".   The name "Sheikh" also occurs in the form "Al sheikh" - and hence the family name "Al Alsheikh".   However, when faced with the family name "Alsheikh" is the "Al" a prefix for the family name or not ?   Sometimes a famous ancestor is singled out as the giver of the family name: this could be the father, with further names written after it.   Furthermore, sometimes the name can be prefixed by 'Al', but not be a family name.   There is also some variability in the way the prefix 'Al' is used.   It is usually treated as a simple prefix, being joined to the family name, but in some cases can be treated as a separate word.

People are often simply known as the son of their father, as in "Ibn Saud".   This can also happen with the mother's name, for example, "bin Kabina" (Thessiger).   In passports in the Arab world it is usual to include the mothers name (only the given name), as an extra aid to the identification in cases where otherwise the names would be the same.

It is quite common for people to become known as "father of" as in example (c).   Traditionally this was more than just a nickname, for the person becomes known by this name and the names "Abu Ja'far" etc. will be quoted in the lineage and in modern usage can become part of the family name.   In current usage however the Abu name is a nickname in situations where courtesy is called for.   In various areas of the Arab world there are alternatives to "Abu" - for example, "Aba", "Ba", or "Bo".   For women an equivalent "mother of" nickname can be used.

The prefix "Abu" in example (c) is an example of a comparatively frequent occurrence - a single name composed of several words.   The most common version of this is those names starting with "Abd" meaning "servant of".   Example (d) shows a name of three components but seven words.

As a final example, an added complexity occurs occasionally where a person is given a double name as in Example (e): the father Mansour gave his son the double name "Mohammed Haffez" and the son is known variously as "Mohammed", as "Haffez", as Mohammed Haffez".

Most of the phenomena noted above do also occur in European names.   Thus in European
names it is not always obvious which is the family name especially since frequently
this is written first.   It is also common for names to consist of several words,
especially for family names (e.g. "van der Merwe"), while double names are also
frequent (e.g. Jean-Louise).   In European names one also has the extra complication
of abbreviations.   Currently abbreviations in Arabic are infrequent, and do not happen with names, excepting that prefixes such as "Abd" are sometimes dropped in European circles.

Note that in Islamic tradition at marriage women do not adopt the name of their hus-
band, but retain their own name with their own paternal lineage.   However in European
influenced Islamic societies, at marriage women do change their name.

2.3.2 How Titles are used in Arabic

It is not customary for Arab people to address each other using their pure names.
We have already seen the use of "Abu", and in formal situations one would insert a
title into the name of the person addressed.

Normally titles precede the name they qualify.   Such are titles like AMIR (price),
SHEIKH, SHERIF (nobleman), DOCTOR, USTAZ (teacher, professor) and HAJ (HAJJI).   Only
AMIR, DOCTOR AND SHERIF always accompany the name of the person.

The confusing titles are SHEIKH and USTAZ.   Although USTAZ us basically used for
teacher at an educational centre, it can be used for any educated man.   The title
SHEIKH is still more confusing.   Originally it was used for the head of a bedouin
tribe, but nowadays the word SHEIKH is used for middle-aged (or older) people who
are usually very rich and enjoy a very high rank whether official, commercial or
social.

A military title is given by a very high authority, and one cannot change it.   Even
when pensioned off one will still have the right to use it as one's title (usually
it is denoted that the involved person is retired).   These military titles become
an integral part of the person's legal name.

Sometimes a man would get a title after his profession.   In such a case this title
is added to the end of the name.   Here we cannot guarantee that the title will appear with the person's name on all occasions.   Examples of such titles are: najjar
(carpenter), Haddad (blacksmith), LaHHam (butcher) and kattan (cotton dealer).

There are a few titles which have become rather archaic nowadays.   These are titles
like Basha, Beik and Afandi.   These titles used to be given to government employees,
Basha being the highest then Beik and the last is Afandi.   Such titles are usually
placed directly after the first name of a person, and the first name plus title were
treated as the full name of the person.

3. ARABIC TEXT IN THE COMPUTER

In the design of a computer system to handle Arabic, we have several related decisions
to make.   We must design the peripherals, displays and printers, and decide in the
external representation of the Arabic characters on each of the devices.   We must
decide on the internal representation of the Arabic script.   And we must also design
the algorithms for use in the software for matching, searching and sorting.

Some discussions of these problems have been made in the literature (e.g. the papers
of Tawakol (15) and of Seflan (14) focussing primarily on the problems associated with
the character sets offered by one particular manufacturer).

3.1 The Coding of Arabic Text

We wish to store inside the computer a representation of Arabic text.   But what is
Arabic text ?   Is it a sequence of letters divided into words, or is it a sequence
of characters, letters in their various forms, with no distinct words, or what is it ?
The answer to this will determine how we construct our code for Arabic.   Do we code
letters only, leaving the conversion to the various forms to a transformation just
prior to printing, or do we code the individual forms of the letter ?

We could choose one distinct code per letter, and generate the various external forms
of characters automatically - the particular form depends at most upon the letters on
either side, assuming that a special symbol is always used to separate words.   The
generation of correct printed form would necessarily be expensive.   Note that this
transformation could actually happen in software if some other code was used for interchange between peripheral and computer, but would most sensibly be done in a micro-
computer in the peripheral.   With a single code per letter, matching for identity
is trivial.   Ehlers (6) has reported a system for computer typesetting of Arabic
with an input keyboard for letters rather than forms and an output character set of
many hundreds of forms.

As the alternative to a code per letter, we could choose one code for each printed
form of each character.   With this strategy, there is the cost of added complexity
of matching and comparing in the computer, though this cost is small.   With this
course there could be an interest (a disadvantage) in reducing the number of forms,
since the number of codes available might be limited.   The character set may have to
accommodate other sets such as Latin capitals.   We would then have to decide how
many forms of each letter are necessary and what the widths of all the forms are.
Having too few forms leads to loss of legibility, but having too many permits alter-
nates choices for a given position (e.g. medial and initial   ) and thus extra
complexity in matching.

Independently of the decision to code letters or forms is the decision concerning
fixed or variable length codes.

If a variable length code was used, it could be convenient to make the lengths of the code (measured in bits or bytes) proportional to the length of the printed character. This would then have the advantage that space requirement for printing a string of text would be proportional to the internal storage requirement. The price for this is a small increase in complexity of matching and comparison algorithms; a single byte by byte comparison from the right hand end still works, but different forms must be recognised as equivalent and only if the different forms were different lengths would any significant extra complexity arise. However, this variable length strategy would be paid for in extra internal storage requirements, possibly leading to the use of too few multiple length forms and degraded legibility.

Another reason for moving to variable length codes could be that the number of codes of fixed length available is less than the number of characters or forms to be represented. In this case it would be sensible to choose the longer codes on the basis of minimising storage for less frequent codes. This would most likely be for non-Arabic characters and an idea worth considering is automatic shift and unshift characters for sequence of non-Arabic characters. One can readily envisage an Arabic - ASCII code to complement the current Latin-ASCII, where in the Latin - ASCII there is a shift-to-Arabic code, while in Arabic - ASCII there is a shift to the Latin code. This strategy is (almost) open-ended, and codes for other scripts could easily be added.

Using a fixed length code with a variable length external representation should, in fact, give remarkably little difficulty, though it would be thought novel for those used to the usual Latin peripherals. In a simple storage and retrieval application, one would capture the data in some predetermined space - and would be able to retrieve to that same size space. What goes in comes out. Now however there is uncertainty about how much internal storage will be required for a given amount of printing space, but an upper bound is easily estimated. There should be little or no storage cost and if an estimate of printing space is required for a given internal string, it is readily calculated by scanning the text. To assist with alignment of columns in printout of tables, a TAB facility on all output peripherals would be very useful.

In all cases, the code sort orders as binary numbers should be the same as the normal Arabic character sort order, with some convention concerning different forms, for example isolated before final before medial before initial.

A problem of some significance is concerned with the diacritics. Which of these are coded ? Of those that are coded, would the code for the diacritic precede or follow that of the letter which it qualifies ? The answer to this second question may well depend upon consideration of peripheral design, addressed in the next section. If diacritics are included, we may have complications in the matching algorithms since in some cases the use of diacritics is optional.

3.2 Peripheral Design

Arabic text stored inside the computer needs to be displayed or printed for the user.

It is also necessary to be able to obtain information in Arabic from the user.   The peripheral devices for this input and output need to be designed: the part of their design that is of interest here is the form the Arabic text takes on paper or visual display screen, and the keyboard used for input.

We assume here that the discourse between peripheral and computer will use the code discussed in section 3.1.   We saw there how the design of this code depended upon considerations of what would be printed - how many different forms of characters are really necessary and how their width would vary.

Arabic characters vary so tremendously in width in normal writing and printing (3 to 5 times) that at least two widths must be considered.   Having recognised this inevitability, it is necessary to decide on the division of the characters into single and double lengths.   The considerations here are legibility and printing density.   If too many characters are forced into single lengths, the basic lengths of these must increase or legibility will suffer.   This leads to reduced packing, with the text requiring more space than necessary upon the screen or paper.   Note that there is an alternative course of action: it might be possible to accommodate all characters in a single width if suitably redesigned, though this may lead to too large a departure tradition to be acceptable.

Other considerations of significance are diacritics and the TAB facility discussed in the last section.   Keyboard layout is also important, and clearly a layout similar to that of typewriters is desirable.

3.3 Coding Forms

When Arabic is typed from a hand-written original, the exact form of the outcome is unpredictable, since the typist has a lot of discretion about layout, and extension bars and spaces can be introduced, and spaces may even be omitted.   This same thing could happen during the entry of data into a computer system, unless we take steps to avoid it.

Given a background of European computing it seems reasonable to use coding forms, with each character written in a separate box, so that a precise character by character specification will be given.   But, coding forms for Arabic may not be a reasonable facility to use.

As has been described above, Arabic uses a cursive script in which characters vary up to five times in width.   There is no hand-written block printing, and even in learning to write Arabic, one progresses almost immediately to the writing of cursive words.   Thus a coding form for Arabic would have to be filled in using cursive writing: to see that this has problems, imagine filling in a coding form in cursive English!

We have conducted a small trial using a standard English coding form for Arabic text, and found that the tradition of extending the separation between characters by using

a horizontal bar is now an asset, and placing letters firmly into boxes is easy apart
from four very broad letters (Saad, Daad, Seen and Sheen) which seem to demand two
boxes ! For these letters, we would need to adopt special conventions of writing
(as is done for some Latin characters, such as zed, to avoid confusions), so that a
one box one keystroke identity can be preserved.

In the context of the design of a single complete computer system, the introduction of
coding forms with rigid blocks for letters could involve a considerable cost in training and in error - correction post data-entry. The alternative is to accept the
traditions of written and printed Arabic and arrange for the computer to handle the
traditional form of Arabic. It will be seen in the following sections that the cost
in software is by comparison very small.

Of course, the use of coding boxes for numerical data in Arabic in perfectly reasonable.

3.4 Matching and Sorting

Let us start by considering the comparison of two strings of Arabic characters. Let
us take as a basic algorithm, a character by character comparison starting from the
right. Then a simple-minded method for comparison would be to say the strings are
identical if each pair of characters were identical, while they would be different if
one or more pairs were different (or if one had more characters than the other) with
the sort order being determined by the rightmost non-matching character. This is
precisely the method we would use based on experience with European languages, but
does it work for Arabic ? We must be clear what we are trying to achieve when comparing two text strings. We are not simply looking for superficial resemblances, we
are asking whether they are the same strings as sequences of symbols in the Arabic
language.

If the coding of Arabic text is at the abstract level, with letters coded and grouped
together as words, then already the nature of Arabic has been taken into account, so
the simple-minded comparison methods described above will suffice. However, the
coding methods that we meet will almost invariably be an encoding of letter forms, as
discussed in section 3.1.

The most obvious variability in Arabic words occur with the optional use of the extension bar and space characters. We could avoid this problem by imposing upon the user
a discipline in the use of these symbols, so that he always uses them in the same way.
At its simplest, this discipline could take the form of disabling the keys for these
characters so that they can never be input. This may be acceptable for the extension
bar, but for spaces we would most likely have to train him in the 'correct' use of
the space character. The alternative is to allow free use of these characters, so
that text can be formatted as desired by the user, but take compensatory action in
software. What we want to do is to compare the strings of text without any extension
bars (or spaces, if we allow freedom with those). The comparison algorithm could

either work directly upon the original strings, skipping any bars or spaces, or one or both strings could be reduced to canonical form by stripping all bars and spaces before making the comparison. If indexes are used, then the index must be built from the canonical form with all extraneous characters removed.

It is possible in typed or printed Arabic for different forms of the same letter to be used in the same position in the same word: this can only happen with particular letters, usually where the initial form can be used in a medial position. There are two ways out: either the peripherals and code could not distinguish between the forms, or the matching algorithms could take this into account. Since there is very little loss in legibility by removing the medial form, and a lot to gain in matching simplicity, it is clearly worth making this simplification in the character set.

If variable length codes are used (for example, some letter forms involving two or more bytes of code, or shift characters present), no further complications arise for exact matches, but for sorting comparisons extra complications can arise because of the variable length, especially if different forms of the same letter have different lengths.

It could happen that in performing a match, in searching for a particular word, one wants to find all grammatical variants of a particular word. To do this in English, you would use the stem or root of the word: for example to match the words "computer" and "computation" you would use the stem "comput-". In English it is adequate to provide the facility to match a string embedded in another string, since the grammatical variants almost always occur with prefices and suffices. Unfortunately, in Arabic the grammatical variants often insert extra letters into the middle of the root of the word, as well as adding suffices and prefices. (See section 2.2.2). Note that within the words derived from a root, the letters of the root can appear in any of their possible character forms and long vowels can actually change into other long vowels. Note also that it is not possible to automatically construct the consonantal root of an Arabic word from the word itself. Thus the user would need to supply the root, or the system would extract all possible candidates, and the user would have to tolerate the retrieval of many nonsense matches.

This discussion of consonantal root matching is one aspect of a more general system's requirement, that of retrieval of information given partial information, or imprecise information. The simplest form is the use of a key, or several keys, which are used to retrieve information. But often the exact form of the keys is not known - hence the consonantal root method. But partial knowledge manifests itself in other ways, and in section 3.6 on names, we will see partial knowledge in the form of parts of names, and variations in transliteration.

3.5 Text Analysis

Often it happens that it is necessary to analyse a piece of text and break it up into

its constituent parts.   This takes us into the realms of linguistic analysis, and
while it is not our intention to discuss this subject in depth, nevertheless it is
necessary to say something if only as background to the examination of names made in
the next section.

One of the simplest forms of analysis is the subdivision of the text into words.
In English this can readily be accomplished by finding spaces, since it is an invari-
able rule that successive words are separated by a space, and possibly also some
punctuation symbol or symbols.   In some programming languages, spaces are also used
to separate words ("language keywords") from each other.   But in Arabic rules for
the insertion of spaces are not invariably applied, as we have seen in section 2.
And when they are, the rules are certainly very different from the conventions of
English (for example, the inclusion of ' و ' meaning "and" into the word that follows).   Thus, the segmentation of Arabic text into words can only be undertaken using
spaces as separators if it is known that a consistent discipline concerning the use
of spaces has been employed.   In general, the segmentation of Arabic text into words
must use some linguistic structure.   However, this does not mean the full linguistic
structure of the language, but simply enough structure for an analysis like the
lexical analysis of compilers (e.g. Gries, 7).

Spaces are often used as separators in programming languages: their use in Arabic
programming languages seems quite defensible, so one could imagine Arabic versions of
languages like PL/I, as well as Arabic versions of ALGOL and FORTRAN where spaces are
(generally) non-significant.

Another simple form of analysis is the searching of a string for a particular substring.   Here all the matching considerations of section 3.4 have to be taken into
account, together with the fact that if the search is for a word, bounding spaces
cannot be assumed.   In particular, this means that sophisticated methods for substring searching (e.g. Boyer and Moor (4)) may in inappropriate or at best in need
of considerable modification.

3.6 Storage and Retrieval of Arabic Names

When Arabic names are to be stored in an information system, there are three levels
of use to which these can be out.   Firstly, they can be used passively, being retrieved only when something else is retrieved.   They are not used as a basis for retrieval.   Alternatively, the names can be used actively, as a basis for retrieval, in
which case, secondly, only the complete name is used, or thirdly components of the
name can be used.

For passive storage and retrieval, the name can simply be stored as text, and there
are no problems.   Any analysis of the name would be done by the users of the system
following retrieval of the name.

If the name in full is to be used actively for retrieval, again it can be treated as as a single text item. For fast retrieval an index could be built: there are many techniques available, some of which may involve truncation of the name (see for example Knuth or Lefkovitz): the matching techniques discussed in section 3.4. are appropriate.

It is only when the name is to be used actively with components of the name being used as the basis for retrieval that we become really concerned with the internal structure of names as explored in section 2. The basic requirement is for one or more components to be specified and for all names having these components to be retrieved together with any associated information stored in the system (such as telephone number, address, date of birth). The retrieval request will most likely be able to specify for each name component the role that it plays - first name, father's name, grandfather's name, family name. We may also wish to use other components of a person's name, such as title either formal or informal, or the name of the eldest son as used in nick-names of the Abu variety. How many name components are used, and what their functions are, are important issues which must be decided and once decided used consistently. However, what these decisions are does not affect the considerations of computer software that are made below.

The primary systems design question is how we store the names so that the required retrieval is efficient. The secondary question is how these names are input at initial capture and at retrieval. Two distinct approaches to storage present themselves. The name components can be separately indexed for fast retrieval on separate name components; or the words forming the name can be separately indexed, with retrieval using several words in combination. In both cases the full name may also be stored as text.

If name components are separately indexed, we have two choices, either to have a single index of all names, or to have one index for each distinct name-role (first, family, etc.). In the first case, the partial name for retrieval would have to be divided into components, each name being used to retrieve a set of record indentifiers; these sets would be intersected to find those records that contain all the specified names; finally the records would be retrieved and the full names checked that the sequencing of the names and their positions is correct. In the second case, the partial names for retrieval would have to have their roles identified; the appropriate indexes would be selected and searched; the sets of records obtained from each index could be intersected to give the required record identifiers. Note that in the first method it is easier to answer a question like "Two of whose names are Mohammed and Ahmed", than by the second method.

With the name components separately indexed, the real problem lies in the capturing of the name. This cannot be input as a single string of normal text and segmented within the computer, because the rules of formation of names are inherently ambiguous, as was seen in section 2 (for example; Said Alkabeer, Alsheik, Mohammed-Haffez).

The computer has to be helped in the segmentation, and either the components have to be separately obtained from the user, or they have to be clearly identified in the text.

Two methods are possible for singling out the name components within text: either single names consisting of several words can be made into a single unit, either running the words together without spaces or using a hyphen convention as in European languages, with the division between name-components marked by a space, or the division between name-components should be explicitly marked with a special symbol.

Where the name components have to be given identified roles, there is the further requirement on data entry that these roles are made explicit. If the components were separately solicited, the role could be declared at that point. However, if the names are obtained as running text with component separators, it is necessary to have a convention for indicating missing names (e.g. a blank between separators) and possibly a convention for declaring the family name (e.g. the last in the sequence, or always the fourth, though the problem of family names which are not the last, seen in section 2, needs to be recognised). A convention regarding connectives like "bin" and "Al" needs to be firmly established.

Instead of indexing name components, one could index individual words within a single index, looking up each word, intersecting the sets obtained, and finally inspecting the full text of the name to eliminate those that do not match the sequence and position requested at retrieval. The advantage of this approach is that names could then be captured as free text, with all superfluous words like "bin" and "Al". The text would be segmented into words using a grammar for the formation of names, assuming that a dictionary of names could be produced. Alternatively the use of space characters could become disciplined and these be used for word segmentation with special routines being used for removing prefixes like "Al". This second approach would not require a comprehensive dictionary, and is very attractive.

A variation on this approach would be the indexing of word fragments rather than words, using some simple minded segmentation rule like all word fragments between initial and final forms of letters. Indeed, because of its simplicity in software, and potential retrieval power for the user, this method is extremely attractive. However, again some method needs to be found to handle the optional "Al" prefixes.

In both the above two approaches of indexing of words or word fragments what has happened is that the retrieval requirement of retrieval given part or all of the name and the retrieval requirement of fast retrieval using indexes, have been separated. The complete method is as follows: on storage, store the complete name as a single string, but index it by extracting suitable words or word fragments; in retrieval using a partial name, extract the words or word fragments by the same method used for storage, look up all the indexes indicated, and retrieve all the names appearing in all indexing results, finally checking the actual names against the retrieval names for matching of words (word fragments) in correct order.

Finally let us consider the system where transliterations of names are used.   Transliterations provide problems of variability which are comparable to writing down names
from dictation in a noisy environment.   An exact match may not be found because a
different transliteration convention was used, and what is really required is a best
match.   Best matches in the context of name retrieval in European systems have been
well studied and a matching technique based on dynamic programming (Wagner and Fisher
18) based on clusters (e.g. 13) would be appropriate: this method is a simple gener-
alisation of indexed sequential methods of indexing.   However it must be pointed out
that these solutions are essentially novel, and though the algorithms involved are
not particularly complex, they have been developed in a research environment and are
not yet part of the software engineers standard collection of methods.

4. CONCLUSIONS

In section 3 we examined aspects of computer system design over a wide front.   We saw
many alternative solutions to the problems posed by Arabic discussed in section 2, and
should now be in a position to choose the best solutions.   But all applications are
different so what is best in one application is not necessarily best in another.
Nevertheless it is of general advantage to adopt a common approach in some aspects
of system design, regardless of application.   Some standardisation is necessary.

If standards are available, this simplifies the task of the implementor of the system.
Instead of having to investigate factors like code design, or peripherals font design,
he knows the methods embodied in the standard are good, and can be used with the confidence that he will not have problems later from this source.   The designer may
also find he has a choice of plug compatible modules to use, and thus save development costs in these areas.

All these advantages for the implementor should also mean a cheaper product for the
customer.   In addition the customer has the security of knowing that the product he
is purchasing attains that level of quality which the standards ensure.

How should we approach standardisation ?   The most appropriate strategy seems to be
to standardise those aspects of computing where standardisation is already accepted
for European computing, and to provide guidelines for other areas.

The prime area for standardisation is undoubtedly character codes, with all the ramifications these imply for standardising character fonts (at least their widths).
This alone would ensure a plug-compatible peripheral market, and there is clearly a
need for free choice on a plug-compatible basis for Arabic peripherals.

A second area for standardisation is programming languages, though there appears to
be little advantage in moving to Arabic programming languages other than a desire to
become culturally self-sufficient.   A programmer, if he is to perform effectively,
needs to be competent in a European language in order to be able to use the documen-

tation of the computer manufacturers and to tap the large store of knowledge about computing available in European languages, principally English.

The area of information storage and retrieval has not yet developed to the point where standardisation is possible. Nevertheless guidelines, particularly in the area of name-handling would be very useful; if these guidelines could be given an authority bordering on that of a standard, a lot would be gained for all concerned.

Surely there is a need for all these matters to be widely discussed throughout the Arab world. But in the area of codes, delay should be avoided at all costs, and the sooner that a standard could be agreed at the International Standards Organisation, the better will be the future of computing in the Arab world.

ACKNOWLEDGEMENTS

In writing this paper we have drawn upon a considerable body of experience within SCICON. Part of this experience is embodied in SCICON's special Arabic peripherals and computing systems that we have studied, and part of this experience is embodied in the colleagues with whom we have discussed these issues. We are very grateful for having been able to tap this experience.

REFERENCES

1. ALBERGA, C.N. "String similarity and Misspellings". CACM, vol. 10, no. 5, pp 302 - 313. May, 1967.
2. AMAJID,Abdullah "The Arab Efforts to Simplify Arabic Writing". Addarah Journal, volume IV (3) 1978. King Abdul Aziz Research Centre, Riyadh, Saudi Arabia (in Arabic).
3. BEESTON A.F.L. "Written Arabic" Cambridge University Press 1968.
4. BOYER R.F. and MOORE J.S. "A Fast String Searching Algorithm" Comm. ACM October 1977 Volume 20 Number 10. pp 762 - 772.
5. DEHLAWI F. "Computer-assisted Instruction in Arabic" 4th Saudi Arabian Computer Conference, University of Riyadh, Saudi Arabia. March 1978.
6. EHLERS G. "Production of High Quality Arabic Texts on a CRT Filmsetting Machine" 4th Saudi Arabian Computer Conference, University of Riyadh, Saudi Arabia. March, 1978.
7. GRIES D. "Compiler Construction for Digital Computers". Wiley 1971.
8. KNUTH D. "The Art of Computer Programming. Vol. 3. Searching and Sorting". Addison-Wesley 1973.
9. LEFKOVITZ D. "File Structures for On-Line Systems". Sparton MacMillan 1969.
10. LIESER H.J. "The Use of Computers to Improve Readability of Print". 4th Saudi Arabian Computer Conference, University of Riyadh, Saudi Arabia. March, 1978.
11. MITCHELL T.F. "Writing Arabic" Oxford University Press 1953.
12. MITCHELL T.F. and BARBER D. "Introduction to Arabic" BBC 1972.
13. SALTON G. "Automatic Information Organisation and Retrieval" McGraw-Hill, 1968.
14. SEFLAN Ali Mohammed "The Need for a Standard Arabic Character Set", Second Saudi Arabian Computer Conference, Riyadh, Saudi Arabia. November 1975.
15. TAWAKOL Samy M.Y. "Arabic Character Graphic Representation" Second Saudi Arabian Computer Conference, Riyadh, Saudi Arabia, November 1975.
16. THESSIGER W. "Arabian Sands".
17. TRITTON A.S. "Arabic" Teach Yourself Books 1943 Introduction pvii.
18. WAGNER R.A. and FISCHER M.J. "The String-to-String Correction Problem" J. ACM, Vol 21, no. 1, pp 168-173. January 1974.

OPERATION AND EVOLUTION OF ORGANIZATIONAL INFORMATION SYSTEMS

Gerald P. Learmonth and Alan G. Merten
Database Systems Research Group
Graduate School of Business Administration
University of Michigan
Ann Arbor, Michigan 48109/USA

1.0 Introduction

In recent years, users and data processing personnel have become increasingly aware of the cost of building and maintaining computer-
based information systems. In addition to recognizing the cost and
complexity of these systems, organizations have recognized that these
systems are built and maintained over a long period of time and that
the life of a computer-based information system can be viewed as consisting of a series of steps or phases. During each of these phases
in the life cycle of a system, different activities are performed, various user and data processing personnel are involved, and different
degrees of time and materials are expended.

Traditionally, the construction or programming phase of the life cycle
received the most attention of practitioners and researchers. Because
of the failure of many large systems to meet the requirements of the
end users, much attention during the past few years has also been directed to the requirements analysis and design phases. Software engineering, structured design, and structured walkthroughs are but a few
of the techniques that have been discussed, developed, and used in an
attempt to improve the quality of these pre-implementation phases.

This shift in emphasis from programming to requirements and design
will most certainly effect the quality of systems. However, in spite
of how systems are built in the future, the importance of post-implementation activities will continue. These phases and activities are
important because of the large number of complex systems in existence
and the large number that are about to become operational (most of
which are being built without the benefit of structured design, sufficient documentation, etc.). Daly [1977] characterized these post-
implementation activities as follows: "Although many feel that the
management of large software development is mysterious, or at least
little understood, the long term control and maintenance of large programs is even more mysterious."

The purpose of this paper is to survey the phases of the life cycle
which occur after the design, construction, and implementation of a
system. These phases will be referred to as the operation phase and

the <u>maintenance and modification phase</u>.

Each of these phases consists of a number of sub-activities. With respect to some of these sub-activities, the paper will address the following:

a) the main problems that occur

b) procedures and techniques used by organizations during these activities to reduce the impact of the problems

c) research results and approaches which could effect the activities of these phases

d) activities which may be performed in earlier life cycle phases to reduce the number of problems during operation and during maintenance and modification

Section 2 will formally define the problem area and highlight the difficulties currently encountered during these two phases. Section 3 will present, discuss, and classify the various types of observation and analysis that occur during the operation phase. Section 4 will discuss maintenance and modification activities. Section 5 will present a summary and conclusion.

2.1 Problem Definition

The phases of the life cycle of a computer-based information system have been defined by many authors and operating organizations. Depending on the originator of the definition, the number of phases, their individual names, and the sub-activities differ widely. For the purpose of this paper, we will assume that the life cycle phases are specified as follows:

a) Specification of General Need

b) Feasibility Study

c) Collection and Analysis of User Needs

d) External Design (General Systems Design)

e) Internal Design (Detailed Systems Design)

f) Construction and Conversion

g) Testing

h) Operation

i) Maintenance and Modification

The detailed description of the sub-activities, problems, and solutions associated with phases a) through g) is the subject of many papers. While the amount of resources expended in these phases is often large, many people now believe the major portion of software manpower is spent in maintenance rather than in development (Boehm [1973], Branscomb [1976], Lientz, et al. [1978]). Similarly, because of the many large systems now in development that will soon transition to the operation/maintenance/modification phases, it is expected that efforts in both of these phases will increase, at least in the short and mid-term range. Coupled with the high system longevity, some predict a five or ten to one ratio of operation/maintenance/modification cost to development cost when viewed over the total life cycle (Gansler [1976]).

The complexity of the post-implementation phases is de-emphasized in the "linear" presentation of the life cycle steps as presented earlier in this section and as presented in most system development manuals. A more representative description is given in Figure 1. The operation phase and the maintenance and modification phase are on-going, cyclic activities. In addition (but not represented in Figure 1), these phases often occur simultaneously, that is, while the system is in operation, maintenance and modification activities take place. The figure explicitly identifies the "analyze" activity and the "obser-vations" and "specification of changes" entities. These entities and activities are important components of the transition between opera-tions and maintenance/modification but are often ignored in the life cycle management of organizational information systems.

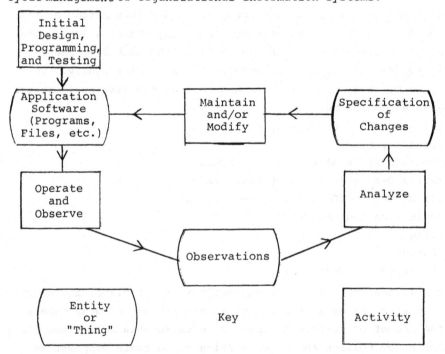

Figure 1. Cyclic Behavior of the Post-Implementation Activities of the System Life Cycle

3.0 Observation and Analysis

3.1 Introduction

To effectively manage a computer-based information system, thorough
and careful observation and analysis of system behavior and performance must be undertaken. This section will review existing methodologies for observing and analyzing a system. Both external, off-line
observation and analysis tools as well as internal, on-line, tools
will be discussed. Several promising new ideas will also be examined.

3.2 Performance Management

With the ever growing complexity of computer systems, the problems
related to computer performance measurement and evaluation have received considerable attention from researchers and practitioners alike.
Most of the concern in this area, however, is related to issues of
hardware and system software performance. Measurement and evaluation
of such criteria as CPU utilization and scheduling policies are aimed
at decisions regarding hardware enhancements or system software tuning.

The effective operation and evolution of an information system requires
observation and analysis which transcends the hardware and system software. The term "performance management" will be used to describe the
activities of observation and analysis both off-line and on-line which
are necessary for managing an information system throughout its operational life-time.

Observation requires the systematic recording of pertinent performance
measures. The analysis activity involves comparing these observed performance measures with certain predefined standards of performance.
If these standards are not met, then maintenance and modification activities are undertaken.

Traditional computer performance measurement and evaluation has developed a common set of performance metrics which are easily observed
and analyzed. When considering performance management of an information system, standards of performance include not only the quantifiable
measures but also a considerable number of intangibles. It is extremely
difficult, if not impossible, to measure and quantify such standards
as user satisfaction, system security, data integrity, and system re-
coverability after unforeseen hardware or system software failures.

The next two sections will address existing tools and highlight some
novel approaches to implementing an effective performance management
program for an operational and evolving information system.

3.3 External Performance Measurement

In the development phase of an information system, alternative system designs are generally compared and the design with the least cost and greatest potential benefit is chosen. To make this decision, various organizational costs are identified and estimated. Among these are personnel costs, equipment costs, cost of vendor software, file conversion costs, and perhaps the opportunity cost of undertaking different projects. To the extent that these costs are able to be estimated, they provide quantifiable standards against which development progress can be measured.

A more difficult task is to quantify benefits which will accrue from an information system. Certainly some benefits can be assigned monetary values such as a reduced billing cycle and improved inventory management. Information systems are, however, often designed and implemented to provide certain intangible benefits to the organization. Improved managerial control, increased data security, improved employee morale are all benefits which are immeasurable yet constitute design goals for an information system.

The methodology of cost/benefit analysis is well defined and widely used in the development of many corporate and social systems. In a recent survey paper, King and Schrems [1978] have adapted cost/benefit analysis to information systems development and operation. Their survey carefully describes the stages of cost/benefit analysis as it would apply to an information system. For information system development, they enumerate many of the costs, both obvious and "hidden," which must be considered. They also provide suggestions for assigning values to many of the heretofore intangible benefits of information systems.

A comprehensive cost/benefit analysis of an information system provides standards against which the development of the system can be measured. More importantly, the cost/benefit analysis performed during system development can provide the standards for the performance management of an operational system.

The environment in which an operational information system exists during its life-time is a changing one. Workload requirements change, data files grow, and hardware is enhanced or replaced. By periodically reassessing the identified costs and benefits of the information system, maintenance and modification tasks may be initiated to ensure satisfactory system performance in light of a changing environment. This application of cost/benefit analysis as an auditing tool can provide significant external information to the process of performance management.

Lastly, cost/benefit analysis as an on-going procedure can provide the necessary information for the all important decision of when to initiate a complete system redesign. Information systems should enjoy only finite life-times. Costs of operation and associated benefits to the organization will eventually exceed reasonable limits and maintenance and modification activities will be insufficient to correct these problems. At such a time, the life-cycle of an information system must be reinitiated.

A novel approach to the observation and analysis of a continuing system is provided by Grochow [1971] who attempted to measure and tune the performance of a time-sharing operating system. Grochow's approach was not to tune the system to maximize CPU utilization or minimize page faults, but rather to maximize user satisfaction. To accomplish this, he devised a utility function which addressed the user-oriented issues of system availability, system reliability, and response time to different classes of tasks. He estimated the shape of the utility function by interviewing users of the system and was able to assess their levels of satisfaction over parameter ranges of three variables. With this empirical utility function, anticipated modifications to the time-sharing system could be measured with respect to their impact upon the user community. In much the same way, the performance management of an operational information system must be sensitive to the user community. External, off-line, measures of user satisfaction with the information system are important factors to consider during operation and may, after analysis, initiate maintenance and/or modification activities to insure user acceptance.

3.4 Internal Performance Management

Effective performance management of an operational information system must combine external and internal measures. In contrast to the external performance measures which are somewhat ad hoc in nature, the internal performance measures discussed in this section are more highly developed, readily available, and easily observed and analyzed.

Widespread interest in computer performance measurement and evaluation has resulted in the development of numerous methods for the collection and analysis of computer performance data (General Services Administration [1977]). These methods vary from sophisticated and expensive hardware attachments to vendor supplied software monitors and lastly to the often overlooked, operator's log and system malfunction reports.

Caution should be exercised in the use of these measures as it has been demonstrated by Stevens [1978] that many of the indicators in widespread use are actually misleading.

Hardware monitors are stand-alone electronic devices which are attached at various points to the main computer system hardware. They have several advantages over software monitors in that they do not induce any unwarranted overhead to the operation of the computer system and they can probe aspects of the hardware performance which would otherwise be unobservable.

Hardware monitors are used to observe device utilization, such as the CPU, channels, and secondary storage devices. They do not observe individual tasks operating in a multiprogrammed environment. They can, however, still be useful in the observation and analysis of an operational information system. It can safely be conjectured that a limiting factor in the performance of an information system will be its efficiency in I/O performance. Possible uses for hardware monitors then might include observation of activity to devices upon which information system files are stored. Maintenance activities might involve a redistribution of files across secondary storage devices or channels to decrease device/channel contention within the information system. Additionally, for on-line information systems supporting a large number of remote terminals, hardware monitors may be used in the analysis of communication bottlenecks. If such problems exist, hardware and/or software modifications may be made to alleviate the problems, thereby prolonging the operational effectiveness of the information system.

Software monitors differ from hardware monitors in that they are typically event driven software systems which log various occurrences during the operation of application programs. These monitors are often supplied by the computer system vendor as part of the operating system. The individual installation has the option of using a software monitor and may select those events which are to be logged. Typical events which may be recorded are job initiation, job termination, interrupts, and various input/output operations.

Software monitors do incur a penalty of increased operating system overhead and may be a source of performance degradation. Most software monitor systems may be turned "on" and "off" to afford periodic snapshots of application program performance with minimum interference.

The analysis of software monitor logged data should be made by an experienced systems programmer. In spite of the most careful detailed system design and testing, the true operational characteristics of an information system will not become evident until the system is in full operation. In analyzing software monitor data, various maintenance and modification projects may be deemed appropriate. For example, different subsystems within the information system may access the same files and consequently cause a deadlocking situation to arise. If the subsystems can be scheduled independently, the problem may be avoided. Additionally, in virtual memory computer systems excessive paging may be a source of overall performance degradation. A simple re-linking of system components may alleviate this problem.

As part of the performance management program for observing and analyzing information systems, these performance monitors, if available, can provide a great deal of insight into actual system performance. Because the environment of an operational information system is continually changing, periodic observation and analysis should be carried out.

A final source of information to a performance management program is the operator's log. This manually maintained data is often overlooked and the information it contains may not be reported to the individual responsible for the information system. In most computer centers, operators record events such as system crashes, disk volume mounts, tape mounts, and hardware malfunctions. If any of these events concern the operational information system, then maintenance and modification activities may be warranted.

The internal performance management tools described above can and should be used to continually assess the overall performance of an operational information system. Recently, attention has been focused on analyzing the performance of a part of a total information system -- the database. It has been recognized that the most volatile aspect of an information system is its data files. While user requirements are sure to change over time, they do so relatively slowly with respect to the changes that take place in files through insertion, deletion, and update of stored records.

One of the most common maintenance activities performed during the operational life-time of an information system is the periodic reorganization of a database. In many information systems, the storage structure and access methods used handle insertion of new record occurrences by means of overflow areas on secondary storage devices. As these overflow areas begin to fill with inserted records, system performance for retrieval applications degrades. Similarly, deleted records are often merely flagged and the space is not recovered. An

important consideration for file systems of this type is when should
the data files be reorganized in order to move inserted records to
their proper positions and recover deleted space.

During observation of an operational information system this problem
may be detected in a number of different ways. Externally, users of
the information system may express dissatisfaction with progressively
deteriorating response time to queries or increased run times for application programs. Internally, performance monitors may indicate
poor performance by recording excessive numbers of I/O requests as
the overflow areas become more heavily used. Additionally, many file
systems maintain logs of all insertion, deletion, and update transaction for backup purposes. Analysis of the log tape will indicate
the volume of insertions and deletions which may be used as a measure
of file degradation.

This problem has been known for a long time and affects many operational information systems independently of their particular application. The solution to the problem is proceeding along two different
fronts. First, there is a cost associated with reorganization of a
file and there is a cost incurred by user applications when overflow
areas are used. The solution then becomes one of determining when
these costs are equal. Shneiderman [1973], Yao, Das, and Teorey [1976],
and Tuel [1978] have examined the problem and have postulated appropriate cost functions. With these cost functions they analytically
derive optimum reorganization points for data files. To adapt their
work to a continuing performance management program, the various cost
functions must be parameterized to reflect the particular information
system, file system, and hardware being observed.

The second approach to solving this operational problem is the more
practical in that it removes the problem completely. Recent advances
in storage structure and access method design allow for efficient insertion of record occurrences in their proper places and recover deleted space automatically. Storage structures based on the B*-tree
developed by Bayer and McCreight [1972] represent approaches to the
solution of the problem.

During the internal design stage of information system development,
decisions are made regarding the logical structure of a database,
the physical placement of record occurrences, and the determination
of which access paths will be maintained. These design decisions impact the operation of the information system throughout its life.
System designers have at their disposal certain tools to assist in
physical design. Among these are the File Design Analyzer (Teorey

and Das [1976]), IBM's Database Design Aid (Hubbard and Raver [1975]), and the Database Design Evaluator (Teorey and Oberlander [1978]). The designer estimates system parameters and anticipated user activity and a proposed design is produced. During actual system operation, it may become apparent that certain parameters were estimated incorrectly or that important considerations were overlooked. Also, information systems are subject to change over time. Files grow in size and patterns of usage change thereby invalidating the basis of the original design. A very promising area being investigated now deals with the question: What can be done during an information system's operational life-time to aid the system in adapting to a changing environment? The answer lies in the continuing observation and analysis of the operational system's characteristics and the use of the results of the analysis to guide the evolution of the information system.

The extent to which an information system may respond to a changing environment partially depends upon the level of logical and physical data independence. With a high degree of physical data independence, self-organizing file schemes may prove beneficial in organizing the "active" portion of the file into a small number of contiguous blocks thereby reducing costly I/O activity. If and when user activity shifts to a different portion of a file, the filing scheme will respond accordingly.

At a higher level of consideration, the maintenance and use of secondary access paths to shared records affects system performance. Schkolnick [1974], King [1974], and others provide design aids which suggest optimum sets of access paths. A cost model which balances the cost of secondary index maintenance with the benefits of reduced I/O activity is used. These suggested designs also tend to lose their effectiveness after the information system has been in operation.

Continuous observation and analysis of an operational system can yield valuable insight into changing requirements. Hammer [1976] has proposed a methodology for collecting and analyzing access information and then allowing for the dynamic creation and deletion of secondary access paths. This methodology is quite promising for future systems. The present drawback is that this type of self-adaptive behavior demands a very high level of data independence. Present day database management systems typically allow the user to be aware of secondary access paths, and any scheme which dynamically creates and deletes these paths would seriously affect applications.

The last topic for discussion concerning observation and analysis of operational information systems deals with the maintenance of software.

Recently, Halstead [1977], Kolence [1975], and others have investigated the measurement and prediction of software development and quality. An excellent review of this recent work is given by Fitzsimmons and Love [1978].

Halstead's theory of software science is particularly applicable to the consideration of large-scale software development as well as maintenance. By cleverly adopting long known physical laws, particularly from thermodynamics, Halstead has postulated a set of "laws" governing program length, programming time, programming effort, and a predictor of program design bugs.

In the survey by Fitzsimmons and Love, numerous experiments are reviewed and a remarkable correlation between Halstead's predictions and actual software practice are cited. Perhaps the most interesting result concerns the prediction of the number of bugs to be expected in operational software systems. The predictors involve easily measurable quantities derived during system development. At the time of system implementation, the number of residual bugs may be estimated and may be used as a measure of the maintenance and modification effort which will be required during the remainder of the system life cycle.

4.0 Maintenance and Modification

4.1 Introduction

The cost and manpower effort expended on maintenance and modification of software system has often been much larger than expected, hidden in various budgets, and hard to estimate beforehand and/or determine after the fact. It is now commonly recognized that the cost of maintenance and modification exceeds that of initial design and construction. Estimates of the rate of maintenance and modification cost to design and construction cost range from two to fifty.

Lientz, et al. [1978] described the results of a survey of 69 organizations regarding maintenance and modification activities. The results of their survey were as follows:

a) maintenance and modification do consume much of the total resources of systems and programming groups

b) maintenance and modification tend to be viewed by management as at least somewhat more important than new applications software development

c) problems of management orientation tend to be more significant than those of technical orientation

d) user demands for modifications constitute the most important management problem area

The purpose of this section is to address the issue of maintenance and
modification in the following manner:

a) present working definitions for "maintenance" and "modification"
b) describe some reasons why the maintenance activity is viewed as
   undesirable by programmers
c) present some suggestions to improve the maintenance function
d) describe what may be done in earlier phases of the life cycle to
   reduce maintenance and minimize the impact of modification

4.2 Definition of Maintenance and Modification

The dictionary provides the following general definitions:
maintain - to keep in existence or continuance; preserve; retain
modify   - to change somewhat the form or qualities of; alter partially

With respect to computer-based information systems, the distinction
between "maintain" and "modify" is often not as precise as may be indicated by the general definitions. When an activity ceases to be
maintenance and becomes modification is not always clear, and what one
organization calls maintenance may be called modification in another.
One area of disagreement is in the classification of changes which improve the performance of a system. Most often, maintenance activities
are those which correct programming and design flaws and those which
change portions of a module or program in response to a requirement
change. Swanson [1976] provided a classification of three types of
maintenance activity. Briefly, these activities are: corrective
maintenance (performed in response to the assessment of failures); a-
daptive maintenance (performed in anticipation of change within the
data or processing environments); and perfective maintenance (performed
to eliminate inefficiencies, enhance performance, or improve maintainability). Modification activities usually include sub-system or system design changes and result in the addition of functional capability.

System maintenance and modification involves users, programmers, and
analysts, the same people that are involved in development activities.
For many years, the data processing industry has implicitly operated
as if development was hard and challenging, and maintenance and modification were easy. The definitions given above and past experience
point out that it is not the magnitude or complexity which determines
whether something is development or maintenance/modification, rather
it is the purpose and/or point of initiation.

Just as organizations do not always agree as to the distinction between

maintenance and modification, they do not agree as to the desirability
of a separate unit within the systems and data processing department
for the performance of maintenance and/or modification functions.
Some organizations require that the developer performs maintenance
but have a separate group perform modifications. In most cases, when
more than one group is involved, the maintenance and/or modification
group as well as the user signs off on the system prior to implementation.

4.3 The Image of Maintenance

Historically, the task of software maintenance has been viewed as an
undesired duty and responsibility by analysts and programmers. Some
of the reasons for this attitude are real, some are only perceived.
The reasons include the following:

a) the name of the function or activity. "Maintenance" implies a
   slight adjustment to something which someone else created. The
   adjustment is most often perceived to be only necessary to correct
   an error in someone else's work.

b) maintenance activities occur in a crisis environment. The system
   that needs to be maintained has been operational and therefore
   users are depending on it. While users are somewhat patient during modification because of their perception of the complexity of
   the task, maintenance is perceived by the user as easy and therefore to be completed immediately.

c) the system to be maintained is most often poorly documented. Liu
   [1976] claims that documentation is either non-existent, misleading,
   or insufficient.

d) both development and maintenance programmers most often over-empha-
   size program efficiency as a design or redesign criteria. This
   situation can have two types of negative effects. On one hand,
   the maintenance programmer is hindered by the fact that since
   "efficiency" and "maintainability" are often conflicting objectives,
   the program to be maintained is difficult to understand. On the
   other hand, maintenance programmers are often frustrated by inefficient programs and often spend too much time attempting to increase the efficiency of the program as opposed to concentrating
   on the task of maintenance.

e) the maintenance function is often viewed by management as a good
   place to provide "on-the-job" training for programmers. Because
   of this, the background and experience of maintenance programmers
   is too often limited. Consequently, many members of a maintenance
   team are often unaware of such issues as the long-term objectives

of the organization, the relationship of a system to others within
the organization, and the evolution of a particular system.

In some sense, the current emphasis in organizations and by researchers
on such techniques as software engineering and structured design is
having a negative effect on the awareness of the importance of maintenance and modification activities. Some of the claims for these
techniques lead one to believe that maintenance and modification problems will automatically be substantially reduced if the system is
properly designed and constructed. However, in many cases in the past,
it was not clear whether a system that was difficult to maintain or
modify was in that condition because of events that occurred during
design and development or because of events that occurred after initial
operation.

In addition to reducing the cost and complexity of maintenance and
modification through improving the techniques and procedures used in
developing and testing (see Section 4.5), it is also possible to reduce
this cost and complexity by improving the techniques and procedures
used in the operation phase and the maintenance and modification phase
themselves. The interest in changing the activities of these phases
is motivated by the fact that many systems are often difficult to maintain because of events in an earlier iteration of the operations phase
or the maintenance and modification phase. Often programming and documentation standards that were followed prior to initial operation are
not followed during maintenance and modification. Similarly a system
may have been easy to maintain when it became operational, but when
efficiency problems arose during operation, the system was changed
without regard to future maintainability.

4.4   Improvements to the Maintenance Function

These problems and negative attitudes toward the maintenance function
have been addressed by some organizations (Daly [1977], Liu [1976],
Izzo [1978], Podalsky [1977]). Some suggestions include the following:

   a) the name of this function should be changed to something more
      positive such as "product support," "application system support,"
      or "software support." This type of name more accurately conveys
      the importance and the need for creativity in this organizational
      function.
   b) senior programmers should constitute at least 30% of the mainten-
      ance staff.
   c) assign maintenance programmers to active design tasks for 50% of
      their working time. This suggestion and the previous one would

tend to increase the expertise of the maintenance staff as well
as potentially lead to systems that would be designed with maintainability as an objective.

d) the project should not be turned over to the maintenance staff
until it has stabilized through a number of iterations. During
these first few iterations, the quality of the system and its
documentation would hopefully be improved by the designers themselves. Placing them in a maintenance function for this short
period of time may help them to realize that while the system is
functionally correct it is also difficult to maintain.

e) develop explicit procedures for testing a system after it has
been "maintained" but before it has been put back into operation.
These testing procedures should address the unmodified portion
of the system and the documentation as well as the modified portions of the system.

These suggestions and others are directed to the staffing and execution of this function. A broader management concept has been proposed,
called software configuration management, which suggests that, beginning
with the life cycle phase of testing, the system be under the control
of a separate organizational function (Daly [1977]). Under configuration management, all code and documentation (original and proposed
changes) must be accompanied by the proper supervisory approval and
must include a test plan. This type of management structure in some
sense applies the quality control normally used by hardware and system
software manufacturers to manage system software to the management of
application software within user organizations.

4.5   Changes to Previous Life Cycle Steps

The size and complexity of the maintenance and modification function
have led many organizations and some researchers to address what can
be done to reduce the required maintenance and modification of a
computer-based system. Much of the work of structural design and
programming (Peters and Tripp [1977]) and that of requirements definition and analysis are directed toward this goal.

Many of the current problems with computer systems can be traced back
to the design objectives of a system. Most often these objectives
are one or more of the following: produce a system that meets the
functional requirements with minimum expenditure of resources, and/or
produce a system that, when operating, consumes a minimum amount of
machine resources. In order to produce systems that are easier to
maintain and/or modify, the design objectives may have to change

(Daly [1977]). Two alternate design objectives are as follows:

   a) perform the desired functional operations and minimize expected
      on-site maintenance costs

   b) perform the desired function operations with minimal expenditure
      of resources (manpower and material) over the entire system life
      cycle.

The high cost of the maintenance and modification activities, have led
many organizations and authors to question the basic structure of the
life cycle as illustrated in Section 2 of this paper. Brooks [1975]
and Podalsky [1977] contend that the life cycle is itself recursive
and that, in addition, organizations often throw away systems sometime
during the life cycle and start all over. While many organizations
recognize these and similar facts, very few organizations have revised
these life cycle management policies and procedures to reflect them.
A related topic of increasing interest to organizations is the concept
of "bread-boarding" or prototype development and use. The experience
with these approaches is limited. However, their effect on the life
cycle in general and maintenance and modification in particular could
be substantial.

5.0 Summary and Conclusion

The cost and complexity of the process of building and maintaining
computer-based information systems is now receiving widespread acknowl-
edgment. Initially, each system was built and maintained based solely
on the experience of the specific users, managers, analysts, and programmers assigned to the system.

Gradually, certain management, engineering, and scientific principles
have been and are being developed by the data processing community or
by a given operating organization which have the effect of bringing
more structured, systematic approaches to the entire life cycle. These
systematic approaces initially appeared in the middle phases of the
life cycle, then in the early phases, and now, to some degree, in the
later phases.

The changes within an organization with respect to life cycle activities are themselves often subject to an evolutionary cycle or series
of phases. First, organizations recognize the problem in terms of
excessive costs, delays, or redundant efforts. For example, in the
early days of computing, the manpower inefficiency of writing programs
in machine languages was recognized. Second, generally accepted rules,
policies, and procedures are developed to organize or structure the
specific life cycle activities. Third, formal methodologies and, in

some cases, computer-aids are developed and become the tools of the user and/or data processing personnel involved in the step. With respect to the programming phase, compilers and operating systems clearly place the industry in the third phase. With respect to design (internal and external), software engineering approaches moved the industry to the second phase, and recent advances indicated a possible transition to the third phase. With respect to operation, maintenance, and modification, the industry appears to be in the second phase although certain computer-aids are under development in the research laboratories (Su and Reynolds [1978] and Swartwout, Novak, and Fry [1977]).

Finally, just as automobiles and hardware have an identifiable useful life span, all software systems will at some point in time be no longer cost justifiable. The transition from the maintenance and modification phase of one system to the specification of user need for another system must be more systematically addressed. Recent work on both problem recognition and proposed solutions has appeared (Oliver [1978] and Fry, et al. [1978]). Formal approaches and, possibly computer-aids, hopefully will soon evolve both for more of the operation/maintenance/modification activities and the inevitable conversion step.

REFERENCES

Bayer, R. and McCreight, E. M. [1972]
"Organization and Maintenance of Large Ordered Indexes", Acta
Informatica, Volume 1.

Boehm, B. W. [1973]
"The High Cost of Software", Proceedings of the Symposium on the
High Cost of Software, Monterey, California, 1973, pp. 27-40.

Branscomb, L. [1976]
"The Everest of Software", Proceedings of the Symposium on Computer
Software Engineering, New York, New York, 1976, pp. xvii-xx.

Brooks, F. P., Jr. [1975]
The Mythical Man-Month, Addison-Wesley, Reading, Massachusetts, 1975.

Daly, Edmund [1977]
"Management of Software Development", IEE Transactions on Software
Engineering, pp. 231-242.

Fitzsimmons, A. and Love, T. [1978]
"A Review and Evaluation of Software Science", ACM Computing
Surveys, Volume 10, Number 1, March 1978, pp. 3-18.

Fry, et al. [1978]
"An Assessment of the Technology for Data - and Program-related
Conversion", AFIPS Proceedings of the National Computer Conference
Volume 47, pp. 887-907.

Gansler, J. [1976]
"Keynote: Software Management", Proceedings of the Symposium on
Computer Software Engineering, New York, New York, pp. 1-9.

General Services Administration (GSA) [1977]
Management Guidance for Developing and Installing an ADP Performance Management Program, Washington, D.C., July 1977.

Grochow, J. [1972]
"A Utility Theoretical Approach to Evaluation of a Time-Sharing
System", Statistical Computer Performance Evaluation, edited by
W. Freiberger, Academic Press, New York, New York.

Halstead, M. H. [1977]
Elements of Software Science, Elsevier North-Holland, New York.

Hammer, M.
"Self-adaptive Automatic Database Design", AFIPS Proceedings of
the National Computer Conference, Volume 46, 1977, pp. 123-129.

Hubbard, G. and Raver, N. [1975]
"Automating Logical File Design", Proceedings of the First International Conference in Very Large Databases, Framingham, Massachusetts, pp. 227-253.

Izzo, J. [1978], personal communication.

King, J. L., and Schrems, E. L. [1978]
"Cost-benefit Analysis in Information Systems Development and
Operation", ACM Computing Surveys, Volume 10, Number 1, March
1978, pp. 19-34.

King, W. F.   [1974]
  "On the Selection of Indices for a File", IBM Research, Report
  RJ1341, San Jose, California, January 1974.

Kolence, K. W.   [1975]
  "Software Physics", Datamation, Volume 21, Number 6, June 1975,
  pp. 48-51.

Lientz, B., Swanson, and Tompkins, G.   [1978]
  "Characteristics of Application Software Maintenance", Communications
  of the ACM, Volume 21, Number 6, June 1978, pp. 466-471.

Liu, Chester C.   [1976]
  "A Look At Software Maintenance", Datamation, Volume 22, Number 11,
  pp. 51-55.

Merten, A. and Fry, J.   [1974]
  "A Data Description Approach to File Translation", Proceedings of
  the ACM SIGMOD Workshop on Data Description, Access and Control,
  Ann Arbor, Michigan pp. 191-205.

Oliver, P.   [1978]
  "Guidelines to Software Conversion", AFIPS Proceedings of the
  National Computer Conference, Volume 47, pp. 877-886.

Peters, L. J. and Tripp, L. L.   [1977]
  "Comparing Software Design Methods", Datamation, Volume 23, Number 11,
  pp. 89-94.

Podalsky, J.   [1977]
  "Horace Builds a Cycle", Datamation, Volume 23, Number 11, pp. 162-168.

Schkolnick, M.   [1974]
  "The Optimum Selection of Secondary Indices for Files", Research
  Report, Department of Computer Science, Carnegie-Mellon University,
  November, 1974.

Shneiderman, B.   [1973]
  "Optimum Database Reorganization Points", Communications of the ACM,
  Volume 16, Number 6, June 1973, pp. 362-365.

Stevens, D. F.   [1978]
  "How to Improve Your Performance Through Obfuscatory Measurement",
  AFIPS Proceedings of the National Computer Conference, Volume 47,
  1978, pp. 425-431.

Su, S. and Reynolds, M.   [1978]
  "Conversion of High Level Sublanguage Queries to Account for
  Database Changes", AFIPS Proceedings of the National Computer
  Conference, Volume 47, pp. 857-875.

Swanson, E. B. [1976]
  "The Dimension of Maintenance", Proceedings of the Second International Conference on Software Engineering, October, 1976,
  pp. 492-497.

Swartwout, D., Deppe, M., and Fry, J.   [1977]
  "Operational Software for Restructuring Network Databases", AFIPS
  Proceedings of the National Computer Conference, Volume 46,
  pp. 499-508.

Teorey, T. J. and Das, K. S., [1976]
"Application of An Analytical Model to Evaluate Storage Structures",
ACM SIGMOD International Conference on Management of Data, 1976,
pp. 9-19.

Teorey, T. J., and Oberlander, L. B. [1978]
"Network Database Evaluation Using Analytical Modeling", AFIPS
Proceedings of the National Computer Conference, Volume 47, 1978,
pp. 833-842.

Tuel, W. G. [1978]
"Optimum Reorganization Points for Linearly Growing Files",
ACM Transactions on Database Systems, Volume 3, Number 1,
March 1978, pp. 32-40.

Yao, S. B., Das, K. S., and Teorey, T. J. [1976]
"A Dynamic Database Reorganization Algorithm", ACM Transactions
on Database Systems, Volume 1, Number 2, June 1976, pp. 159-174.

HUMAN RESOURCES SYSTEMATICALLY APPLIED TO ENSURE COMPUTER SECURITY

by

V.P. Lane and F.G. Wright
Southern Water Authority
Worthing, West Sussex,
UK

ABSTRACT

In order to maintain the integrity of a computing service, it is essential to assume
that there is no limit to the ingenuity of men who wish to break the service's security
measures and no limit to the carelessness of those parties given responsibility for
maintaining its integrity.

From a statement of security requirements from the point of view of the user, the
paper discusses the practicalities of satisfying these objectives. This is an extremely demanding task. However, if it is approached in a systematic manner it can
often be achieved without any great increase in operating costs.

A great deal of attention has been given to security aspects of hardware and systems
software, and consequently the personnel, a vulnerable area, has received little or
no attention. The paper explores the contribution that end-user and computer personnel
make to computer security through their routine duties. The paper identifies the
need for, and the methods to achieve, a logically developed approach to security, to
enable those responsible for computing, both end-users and data processing professionals, to identify, evaluate, and deal effectively with their own security requirements.
The basic philosophy of physical, document and personnel security must be applied in
concert, for if they are applied independently they are ineffective.

1. THE COMPUTING SERVICE AND ITS SECURITY

The data processing department is a production unit which receives raw material from
the user in the form of data and converts this into information which is returned to
the user or other users to form the basis of their decision making. The basic requirements of the user are that the service should provide results which are correct
and to an agreed timetable not affected by outside influences, and such that the
service cannot be used or influenced by unauthorised personnel.

The computing service as illustrated in figure 1 consists of four components, i.e.
equipment, software, data and personnel. All four components require protection.
Management generally appreciate that the computer system and the data comprise a
major asset of an organisation and therefore that the equipment, the software and the
data require protection without realising that personnel is the area of greatest
vulnerability. Personnel create two problem areas:-

       (i) data privacy

      (ii) data integrity

Most organisations fall into one of two categories. The first category includes

organisations such as Water Authorities in which only a small part of the work has a
need for high security because of its financial or confidential nature and the bulk
of the work has relatively low security value. The second category is at the other
extreme and includes organisations, like finance houses and Government bodies, for
which the majority of the computing is high risk; however, it is common for this type
of organisation to dedicate itself to a relatively small number of computer applications. In both categories, security is achieved in the same way. First standard
protection is ensured for the equipment, the software and the data. Then in addition
and most important of all, attention is directed to the personnel to create, through
staff training and management supervision, simple but effective security procedures.
All breaches of security being promptly investigated followed by swift disciplinary
action which must be taken against defaulters. The result will be a secure service
which meets the approval and the needs of the user without any significant increase
in cost for security measures.

2. SECURITY - THE USER SPECIFICATION

The specification of the user's needs originates from his computing service requirements. As illustrated in figure 1, the computing service is made from four components,
i.e., the equipment, the software, the data and the personnel. Each of these may be
subject to threat and therefore requires protection.

The security specification from the user's point of view is that the computing service
must be protected against a diverse range of unwelcome events which may be natural or
man-made and planned or accidental. The service must be protected against:-

   (i)   Acts of God i.e., the service should be flood-proof, fire-proof, etc.

   (ii) Accidental man-made disasters i.e., it should be safe:-

         (a) from machine failures such as from power cuts, mal-functioning
              of equipment, disc crash or disc dropped, tape torn, etc. and

         (b) from human error causing bugs in systems or application software

   (iii) Planned man-made disasters from outside the organisation. This is generally
         directed at the premises although it might be aimed at the computer and
         its ancillary technology

   (iv) Planned attack from inside the organisation e.g., program "time-bombs" to
         destroy data files

   (v)   Theft and espionage - crimes may range from selling time of the computer,
         or mailing lists, or application software; fraud or embezzlement (there is
         the incident of a credit card company having its list of members stolen
         together with a set of plastic cards which resulted in a loss of over £1
         million)

   (vi) Espionage - information may be wrongly disclosed within the organisation
         and although this is a breach of security it may not be so significant as
         cases where the privacy leak results in wrongful disclosure to an external
         party

In the final analysis, however, the user is only really interested in the integrity of his data, the confidentiality of information and in a service to an agreed timetable. The features of security which are not evident in this simple approach is of little interest to him as this is the responsibility of the Computer Manager.

3. A SECURITY STRATEGY

It is apparent that few companies could hope to justify in economic terms complete protection against the threats listed above. A government department with a high proportion of top secret data is one of the few installations attempting to achieve perfect security. Fortunately for the average data processing installation with a relatively small proportion of security problems, the task is less daunting, more likely to be satisfied, and correspondingly less expensive. A satisfactory service can only be achieved through an effective use of resources by concentrating on those areas of the business with a real security need [1] .

Responsibility, risk evaluation and defence priorities

The computer services manager is responsible for all aspects of the computing service security. However, as the service becomes less centralised, the users must bear more responsibility for the protection of their part of the system. In order to identify

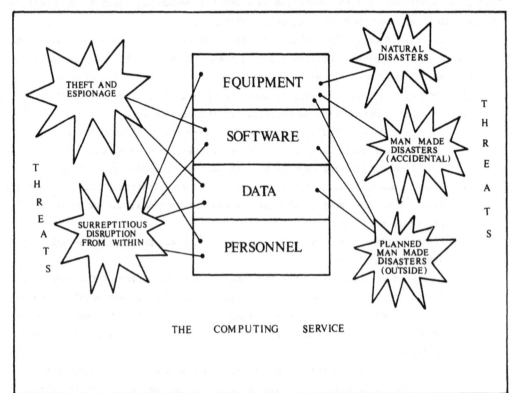

Figure 1   Threats to the computing service

those areas of the business which are worthy of special attention, it is necessary to evaluate the effect of each threat. First one calculates the loss if a breach of security occurred and this will indicate those parts of the business with a high priority in security terms. Then defences are considered, priced and selected. As implied above the computer services manager should complete the evaluation of those parts of the service with which the user does not have any direct contact.

Hardware and systems and application software

The majority of threats indicated in figure 1 are intimately associated with the hardware and software. Fortunately, this is under the direct control of the computer professionals. Organisations are generally sympathetic to funds being used for the computer equipment and its immediate environment. The user therefore expects basic security features without any effect on his budget. The user presumes that the equipment will be protected by fire detectors and extinguishers; that physical security is adequate through the use of locks, bolts and personnel entry control stations; and that the hardware is protected against mal-functioning.

The cost of developing the system software is such that few organisations can contemplate the writing of their own software. Therefore the computer manager and the user are dependent upon the manufacturer and this is generally perfectly satisfactory. In case of catastrophe, back-up on other computers [2] is taken for granted.

The computer professional will provide a service with all the above characteristics and the additional cost to the user, because of the security features, will be insignifcant since their cost is negligible in terms of the total data processing budget. It is common for an organisation to provide this type of service and the user computing service to be anything but secure.

The explanation is simple. Virtually all of the discussion and protection described above has been related to technological areas, whereas the major vulnerable parts are those in which there is significant human involvement. In many respects to talk of physical security of the computer or of back-up services are red-herrings in that they distract attention from the main area of security risk, namely the personnel in the computer department and in the user departments.

With respect to application software, the user expects it to be available not without bugs, but with controls [3] so that errors will be apparent to the user and will not damage the operation of the business.

In the last decade, the majority of reported security breaches have not resulted because of the mal-functioning of security with respect to physical protection of hardware or to systems software, but have resulted from the simple actions of personnel in situations where basic security was lax, and/or controls in the application software or computer installation were non-existent.

4. HUMAN FACTORS AND A SECURE SERVICE

Computers are employed by an organisation in order to assist the user in the operation of the business. The computer and the data it holds are major assets of the company. Although the computer services manager must be responsible for the day-to-day operation of the computer services to a given minimum security standard, it is only the user who can specify what this security standard should be based on, from his practical knowledge of the cost of a breach of security.

The user, as with the design of computer-based systems, must be the major contributor in establishing security standards because of his unique knowledge of the business. Based on this contribution, the organisation will agree and establish security procedures. Although a user will initially state that a security system is required that is impregnable, it must be recognised that complete protection as an objective is impracticable except for high security government departments.

If the user is given the task of evaluating the cost of a security breach, he will quickly establish those parts of the business where security is essential and other areas where it is insignificant. Protection must reflect the sensitivity of the operation and not necessarily the equipment nor the applications. For example, a computer used for real-time control of a water distribution network which in the event of computer failure could be operated manually for hours or even days without any real damage does not justify the same defence as similar equipment used for air traffic control at a major airport - in this case both one hundred percent back-up facility, and contingency plans to ensure the service in the event of a major disruption are essential. Similarly, a payroll system, often considered to be an extremely sensitive area, can be regarded as low risk in an organisation in which the grades of staff are virutally common knowledge and personnel are employed on grades with corresponding salary scales - whereas in an organisation in which salary differentials are not common knowledge then payroll confidentiality is more important.

Following the user analysis of security objectives, it is necessary to recognise the threats. Referring to figure 1, the threats from natural disasters, accidental man-made disasters and from planned man-made disasters from outside the organisation have usually been considered in general defence plans for the computer hardware. It is the threats of theft, confidentiality of information, data integrity, espionage and internal disruption - all human threats - which are the biggest dangers because they have seldom been considered systematically. All computer applications must be examined by the users in terms of risk to attack from personnel. Although protection may be only economically justifiable for a few systems, the security procedures and working practices which ensue will bring benefits and protection to other systems which in themselves may be undeserving.

The user analysis will identify at least three major areas, namely:

    (i) document security and information confidentiality,

    (ii) data integrity - data processed in computer systems must be processed to give correct answers (i.e., accuracy of data) and all the data authorised, no more and no less, must be included (i.e., reliability of data) and

(iii) the need for at least similar levels of security within
       the computer department to those established in the user
       section.

Systematic examination for document security and data privacy

It is surprising that few computer organisations have requested a systematic study of
the threats to company information in the manner outlined above and illustrated in
figure 2. The computer professional is ideally equipped to encourage the user to
embark on this type of study. One can only assume that the reason for this lack of
systematic analysis of security needs is because the study will suggest a concentration
on manual control procedures (rather than computer-based systems) within the user
department and the computer departments and between the two units.

Manual procedures do not have the glamour and therefore the appeal of computer systems.
It is regrettable that basic procedures of prime significance to the business do not
attract the same degree of professional attention as sophisticated computer systems.
The organisation must examine each involvement by a user with the computer, to establish for the systems identified, methods to give protection from user to machine and
from machine to user - this will require procedures for registration of work, for
transferring between machine and ultimate recipient, and for security classification.

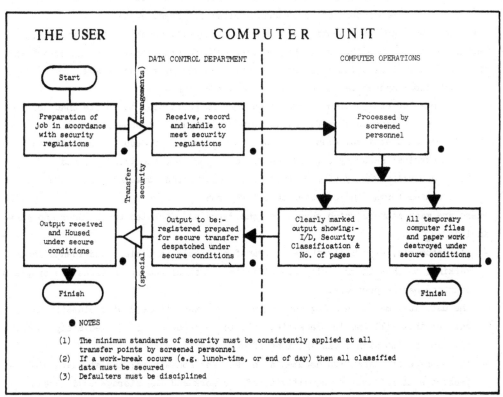

Figure 2 Data protection

It is apparent that security classifications established and operated by the user should be consistent throughout the organisation. The computer unit may give work a higher degree of security than the user requests but common minimum standards of protection are essential if one department is to entrust its work to another on the understanding that it will be adequately protected.

The examination will encompass the following:-

(i)   identification of appropriate security measures for each user application - measures which reflect the sensitivity of the work

(ii) formulation of security and operating instructions

(iii) training plans for operatives and management

(iv) enforcement of procedures and

(v)   provision for regular review of procedures.

User procedures for document security and data privacy

The threats to which the user department are prone include:-

damage to terminal equipment,

unauthorised viewing of information,

unauthorised copying of documents,

the introduction of irregular routing channels and

the introduction of unauthorised or irregular data.

Therefore the user must identify the degree of security which is appropriate to a particular job; agree this with the computer unit in order that they can establish similar or better security standards to that operated by the user; and then proceed in a methodical manner as illustrated in figure 2 with:-

(i)   regular examination of operation logs in particular to locate unauthorised use of documents or for files being referred to at unusual times

(ii) management involvement to ensure that security procedures are followed

(iii) physical security of, and control of access to, the terminal, and

(iv) overall security of documents i.e., input forms containing data and unused forms, and computer output reports.

There are many methods which should be considered and operated by the user and these include:-

- all documents, especially those in transfer, should have clear and identifiable markings to show the security classification; colour coding is effective or plain language may be preferred

- the user must maintain a log showing where and what is held of a classified nature; this will enable classified data to be accounted for immediately at any time and assist with investigations if defences fail

- the storage of classified documents in "secure" containers; the standard of protection will reflect the sensitivity of the work and the storage environment.

- a routine procedure for the destruction under secure conditions of obselete doc-

uments will have a working life of days and others months but all documents at
the end of their useful life must be totally destroyed and
- operational arrangements for special jobs i.e., non-routine work; in such cases
it is imperative that other parties such as the computer manager are given notice
of the user's irregular intentions in order that adequate safeguards can be brought
into operation.

A secure service is based on the commitment of user management. Management must
provide the impetus to set-up working procedures but most important of all they must
give the leadership to make certain that the procedures are followed.

Data control for data integrity

The majority of errors in the use of computer applications is not generated by the
computer system, but by the personnel involved in the system. Humans will always
make mistakes and the purpose of data control is to identify human errors and thence
remove their effect. In addition, data control will help to stop intentional entry
of illegal data for theft.

It is not possible within this paper to describe the full nature of data control but
it must be realised that controls within a computer system are fundamental to security.
As such, data control must be considered in the first stages of the design of a system,
and not added to a system as an afterthought; they must be considered in a comprehensive manner as described in [3] . The objective of data control is to create accurate
and reliable data i.e., data integrity. Therefore controls must be introduced to
ensure:-

(i)   completeness of processing of input data
(ii)   accuracy of processing of input data
(iii) authorisation of input data
(iv)   authorisation of amendments to master files
(v)   completeness and accuracy of processing of amendments to master files and
(vi)   master file balancing

Some of the tasks which are essential to the above may include:-

       raising of input documents
       batching of input documents
       authorisation of batches
       recording of the above tasks
       recording of receipt of computer output and
       reconciliation checks before output is distributed and used.

It is possible to design good data control but still have insecure systems. As indicated earlier, hardware may be at risk because security procedures are not followed
and similarly data control procedures will not be effective unless they are practised.
For effective data control, the pre-requisite is a review and monitoring function by
management.

If the above features together with good accounting practices are implemented, it is relatively easy to achieve data integrity.

The computer services department

The user's expectations may be summed up as a professional approach from the computer personnel to provide a service which includes computer department security procedures consistent with those enforced by the user; duplication and back-up facilities as necessary; the correct labelling of all output being directed to the user with the security classification, job identification, clear indication of beginning and end of output reports, and user identification; and contingency plans for continuation of services in the event of a major disruption. With respect to data held on behalf of the users on magnetic media or in other form. They must be controlled in a similar manner to that indicated in figure 2 to ensure consistent levels of security throughout the organisation.

These procedures are generally "observed" in any professional computer department, but it would be unwise if reference was not made to the weakness which is inherent in any computer system i.e., it is not the technology nor the procedures which are likely to cause danger but the personnel e.g., the major fraud case at Equity Funding Corporation of America [4] where the computer was utilised to cover a fraud involving $110 million. The integrity of the computing service will be maintained only if the computer controller exerts his management influence on his staff.

5. ILLUSTRATIVE EXAMPLES OF SECURITY FAILURES

The following case studies, well documented security failures, are described briefly and are then analysed to illustrate the significance of the security principles described in this paper.

A document security breach and personnel vetting

In this incident, two members of an ICI computer department stole 48 magnetic discs and over 500 tapes containing major company data and the back-up copies. Then they requested £275,000 from the company for the return of the files or threatened the files would be destroyed [5] .

This example illustrates the failure of the "document" security procedure. The procedure, shown in figure 2, should be followed between user department and Computer Unit - but the same procedure must be followed between other departments e.g. departments within the computer unit. The case study shows how simple it is for the elementary rules of security to be ignored. Operators had access to both main files and their back-up copies and the files could be obtained without any other authorisation. In effect the files were unprotected.

There is a further lesson to be learned from this case. The operator involved in the theft whilst awaiting trial was employed by a number of other computer organisations. A clear indication of the negligible extent of personnel vetting done at the interview stage. When it was discovered who he was, the operator was removed immediately, not because of what he had done, but through fear of what he might do.

A data control failure

This second example is taken from a health authority paymaster's department [6] . Expenses incurred by personnel were normally paid at the end of each month with the monthly salary payment. However there was the facility to pay expenses separately via another system, quite independent of the main payroll system - presumably for special cases and for speedy payments. One of the pay-clerks took advantage of this. In general, doctor's expenses were paid via the main payroll to the doctor whilst the second system was used for a duplicate payment of the same expenses claim to the pay-clerk direct. After £13,000 had been stolen, the mal-practice was discovered because of the clerk being sick and absent. A doctor asked for payment of a claim he had submitted earlier in the month which (although he was not aware of it) would have been paid with his monthly salary. This enquiry brought to light the fact that the expenses had already been paid via the secondary system but the payment had gone astray.

This breach raises a number of questions. Was there no overall reconciliation i.e., control figures, between the two systems? Did the person who handled the money have the power to authorise payment? The case indicates (i) the failure to design and/or operate data control and (ii) the neglect of good accounting practices. In addition, it brings out the need for management to monitor procedures continually.

The role of senior management

In the security principles outlined in this paper it has been stated that the chief executive is the prime-mover in security matters. The example of the Equity Funding fraud involving $110 millions [3] involved its president and senior computer personnel. This security breach illustrates the corollary of the security principle i.e., if the chief executive with assistance of colleagues, conspires to breach security there is little hope of effective defense.

However in this example if effective security practices had existed, with respect to the operation of the computer together with comprehensive data control on the application software, prior to the commencement of the fraud, the president might not have fallen to the temptation. Both of these security features must have been lax or non-existent. If they had been operational it would have been difficult for the activities of the president to proceed without staff being at least suspicious if not aware of the irregularities. The result might have been that the crime never started or that the mal-practices were common knowledge earlier and the size of the fraud smaller.

6. MAINTAINING A SECURE SERVICE

It must not be assumed that once a secure method of operation as illustrated in figure 2 has been established that the position can be maintained without effort.

A secure service will only be maintained if it incorporates at its inception realistic methods by which it may evolve. The pre-requisites are:-

(i) simple staff training; supervision and defined disciplinary correctives for staff who fail to observe established procedures; disciplinary action must be enforced swiftly and strictly initially, for incidences such as general breaches of professional standards, circumventing despatch control procedures, or disclosure of passwords

(ii) report procedures for immediately informing management of all suspected breaches of security

(iii) a routine review (annually and also following each breach) of the procedures and a continuous programme of re-assessing security classifications; a good secure service depends upon the proportion of top security work being kept to a minimum and this pre-supposes that some high security work may at a later stage be downgraded, and

(iv) security exercise tests both as part of the regular audit of the organisation and its systems, and random unscheduled exercises being regularly carried out.

7. CONCLUSION - DEMANDING BUT NOT EXPENSIVE

Security is an essential feature of an efficient organisation. The introduction of security measures into a computing service which has grown negligent is difficult, can be costly in time and effort and be a traumatic experience for the personnel. All this inconvenience can be avoided by designing at the system inception total computer systems, from user back to user, with security as an essential and integral component.

Personnel are both the strength and the weakness of all security systems. Therefore every person from user through to computer operator must be made to realise the part they have to play in maintaining the security and the integrity of the computing service and must be trained to appreciate his individual responsibilities. If this is achieved the benefits of security to the efficiency of the company will accrue with the minimum of effort and corrective action.

The cost of devising security procedures is not expensive in terms of data-processing budgets. The significant cost is that required to ensure that the written security procedures become a reality. This is achieved through audit-checks and surprise visits to test the effectiveness of the defences. Although technological development may assist in security, the major threats are caused by human weaknesses and the best defence is good management.

ACKNOWLEDGEMENTS

The authors wish to acknowledge the help give to them by many previous colleagues and by Southern Water Authority. The opinions expressed in the paper are those of the authors and do not necessarily represent those of the Southern Water Authority.

REFERENCES

1. Kluwer Harrap Handbooks. 'Handbook of security'. Kluwer Harrap, Netherlands, 1977.
2. Hemphill, C.F., and J.M. 'Security procedures for computer systems'. Dow Jones-Irwin, Illionois, USA.
3. Sharrat, J.R. 'Data control guidelines'. National Computing Centre, Manchester, England, 1974.
4. Hamilton, P. 'Computer security' Cassell Associated Business Programmes, 1972.
5. Hampshire Regional Health Authority. 'Report of members enquiry into salary misappropriation at area headquarters'. Hampshire Regional Health Authority, UK, June 1977.
6. Computerview. 'The ICI ransom case: some lessons to be learnt'. Computer Weekly, IPC Business Press, 9 Feb., 1978, pp2.

List of Authors

P. Aanstad

M. Adiba

P. Ancilotti

M. Badel

S. Baragli

W. Bartussek

M. Boari

W. Cellary

C. Ciborra

J.-P. De Blasis

C. Delobel

M. Demuynck

M. Di Manzo

A. Endres

A. Flory

A.L. Frisiani

G. Gasbarri

W. Glatthaar

P.A.V. Hall

J. Hawgood

I.A. Hussein

T. Johansen

J. Klonk

F. Kolf

J. Kouloumdjian

S. Krakowiak

D. Kropp

F. Land

V.P. Lane

B. Langefors

G.P. Learmonth

J. Leroudier

F. Lesh

W.K. Liebmann

N. Lijtmaer

P.C. Maggiolini

A.G. Merten

H.D. Mills

P. Moulin

E. Mumford

G. Olimpo

H.J. Oppelland

D.L. Parnas

E. Pichat

W.A. Potas

P.L. Reichertz

M. Ricciardi

N.L. Richards

K. Sauter

H.-J. Schek

D. Seibt

G. Skylstad

R. Słowiński

A. Sølvberg

R.K. Stamper

N. Szyperski

P.R. Torrigiani

J.M. Triance

S. Valvo

S. Vinson

W. Weingarten

F.G. Wright

H. Wrobel

Vol. 49: Interactive Systems. Proceedings 1976. Edited by A. Blaser and C. Hackl. VI, 380 pages. 1976.

Vol. 50: A. C. Hartmann, A Concurrent Pascal Compiler for Minicomputers. VI, 119 pages. 1977.

Vol. 51: B. S. Garbow, Matrix Eigensystem Routines – Eispack Guide Extension. VIII, 343 pages. 1977.

Vol. 52: Automata, Languages and Programming. Fourth Colloquium, University of Turku, July 1977. Edited by A. Salomaa and M. Steinby. X, 569 pages. 1977.

Vol. 53: Mathematical Foundations of Computer Science. Proceedings 1977. Edited by J. Gruska. XII, 608 pages. 1977.

Vol. 54: Design and Implementation of Programming Languages. Proceedings 1976. Edited by J. H. Williams and D. A. Fisher. X, 496 pages. 1977.

Vol. 55: A. Gerbier, Mes premières constructions de programmes. XII, 256 pages. 1977.

Vol. 56: Fundamentals of Computation Theory. Proceedings 1977. Edited by M. Karpiński. XII, 542 pages. 1977.

Vol. 57: Portability of Numerical Software. Proceedings 1976. Edited by W. Cowell. VIII, 539 pages. 1977.

Vol. 58: M. J. O'Donnell, Computing in Systems Described by Equations. XIV, 111 pages. 1977.

Vol. 59: E. Hill, Jr., A Comparative Study of Very Large Data Bases. X, 140 pages. 1978.

Vol. 60: Operating Systems, An Advanced Course. Edited by R. Bayer, R. M. Graham, and G. Seegmüller. X, 593 pages. 1978.

Vol. 61: The Vienna Development Method: The Meta-Language. Edited by D. Bjørner and C. B. Jones. XVIII, 382 pages. 1978.

Vol. 62: Automata, Languages and Programming. Proceedings 1978. Edited by G. Ausiello and C. Böhm. VIII, 508 pages. 1978.

Vol. 63: Natural Language Communication with Computers. Edited by Leonard Bolc. VI, 292 pages. 1978.

Vol. 64: Mathematical Foundations of Computer Science. Proceedings 1978. Edited by J. Winkowski. X, 551 pages. 1978.

Vol. 65: Information Systems Methodology, Proceedings, 1978. Edited by G. Bracchi and P. C. Lockemann. XII, 696 pages. 1978.

This series reports new developments in computer science research and teaching — quickly, informally and at a high level. The type of material considered for publication includes:

1. Preliminary drafts of original papers and monographs
2. Lectures on a new field or presentations of a new angle in a classical field
3. Seminar work-outs
4. Reports of meetings, provided they are
   a) of exceptional interest and
   b) devoted to a single topic.

Texts which are out of print but still in demand may also be considered if they fall within these categories.

The timeliness of a manuscript is more important than its form, which may be unfinished or tentative. Thus, in some instances, proofs may be merely outlined and results presented which have been or will later be published elsewhere. If possible, a subject index should be included. Publication of Lecture Notes is intended as a service to the international computer science community, in that a commercial publisher, Springer-Verlag, can offer a wide distribution of documents which would otherwise have a restricted readership. Once published and copyrighted, they can be documented in the scientific literature.

Manuscripts

Manuscripts should be no less than 100 and preferably no more than 500 pages in length.
They are reproduced by a photographic process and therefore must be typed with extreme care. Symbols not on the typewriter should be inserted by hand in indelible black ink. Corrections to the typescript should be made by pasting in the new text or painting out errors with white correction fluid. Authors receive 75 free copies and are free to use the material in other publications. The typescript is reduced slightly in size during reproduction; best results will not be obtained unless the text on any one page is kept within the overall limit of 18 x 26.5 cm (7 x 10½ inches). On request, the publisher will supply special paper with the typing area outlined.
Manuscripts should be sent to Prof. G. Goos, Institut für Informatik, Universität Karlsruhe, Zirkel 2, 7500 Karlsruhe/Germany, Prof. J. Hartmanis, Cornell University, Dept. of Computer-Science, Ithaca, NY/USA 14850. or directly to Springer-Verlag Heidelberg.

Springer-Verlag, Heidelberger Platz 3, D-1000 Berlin 33
Springer-Verlag, Neuenheimer Landstraße 28–30, D-6900 Heidelberg 1
Springer-Verlag, 175 Fifth Avenue, New York, NY 10010/USA

ISBN 3-540-08934-9
ISBN 0-387-08934-9